# Introduction
## to
# Engineering
# Materials

# Introduction
# to
# Engineering
# Materials

Behavior, Properties, and Selection

# MATERIALS ENGINEERING

*Additional Volumes in Preparation*

# Introduction
# to
# Engineering
# Materials

Behavior, Properties, and Selection

# G. T. Murray

*California Polytechnic State University*
*San Luis Obispo, California*

**Taylor & Francis**
Taylor & Francis Group

Boca Raton   London   New York   Singapore

A CRC title, part of the Taylor & Francis imprint, a member of the
Taylor & Francis Group, the academic division of T&F Informa plc.

Published in 1993 by
CRC Press
Taylor & Francis Group
6000 Broken Sound Parkway NW, Suite 300
Boca Raton, FL 33487-2742

International Standard Book Number- 0-8247-8965-2 (Hardcover)
Library of Congress Card Number 93-12660

**Library of Congress Cataloging-in-Publication Data**

---

Murray, G. T.
  Introduction to engineering materials : behavior, properties, and selection / G. T. Murray.
      p. cm. (Engineered materisl ; 2)
  Includes bibliographical references and index.
  ISBN 0-8247-8965-2 (alk. paper)
  1. Materials.   I. Title. II. Series.
TA403.M84   1993
629,1'1—dc20                                                                     93-12660

---

Taylor & Francis Group
is the Academic Division of T&F Informa plc.

Visit the Taylor & Francis Web site at
http://www.taylorandfrancis.com

and the CRC Press Web site at
http://www.crcpress.com

*To Karen and Mike*

# Preface

This book was written for students who are not materials engineering majors, although it could serve as an introductory-level text for materials engineering majors. Throughout the book the comparative behavior and properties of the four major classes of materials—namely, metals, ceramics, polymers, and composites—are emphasized and repeated, even in the introductory basic science chapters. Separate chapters on each of these classes of materials follow the basic science chapters, and again comparisons are intentionally repeated. After presentation of chapters on electronic materials and environmental degradation of materials, where again appropriate comparisons are made, a chapter entitled "Comparative Properties" is introduced that summarizes the property differences of the various classes and leads into the final chapter on materials selection. Probably the chief difference between this and the many excellent materials science and materials engineering texts now available is that this text places more emphasis on comparative properties and materials selection. It is my opinion that non-materials majors are often presented with materials selection problems when they enter industry. In fact, in the curriculum of a few schools, a two-course approach has been taken—one on the fundamentals, followed by one on materials selection. We may see more of these two-course sequences in the future because of the vast and continuing increase in the number of new materials becoming available to design engineers. The science content has been condensed somewhat to allow more space for materials properties and selection.

As in all textbooks of a survey nature, considerable material was obtained from other sources. Where known, these sources have been appropriately identified and credited. The ASM, Int., handbooks were an invaluable source of information and I appreciate receiving permission to use material from them. I also would like to thank my colleagues at Cal Poly, Robert Heidersbach, William Forgeng, Robert Leonesio, Anny Morrobel-Sosa, Linda Vanusupa, and Dan Walsh for their reviews of the manuscript, their many helpful comments, and material supplied. Bill Forgeng and his Physical Metallurgy and Metallography students deserve special credit for the photographs they provided. Credit to the involved students has been given in the figure legends. The typing services of several students and the drawing services of Jorgensen Illustrations are gratefully acknowledged. Finally, the patience and assistance of my wife Tonny during the nearly two-year writing period of this book, during which she could be described as a book widow, are deeply appreciated. I am sure that she is ready to return to her former and less demanding role as a golf widow.

*G. T. Murray*

# Contents

**Figure 1.3**  *Voyager* attained its light weight through the use of graphite–epoxy composites in conjunction with a honeycomb structure. (Courtesy of Voyager Enterprises, Inc.)

over the years as a result of new alloy development (Figure 1.5), advances in these alloys are currently limited because operating temperatures are approaching their melting point. One alternative would be to go to higher-melting-point metals such as tantalum, niobium, molybdenum, and tungsten. But these metals have the disadvantage of being very reactive to gases at elevated temperatures, hence requiring some kind of protective surface layer. Experiments with ceramic coatings have not been successful. Therefore, materials engineers are currently working on the development of ceramic and ceramic composite materials for use in the hottest parts of the engine. Useful alloys developed to date are limited to about 1040°C (1900°F), while ceramics can retain their strength to 1650°C (3002°F) and are potentially useful to 2760°C (5000°F).

Automotive engineers are also participating in the development of ceramic and ceramic composite materials. Approximately one-third of the energy generated in an internal combustion engine is lost in the cooling process necessary to prevent distortion of the metallic parts. Piston rods and ceramic cylinder liners, predicted by some to be used before the year 2000, would lessen the amount of cooling required, while a ceramic engine block would eliminate water cooling altogether. The ceramic engine block must, however, await the development of less brittle ceramic materials.

Materials engineers in the automotive industry have also developed stronger steels that result in less weight and better fuel efficiency. These steels are generally used in the higher-load-bearing components. For the sheet steel used in auto body construction, engineers are examining polymers and polymer com-

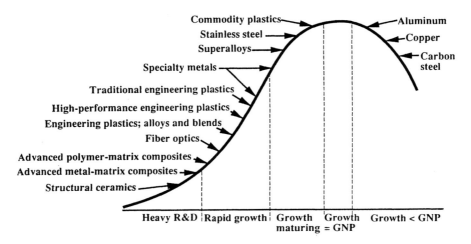

**Figure 1.2** Materials maturity curve (From Louis J. Sousa, *Problems and Opportunities in Metals and Materials: An Integrated Perspective,* U.S. Dept. of the Interior, Bureau of Mines, Washington, D.C., 1987.)

riveted aluminum alloy skin was first used for airframe construction and was the predominant fuselage material until the advent of supersonic aircraft. The Boeing 757, for example, contains 80% aluminum, 12% steel, and 3% composites. But aluminum alloys are limited to temperatures of about 175°C. The Concorde, the only supersonic commercial aircraft in use today, and military aircraft such as the SR-71 surveillance aircraft (Mach 3), have titanium alloy fuselages. Starting with the Harrier, an aircraft designed in the 1970s at McDonnel Aircraft Company, much of the aluminum support structure was replaced by lighter and stronger carbon–epoxy composites. Strong, lightweight composites are currently playing a major role in nonmilitary aircraft. The *Voyager* (Figure 1.3) achieved its global circuit without refueling, in large part due to the use of graphite–epoxy composites in conjunction with honeycomb structures. The Beech Starship Model 2000 (Figure 1.4) is the first all-composite aircraft to be certified by the Federal Aviation Administration (FAA). Predictions are that we will see many more composites in the aircraft of the twenty-first century, not only in the fuselage and support structures but in the jet engine itself.

When we examine the structure of the current jet engine we find that advances made in the alloys of nickel, cobalt, and chromium, called *superalloys,* have permitted the increase in operating temperature necessary for increased engine efficiency, which translates into faster travel speeds and increased fuel economy. Even though the operating temperature has increased

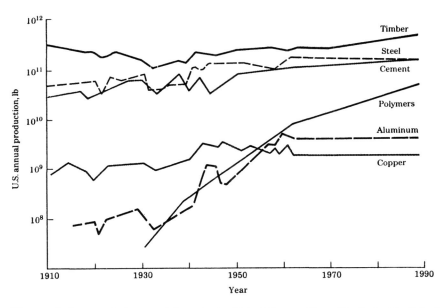

**Figure 1.1** Trends in material use in the United States. (From W. F. Smith, *Principles of Materials Science and Engineering,* 2nd ed., McGraw-Hill, New York, 1990.)

material developments can be transferred to the older, more mature industries. Figure 1.2 depicts a *maturity curve* for a variety of materials. This curve is, to a certain degree, based on projected use and should be considered in that light. The important point that is illustrated here is that in the future we can expect to see more and more applications of polymers, ceramics, and composites. Composites are a combination of two or more materials. Fiberglass and concrete are familiar examples. Composites now fall into a new and distinct category that is being referred to by technologists as *engineered materials.* In the past we have selected existing materials to fit a design or application. If the fit did not work, the design was altered or different materials were selected. In the future we can expect to see more materials that are engineered to suit an application. As stated in a *Scientific American* review article (2), we are now leaving the *basic materials age* (i.e., the stone, bronze, and iron ages) and entering the *era of engineered materials.*

It is of both historical and technical interest to follow the development of materials for aircraft construction, because in no other situation have the properties of strength, weight, and operating temperature been so interrelated and important. The first all-metal monoplane, designed by Hans Reissner of Germany which first flew in 1912, had wings of pure aluminum. In the early 1930s

# 1

# Classification of Materials

## 1.1 INTRODUCTION

We currently have available for potential use in engineering structures and electronic devices about 45,000 different metallic alloys, of the order of 15,000 different polymers, and hundreds, if not thousands, of other materials that fall in the categories of wood, ceramics, fabrics, and semiconductors. An automobile alone contains several hundred different materials. A 1986 Mercedes-Benz contains 67 wt % cast iron and steel, 12 wt % fabrics, 12 wt % polymers, and 4 wt % aluminum alloys, the balance being composed of glass and other non-ferrous alloys (1). Within each of these categories there are a number of specific materials that are used and many more that must be considered before reaching a decision on the material specifications for that particular model. These decisions are really not entirely final; they must be reviewed and updated periodically as new materials are developed and design changes are made. Mercedes-Benz engineers project that by 1996, a span of only 10 years, iron and steel use will fall to 62 wt %, while polymers and aluminum alloy content will increase to 18 and 6 wt %, respectively. Throughout all industries, the trend to more polymer use is evident (Figure 1.1). As you are probably aware, polymers have replaced metals in many household functions. It is thus not only the current materials available with which we must be concerned in the material selection process but also those on the horizon, particularly the new developments in the faster-growing technologies. Frequently, such

*1*

**Figure 1.4** Beech Aircraft Corporation Starship: graphite–epoxy composite used for airframe construction. (Courtesy of Beech Aircraft Corp.)

posites as possible replacement materials in an effort to reduce the weight further. The General Motors Pontiac Fiero was the first mass-produced car that incorporated plastics in all its outer panels. Although the jury is still out on this, the use of reinforced polymers for auto exterior parts has increased steadily since the mid-1970s.

## 1.2 METALS

Metals have been widely used for several thousand years, beginning in the Bronze Age, which historians date to about 3000 to 1100 B.C. The Bronze Age was supposedly replaced by the Iron Age, which we are currently experiencing. Actually, we still use a considerable quantity of bronze, but steel use is many times that of the bronzes.

Metals have been defined over the years according to their characteristics: They are hard; reflect light, giving a metallic luster; and have good thermal

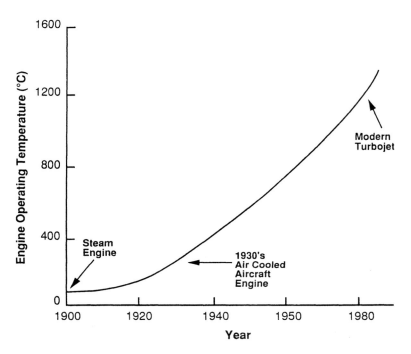

**Figure 1.5**  Aircraft engine operating temperatures have increased with the development of new alloys. (From a report on the National Research Council's study on materials science and engineering, 1989.)

and electrical conductivities. However, with the advent of semiconductors and superconductors, metals can best be described by comparing their electrical conductivities with these materials and with insulating materials (Section 2.3.1). Another characteristic of metals is that in the solid state they are crystalline in form; that is, the atoms within the body occupy well-defined positions, in contrast to noncrystalline amorphous solids such as glasses, wood, and many polymers.

Metals have been classified by technologists into two categories, ferrous and nonferrous, the term *ferrous* being derived from the Latin word *ferrum,* meaning iron. This distinction arose for two reasons. First, the steels (i.e., iron-based alloys) have been used in quantities exceeding all the others combined, and thus to some extent warranted a separate category based on quantity of use. Production of iron and steel on a weight basis in 1988 exceeded that of all other metals combined by a ratio of approximately 7:1. Second, the process of extracting iron from its ore is a complex process and somewhat different from the processes used, for example, for copper, nickel, and aluminum extraction. It was these extractive metallurgists that first separated metals into the ferrous and nonferrous categories. The physical metallurgy of alloys, which includes the composition and processing of such alloys, is basically the same in principle for all metals. Many physical metallurgists are not convinced that there is a need for these two separate categories of metals. We will follow convention, however, and begin our classification of metals with the ferrous metals and follow with a separate section on nonferrous alloys.

## 1.2.1  Iron and Steel

Iron-based alloys (i.e., where iron is the major constituent) include the cast irons, a number of steels, and a few iron-based alloys that are not called steels. Some wrought iron, which consists of iron silicate fibers in an iron matrix, is in use today in the form of pipe, grills, and decorative objects.

Tools made of iron appeared around 1300 B.C. in Palestine, where an iron furnace has been found. Smelting of iron to extract it from its ore was believed to have begun about this time. Both steels and cast irons are basically alloys of iron and carbon, with the former containing up to about 2 wt % carbon and the latter about 2 to 4 wt % carbon. Why are the higher-carbon alloys called "cast irons"? Simply because they are too brittle to permit forming to the desired shape and therefore must be cast to shape. Cast iron first appeared in China around 200 B.C. It is widely used in large, intricately shaped structures that cannot be machined or forged to shape. It comprises approximately 95% of the weight of a typical automobile engine in the form of the engine block, head, camshaft, piston rings, lifters, and manifolds. Some competition exists in the form of aluminum alloy engine blocks, and since weight reduction is important, we can expect that some future cast iron applications will be lost to composites. There are several types of cast iron from which to choose, referred to as either white, gray, ductile, or malleable iron. We examine these in more detail in Chapter 6.

Steel, because of its strength, formability, and abundance, and therefore lower cost, is still and will be for some time to come the dominant metal used for structural applications. Numerous steel alloys have emerged over the years, each class of such alloys having its own niche among the myriad of modern designs. These classes are listed and discussed briefly in the following, but first we need to examine the steel numbering system.

Steel Numbering System

There are many numbering systems used to designate the composition of steels. The most common is that used by the Society of Automobile Engineers (SAE). Examples are given in the following sections. The American Society of Testing and Materials (ASTM) has formulated a system that is becoming more widely used, especially for procurement specifications. An ASTM specification may include more than one SAE steel since a given ASTM specification number frequently designates a range of properties plus the finishing process (e.g., hot-rolled plate, cold-rolled strip, etc.). There also exist many foreign numbering systems. The Germans use a DIN number, the "D" standing for Deutsch; the Japanese a JIS number; and the British a BS number, the "BS" standing for British steel. Fortunately, a unified number system (UNS) has been developed by ASTM, SAE, and other technical societies, and the U.S. government. The various numbering systems have been compared in handbooks. The recently published *Metals and Alloys in the UNS System, 5th ed., 1989,* by ASTM and SAE is updated every few years. The *Metals Handbook,* Vol. 1, 10th ed., by the American Society for Materials, International, Metals Park, Ohio, 1990, is also an excellent source, not only for clarifying the numbering system, but for the properties, processing, and selection of all ferrous alloys.

Plain Carbon Steels

As the term *plain carbon steels* implies, these are alloys of iron and carbon. Some small quantities of manganese and silicon may be present, but carbon is the alloying element that dictates the properties. They were the first steels developed, are the least expensive, and have the widest range of applications. According to the SAE system, plain carbon steels are identified by a four-digit number, the first two digits being a one and a zero, and the second two digits indicating the wt % carbon content. For example, a 1045 steel is a plain carbon alloy containing 0.45 wt % carbon. The corresponding UNS number is G10450.

Alloy Steels

The alloy steels contain small amounts, on the order of 2 to 5 wt % alloying elements in addition to carbon. Sometimes these steels are called low-alloy steels, to distinguish them from high-strength steels such as maraging steels, which contain about 30 wt % of alloying elements. The most common alloying elements in the low-alloy steels are nickel, chromium, and molybdenum. These alloys permit the hardening of the steel to greater depths when quenched from elevated temperatures. Whereas plain carbon steels can achieve only about 1/8 in. depth of hardening, the alloy steels can be hardened to depths of 2 in. This means that a shaft of 4-in. diameter could be hardened throughout the entire cross section. As you might expect, these steels are used in large-diameter shafts such as crankshafts, axles, rolls, and large bolts. In the heat-treated condition their strengths are usually in the vicinity of 1035 MPa (150 ksi) to 1380 MPa (200 ksi). Plain carbon steels can also attain this strength level, but only in thin sections.

SAE numbers for alloy steels are also four-digit numbers, with the first two digits specifying the alloying elements and their percentage and the last two digits their carbon content. One cannot look at this number and immediately know the alloy content, unless it has been memorized. To find the alloy content we must refer to handbooks. The last two digits tells us the carbon content, just as in the case of the plain carbon steels. If we look on page 152 of the *Metals Handbook,* Vol. 1, 10th ed. (ASM, Metals Park, Ohio) we find that an SAE 4340 steel contains 0.40 wt % carbon, 1.65 to 2.00 wt % nickel, 0.70 to 0.90 wt % chromium, 0.20 to 0.30 wt % molybdenum, 0.65 to 0.85 wt % manganese, and small amounts of silicon, phosphorus, and sulfur. We also find that the corresponding UNS number is G43400.

High-Strength Low-Alloy Steels

The term high-strength low-alloy steels is a misnomer since the name implies that they are of higher strength than the alloy steels, but such is not the case. They are stronger than the plain carbon steels and have a lower alloy content than that of the alloy steels. They are primarily structural steels used in large buildings, bridges, ships, oil and gas pipelines, automobiles, and pressure vessels. They are sometimes called *microalloyed steels,* since they contain minute quantities of niobium, titanium, vanadium, and/or molybdenum. Their exact compositions and processing methods are often proprietary information. They contain less than 0.2 wt % carbon, such a low carbon content being necessary for welding operations. The SAE numbering system described above is not used for these steels. They were identified by trade names as they were developed, but the trend today is to use ASTM designations and/or UNS numbers. ASTM specification A572 covers an important group of these steels. Their strength levels are on the order of 552 MPa (80 ksi).

Specialty Steels

The three categories of steel listed above probably account for over 90% of steel use. The remainder include a group of steels that we call specialty steels. The most prominent steels of this group of special steels are the stainless steels. Another class of specialty steels comprises the tool steels.

Stainless steels require a minimum of 10.5% chromium to be stainless (i.e., to be very highly resistance to corrosive environments). They achieve their stainless properties through the formation of a thin adherent chromium oxide surface layer. Other elements are added for a variety of reasons, which we consider in more detail in Chapter 6. The stainless steels are classified as martensitic, ferritic, austenitic, and precipitation-hardening stainless steels. Their numbering system is somewhat confusing and fits into no particular pattern as we had for the plain carbon and alloy steels. They can also be identified by a UNS number, but this method has not yet caught on.

Austenitic stainless, the most common and widely used, falls into the 300 series. Type 304 is best known because it is used for most household applications. The ferritic and martensitic groups both have a 400 series of numbers.

Of the martensitic steels, the 410 and 440 stainless are the most popular. The precipitation-hardened stainless steels have been numbered according to the percent chromium and nickel content. PH 17-4, the best known steel in this group, probably because they are used for golf club heads and other consumer products, contains 17 wt % chromium and 4 wt % nickel. Other such steels, the PH 15-5 and 17-7, are similarly numbered according to chromium and nickel content. A relatively recent addition to the family of stainless steels are the duplex stainless steels, which are really a mixture of the ferritic and austenitic types. We will find out later how these names emerged during the development of stainless steel metallurgy. The composition of these stainless steels, using both the numbering system above and the UNS system, can be found on page 843 of Vol. 1 of the *Metals Handbook,* 10th ed.

Tool steels, which are used for cutting and forming operations such as shears and dies, are a subject by themselves. Another long-standing reference of a type similar to the ASM *Handbook* deals only with tool steels (3). The tools steels are normally divided into four subgroups: air-hardened, oil-hardened, shock-resistant, and high-speed steels. The first two groups are so named because of the quenching media used in heat treatment. The shock-resistant tool steels are used for chisels, punches, and driver bits, and other applications where the tool is subjected to impact loading. The high-speed group consists of those that experience very high temperatures during use, such as machine tools and hot-working dies. The high-speed steels are designated by a significant letter preceding the number. The molybdenum-containing high-speed tool steels have the letter M preceding the number, the tungsten-containing high-speed steels have the letter T preceding their number, and those used at very high temperatures for hot forming of metals have the letter H preceding their number independent of their major alloying element. The letters preceding the numbers for the air-hardened, oil-hardened, and shock-resistant steels are A, O, and S, respectively. These designations, together with their corresponding UNS number, can be found on page 758 of the *Metals Handbook.*

## 1.2.2  Nonferrous Alloys

The principal nonferrous alloys are those in which the base metal (i.e., the major constituent) consists of either aluminum, copper, nickel, magnesium, titanium, or zinc. Alloys of these six metals account for over 90% of nonferrous alloy use. World production of these primary metals from their ores and their 1988 prices are listed in Table 1.1. Iron is included for comparative purposes. About 30% of aluminum, copper, and magnesium is recycled, which makes their consumption somewhat higher than the primary metal production numbers. Note that primary lead production exceeds that of nickel, mag-

**Table 1.1** World Production and Prices of Primary Metals from Their Ores

| Metal | Production (thousands of tons) | 1990 Price ($U.S./ton) |
|---|---|---|
| Pig iron | 656,000 | 187 |
| Aluminum | 18,256 | 1,281 |
| Copper | 9,722 | 1,914 |
| Zinc | 8,075 | 1,112 |
| Lead | 3,830 | 623 |
| Nickel | 884 | 5,340 |
| Magnesium | 363 | 3,204 |
| Tin | 207 | 5,607 |
| Titanium | 24 | 10,000 |

*Source*: H. Spoel, "The Current Status of Scrap Metal Recycling," *J. Miner. Met. Mater.*, Apr. 1990.

nesium, and tin. Except for solders and lead alloy bearings (babbitts), lead is not used much in alloy form. In nonalloy form it is used for radiation shielding and in batteries.

Just as there is a *Metals Handbook* for iron and steel, there is also one for the nonferrous alloys. The 1990 *Metals Handbook,* Vol. 2, 10th ed., by ASM International includes properties, processing, selection, and numbering systems for all prominent nonferrous alloys. The numbering system for nonferrous alloys is just as confusing as that for ferrous alloys. We have the old numbering system, mostly developed by alloy manufacturers, plus the UNS numbers and ASTM specifications that may include more than one alloy.

Aluminum Alloys

Aluminum is produced by the electrolytic reduction of aluminum oxide. In 1900 only 6384 tons of aluminum was produced, compared to 18.3 million tons in 1988. The advent of the aircraft industry early in the twentieth century propelled aluminum to the top of the list of nonferrous metal use. Subsequently, many other uses for aluminum were developed. About 1.3 million tons was used in 1988 for beverage cans alone, of which about 60% (1989) was recycled. The percentage recycled is expected to grow rapidly with the introduction of recent conservation and environmental regulations. The energy required for recycling is about 5% of that required to extract aluminum from aluminum oxide, a tremendous savings in dollars and energy.

The attractiveness of aluminum as a structural metal resides in its lightweight. Its density is only 2.7 $g/cm^3$, compared to 7.87 $g/cm^3$ for iron. Aluminum alloys can be heat treated to attain yield strengths in the vicinity of 550

MPa (80 ksi), compared to high-strength steels of about 1550 MPa (225 ksi).
On a strength-to-density ratio (sometimes referred to as *specific strength*)
aluminum alloys come out ahead, being approximately 204, while the higher-
strength steels have a ratio of 197. These high-strength steels are most often
used in gears, shafts, axles, and the like. More appropriately, we should compare the strength-to-density ratios of aluminum alloys to those for structural
steels. The latter have a strength-to-density ratio of about 70. In this case
aluminum alloys come out on top by a factor of about 3. It is rather obvious
why aluminum became the structural material for aircraft fuselages, wings, and
support structures. A bonus also arises when comparing the corrosion resistance of aluminum to steels (other than stainless steels, which, aside from
weight problems, are also quite expensive). The most common aluminum
alloys are those containing copper and zinc in quantities of 3 to 4%. These
alloys are in the 2000 and 7000 series in the four-digit numbering system.
Aluminum alloys produced in wrought form, such as sheet, rod, wire, and
extruded shapes, are classified according to their content of major alloying elements. The first digit indicates the alloy group, while the other digits refer to
modifications of the original alloy. The second digit originally had something
to do with aluminum purity, but since all alloys are now made from essentially
the same commercial purity aluminum, this designation has lost its
significance. The numbers are listed according to alloying element in Table
1.2. These alloys can also be specified by a UNS number.

In recent years 2024 (UNS A92024) and 7075 (UNS A97075) have been the
most widely used wrought alloys where specific strength is important. Alloy
3003, which contains 1.2 wt % manganese, is used for beverage cans. Alloy
1350, which is 99.5% aluminum, makes up most electrical conductors.

Aluminum casting alloys now have a four-digit series numbering system
similar to that for wrought alloys. Their numbering system is listed in Table
1.3. Aluminum casting alloys were for many years designated by a three-digit

**Table 1.2**  Wrought Aluminum Alloy Numbering System

| | |
|---|---|
| Aluminum—99.0% minimum aluminum | 1xxx |
| Alloys by major element | |
| Copper | 2xxx |
| Manganese | 3xxx |
| Silicon | 4xxx |
| Magnesium | 5xxx |
| Magnesium and silicon | 6xxx |
| Zinc | 7xxx |
| Other elements | 8xxx |

**Table 1.3** Cast Aluminum Alloy Numbering System

| | |
|---|---|
| Aluminum—99.0% minimum aluminum | 1xx.x |
| Alloys by major element | |
| Copper | 2xx.x |
| Silicon with copper and magnesium | 3xx.x |
| Silicon | 4xx.x |
| Magnesium | 5xx.x |
| Zinc | 7xx.x |
| Tin | 8xx.x |
| Unused series | 6xx.x |

number, sometimes with a letter preceding the number. Alloy 242, for example, is a sand-cast aluminum–copper alloy that now has the number 2420 (UNS A02420).

Copper Alloys

Copper ores found in the United States contain about 1% copper in the form of copper sulfide. Copper sulfide concentrates are smelted to yield a matte that is a mixture of copper and iron sulfides. The copper-containing mattes are melted again in a converter. Air is blown through the mattes to oxidize the remaining sulfur. The remaining copper is called *blister copper* and is about 99% pure. The blister copper is further fire-refined to remove other impurities, leaving a *tough-pitch* copper of around 99.5% purity. This is suitable for many alloys, but for some a higher purity of 99.9% is obtained by electrolytic refinement of the tough-pitch copper.

Copper–zinc brasses are the most frequently used copper alloys. The zinc content can vary from 5 to 40 wt % and range in yield strengths from 100 MPa (14.5 ksi) to about 345 MPa (50 ksi), depending on zinc content and amount of cold work. Known as cartridge brass, 70% CU–30% Zn is also used for radiator cores, plumbing accessories, nuts, bolts, fasteners, heat exchangers, lamp fixtures, and a host of small parts.

Copper–tin alloys are the original bronzes for which the Bronze Age was named. They are sometimes referred to as tin bronzes, containing up to 10 wt % tin, but also known as phosphor bronzes because of the addition of phosphorus to remove oxygen. Only a small amount of phosphorus remains, in the form of $Cu_3P$. These bronzes can be cold worked to increase strength and in this form are stronger but more expensive than the brasses. They are used in the as-cast condition for gears, bells, and bearings. Wrought tin bronzes have strengths up to 840 MPa (120 ksi), while the as-cast parts are much weaker, being on the order of 140 MPa (20 ksi).

Other copper alloys of significance include the copper–beryllium alloys, known as beryllium bronze, which can be heat treated to exceptionally high strength levels, on the order of 1070 MPa (155 ksi). These alloys are the strongest of the copper-based alloys and require only about 2 wt % beryllium to achieve this strength level. They are used for springs, electrical switch contacts, diaphragms, and fasteners in the cold-worked state and to some extent as molds, bushings, gears, and valves in the cast state. Copper–aluminum alloys, usually containing about 6 to 8 wt % aluminum, are known as aluminum bronzes and are also relatively high-strength copper alloys achieving about 517 MPa (75 ksi) strength levels in the wrought cold-worked state and around 370 MPa (54 ksi) in the cast state. They also have good corrosion resistance, due to the aluminum content, and are frequently used for gears, bearings, bushings, valves, and parts that may be exposed to marine environments. Copper alloys now have a UNS numbering system, which was derived from the Copper Development Association numbers, but as with other alloys can also be described by an ASTM specification.

Other Nonferrous Alloys

Zinc is used more as an alloying element and in galvanizing steel than as a base metal. A few zinc alloys containing a small percentage of aluminum are used for die castings, a process whereby the molten alloy is squeezed into a die. Zinc alloys also have a UNS numbering system in addition to their manufacturer's number and ASTM specifications.

Magnesium alloys are in competition with aluminum alloys because of their low density (1.74 g/cm). They are used in the aircraft industry and are competing with aluminum for several applications in the automotive industry. Their disadvantages are their brittleness, requiring hot- rather then cold-forming procedures, and their relatively high cost, being about 2.5 times that of aluminum. Their strengths are on the order of 200 MPa (29 ksi) and are not competitive with aluminum on this basis. They are used in cast, extruded, and forged conditions. Magnesium alloys are identified by a UNS number in addition to their original numbering method, which included both letters and numbers. An AZ92A alloy contains 9 wt % aluminum and 2 wt % zinc and has a UNS number of M11920.

Nickel-based alloys, although in fifth place on the nonferrous metals production list, are far more important than this statistic suggests. Their claim to fame is their good high-temperature strength and corrosion resistance. Both nickel- and cobalt-based alloys are used almost exclusively for the components in the hottest regions of jet engines (i.e., the turbine blades and compressor parts). For this reason, they have been termed *superalloys*. Although cobalt-based alloys were the front-runners here, nickel alloys have retaken the lead, due primarily to the scarcity and uncertainty of the cobalt supply. All of these alloys contain up to 25 wt % of chromium for high-temperature oxidation resistance. They are known by trade names such as Inconel, Incalloy, Waspalloy, Udimet, Hastelloy, and Nimonic as well as by UNS numbers. Inconel 625 is designated as N06625 in the UNS system.

The room-temperature use of nickel alloys includes the use of some of the Inconels, but most are nickel–copper alloys, known as Monels. These contain about 30% copper and have a room-temperature strength on the order of 262 MPa (38 ksi), although one alloy can be heat treated to strengths nearly twice this level. They have good corrosion resistance and find frequent application as valves, pumps, springs, and so on, in the marine and petroleum industries. To some extent they are in competition with the aluminum bronzes (copper-aluminum), and in most case would win out if it were not for the high cost of nickel, which is nearly three times that of copper.

When it first emerged in the 1940s, titanium was called the *wonder metal* because it appeared to have all the desirable characteristics of strength, low density, corrosion resistance, and formability. In has not quite lived up to the forecasts but is, nevertheless, still in the running to replace certain other metals. The density of titanium, 4.5 g/cm$^3$, falls between that of aluminum and iron. Alloys have been developed with strength levels of about 1170 MPa (170 ksi), and on a strength-to-density ratio beat both aluminum and iron alloys. Because of this characteristic, plus the fact that it retains a relatively high strength at temperatures reaching 500°C (930°F), titanium is the obvious metal for the fuselage structure of supersonic aircraft. It cannot be used at temperatures above 500°C because of its high reactivity with gases. Thus it cannot be used in jet engines unless a suitable protection coating is developed. Also because of the high reactivity, precautions must be taken during melting and hot-forming operations. Such difficulties, along with the high cost of extracting it from its ore, has priced titanium out of reach for applications outside the aerospace industry. However, titanium is finding a niche in body implants (prostheses) and other specialty applications, where its strength and corrosion resistance outweigh the cost factor. The numbering system for titanium alloys created by the manufacturers is based on the percent alloying element. The popular alloy titanium–6% aluminum–4% vanadium is known Ti-6-4. The corresponding UNS number is R56400.

## 1.3 CERAMICS

Ceramics consist of the combination of one or more metals with a nonmetal. They have also been defined as inorganic nonmetallic materials. They are often classified according to the nonmetallic element (e.g., as oxides, carbides, nitrides, and hydrides), depending on whether the metal is combined with oxygen, nitrogen, carbon, or hydrogen, respectively. The halides, such as the chlorides, bromides, iodides, and fluorides, are today considered to be a special type of ceramic. Our common table salt, sodium chloride, is typical of this group. We usually do not think of these materials as ceramics because of their principal use as compounds in some type of solution, whereas the common *ceramic* term has been applied primarily to such items as bricks, concrete, and pottery. The halide compounds do have some limited use in the solid state as optical materials. There are also some less common metal–nonmetal compounds, such as cadmium–telluride, cadmium–sulfide, gallium–arsenide, and so on, which, by our definition above are ceramic materials. But these are also semiconductors and hence will be treated under that heading. All of the compounds above are crystalline in their solid form. The atoms take up specific positions in a type of lattice arrangement that is similar but often more complex in pattern than that found in metals. The largest group of ceramics found in nature are the silicates. Silicon and oxygen are the most abundant elements in the earth's crust. When silicon combines with oxygen we have the compound silicon dioxide, which is called silica and written as $SiO_2$. It is crystalline in its natural form, the most common arrangement being that found in crystals of beach sand. When silica is heated to about 1700°C it becomes a liquid, which, as all liquids, is noncrystalline. Unless the molten mass is cooled very slowly, the atoms do not have time to take up their specific positions. This solid form of silica is a glass. $SiO_2$ is the main ingredient in glasses, but it is still a ceramic material. Hence ceramics may be crystalline or noncrystalline materials, the noncrystalline ceramics being the glasses.

## 1.4  POLYMERS

Polymers are better known to the general public as *plastics*. The term *plastic* is defined in the dictionary as "capable of being easily molded," such as putty or wet clay. It was adopted to describe the early polymeric materials because such materials could easily be molded. This was an unfortunate choice since the term *plastics* now is used for all polymers, many of which are quite brittle and incapable of being molded once they are formed. Actually, most metals are more plastic than are these brittle plastics. Therefore, hereafter we will not use the term *plastic* to describe polymers, and hope that students will do the same. Perhaps together we can reverse the trend, even though the term *plastic* is well entrenched in the popular literature. Credit cards, for example, are frequently called *plastic money*. Have you ever tried to mold or reshape a credit card?

Polymers can generally be classified into three categories: thermoplastic polymers, commonly called *thermoplasts; thermoseting* polymers, called *thermosets;* and *elastomers,* better known as rubbers. Thermoplasts are long-chain linear molecules that can easily be formed by heat and pressure at temperatures above a critical temperature referred to as the *glass temperature*. This term was first applied to glass to represent the temperature at which glass became plastic and easily formed. The glass temperature for many polymers is below room temperature, and hence these polymers are brittle at room temperature. But they can be reheated and re-formed into new shapes, and thus can be recycled. Polyethylene is the most common thermoplast and consists of a long chain of carbon atoms with hydrogen atoms being attached to each side of the carbon atoms. It has a nominal glass transition temperature of $-100°C$. We use the term *nominal* here because the glass temperature can vary somewhat depending on the density of the polymer and the various additives, such as plasticizers and fillers, that are used to attain specific desired properties. All polyethylene, however, is easily formed at room temperature and is used in film form as covering material and in sheet form for the manufacture of bottles, housewares, and chemical tubing, to name a few applications. Thermoplasts can be either crystalline or amorphous, depending on composition and processing methods uses.

Thermosets are polymers that take on a permanent shape or *set* when heated, although some will set at room temperature. An example of the latter are epoxies that result from combining an epoxy polymer with a curing agent or catalyst at room temperature. Thermosets consist of a three-dimensional network of atoms rather than being a long-chain molecule. They decompose on heating and thus cannot be re-formed or recycled. Thermosets are amorphous polymers.

Elastomers are polymeric materials whose dimensions can be changed drastically by applying a relatively modest force, but which return to their original values when the force is released. The molecules are extensively kinked such that when a force is applied, they unkink or uncoil and can be extended in length up to about 1000% with minimal force, and return to their original shape when the force is released. In general, they must be cooled below room temperature to be made brittle (i.e., their glass temperature is below room temperature).

Natural rubber, polyisoprene, is an elastomer that consists of long-chain molecules made up of carbon and hydrogen atoms arranged in a somewhat more complex form than that of the simple polyethylene arrangement. Natural rubber as obtained from rubber trees is a sticky, gumlike material which in addition to the isoprene molecule contains small amounts of liquids, proteins, and inorganic salts. To be useful it must be processed further. In 1839, Goodyear developed the vulcanization process, whereby the rubber is changed from a thermoplastic molecule to an elastomer. In this process natural rubber is heated in the presence of sulfur and lead carbonate, the sulfur atoms causing the long-chain molecules to become cross-linked by acting as pinning points between the long-chain molecules. When a force is applied, the molecules uncoil. When the force is released, the sulfur cross-links pull the molecules back to their original positions.

Polymers are usually designated by their name or an abbreviation. Polyvinyl chloride is known as PVC. The Society of Plastics Engineers (SPE) has established SPE identification numbers, which identify the material and its form. Flexible PVC has an SPE number of 29, and rigid PVC has the number 30.

## 1.5 COMPOSITES

A composite is somewhat difficult to define. On a microscale many materials could be thought of as composites since they are made up of different atoms. But we will be dealing with composites on a macroscale and thus define them as a mixture of two or more materials that are distinct in composition and form, each being present in significant quantities (e.g., greater than 5 vol %). By this definition, microalloyed steel would not be a composite, even though we can see distinctly different substances in the electron microscope at very high magnifications (e.g., × 100,000). The microalloy constituents are present in quantities much less than 1% by weight or volume and, most often, less than 0.1%.

Wood is actually a composite of cellulose, lignin, and other organic compounds. One of the earliest engineered composites was that of glass fibers in a polymer matrix, known as fiberglass. The possibilities of material combinations are quite extensive. Some of the more advanced composites include polymer fibers in a polymer matrix, such as Kevlar fibers, introduced in 1972 by DuPont; ceramic fibers in a metal matrix; metal fibers in a polymer matrix; ceramic particles in a metal matrix; and carbon fibers in several matrices, including a carbon matrix. NASA has an ongoing high-temperature-materials program with goals to achieve a high-temperature composite material: to 425°C (800°F) for polymer matrix composites, to 1250°C (2280°F) for metal and intermetallic compound matrix composites, and to as high as 1650°C (3000°F) for ceramic matrix composites. The object of all composite materials is to achieve properties in composite form that exceed those of their individual components alone. Sometimes developing these properties involves a trade-off. Combining a strong but brittle ceramic fiber in a ductile and a weaker metal matrix results in a composite whose strength lies somewhere between the strength of the ceramic fiber and that of the metal matrix, but which at the same time is not as brittle as the ceramic alone. Such composites are true engineered materials.

## DEFINITIONS

*Alloy*:   metallic material composed of two or more metallic elements. *Note*: This is the usual dictionary definition and the one used by metallurgists.

Recently, however, materials engineers have started using such terms as *polymer alloys* and *ceramic alloys*, meaning combinations of polymers and combination of ceramics.

*Amorphous solid*: noncrystalline solid.

*Bauxite*: aluminum-containing compound from which alumina is extracted.

*Blister copper*: copper product obtained from the smelting of copper ores, approximately 99 wt % copper.

*Brass*: alloy of copper and zinc.

*Bronze*: alloy of copper with either tin, aluminum, beryllium, or manganese.

*Cast iron*: alloys of iron and carbon containing 2 to 4 wt % carbon.

*Crystalline*: describes a solid structure where atoms take up precise positions in a repeat pattern.

*Density*: weight per unit volume of a substance.

*Elastomer*: polymeric material whose dimensions can easily and extensively be changed by force and which return to their original dimensions when the force is released.

*Extractive metallurgy*: metallurgy that involves extracting the metal from its ores.

*Ferrous*: iron-based.

*Fiberglass*: composite of glass fibers in a polymeric matrix.

*Glass temperature*: temperature below which glass and thermoplastic polymers become brittle.

*Heat treatment*: process whereby the properties of alloys can be changed drastically by heating to certain temperatures and cooling at certain rates. Some processes involve reheating at lower temperatures. The reheating process is often referred to as *tempering* for steels and as *aging* for nonferrous alloys.

*Kevlar fiber*: aramid polymer fiber introduced by DuPont in 1972. Composites made from this fiber are often called Kevlar, even though the name actually applies only to the fibers.

*ksi*: unit of stress measurement: thousands of pounds per square inch; 1 ksi = 1000 pounds per square inch (psi); 1 ksi = 0.145 MPa.

*Macroscale*: Dealing with constituents that can be seen by the eye or at relatively low optical magnifications (e.g., 10 times).

*Matrix*: continuous and/or major constituent of a combination of materials.

*Microalloy*: steel that contains less than 1.0 wt % and often less than 0.1 wt % of alloying elements, usually vanadium or niobium.

*Microscale*: dealing with constituents on a scale requiring high magnifications, on the order of 1000 times by optical microscopy or up to 100,000 times by electron microscopy.

*MPa*: megapascal; 1 MPa = 6.9 ksi.

*Mach*: speed of sound; Mach 3 = 3 times the speed of sound ($\sim$3300 meters per second in air).

*Nonferrous* alloys: alloys that contain little or no iron.

*Pig iron*: iron resulting from smelting of iron ores, usually about 92 wt % iron and 4 wt % carbon.

*Phosphor bronze*: copper–tin bronzes to which phosphorus has been added to remove oxygen.

*Physical metallurgy*: science and practice of metallurgy, which deals with the development of alloys and their resulting properties.

*Polymers*: literally "many-mers," where a "mer" is the basic building block of the giant molecule.

*Refractory metals*: high-melting-point metals, usually niobium, molybdenum, tantalum, and tungsten.

*Semiconductor*: solid such as silicon or germanium, including many compounds that have conductivities in the range somewhere between those of conductors and insulators.

*Silicates*: compounds of silicon and oxygen, and usually also with metals; they constitute the most abundant ceramic materials in the earth's crust.

*Silicon dioxide*: compound of silicon and oxygen called silica, which is the major constituent of glasses and silicates.

*Superconductors*: solids that have extremely high conductivities, about $10^{13}$ times that of metals, at temperatures below a certain critical transition temperature.

*Tempering*: process of heating quenched steel to relatively low temperatures, approximately 800°F, to reduce their brittleness.

*Thermoplasts*: long-chain polymer molecules that can easily be molded on heating and can be remolded to a new shape by reheating. They are recyclable.

*Thermosets*: polymers that have three-dimensional atom arrangements. They cannot be reshaped when reheated and cannot be recycled.

*Tough-pitch copper*: result of fire-refining blister copper. It is about 99.5 wt % copper.

*Vulcanization*: process of combining natural rubber with sulfur to form the main class of commercial rubbers.

## QUESTIONS AND PROBLEMS

The ASM *Metals Handbook* (ASM International, Metals Park, Ohio) and *Smithells' Metals Reference Book* (E. A. Brandon, Ed., Butterworth, Stoneham, Mass., 1983) contain an abundance of information pertaining to composition, properties, and applications of metals and materials commonly used in industry.

1. What are the differences between cast iron and steel? Do we have or should we have a category called "cast steels"?

2. What is the term most commonly used for polymers? Why is this term misleading?
3. Convert the following Fahrenheit temperatures to Celsius.
   (a) $-32°F$      (b) $32°F$      (c) $212°F$      (d) $1500°F$
4. Define a metal. What is one property of a metal which more than any other property distinguishes a metal from a polymeric or ceramic material?
5. What do the following groups of letters mean? SAE; UNS; ASTM.
6. What carbon content does an SAE 1010 steel have? What other name is often used for low-carbon steels that contain only iron and carbon?
7. What is the composition of a 4340 steel? Where could you find the composition of a 4140 steel?
8. What is the major alloying element in a 2024 aluminum alloy? For what aircraft component or part is this alloy frequently used?
9. A steel with the designation 17-7 PH is what type of steel? What are the two major alloying elements and the amount of each present?
10. Maraging steels are the strongest of all commercial steels. A common maraging steel has the designation 18 Ni (300). It contains 18% nickel. Can you guess what the 300 number signifies?
11. What is the approximate copper content in an aluminum bronze? Are they stronger or weaker than beryllium bronzes? What alloy does the term "bronze" suggest to you in reference to the alloy composition?
12. What is a common theromoplastic polymer? How does it differ in properties and behavior from the elastomers and the thermoset polymers?
13. During the 1980s a new type of material was introduced for golf club shafts. Golfers referred to them as graphite shafts. What were the components of this composite material?

## REFERENCES

1. Razin, C., and Kanuit, C., "Materials-Selection Trends at Mercedes-Benz," *Adv. Mater. Process.,* Vol. 137, 1990, p. 43.
2. "Materials for Economic Growth," *Sci. Am.,* No. 4, 1986, p. 255.
3. Roberts, G. A., and Gary, R. A., *Tool Steels,* 4th ed., ASM International, Metals Park, Ohio, 1980.

# 2

# Properties and Their Measurement

## 2.1 INTRODUCTION

Prior to a consideration of the mechanical and physical behavior of materials it is important that we understand the methodology underlying the determination of their properties. Mechanical behavior deals with the reaction of the body to a load or force, whereas physical behavior deals with electrical, optical, magnetic, and thermal properties. One could also discuss chemical properties, but the only chemical characteristics that we will study are those pertaining to environmental effects, and we deal with those measurements in the appropriate chapter. For some applications mechanical properties will be of prime interest, while for others the physical properties will be of major concern. In most metallic applications, corrosion resistance must be taken into consideration.

## 2.2 MECHANICAL PROPERTIES

The mechanical properties of most interest and the ones that we will consider here include yield strength, ultimate tensile strength, modulus of elasticity, ductility, hardness, fatigue life, fatigue limit, creep, and fracture toughness. The strength, modulus, and ductility are of prime importance and are most often determined in a uniaxial tension test. There are a few exceptions, however. Brittle materials are frequently tested in uniaxial compression since it is difficult to grip a brittle specimen for a tensile or pulling load. In a compression test we can apply a load without having to grip the specimen. In general, properties determined in compression will be similar to those determined in tension. Another test that is occasionally used for strength and ductility is the bend test. Sometimes a bend test is easier to perform than a tension or compression test. In a bend test the outer portion of the specimen is in tension while the inner part is in compression. The torsion test, in which a twisting force is applied, is used to determine the torsion yield strength and the torsion modulus, which are not equal to the yield strength and modulus values obtained in tension. This is a very important point to remember and we emphasize it more strongly later. In selecting a material for a certain design it is important to know the type of loading to which the material will be subjected. Generally, the yield strength under a torque load (torsion) will be about one-half of that in tension, and the corresponding torsion modulus of elasticity, sometimes called the shear modulus, will be about 40% of that measured in the uniaxial tension test.

## 2.2.1 Stress–Strain Relations in Tension

All materials deform to some extent when subjected to a force. If the material returns to its original shape, it is said to have deformed elastically. If some permanent change remains after removal of the force, it has been plastically deformed. In the latter case it was also elastically deformed since, for engineering materials, some elastic deformation precedes the plastic deformation. The elastic part, however, disappears when the force is removed. It is common engineering practice to use the term *stress,* which is defined as the applied force per unit area of the specimen, and the resulting deformation as *strain,* which can be defined as change in volume per unit volume, or as is more often done for convenience, as change in length per unit length and/or change in cross-sectional area per unit area. The mechanical behavior can to a large extent be described in terms of the relationship between stress and strain during deformation. This relationship is often portrayed in the form of a stress–strain curve in uniaxial tension.

In a uniaxial tension test a force is applied, usually hydraulically or by means of an electric motor, coaxially to a cylindrical bar of material. There are situations where the specimen might be cut from sheet material, in which case a cylinder is not used, but the same principles apply. The American Society for Testing and Materials (ASTM) has specified the shape and dimensions of specimens for the standard tests, the most often used being a cylindrical bar of 0.505 in. diameter and a gauge length of 2.00 in. The specification number is ASTM E8-90 and the specimen dimensions are shown in Figure 2.1. This specimen cross-sectional area is 0.20 in.$^2$ if we use a 0.505-in.-diameter bar, which is well within ASTM specifications. Since we are usually measuring the

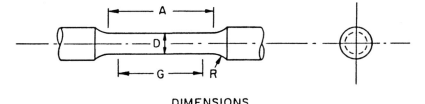

| DIMENSIONS | | |
|---|---|---|
| | STANDARD SPECIMEN | |
| | in. | mm |
| NOMINAL DIAMETER | 0.500 | 12.5 |
| G  GAGE  LENGTH | 2.000 ±<br>0.005 | 50.0±<br>0.10 |
| D  DIAMETER (NOTE 1) | 0.500 ±<br>0.010 | 12.5±<br>0.25 |
| R  Radius of  fillet, min | 1/8 | 10 |
| A  Length of  reduced  section | 2-1/4 | 60 |

**Figure 2.1**  Tensile test specimen dimensions per ASTM E8-90 specifications.

force applied to the bar, the stress can easily be computed during the test since the stress is given by

$$\sigma = \frac{F}{A} = \frac{F}{(1/5)\text{in.}^2} = 5 \times F \qquad (2.1)$$

This was advantageous years ago because the test equipment used at that time showed the applied force via a moving needle, much as in a large dial gauge, or as an indicator on a strip-chart recorder. Without bothering with the now-obsolete slide rule or a calculator, one could simply mentally multiply the force by the number 5 to obtain the stress. Much of this older equipment is still in use today; however, the more up-to-date test equipment is computerized such that the stress appears without the necessity of the arithmetic operation of equation (2.1). During the test the specimen increases in length and decreases in area. Thus the specimen dimensions are constantly changing. When the force is divided by the original area, $A_0$, the expression for stress becomes

$$\sigma = \frac{F}{A_0} \qquad (2.2)$$

This is the more common way of describing stress and is called the *conventional* or *engineering stress*. To know the true stress, the specimen diameter must be measured at the specific point of interest in the test. The true stress is defined as

$$\sigma = \frac{F}{A_i} \tag{2.3}$$

where $A_i$ is the instantaneous area. The units of stress can be expressed in the English system as pounds per square inch (psi), in the metric system as kilograms per square meter, and in the System International as pascal. We frequently use ksi for psi $\times 10^3$.

The strain, $\epsilon$, in uniaxial tension is measured by either the change in length per unit length or by the change in area per unit area and is expressed in terms of conventional or engineering strain by the following equations:

$$\epsilon = \frac{l - l_0}{l_0} = \frac{\Delta l}{l_0} \tag{2.4}$$

$$\epsilon = \frac{A_0 - A}{A_0} = \frac{\Delta A}{A_0} \tag{2.5}$$

where $l_0$ and $A_0$ are the original length and the original area, respectively, and $l$ and $A$ are any dimensions of interest during the test. Again we are using the original length and area even though the length is increasing during the test while the area is decreasing. To know the true strain at any time during the test, we must know the specimen dimensions at that specific point in the test. The true strain is defined as

$$\epsilon_{\text{true}} = \frac{\Delta l}{l_i} \tag{2.6}$$

where $l_i$ is the instantaneous length and $\Delta l$ is a small incremental change in length at $l = l_i$. In terms of area changes the true strain becomes

$$\epsilon_{\text{true}} = \frac{\Delta A}{A_i} \tag{2.7}$$

The change in length during the test is detected by an extensometer, which is attached to the specimen. An iron slug, which is contained within an induction coil, moves with the specimen during the test and feeds an electrical signal to a strip-chart recorder, which displays it as a change in length. There are also other methods for detecting length changes (e.g., the bonded wire strain gauge technique). In this test an increase in length and a decrease in area of a fine wire bonded to the specimen causes an increase in electrical resistance and a corresponding electrical signal to be fed to a sensitive recording device. Strain, being a change in length per unit length, is a dimensionless quantity. If convenient, it could be expressed as miles per mile. More often, it is expressed as inches per inch or centimeters per centimeter, or in the case of interatomic distances as angstrom per angstrom Å). Typical interatomic distances are on the order of 2 to 5 Å ($1\,\text{Å} = 10^{-8}$ cm). Of course, since strain is dimensionless, you may often find only a number for strain, either as a number or as a percent change. Again in the case of computerized equipment, the program can convert extension to strain automatically. One of the modern uniaxial test machines is shown in Figure 2.2, illustrating the stress–strain curve being plotted as the test takes place.

Elastic Deformation

For a crystalline solid the force or stress can be viewed as extending to each row of atoms and that the elongation of the bar takes place by separation of the

**Figure 2.2**  Modern tensile test apparatus. (Courtesy of Instron Corporation.)

atoms in the direction of loading. On an atomistic basis the strain is the change in the interatomic spacing per unstrained equilibrium interatomic distance. The *elastic strength* of the material is a function of this atom bond strength. If the force is removed, the atoms snap back to their original position much as an extended spring would do. Elastic deformation is reversible. For polymers the elastic deformation often consists of the unkinking of the long-chain molecules. Some atom separation may be involved, particularly in those polymers that do not consist of long chains but of a more three-dimensional network of atoms. But again the elastic deformation is reversible. In some polymers there may be a time delay for the atoms to return to their original position, resulting in a mechanical hysteresis.

Although the *elastic strength* is an important material characteristic, since it is a measure of the atom bond strength, you will not find this term used in handbooks or in most texts. Instead, we use *modulus of elasticity*. Figure 2.3 shows how this property is determined. It is simply the slope of the line that represents the elastic portion of the stress–strain graph (i.e., it is the stress required to produce unit strain). In crystalline materials it is a good indicator of the atom bond strength. The uniaxial modulus of elasticity is often referred to as Young's modulus. The moduli of some common engineering materials are listed in Table 2.1. Since the atom bond strength decreases with increasing temperature the moduli also decrease as temperature increases (Figure 2.4). Since strain is dimensionless, the modulus has the same dimensions as stress. Conversion units are listed in Appendix A.

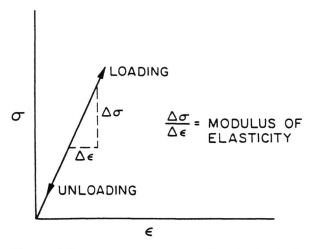

**Figure 2.3** Elastic stress–strain relationship showing how the modulus of elasticity is determined.

**Table 2.1**  Typical Moduli and Poisson's Ratio at 20°C

| Material | Average elastic modulus,[a] $E$ | | Poisson's ratio |
|---|---|---|---|
| | MPa $\times$ 10$^5$ | psi $\times$ 10$^6$ | |
| Aluminum alloys | 0.71 | 10.3 | 0.31 |
| Plain carbon steels | 2.0 | 29 | 0.33 |
| Copper | 1.1 | 16 | 0.33 |
| Titanium | 1.17 | 17 | 0.31 |
| Tungsten | 4.0 | 58 | 0.27 |
| MgO (magnesia) | 2.07 | 30 | 0.36 |
| $Si_3N_4$ | 3.04 | 44 | 0.24 |
| $Al_2O_3$ (alumina) | 3.80 | 55 | 0.26 |
| BeO | 3.11 | 45 | 0.34 |
| Plate glass | 0.69 | 10 | 0.25 |
| Diamond | 10.35 | 150 | — |
| Nylon | 0.025 | 0.4 | — |
| Polymethyl methacrylate (plexiglass) | 0.035 | 0.5 | — |

[a] The moduli vary with direction of measurement and with composition, the latter being a small effect and the former being a small effect in most large polycrystalline bodies.

Any force, no matter how small, will extend the atom spacing if it is a tensile force and compress the spacing for a compressive force. As long as the elastic condition is maintained, the strain is linearly proportional to the stress as expressed by Hooke's law, that is,

$$\sigma = E\epsilon \tag{2.8}$$

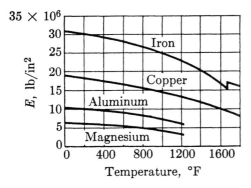

**Figure 2.4**  Elastic modulus versus temperature for some common metals. (From A. G. Guy, *Elements of Physical Metallurgy*, Addison-Wesley, Reading, Mass., 1959.)

The proportionality constant relating stress to elastic strain is $E$, *Young's modulus of elasticity.* It will be the same value in both tension and compression. It is of interest now to examine the magnitude of the changes in length that one might expect when subjecting a common engineering material to a wide range of forces.

   *Example 2.1.* For a cylindrical steel bar of 2-in. gauge length and 0.505 in. diameter (ASTM E8-90 specifications), compute the stresses and strains for applied forces of 6 and 6000 lb.

   Solution.   Since this bar has a cross-sectional area of 0.2 $in.^2$, the stress for 6 lb becomes

$$\sigma = \frac{6\,lb}{0.2\,in.^2} = 30\,psi = 0.2\,MPa$$

From Hooke's law (modulus from Table 2.1),

$$\epsilon = \frac{\sigma}{E} = \frac{30\,psi}{30 \times 10^6} = 10^{-6}$$

Remember that strain is a dimensionless quantity. Therefore, for a 2-in. gauge length, the total length change, $\Delta l$, becomes

$$\Delta l = 2 \times 10^{-6}\,in.$$

Thus, even a small force, barely measurable in common engineering tests, will elongate the specimen 2 $\mu$in., also barely measurable on most test equipment. For a 6000-lb force, the strain will be 1000 times greater (i.e., $10^{-3}$) and the corresponding elongation will be $2 \times 10^{-3}$ in. This strain value is easily measured and the order of those commonly experienced in modulus determinations.

   In the elastic region of a crystalline material a lateral contraction also occurs during extension. The ratio of lateral contraction strains to longitudinal extension strains is called *Poisson's ratio,* which can be expressed by the following:

$$\frac{\Delta d}{d} = -\epsilon_x = -\epsilon_y = \mu\epsilon_z \qquad (2.9)$$

where $d$ = bar diameter
   $\epsilon_x, \epsilon_y$ = lateral contraction strains (negative)
      $\epsilon_z$ = positive longitudinal strain
       $\mu$ = Poisson's ratio

This lateral contraction arises due to the nature of the atom bond. The equilibrium distance is a balance between attractive and repulsive forces. When this balance is disturbed in one direction a disturbance will be necessitated in other crystallographic directions in order to balance the repulsive and attractive forces with the external forces. For an isotropic solid (i.e., one whose mechanical properties are independent of direction), $\mu = 0.25$. However, crystalline solids are not isotropic and neither are most noncrystalline solids. A metal or ceramic body is made of many grains (crystals) of varying orientation such that we are actually dealing with average values when measuring directionally dependent properties of polycrystalline bodies. For most polycrystalline metals $\mu$ is found to be approximately 0.3, and for many glasses $\mu = 0.25$ (i.e., isotropic behavior). It is of little or no practical significance, but it should be pointed out that in the elastic region there will be a slight change in volume when $\mu > 0.5$, and it is always a positive number, which can be expressed by the equation

$$\frac{\Delta V}{V} = \epsilon_z(1 - 2\mu) \tag{2.10}$$

The volume change is so small that for practical purposes the modulus can be computed by measuring either the change in length or area. Generally, we use equation (2.4) for the strain value in moduli determinations since the extension is easier to measure accurately than is the reduction in area.

Plastic Deformation

As stress is continually applied to a material body, a point will be reached where stress and strain are no longer related in a linear fashion (i.e., Hooke's law is no longer obeyed). Furthermore, if the force is released, the bar will no longer return to its original length or shape. Some permanent deformation has occurred. Figure 2.5 shows schematically an engineering stress–strain curve for a relatively easily deformed metal. The elastic region is repeated here and

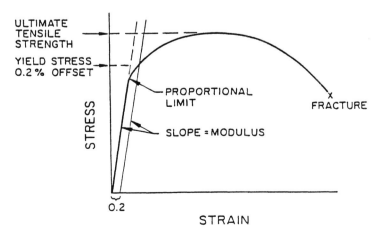

**Figure 2.5** Typical stress–strain curve for a ductile metal.

is shown to extend from the origin to the proportional limit, where departure from a linear relationship between stress and strain occurs. Often this point of departure is difficult to measure, so engineers devised a scheme whereby a line is constructed parallel to the elastic line, but is offset by a strain of 0.2% (numerically = 0.002) on the abscissa or strain axis. This 0.2% offset is defined by ASTM Specification E8-90. The point where the constructed line intersects the actual stress–strain curve is called the *yield stress*. Other offsets may sometimes be used, but the 0.2% is common because it is sufficient in most cases to clearly define the departure from the linear relationship, and also provides a tolerable amount of plastic deformation for most designs. If the offset is not given, it may be assumed to be 0.2%. The proportional limit is almost never used to define the yield stress, although this is actually the point at which plastic deformation (yielding) on a macroscale scale begins. As the stress is increased further, the shape of the curve depends on the plastic behavior of the material being tested. For most metals the stress to maintain plastic flow increases due to strain hardening, a concept we explore in Chapter 5. Therefore, the stress must increase with increasing plastic strain and the curve rises to its maximum value called the *ultimate tensile strength*, or as it is more commonly called, the *tensile strength*.

From the point of yielding on the curve in Figure 2.5 to the maximum in the curve, denoted by ultimate tensile strength, the specimen strain hardens and elongates uniformly. The volume remains constant during plastic deformation; therefore, the diameter must be reduced to maintain this constancy of volume. One could use equation (2.9) with a value of $\mu = 0.5$ to describe this region. However, the term *Poisson's ratio* is normally used to define elastic contraction, where there is a volume change dependent on the value of $\mu$. It is not appropriate to use it for the plastic situation where the contraction involves no change in volume and is not dependent on material constants. In this region between yielding and the ultimate tensile strength of the stress–strain curve, it should be realized that the plastic strains as expressed by equations (2.4) and (2.5) are not equal. This is strictly a consequence of the mathematical expressions for the two strains. In the case of length change we are stating a percent increase in length, whereas for the change in area definition we are expressing the strain as a percent decrease in area.

If we elected in the latter case to define strain as $\epsilon = (A_f - A_O)/A_f$, the two strains would be equivalent. However, this is not used since we elected to express strain as a percent change from original dimensions.

*Example 2.2.* A 0.5-in.-diameter test bar of a 2-in. gauge length was elongated to a permanent length of 2.3 in. The ultimate strength was not attained. Compute the plastic engineering strain by (a) change in length and (b) reduction in area.

Solution. (a) By change in length:

$$\epsilon = \frac{\Delta l}{l_0} = \frac{2.3 - 2.0}{2.0} = 0.15 = 15\%$$

(b) By change in area:

$$\epsilon = \frac{\Delta A}{A_0}$$

From constancy of volume (i.e., $A_0 l_0 = A_1 l_1$),

$$A_1 = \frac{A_0 l_0}{l_1} = \frac{0.2 \times 2.0}{2.3} = 0.174 \text{ in.}^2 \qquad \text{(area at a permanent extension of 2.3 in.)}$$

$$\epsilon = \frac{0.2 - 0.174}{0.2} = 0.13 = 13\%$$

Which method should be used in reporting? Actually, if the deformation is uniform throughout the length of the bar, it does not matter as long as the method used is stated and that we are aware that there is a difference in the values obtained by the two methods.

In Figure 2.6a a stress–strain curve is plotted where the test has been halted prior to fracture and also prior to attaining the ultimate tensile strength. When the load is released, as indicated by the downward arrow, the plot follows a

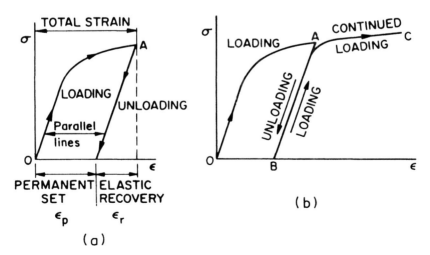

**Figure 2.6** Strain hardening causes the stress required to maintain plastic flow to increase but does not affect the modulus of elasticity.

line parallel to the original elastic region. The strain remaining, $\epsilon_p$, is permanent plastic deformation. The recovered elastic strain, $\epsilon_r$, is actually larger than the original elastic strain at the point of yielding. During plastic deformation some elastic strain is introduced, which adds to that previously introduced prior to yielding. Now if we reload the specimen (Figure 2.6b), the new yield strength becomes that at the point we released the load. The material has become stronger via the strain-hardening process. This is a common method of strengthening metals, although we do not use a tensile test for accomplishing the strengthening; rather, the metal is deformed by rolling for both sheet and rod material, or by drawing in the case of wire. Sometimes the metal is deformed in bending, torsion, compression, and tension, or combinations of these, to obtain a desired shape, as for example in manufacturing automobile bodies. The increased strength may be an added advantage. However, in many cases the elastic strains resulting from complex forming operations are not fully recovered when the load is released, due to constraints placed on the body in some regions. A region experiencing elastic tensile strains adjacent to one of elastic compressive strains may result in a net localized zero change in dimensions when the load is released, yet the strains remain. These residual strains are accompanied by residual stresses. Stresses of this type are often a problem. Machining a part that has residual stresses may result in warping due to stress release during the machining operation. Residual stresses can also promote corrosion. Frequently, a deformed part is heated to remove these stresses. This process, called a stress relief anneal, is discussed further in Chapter 5. Note in Figure 2.6a that the slope of the unloading line is the same as that of the original loading line. The forces that pull the atoms back together when the load is removed are the same interatomic forces that had to be overcome in the first elastic elongation. Thus the modulus is not changed by plastic deformation.

There is one more part of the engineering stress–strain curve that must be examined, and that is the region between the maximum in the curve and the fracture point. Why does the stress decrease at the point of the maximum in the curve? This would imply that strain hardening ceases or even that strain softening may be occurring. Actually, softening does not occur. The metal still strain hardens, but the deformation becomes localized in a small-"necked" region (Figure 2.7). The diameter in this region begins to reduce faster than in the other regions. This causes the actual or true stress in this region to increase because the true stress is defined by equation (2.3), and $A_i$ is constantly becoming smaller in the necked region. To measure the stress in this region we must measure the actual diameter during the test. Sometimes this may be done, but most often the true stress is ignored since, for design purposes, the yield strength is usually the significant number. When the maximum in the curve is reached and necking begins, the true stress increase resulting from the reduc-

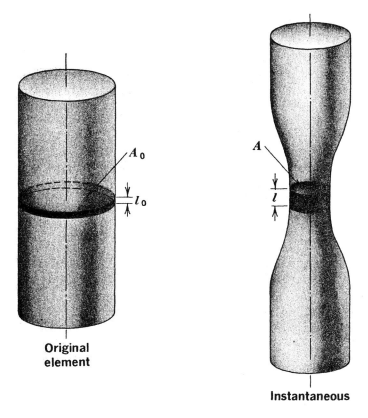

**Figure 2.7** Localized plastic flow causes "necking" to occur in a tensile specimen.

tion in area occurs faster than strain hardening. This behavior has been described as *plastic instability,* which causes the material to go rapidly to fracture. On our engineering stress–strain curve, however, the stress appears to decrease, but this is not a true decrease. It occurs only because we elected to define stress as the force over the original area, the latter number being a constant. The actual force required to maintain extension decreases after the maximum in the engineering curve is reached, and since this force is divided by a constant, $A_0$, the engineering stress decreases. Engineering strain also differs from the true strain since again we are dividing by a constant, $A_0$ or $l_0$, in defining engineering strain. True strain is obtained by measuring the change in length or area and dividing this number by the actual length or area at the time of the incremental measurement. Mathematically, the true strain is expressed as

$$d\epsilon = \frac{dl}{l} \tag{2.11}$$

The total true strain to fracture, $\epsilon_f$, can be obtained by integration over the entire change in length, resulting in

$$\epsilon_{\text{true}} = \int_{l_0}^{l_f} \frac{dl}{l} = \ln \frac{l_f}{l_0} \tag{2.12}$$

Since $A_0 l_0 = A_f l_f$ from the constancy of volume requirement, the true strain to fracture as measured by reduction in area becomes

$$\epsilon_{\text{true}} = \ln \frac{A_0}{A_f} \tag{2.13}$$

Figure 2.8 shows a comparison of engineering and true stress–strain curves for actual experimental data obtained for hot-rolled 1018 steel. The experimental data points for the two curves are connected by the dashed line.

*Example 2.3.* A 0.505-in.-diameter test bar was elongated to a permanent length of 2.3 in. (see Example 2.2).

(a) What was the total true strain obtained as measured by the change in length?

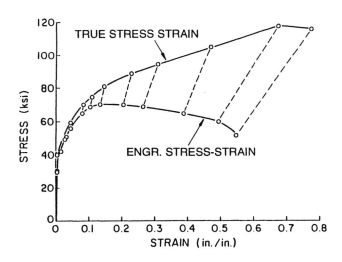

**Figure 2.8** Comparison of engineering and true stress–strain curves for hot-rolled 1018 steel.

$$\epsilon_{true} = \ln\frac{l_f}{l_0} = \ln\frac{2.3}{2.0} = 0.14$$

(*Note*: From Example 2.2 the engineering strain was 0.15.)

(b) What was the total true strain as measured by reduction in area?

$$\epsilon_{true} = \ln\frac{A_0}{A_f} = \ln\frac{0.2}{0.174} = 0.14$$

(*Note*: From Example 2.2 the engineering strain was 0.13.)

The differences in true and engineering strain values become much larger at higher values of strain. Also note that the true strain is the same whether measured by change in length or area. This will always be true as long as we are dealing with uniform deformation, but not when localized necking occurs.

(c) If the load on this bar to produce the total strain measured was 1000 lb, what was the final engineering and true stresses in psi and MPa units?

$$\text{Engineering stress} = \frac{1000\ \text{lb}}{0.2\ \text{in.}^2} = 5000\ \text{psi}\ (34.5\ \text{MPa})$$

$$\text{True stress} = \frac{1000\ \text{lb}}{A_f} = \frac{1000\ \text{lb}}{0.174\ \text{in.}^2} = 5747.13\ \text{psi}\ (39.66\ \text{MPa})$$

The strain to fracture (i.e., the total observed plastic strain) is called the *ductility* of the material. It can be measured by either the overall change in length or the overall change in area. It is made on the specimen after fracture by meshing together the two broken sections. Therefore, this strain does not include the elastic strain, which was released at the point of fracture. Although some data may be reported in terms of the engineering strain by using either the original length or the original area, it is recommended that true strains to fracture, as obtained from equations (2.12) or (2.13), be used.

Which is better, the strain to fracture as measured by change in length or by change in area? When extensive necking occurs, the deformation is highly localized and the strain to fracture as measured by change in length becomes a function of the gauge length. Thus the strain for a 2-in. gauge length will be different from that of a 10-in. gauge length (Problem 2.7). The ductility should represent a material property and not be a function of the specimen dimensions. Therefore, strain to fracture as measured by reduction in area is a better measure of ductility.

In the discussion of stress–strain relationships above, we have used metals as an example. The same equations also hold for ceramics, polymers, and composites. Their will be a big difference in properties for the various materials, especially for ductility and modulus of elasticity. Materials that exhibit little or no plastic strain prior to fracture are said to be brittle. Glasses, ceramics, and cast iron are examples of brittle materials.

The room temperature stress–strain curves of some typical materials are compared in Figure 2.9. (See the Suggested Reading for a compilation of stress–strain curves.) Note that in Figure 2.9a, two yield points are evident. This is a characteristic of many steels and some alloys and will be explained in a later section. Also note that only the steel shows strain hardening, a process most often present in metals or metallic components of composites. It will seldom be seen in ceramics or polymers.

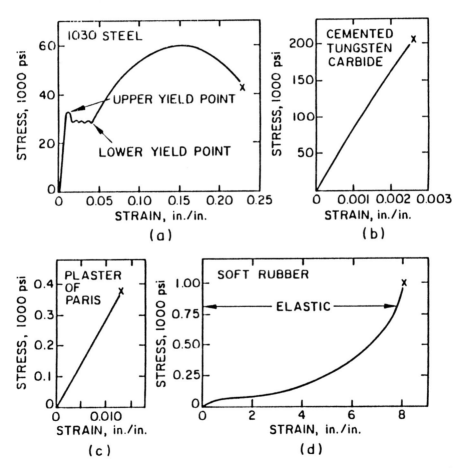

**Figure 2.9**  Comparison of engineering stress–strain curves for various types of materials. (From H. W. Hayden, W. G. Moffatt, and J. Wulff, *The Structure and Properties of Materials,* Wiley, New York, 1965.)

The stress–strain behavior of all materials is strongly temperature dependent. We noted previously that the elastic moduli, and hence the slope of the elastic stress–strain line, decreases with increasing temperature. The yield strength, ultimate tensile strength, and stress to maintain plastic flow also decrease with increasing temperature while the ductility and toughness increase. Generally, one can, by increasing the temperature sufficiently, change a material that is brittle to one that is ductile. The effect of temperature on the shape of the stress–strain curve is depicted in Figure 2.10.

### 2.2.2 Stress–Strain Relations in Torsion

At this point in our discussion we should briefly examine the elastic stress–strain relations in torsion. In Figure 2.11 we show a cylindrical bar being elast-

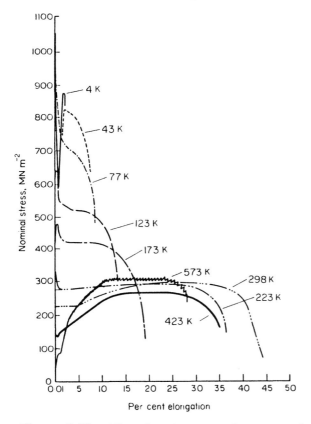

**Figure 2.10** Effect of temperature on the stress–strain curve of tantalum. (From J. H. Bechtold, *Acta Metall.*, *3*, 249, 1955.)

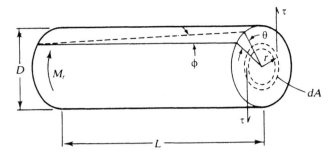

**Figure 2.11** Torsion in a cylindrical bar.

ically twisted through an angle $\phi$ by a stress $\tau$. This is a shear stress, and by convention the symbol $\tau$ is used to distinguish it from normal or uniaxial stress, which are usually denoted by the symbol $\sigma$. The shear strain $\gamma$ is given by $\delta/l$, which is the tangent of $\phi$. The stress is a function of the torque, or twisting moment $M_t$, at the point of interest $r$ and the bar diameter $D$, and can be expressed by

$$\tau = \frac{32M_t}{\pi D^4} r \tag{2.14}$$

The maximum stress occurs on the outer fiber where $r = D/2$. The shear strain is proportional to the shear stress, with the proportionality constant being the shear modulus $G$, that is,

$$\tau = G\gamma \tag{2.15}$$

This is simply Hooke's law in torsion. It can be shown that $G$ and $E$ are related by

$$G = \frac{E}{2(1 + \mu)} \tag{2.16}$$

The torsion test can be extended to the plastic range but is seldom used to determine ductility or flow stress.

### 2.2.3 Hardness

The first method used to measure hardness was that developed by Friedrich Mohs in 1832, in which he placed materials on a hardness scale of 1 to 10. Mohs devised the scale based on which material could produce a scratch on other materials under ordinary hand abrasion techniques. Nothing would scratch diamond, so it was given the number 10. Corundum, a form of aluminum oxide, was given the number 9 and the scale progressed downward to talc, which one could say perhaps could not produce a scratch on any other material. Subsequent to the time of Mohs scale, many hardness tests and hardness scales have been developed that are more quantitative. They all have one thing in common, however. Their hardness numbers are relative. There is no such thing as an absolute hardness number as in yield strength, for example. The Brinell hardness test, one of the older methods, uses the microscopically measured diameter of the indentation, which results from the penetration of a hardened steel or tungsten carbide ball that has been pressed into the sample under a specified load, usually 500 or 3000 kg. The Vickers, or diamond pyramid hardness (DPH), was later developed and used the same principle as the Brinell except that the indenter was diamond shaped and usually made of a diamond crystal. Here the diagonal of the indentation was measured. A closely related method was the Knoop hardness test, which uses a different-shaped diamond for the indenter. The DPH and Knoop techniques are often employed in microhardness tests, where one is interested in the hardness of a very small region (e.g., the depth of hardening for a surface-hardened steel and for thin sheet specimens). The DPH and Knoop scales have one advantage over the others in that they are more linear (i.e., a DPH value of 300 is probably close to three times the hardness of a material that has a measured 100 DPH number). An optical microscope is used to measure the diagonals of the indentation, which is then converted to a hardness number.

The Rockwell scale is undoubtedly the most widely used. It was developed by S. P. Rockwell in 1919. It is not as linear as the others, but it has caught on, probably because of its simplicity. The basic Rockwell tester has a number of scales, which consists of various indenters used in combination with a variety of loads. The most common scales employ either a 1/16-in.-diameter steel sphere, a 1/8-in.-diameter steel sphere, or a diamond indenter. Each of these indenters can be used with a load of 60, 100, or 150 kg, giving a total of nine common scales. The Rockwell tester is so devised that it measures the depth of the indentation, which is then automatically converted to a hardness number without any actual microscopic or other form of measurement of depth of the depression. As such, the scale is strictly arbitrary and very nonlinear. Anytime that you have a measurement near the end of the scale, you must resort to another scale. The Rockwell C scale, 150 kg with the diamond pyramid indenter, and the Rockwell A scale, 60 kg with diamond indenter, are normally used for steels and similar hard alloys. Aluminum alloys are usually measured on the B scale, 100 kg with a 1/16-in.-diameter sphere, while some copper alloys are measured on the K scale.

Polymers are frequently measured on the Rockwell E and M scales, with the M scale being preferred for the harder polymers such as the phenolics. The Barcol hardness test measures the resistance to penetration of a sharp steel point under a spring load. It has a direct-reading scale from 0 to 100 and is often used to measure the degree of cure of a polymer. The Shore hardness test measures the resistance of a material to indentation by a spring-loaded indenter and is normally used for rubbers.

Ceramic materials are relatively brittle and frequently crack if too much load is applied during the test. Nevertheless, their hardnesses are measured. Vickers (the DPH) or the Knoop test are the preferred methods because of their linearity and the fact that cracking can be seen through the microscope. If the material cracks, the hardness should be remeasured with a smaller load. ASTM specifications have been developed for the common tests: namely, ASTM E10 for the Brinell method, ASTM E18 for the Rockwell test, and ASTM E92 for the DPH methods.

Hardness tests consume little time and are very informative. They do not have the precision that other tests do, but one can gain a lot of information in about 1 minute. The hardness is related to strength. The indenter, under load, is plastically deforming the material and thereby pushing it out of its way. The material is strain hardening during the tests, and as deformation continues, the material resists further penetration by the indenter. Usually, fracture does not occur, except in very brittle materials, but if it does occur, the test is not a valid one. The resistance to deformation is the reason that the hardness value obtained can be related to the ultimate tensile strength of the material. It takes into account any strain-hardening effects. Empirical relationships have been developed and published in the form of charts. By far the most widely used is that published by the developers of the Rockwell testers, the Wilson Company. Their chart is reproduced in Table 2.2. One must remember that these comparisons are empirical, but they can give an approximate strength of the material without the time-consuming operations of machining and testing a tensile bar. The hardness test is the most utilized mechanical property test of all methods available. It is one that you should learn to conduct and be able to recognize its advantages and limitations. But this can only be accomplished by actual operation of the tester and is one of the reasons that most material courses require laboratory experiments. If you want to obtain a good return for a few minutes of effort, learn and understand the hardness test. A comparison of the various methods, together with the indenter shape, is shown in Figure 2.12 and a modern digital-readout Rockwell tester is shown in Figure 2.13.

## 2.2.4 Fracture Toughness

Although the ductility of a material is an important property, frequently the fracture toughness of a material will also be specified for design purposes. Generally, but not always, a ductile material will also be a tough material. Ductility is measured in a relatively slow strain rate test. Strain rates on the order of $10^{-3}$ to $10^{-4}$ per second are normally used. In fact, the ASTM E8-90 specification states that the strain rate shall not exceed $8 \times 10^{-3}$ per second. It

**Table 2.2** Hardness Conversion Numbers from a Variety of Test Methods

| C | A | 15-N | 30-N | Vickers | Knoop | Br'l | Tensile Strength |
|---|---|---|---|---|---|---|---|
| 150 kg | 60 kg | 15 kg N | 30 kg N | 10 kg 136° Diamond | 500 Gr. & Over | 3000 kg 10mm Ball | |
| Brale® | Brale | Brale | Brale | | | | |
| Rockwell | Rockwell | Rockwell Superficial | Rockwell Superficial | Vickers | Knoop | Brinell** (Standard Ball) | Thousand lbs. per sq. in. |
| 70 | 86.5 | 94.0 | 86.0 | 1076 | 972 | — | |
| 69 | 86.0 | 93.5 | 85.0 | 1004 | 946 | — | |
| 68 | 85.6 | 93.2 | 84.4 | 940 | 920 | — | |
| 67 | 85.0 | 92.9 | 83.6 | 900 | 895 | — | |
| 66 | 84.5 | 92.5 | 82.8 | 865 | 870 | — | |
| 65 | 83.9 | 92.2 | 81.9 | 832 | 846 | — | Inexact and only for steel |
| 64 | 83.4 | 91.8 | 81.1 | 800 | 822 | — | |
| 63 | 82.8 | 91.4 | 80.1 | 772 | 799 | — | |
| 62 | 82.3 | 91.1 | 79.3 | 746 | 776 | — | |
| 61 | 81.8 | 90.7 | 78.4 | 720 | 754 | — | — |
| 60 | 81.2 | 90.2 | 77.5 | 697 | 732 | — | — |
| 59 | 80.7 | 89.8 | 76.6 | 674 | 710 | 634 | 351 |
| 58 | 80.1 | 89.3 | 75.7 | 653 | 690 | 615 | 338 |
| 57 | 79.6 | 88.9 | 74.8 | 633 | 670 | 595 | 325 |
| 56 | 79.0 | 88.3 | 73.9 | 613 | 650 | 577 | 313 |
| 55 | 78.5 | 87.9 | 73.0 | 595 | 630 | 560 | 301 |
| 54 | 78.0 | 87.4 | 72.0 | 577 | 612 | 543 | 292 |
| 53 | 77.4 | 86.9 | 71.2 | 560 | 594 | 525 | 283 |
| 52 | 76.8 | 86.4 | 70.2 | 544 | 576 | 512 | 273 |
| 51 | 76.3 | 85.9 | 69.4 | 528 | 558 | 496 | 264 |
| 50 | 75.9 | 85.5 | 68.5 | 513 | 542 | 481 | 255 |
| 49 | 75.2 | 85.0 | 67.6 | 498 | 526 | 469 | 246 |
| 48 | 74.7 | 84.5 | 66.7 | 484 | 510 | 451 | 238 |
| 47 | 74.1 | 83.9 | 65.8 | 471 | 495 | 442 | 229 |
| 46 | 73.6 | 83.5 | 64.8 | 458 | 480 | 432 | 221 |
| 45 | 73.1 | 83.0 | 64.0 | 446 | 466 | 421 | 215 |
| 44 | 72.5 | 82.5 | 63.1 | 434 | 452 | 409 | 208 |
| 42 | 71.5 | 81.5 | 61.3 | 412 | 426 | 390 | 194 |
| 40 | 70.4 | 80.4 | 59.5 | 392 | 402 | 371 | 182 |
| 38 | 69.4 | 79.4 | 57.7 | 372 | 380 | 353 | 171 |
| 36 | 68.4 | 78.3 | 55.9 | 354 | 360 | 336 | 161 |
| 34 | 67.4 | 77.2 | 54.2 | 336 | 342 | 319 | 152 |
| 32 | 66.3 | 76.1 | 52.1 | 318 | 326 | 301 | 146 |
| 30 | 65.3 | 75.0 | 50.4 | 302 | 311 | 286 | 138 |
| 28 | 64.3 | 73.9 | 48.6 | 286 | 297 | 271 | 131 |
| 26 | 63.3 | 72.8 | 46.8 | 272 | 284 | 258 | 125 |
| 24 | 62.4 | 71.6 | 45.0 | 260 | 272 | 247 | 119 |
| 22 | 61.5 | 70.5 | 43.2 | 248 | 261 | 237 | 115 |
| 20 | 60.5 | 69.4 | 41.5 | 238 | 251 | 226 | 110 |

**Table 2.2**   Continued

| B 100 kg 1/16" Ball Rockwell | F 60 kg 1/16" Ball Rockwell | 30-T 30 kg 1/16" Ball Rockwell Superficial | E 100 kg 1/8" Ball Rockwell | Knoop 500 Gr. & Over Knoop | Br'l 3000 kg D.P.H. 10 kg Brinell | Tensile Strength Thousand lbs. per sq. in. |
|---|---|---|---|---|---|---|
| 100 | — | 83.1 | — | 251 | 240 | 116 |
| 99 | — | 82.5 | — | 246 | 234 | 114 |
| 98 | — | 81.8 | — | 241 | 228 | 109 |
| 97 | — | 81.1 | — | 236 | 222 | 104 |
| 96 | — | 80.4 | — | 231 | 216 | 102 |
| 95 | — | 79.8 | — | 226 | 210 | 100 |
| 94 | — | 79.1 | — | 221 | 205 | 98 |
| 93 | — | 78.4 | — | 216 | 200 | 94 |
| 92 | — | 77.8 | — | 211 | 195 | 92 |
| 91 | — | 77.1 | — | 206 | 190 | 90 |
| 90 | — | 76.4 | — | 201 | 185 | 89 |
| 89 | — | 75.8 | — | 196 | 180 | 88 |
| 88 | — | 75.1 | — | 192 | 176 | 86 |
| 87 | — | 74.4 | — | 188 | 172 | 84 |
| 86 | — | 73.8 | — | 184 | 169 | 83 |
| 85 | — | 73.1 | — | 180 | 165 | 82 |
| 84 | — | 72.4 | — | 176 | 162 | 81 |
| 83 | — | 71.8 | — | 173 | 159 | 80 |
| 82 | — | 71.1 | — | 170 | 156 | 77 |
| 81 | — | 70.4 | — | 167 | 153 | 73 |
| 80 | — | 69.7 | — | 164 | 150 | 72 |
| 79 | — | 69.1 | — | 161 | 147 | 70 |
| 78 | — | 68.4 | — | 158 | 144 | 69 |
| 77 | — | 67.7 | — | 155 | 141 | 68 |
| 76 | — | 67.1 | — | 152 | 139 | 67 |
| 75 | 99.6 | 66.4 | — | 150 | 137 | 66 |
| 74 | 99.1 | 65.7 | — | 147 | 135 | 65 |
| 72 | 98.0 | 64.4 | — | 143 | 130 | 63 |
| 70 | 96.8 | 63.1 | 99.5 | 139 | 125 | 61 |
| 68 | 95.6 | 61.7 | 98.0 | 135 | 121 | 59 |

**Table 2.2** Continued

| B | F | 30-T | E | Knoop | Br'l | Tensile Strength |
|---|---|---|---|---|---|---|
| 100 kg 1/16″ Ball | 60 kg 1/16″ Ball | 30 kg 1/16″ Ball | 100 kg 1/8″ Ball | 500 Gr. & Over | 3000 kg D.P.H. 10 kg | Thousand lbs. per sq. in. |
| Rockwell | Rockwell | Rockwell Superficial | Rockwell | Knoop | Brinell | |
| 66 | 94.5 | 60.4 | 97.0 | 131 | 117 | 57 |
| 64 | 93.4 | 59.0 | 95.5 | 127 | 114 | |
| 62 | 92.2 | 57.7 | 94.5 | 124 | 110 | |
| 60 | 91.1 | 56.4 | 93.0 | 120 | 107 | |
| 58 | 90.0 | 55.0 | 92.0 | 117 | 104 | |
| 56 | 88.8 | 53.7 | 90.5 | 114 | 101 | |
| 54 | 87.7 | 52.4 | 89.5 | 111 | *87 | |
| 52 | 86.5 | 51.0 | 88.0 | 109 | *85 | |
| 50 | 85.4 | 49.7 | 87.0 | 107 | *83 | |
| 48 | 84.3 | 48.3 | 85.5 | 105 | *81 | |
| 46 | 83.1 | 47.0 | 84.5 | 103 | *79 | |
| 44 | 82.0 | 45.7 | 83.5 | 101 | *78 | |
| 42 | 80.8 | 44.3 | 82.0 | 99 | *76 | |
| 40 | 79.7 | 43.0 | 81.0 | 97 | *74 | |
| 38 | 78.6 | 41.6 | 79.5 | 95 | *73 | |
| 36 | 77.4 | 40.3 | 78.5 | 93 | *71 | |
| 34 | 76.3 | 39.0 | 77.0 | 91 | *70 | |
| 32 | 75.2 | 37.6 | 76.0 | 89 | *68 | |
| 30 | 74.0 | 36.3 | 75.0 | 87 | *67 | |
| 28 | 73.0 | 34.5 | 73.5 | 85 | *66 | |
| 24 | 70.5 | 32.0 | 71.0 | 82 | *64 | |
| 20 | 68.5 | 29.0 | 68.5 | 79 | *62 | |
| 16 | 66.0 | 26.0 | 66.5 | 76 | *60 | |
| 12 | 64.0 | 23.5 | 64.0 | 73 | *58 | |
| 8 | 61.5 | 20.5 | 61.5 | 71 | *56 | |
| 4 | 59.5 | 18.0 | 59.0 | 69 | *55 | |
| 0 | 57.0 | 15.0 | 57.0 | 67 | *53 | |

Even for steel, tensile strength relation to hardness is inexact unless determined for specific material.

*Below Brinell 101 tests were made with only 500 kg load and 10mm ball.

**Above Brinell 451 HB tests were made with 10mm carbide ball.

*Source*: Courtesy of Page-Wilson Corporation.

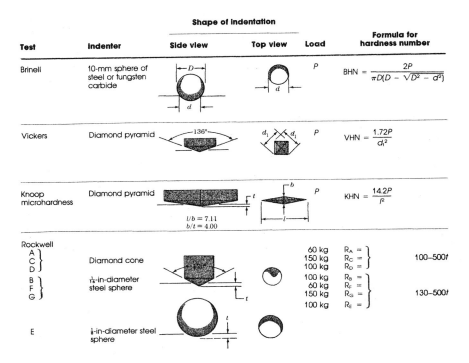

**Figure 2.12** Hardness test indenters. (From H. W. Hayden, W. G. Moffatt, and J. Wulff, *The Structure and Properties of Materials,* Wiley, New York, 1965.)

is possible for a material that is ductile in a slow strain rate test to fracture in a brittle manner in a fast impact test. Toughness is related to the plastic deformation during crack propagation. When a crack propagates, plastic deformation occurs in the vicinity of the crack tip and acts to slow the propagation rate by absorbing some of the energy applied. But at fast strain rates and also at low temperatures this plasticity is inhibited, thereby favoring brittle or fast fracture, the crack propagating through the body in a matter of micro- to milliseconds. Toughness, then, is the ability of a material to resist crack propagation, and it is frequently determined by measuring the energy absorbed during the fracture process while the material is being impacted by a fast-swinging pendulum. Strain rates in the impact test are on the order of $10^2$ to $10^4$ per second. A schematic of a simple impact-testing machine is shown in Figure 2.14. It is often referred to as a Charpy V-notch test. The Charpy specimen shape is shown in the schematic just above that of the impact tester. The specimen is placed on the anvil and the pendulum hammer is released from its cocked starting position, impacting the specimen and thereby causing a fast fracture to

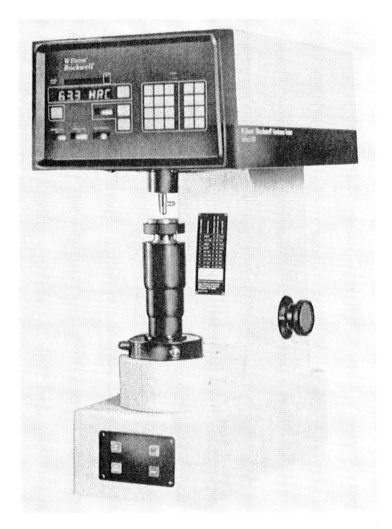

**Figure 2.13**   Digital-readout hardness tester. (Courtesy of Wilson Company.)

occur. Even though this is a relatively fast fracture, it may be of either a brittle or ductile nature or a combination of both, depending on the material history. The higher the position of the hammer after the fracture occurs, the less energy absorbed during the fracture process. A pointer attached to the hammer will show on a calibrated scale the foot-pounds of energy absorbed. Thus a brittle material will absorb very little energy and the pointer will point to the right

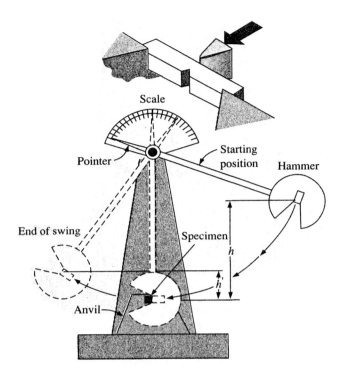

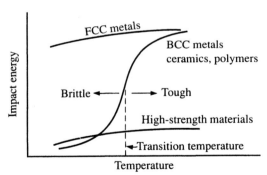

**Figure 2.14**  Schematic drawing of a standard impact testing apparatus. (From H. W. Hayden, W. G. Moffatt, and J. Wulff, *The Structure and Properties of Materials*, Wiley, New York, 1965.)

part of the scale near the zero mark. By some definitions or specifications a reading below 25 ft-lb for steels would be considered unacceptable and the material would not be used if it would experience the same use temperature as that of the test. This number of 25 ft-lb is somewhat arbitrary. Glass in a room-temperature test would show less than 1 ft-lb of energy absorbed during fracture, yet glass is widely used at room temperature. One must just be careful that it does not experience impact loading.

Notice that the word *temperature* was used a number of times in the preceding paragraph. Like most mechanical properties, toughness is temperature dependent. This temperature dependency of the fracture toughness of steels has resulted in some rather tragic failures in the past. One of the most striking observances of this effect was the fractures of the hull of Liberty cargo ships in World War II. These ships had been fabricated by welding relatively low-carbon steel plate. About 25% of the ships developed severe cracks. Subsequent investigations showed that these steels became brittle at temperatures on the order of 40°F. Stress concentrations in the original design also contributed to the failures. A typical impact–temperature curve, for a 1018 steel together with optical and scanning electron microscopic pictures of the fracture surfaces, is illustrated in Figure 2.15. This change in impact energy absorbed during fracture as a function of temperature is known as the brittle–ductile transition, and the temperature where the largest rate of change occurs is often called the *transition temperature*. This transition temperature is not a precise value such as the melting temperature, for example. It has been defined by different persons in different ways, some of which base it on the energy absorbed during fracture, while others might use the fracture appearance. When the material is brittle, the fracture surface appears granular by eye or at low magnifications and shows cleavage markings at higher magnifications in the electron microscope. The crack propagates by a cleaving action much like that the craftsman employs in the cleaving of diamonds to obtain the desired shape and cleavage facets. The ductile fracture appears "fibrous" at low magnifications and as an abundance of pits under the electron microscope. In contrast to cleavage, ductile failures occur more as a tearing action. The transition temperature varies with strain rate, stress concentration, and certain metallurgical variables associated with the processing of the steel. Fast strain rates, low temperatures, and stress concentrations tend to promote brittle behavior. The Charpy V-notch test represents the worst condition [i.e., a very fast strain rate impact on a notched specimen (notches create stress concentrations)]. The impact test tells us at what temperature it is safe for one to use the particular material in question. It does not provide data that can be used for design purposes.

In the late 1950s and early 1960s, a different approach to fracture toughness emerged, the principles of which encompass the subject of fracture mechanics.

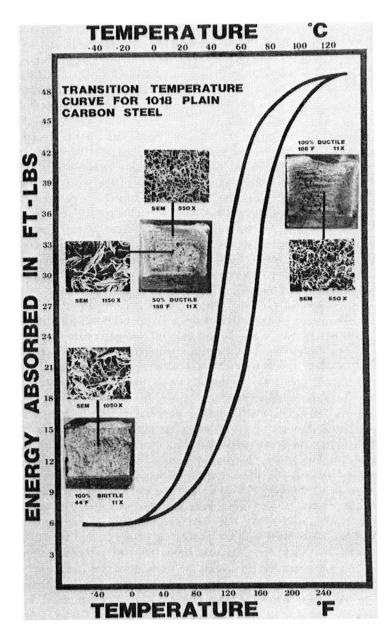

**Figure 2.15**   Brittle–ductile transition temperature determined by the impact test for a 1018 hot-rolled steel. (From senior project of Jon Gergen, Cal Poly.)

The theory underlying the fracture mechanics approach is beyond the scope of this book. It is based on the concept that at existing flaws (cracks or other discontinuities in structure) which act as stress concentrators, a certain flaw shape and size provides sufficient stress concentration that for a given applied stress, the flaw (or crack) will propagate to fracture. The stresses near the tip of a very sharp crack are governed by a parameter called the *stress intensity factor K*, which is a function of the crack length $a$ and the applied stress $\sigma$. It is expressed by the formula

$$K_1 = Y\sigma\sqrt{\pi a} \tag{2.17}$$

where $\sigma$ = applied nominal stress

$a$ = flaw depth for an edge flaw or crack, or half the length of an internal crack

$Y$ = a dimensionless geometric factor that involves the flaw geometry with respect to specimen dimensions (this value has been computed by stress analysts and is available in handbooks for a variety of geometries)

The critical stress intensity that will propagate the crack to failure is called $K_{1c}$ (pronounced "kay-one-see") and is a true material constant. It is the fracture toughness of the material. The fracture toughness has the SI units of MPa $\sqrt{m}$ and English units of ksi $\sqrt{in}$. The conversion factor between the two units is 1.1, so the values in these units for all practical purposes are interchangeable. Attempts have been made to empirically correlate $K_{1c}$ values with Charpy impact tests, but they must be treated with caution. The advantage of the fracture mechanics approach is that one can assume the existence of a certain flaw size, probably just at the limit of that detectable by nondestructive testing techniques, and compute therefrom the applied stress that would produce fracture.

*Example 2.4.* A 1-m-wide aluminum plate has a fracture toughness of 25 ksi $\sqrt{in}$. (27.5 MPa$\sqrt{m}$) and a yield strength of 70 ksi (517.5 MPa) and contains a center crack 0.1 in. (2.54 mm) in length. Calculate the fracture stress assuming that $Y = 1$ (valid when plate width is very large compared to crack length).

Solution.

$$\sigma = \frac{K_{1c}}{\sqrt{\pi a/2}} = \frac{27.5 \text{ MPa} \sqrt{m}}{\sqrt{2.54\pi \times 10^{-3}}} = 435.6 \text{ MPa} = 84\% \text{ of yield strength}$$

We use $a/2$ rather than $a$ for a center crack.

## 2.2.5 Fatigue Life, Fatigue Limit, and Fatigue Strength

In Section 2.1.1 we described the characteristics of the yield strength, a value that is widely used by engineers in their designs. The design engineer should be aware that in situations where alternating stresses are anticipated the yield strength values obtained in a uniaxial tension tests should not be used. Structural members that have been subjected to an alternating stress for long periods of time have been observed to fail at stresses as low as 25% of their uniaxial tensile yield strength. A rule of thumb for steels is that the failure stress in fatigue is about 50% of the ultimate tensile strength. Of course, such rules cannot be applied in design. Some steels show better fatigue resistance than others. Surface condition and the design geometries also play a major role. The presence of stress concentrators such as scratches, sharp corners, and notches will cause failure in fatigue at stresses much lower than that for polished parts and for components where generous radii at reduced sections, threads, and keyways have been provided by the design engineer.

Fatigue has been widely studied in metals because the majority of rotating machine parts and flexing parts, such as springs and beams are usually of metallic construction. It was also first noted in metals. The term *fatigue* was coined in the mid-nineteenth century to describe failures in such things as railroad wheel axles. Some of the early reports described fatigue failures as being due to *crystallization* of the metal. But of course this could not be happening, because metals become crystalline bodies during solidification from the molten state. Nor do metals become *tired* as a result of the stress concentration when subjected to alternating stresses. It is now realized that plastic deformation can take place on a microscale at stresses below the yield stress (i.e., in the normal elastic range). The to-and-fro movement under alternating stresses of dislocations, a type of crystalline imperfection that exists in all metals and that we discuss at length in Chapter 5, can cause permanent deformation in localized regions. This deformation can serve to nucleate cracks. But often microcracks, sufficiently small that an electron microscope is needed for their detection, already exist. Alternating stresses, even though in the normal elastic range macroscopically, can, via stress concentration, cause plastic deformation at these crack tips and thereby cause the cracks to slowly propagate. Over a period of time, which may vary from a few months to a decade, the crack will grow until it reaches a critical size sufficient to cause failure of the entire part. As the crack grows, the remaining good material carries more and more of the load and is thus being subjected to a larger stress, even though the overall nominal applied stress remains constant. At a critical stress level the member fails catastrophically. Because of this long period of time required for failure to occur the process was called *fatigue*. One could say that a metal member as a whole becomes weaker with time as the crack grows, but only in the sense that the load-bearing portion is becoming smaller. Since the stress is equal to force divided by area, if the area decreases the stress must increase.

A typical fracture surface of a shaft that failed by fatigue is shown schematically in Figure 2.16. It is composed of three parts according to the three stages

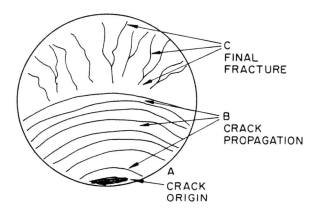

**Figure 2.16** Schematic description of a fatigue fracture surface of a shaft.

of fatigue. The first region, denoted by the letter A, is the region where the crack nucleates, or if it already exists, propagates to a size sufficient for detection. This region is a very small part of the total fracture surface but it may represent a large part of the total life. Region B is the macroscopic propagation stage. It often contains characteristic markings commonly called *beach* markings or *clamshell* markings, for obvious reasons. These markings result from the interruption of the load. The crack then stops propagating, and when the load is resumed at this same value again the crack resumes propagation, probably on a slightly different geometrical level. The crack tip also may become oxidized if the load is removed or reduced below the critical crack propagation level for a period involving hours or days. Such oxidation serves to make the markings more visible. This propagation stage, which may also be a large portion of the part life, requires that a tensile stress be present. Tensile, not compressive stresses, propagate cracks. The absence of the characteristic markings does not necessarily mean that fatigue has not occurred. If the two fracture surfaces rub together during the propagation stage they may smear the markings until they cannot be distinguished. Also, if the load is not interrupted during the propagation stage, the markings will not be present. On the other hand, their presence assures that fatigue was the cause of the failure. Finally, the crack attains the length where the remaining material cannot carry the load and catastrophic failure occurs in a matter of seconds. This region is labeled in Figure 2.16 with the letter C. A large final fracture region indicates that the part was subjected to a rather high stress, while a small region indicates a low overall stress application. An actual fracture surface of a shaft that failed in fatigue is shown in Figure 2.17. Note that the failure initiated in two different points and thus caused two sets of beach markings.

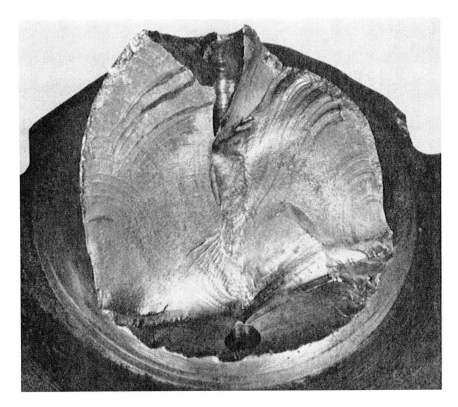

**Figure 2.17**   Fatigue fracture surface of a shaft.

There is another fatigue fracture surface marking that is frequently observed when examining the fracture at high magnifications under a scanning electron microscope (Figure 2.18). These markings are called *striations* and are a result of the plastic deformation that occurs due to the stress concentrations at the advancing crack tip. In most cases each striation represents one cycle of stress application.

Now let us examine how fatigue data are presented and used in engineering design. In laboratory tests, simple alternating tensile and compressives stresses are often used even though in the real world these alternating stresses are superimposed on a mean stress. An aircraft wing, for example, is subjected to a mean or average stress during a smooth flight, but it experiences an additional alternating stress during the periods of turbulence. Similarly, the fuselage is subjected to additional alternating stresses due to pressure changes with altitude. Aircraft failures have been known to occur from the latter situa-

**Figure 2.18** Fatigue striations that occurred during crack advancement in an aluminum alloy.

tion. Although the mean stress plays a role, it has been determined that the crack growth rate is more dependent on $\sigma_a$, the alternating stress amplitude. Fatigue data are usually represented by an *S-N* curve, a plot of the cycles to failure *N* versus the applied stress *S*. Actually, we plot stress versus log *N* since *N* can be a large number. Unfortunately, fatigue is statistical in nature. The distribution and severity or microcracks and/or microdefects cannot be predicted with any degree of certainty. Therefore, a number of tests must be run at each stress level. Figure 2.19 shows the scatter in fatigue limit for 10 specimens tested in identical fashions. Often data will be represented by several *S-N* plots, one curve being the median value, each point on the curve representing the number of cycles for 50% survival. In addition, curves for other percentages of survival, frequently for 5% and 95% survival, will be plotted as shown in Figure 2.20. Fatigue data have not been generated to the same extent that have yield and ultimate strength data. The latter are available from many handbooks and also from manufacturers of the alloys. A compilation of *S-N* curves and crack growth rate curves has been published (see the Suggested

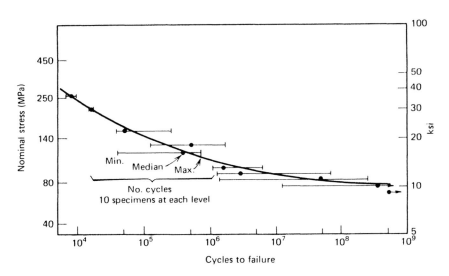

**Figure 2.19**  *S–N* curve obtained for a 7075-T6 aluminum. (From H. F. Hardrath, E. C. Utley, and D. E. Guthrie, *NASA TN D-210,* 1959.)

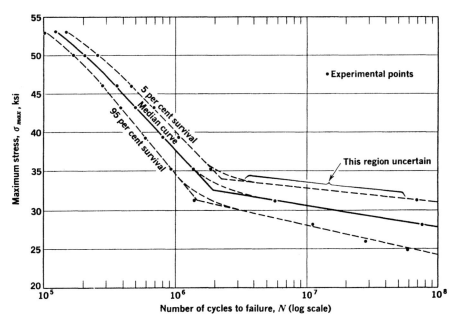

**Figure 2.20**  *S–N* data for phosphor bronze strip. (From M. N. Torey and G. R. Gohn, *ASTM Proceedings,* 1956.)

Reading). It probably represents the best source of fatigue information available at this time.

Some alloys, particularly ferrous ones, have *S–N* curves that tend to level out at a certain value of stress (i.e., the material will last indefinitely at this stress). This stress would then be a safe design value, applying an appropriate factor of safety. Remember that fatigue data must be treated statistically. If the handbook or other published data do not show the scatter obtained, as shown in Figure 2.19 or 2.20, it is assumed that the curve represents that for 50% survival at each stress value. The stress at which the material would last for an infinite number of cycles is called the *fatigue limit* or *endurance limit*, the former being the preferred term. For materials that do not show a well-defined limit, we use the number of cycles that the material will withstand at any specified value of stress. One can call the number of cycles the *fatigue life* for that stress and also speak of the stress level as being the *fatigue strength* for a given life rather than using the fatigue limit term.

During the last 20 years or so we have seen an increase in the use of polymers for machine parts such as gears and bearings. We find that many of these parts fail by fatigue and often show a *S–N* curve similar to that for metals. Many polymers show a fatigue limit like those exhibited by ferrous metals, and as with high-cycle long-life metallic parts, the nucleation of the crack occupies a large fraction of the polymer life. The mechanisms are somewhat different however. In some cases fatigue failure in polymers occurs via large-scale hysteretic heating that leads to softening of the polymer. Hysteretic heating is due to the fact that the stress and strain are not always in phase, giving rise to a mechanical hysteresis heating effect. In other cases cracks nucleate and propagate similar to that in metals except that the crack-tip plastic deformation occurs by a different atomic mechanism in polymers.

Fatigue behavior in composites is very complex due to the variation of the properties with direction. In general, the damage done by alternating stresses is similar to that under short-time static loading except that fatigue at a given stress level causes increasing damage with the increase in number of cycles. Instead of a single crack propagating to failure as in metal fatigue, one speaks of the increase of crack density with the number of cycles. The damage ratio is described as the ratio of crack density at *n* cycles to the crack density at final failure. In fatigue, more cracks occur than during static loading and appear to reach a crack density limit. This limit occurs during the first 20% or so of fatigue life. The rest of the fatigue life is largely occupied by delamination, fiber–matrix debonding, and fiber breakage. Fatigue data for composites are often expressed in *S–N* curves, just as for metals and polymers but with the added information reflecting the orientation of the components.

## 2.2.6  Creep and Stress–Rupture

Creep is a very slow plastic deformation of materials that occurs over a long period of time and is most noticeable at elevated temperatures. Like fatigue, it occurs at stresses far below the yield stress for the temperature in question. Creep can occur in all materials included in our study, although it was first noted and has been most studied in metals. The lower-melting-point metals like lead and tin will creep at room temperature at stresses well below their room-temperature yield strength. This has become a problem in lead–tin solders in electronic circuits. Stress relaxation by creep of a solder joint can result in an open circuit. (Fatigue is also common in solder joints.) One of the more important cases observed of room or ambient temperature creep was that of aluminum cables used in cross-country electrical power distribution. Pure aluminum is needed for high electrical conductivity, but as is the case for most pure metals, has a relatively low strength compared to the metal in alloy form. Some of the longer spans (e.g., in river crossings) would sag under their own weight and thus require frequent tightening. To solve this problem the high-purity aluminum strands were wound around an aluminum alloy core, and for very long spans, around a steel core. Steel does not creep at room temperature and for most aluminum alloys it is barely detectable. The steel-reinforced cables are described in ASTM specification B232. Their strength-to-weight ratio is about two times that of copper for equivalent direct-current resistance.

Creep data are usually presented in the form of creep strain (i.e., change in length per unit length) versus time (Figure 2.21). The curve can be divided into three regions. The first is called *primary creep,* where the creep rate decreases with time. The second and most important region comprises the *secondary creep,* where the creep rate is constant over a long period of time. This region is used for design purposes. When the creep curve is run at the same temperature as the intended application, the life, that is, the time for allowable creep strain to take place, can easily be predicted. Creep data for a given alloy are generally presented at a number of temperatures and extrapolation to other temperatures is permitted with certain limitations. One rule to remember is that the most creep rates increase exponentially with temperature and linearly with time. It is much safer to extrapolate downward to lower temperatures than upward to higher temperatures. (See the Suggested Reading for compilation of creep data.)

*Example 2.5.*  The initial clearance between the ends of the turbine blades and the housing of a steam turbine is 0.003 in. The blades are 8 in. long and their elastic elongation during operation is calculated to be 0.0008 in. (a) If it is desired to hold the final clearance to a 0.001 in. minimum, what is the maximum percent creep that can be allowed in the blades? (b) What will be the life of the turbine blade?

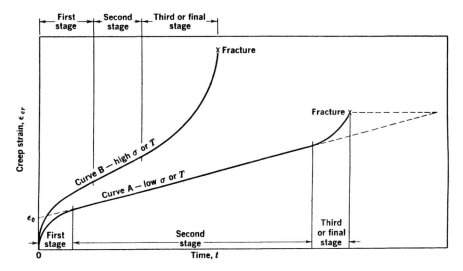

**Figure 2.21** Schematic creep curves showing the three stages of creep. (From C. W. Richards, *Engineering Materials Science,* Wadsworth, Belmont, Calif., 1961.)

Solution. From published data it is found that at the intended operating temperature and stress, this alloy will have a primary creep strain of $2 \times 10^{-5}$ and a secondary creep rate of $6 \times 10^{-6}$ per hour. Allowable creep strain = 0.003 in. − 0.001 in. − 0.0008 in. = 0.0012 in.

(a) 0.0012 in./8.0 in. = $1.5 \times 10^{-4}$ in. = 0.015% max.

(b) 0.0012 in. − 0.00005 in. = 0.00115 in. allowable secondary creep strain. 0.00115 in./0.000006 in./in./hr = 192 hr.

In many high-temperature applications, such as the superalloy components inside a jet aircraft engine (mentioned in Chapter 1), we are more interested in the time to fracture than in the creep rate. When data are presented in this fashion, the process is referred to as *creep-rupture*. Whereas the creep rate increases with increasing temperature according to the Arrhenius relationship

$$\frac{d\epsilon}{dt} = A \exp \left[ -\frac{Q}{RT} \right] \tag{2.18}$$

the time to rupture, $t_r$, decreases with increasing temperature, that is,

$$t_r = A \exp \left[ \frac{Q}{RT} \right] \tag{2.19}$$

In both these expressions $A$ is a proportionality constant, $Q$ is the activation energy for the process causing creep strains or causing the material to rupture,

*R* the gas constant, and *T* the absolute temperature in Kelvin. Remember to use the Kelvin scale because the atoms do not recognize or understand other temperature scales. The activation energy is often that for the movement of atoms from one position to another in the material. This process, called *diffusion,* enables plastic deformation to occur, which is manifested as creep. This associated plastic deformation eventually leads to fracture, just as it did in the uniaxial tension test, but over a much longer period of time. We cover the diffusion process in Chapter 5.

Polymers, especially the thermoplastic ones, are very prone to creep, probably more so than metals. The chief mechanisms of creep in polymers are the uncoiling of polymer chains and the slippage of polymer molecules past one another. These are the same mechanisms that were responsible for plastic flow in the uniaxial tension test. The only difference in creep is that a lower constant load is applied such that the plastic deformation proceeds slowly over a long period of time. Polymer creep shows many of the same creep characteristics as do metals: an instantaneous elastic response to the load, a fast creep rate primary region that decreases with time, and slow secondary creep processes that eventually lead to fracture. Data are also expressed as creep–rupture: the lifetime of a polymer at constant load. A high degree of cross-linking results in a dramatic reduction in molecular mobility and correspondingly, the creep rate. The hard elastomers and thermosets will not exhibit extensive creep but will rupture after a relatively small amount of creep strain.

## 2.3  PHYSICAL PROPERTIES

The physical properties of materials generally include electrical and heat conductivity, thermal expansion, magnetic properties, dielectric strength, optical properties, ferroelectricity, and piezoelectricity. In this section we cover only the determination of the electrical and thermal properties of metals, polymers, and ceramics. The other properties listed above apply only to specific classes of materials and are covered in later sections of the book.*

### 2.3.1  Electrical Conductivity

Electrical conductivity and its reciprocal, electrical resistivity, are the physical properties of most interest to engineers. In fact, we now use these properties to classify materials. Metals, for example, were once described in terms of their characteristics such as metallic luster and hardness. Metals, semiconductors,

---

*Some instructors may prefer to omit this section for now and cover all these measurement methods when discussing the characteristics of the specific class of materials.

and insulators are classified today according to where they are on the electrical conductivity scale of materials. Such a scale for room-temperature values is shown in Figure 2.22. Those materials with the lowest conductivity are at the top of the scale. Note that resistivity is listed on one side of the scale and conductivity on the other. This can easily be accomplished by changing the sign of the exponent since these properties are reciprocals of each other. Also note the wide range of values for all materials, from about $10^{-16}$ $(\Omega \cdot m)^{-1}$ for diamond conductivity to $10^8$ for copper (i.e., copper is $10^{24}$ times more conductive than diamond). Semiconductors, as their name suggests, fall in between these two extremes. If we consider superconductors below their superconducting transition temperature we see even a wider range of conductivity values. Their conductivities are of the order of $10^{25}$ $(\Omega \cdot m)^{-1}$, or a factor of about $10^{40}$ times that of diamond.

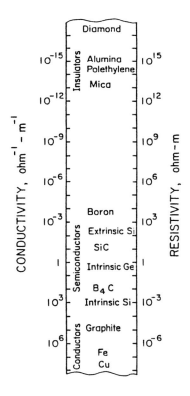

**Figure 2.22**  Range of electrical conductivities for various materials.

Let's examine how the units of conductivity arise. Ohm's law states that the current $I$ passing through a material that is subjected to a potential gradient $V$ is proportional to $V$ and inversely proportional to the materials resistance $R$, or

$$I = \frac{V}{R} \tag{2.20}$$

The resistance of a material is dependent on its dimensions. If we consider a wire or rectangular-shaped specimens, the larger the cross-sectional area, the larger the current flow for a given value of $V$, and the longer the length of the conductor, the less the current flow. This can be expressed in equation form as

$$R = \rho \frac{l}{A} \tag{2.21}$$

where $R$ = resistance
       $A$ = area
       $l$ = length
       $\rho$ = a proportionality constant called resistivity

The resistivity is not a function of specimen dimensions. Since $R$ is expressed in ohms and the dimensions in meters, the units for $\rho$ become

$$\rho = \frac{RA}{l} = \frac{\Omega \cdot m^2}{m} = \Omega \cdot m \tag{2.22}$$

The conductivity, $\sigma$, is simply the reciprocal of resistivity; thus

$$\sigma = \frac{1}{\rho} (\Omega \cdot m)^{-1} \tag{2.23}$$

The reciprocal of the resistance is called the conductance, $G$:

$$G = \frac{1}{R} = \Omega^{-1} = \text{seimens}$$

The conductance can also be expressed as

$$G = \frac{\sigma A}{l} \tag{2.25}$$

Like resistance, conductance is a function of specimen dimensions.

The conductivity and resistivity of metals are generally measured by the four-point probe technique, where a current of a few amperes is fed through the two outside leads, often to a cylindrically shaped specimen. The voltage drop is picked off via the two inside leads that contact the specimen over some known specimen length between the two current leads. To measure resistivity accurately the specimen dimensions must be known accurately. Knife edges for the voltage contact points minimize the error in measurement of specimen length. The resistance is determined from the known $I$ and $V$ values using equation (2.20), and then the resistivity can be computed via equation (2.22). The voltage drop is small but can be measured with a microvolt meter or, for high accuracy, a Wheatstone bridge.

The four-point probe method can also be used to measure the sheet resistivity of silicon wafers. The four points—the two current and the two voltage contacts—rest on the surface of the silicon as shown in Figure 2.23. Since the current path is not directly through the wafer thickness, the value measured is called sheet resistivity rather than bulk or volume resistivity. However, when certain correction factors are applied to account for the wafer thickness and current path, the resistivity value obtained is valid for circuit design purposes. The correction factors are obtained from the solution of a partial differential equation for the current flow under certain boundary conditions. Two assumptions are made: (1) minority carrier injection at the metal–semiconductor current contacts recombine near the electrode, and (2) the boundary region between the current contacts and the silicon is hemispherical in shape and small in diameter compared to the distance between the probes. A satisfactory arrangement used in our laboratories consists of a probe head that contains four tungsten carbide points spaced 0.1 cm apart (probe head and specimen holder were obtained from Signatone Corporation). The outer two carbide points supply the current, usually about 1 mA, while the two inner probes pick up the voltage drop and feed it to a millivolt range meter. Gallium arsenide resistivity measurements are somewhat more difficult than those for silicon, due to the high ohmic contact values at the probe–semiconductor interface. Indium solder

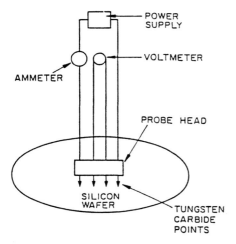

**Figure 2.23** Four-point probe method of measuring the sheet resistivity of semiconductor wafers.

joints minimize these high-resistance contacts. When the appropriate boundary conditions are applied, the solution to the partial differential equation results in the following simple equation for the resistivity:

$$\rho = \frac{\pi V W}{I \ln 2} \qquad (2.26)$$

where $W$ is the wafer thickness.

The electrical resistivity of insulators is more difficult to measure than that for metals and semiconductors. The dc resistance for ceramic insulators of less than $10^{19}$ $\Omega \cdot$ m can be computed from measurements of the voltage drop through both the specimen and a standard resistor. The current values are small and require an electrometer, which is essentially a sensitive galvanometer, and in some cases a dc amplifier. The details of circuitry and apparatus can be found in the ASTM method number D257-78 (reapproved 1983). Insulating polymers of resistivities on the order of $10^{16}$ $\Omega \cdot$ m have been measured in our laboratories using a special adapter (Keithley model 6105) and an electrometer (Keithley model 617). The adapter can hold specimens to 4 in. in diameter and thickness to ¼ in.

## 2.3.2 Thermal Properties

Temperature Scales

In Chapter 1 we frequently listed temperature in both °C and °F. If the student is not already familiar with the various temperature scales that are used, the following discussion should be helpful. Perhaps a little history will assist in our understanding and comparison of the scales. Gabriel Daniel Fahrenheit (1686–1736), a German physicist, developed the scale known by his name. He used three major points on his scale: °0 for the freezing point of a salt solution, 32°F for the freezing point of pure water, and 96°F for the normal temperature of the human body (later found to be 98.6°F). Unfortunately for U.S. residents, Anders Celsius was born after Fahrenheit. If it had been the other way around, the Fahrenheit scale would never have existed and we would not be subjected to the anguish of having to convert from Fahrenheit to Celsius, and vice versa. Celsius devised the centigrade scale, which is now officially called the Celsius scale, in 1742. He essentially had two major points on his scale, that of the freezing and boiling points of pure water. The scale contained 100 divisions between these two points, a much more logical approach. This scale is used just about everywhere but the United States and is used internationally by all scientists. Manufacturers of temperature controllers and recorders for furnaces, refrigerators, and the like, should, in my opinion, be required to use the Celsius scale. Then if weather reports were given in °C (perhaps in both °C and °F for a short transition period), the conversion to the Celsius scale would be quick and painless. The same approach could be used in converting to the metric system. A few years ago many gasoline stations in the United States converted their pumps from gallons to liters, but the experiment failed because the automobile manufacturers still reported mileage for their cars in miles per gallon. We should learn to think metric and Celsius.

Another scale that is even more logical than the Celsius scale is that suggested by William Thomson Kelvin in the late nineteenth century and bears his name. The Kelvin or absolute temperature scale is widely used by scientists and could, and perhaps should, be adopted by engineers. A unit of measurement on the Kelvin scale is the same as that on the Celsius scale, but the zero point, theoretically the lowest temperature attainable, is $-273°C$. Shortly thereafter, William Rankine introduced the Rankine absolute scale. It is similar to the Kelvin scale except that the unit of measurement is the same as on the Fahrenheit scale. Absolute zero now becomes $-459.69°F$, and the freezing and boiling points of pure water become $491.69°R$ and $671.69°R$, respectively. If we could do away with the Fahrenheit scale, the Rankine would also cease to exist. These temperature scales are compared in Figure 2.24.

The thermal properties of most interest are thermal conductivity and thermal expansion. The temperature coefficient of electrical resistivity (i.e., the change of resistivity with temperature) is a very important thermal–electrical characteristic of both metals and semiconductors, but we hold this subject for a later section.

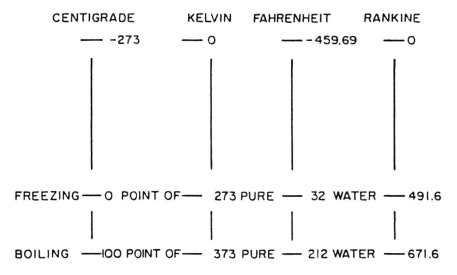

**Figure 2.24** Comparative temperature scales.

Thermal Conductivity

The heat flow through a wall per unit of area is called the thermal flux $J$, which is proportional to the thermal gradient, the proportionality constant being the thermal conductivity. In equation form we have

$$J = K\frac{\Delta T}{\Delta X} \tag{2.27}$$

where $\Delta T/\Delta X$ is the thermal gradient (i.e., temperature change per unit thickness, $X$) and $K$ is the thermal conductivity. The thermal conductivity is expressed in English units as BTU $-$ ft/(hr $-$ ft$^2$ $-$ °F) and in the metric system as kcal $\cdot$ m/(s $\cdot$ m$^2$ $\cdot$ K). In word form, the thermal conductivity is the quantity of heat flow through unit thickness and unit area per unit time.

Thermal Expansion Coefficient

This coefficient represents a change in dimension per unit temperature change. Just as in mechanical strain, thermal expansion and contraction can be expressed by a change in either volume, area, or length, with the latter being the most frequently used. The volume coefficient of expansion is three times the linear coefficient. Linear handbook values will usually be given in cm/cm/ °C or in./in./°F.

As the temperature of a material is increased from absolute zero on the Kelvin scale, the atoms begin vibrating about their equilibrium positions and the amplitude of vibration increases with temperature. This amplitude of vibration is not symmetrical, however. As we discover later, a balance of both attractive and repulsive forces between atoms determines the equilibrium position. The repulsive forces increase as atoms are pushed together more drastically than do the attractive forces as atoms are pulled apart. In other words, it is easier to pull atoms apart than to push them together. Thus the amplitude of vibration is nonsymmetrical, resulting in a net atom positive displacement, the magnitude of which increases with increasing temperature. There are a few exceptions to this behavior. Uranium, for example, has a negative temperature coefficient of expansion; that is, it contracts, in one crystallographic direction. This is a result of the directionality of the bonding forces.

In many crystalline materials there also exists a very sharp discontinuous change in dimensions at certain very precise temperatures. As the atoms gain energy they find that at certain well-defined temperatures they prefer to alter their geometrical arrangement from one type of atom architecture to another. This dimensional change (called a *phase change*) is not related to the normal thermal expansions and contractions and will be better understood after we have examined the various atom arrangements that can exist in crystalline structures. Thermal expansion data are normally reported to three significant figures and can be measured accurately by a number of methods. Dilatometers can be readily purchased for this purpose. Some of the simpler ones mechanically transmit the linear dimensional change from a specimen in a furnace via a zero- or low-coefficient-expansion material to a dial gauge or an electronic sensor. It should be realized that the coefficient varies somewhat with temperature and is frequently reported to be that over a fairly narrow temperature (e.g., from 0 to 100°C).

## DEFINITIONS

*Activation energy*: energy required to initiate a process or chemical reaction.

*Brinell hardness number*: hardness number obtained by measuring the diameter of the impression made by a specific-size hard ball under a specific load.

*Brittle*: describes a material that shows little or no plastic deformation prior to and during fracture.

*Charpy V-notch test*: impact test in which the specimen has a V-shaped notch and in which the energy absorbed during fracture is measured at a specific temperature.

*Diamond pyramid hardness (DPH)*: hardness value obtained when a pyramidal-shaped diamond makes an impression in a material under a specified load, and in which the hardness number is determined from the length of the diagonals of the impression.

*Ductility*: strain obtained prior to fracture, usually in an uniaxial tension test.

*Elastic deformation*: recoverable deformation (change in shape) obtained after the force is removed.

*Elastomer*: polymer material that essentially exhibits only elastic deformation when stressed.

*Electrical conductance*: reciprocal of electrical resistance. It is a function of specimen dimensions.

*Electrical conductivity*: measure of the ease of the passage of electrical current. It is the reciprocal of electrical resistivity and it is independent of specimen dimensions.

*Electrical resistance*: resistance to current flow. It is the reciprocal of electrical conductance and is a function of specimen dimensions.

*Electrical resistivity*: inherent ability of a material to resist current flow. It is the reciprocal of conductivity and is independent of specimen dimensions.

*Engineering strain*: change in length or area divided by the original length or area.

*Engineering stress*: force (or load) on a body divided by the original cross-sectional area.

*Fatigue life*: number of cycles a material will survive at a specified alternating stress.

*Fatigue limit*: number of cycles in a fatigue *S–N* curve where the curve becomes horizontal (i.e., at this stress value the material has an infinite life).

*Fatigue strength*: stress level corresponding to a specific fatigue life.

*Fracture toughness*: resistance to crack propagation. It is an inherent property of the material.

*Hooke's law*: law that states that strain is a linear function of stress in the elastic condition.

*Modulus (shear)*: proportionality constant between shear stress and shear strain during elastic behavior.

*Modulus (Young's)*: proportionality constant between normal stress and normal strain in the elastic state. It is determined from the slope of the elastic stress–strain line.

*Plastic deformation*: nonrecoverable permanent deformation. It remains when the force is released.

*Poisson's ratio*: ratio of the lateral contraction to longitudinal extension when a body is tested in uniaxial tension in the elastic region of behavior.

*Proportional limit*: point on the stress–strain diagram where Hooke's law ceases to be obeyed and the graph becomes nonlinear.

*Shear strain*: tangent of the angle of twist in torsion; the shear displacement divided by the distance over which it acts.

*True strain*: natural logarithm of the ratio of the instantaneous length of a specimen to its original length; also in terms of the area, it is the natural logarithm of the original area divided by its instantaneous area.

*True stress*: force on a body divided by the instantaneous cross-sectional area on which it acts. *Note*: Stress at a point is defined by six independent parameters, of which three act normal to the faces of an infinitely small cube; the three shear stresses act parallel to these faces.

*Ultimate tensile strength*: highest point in the engineering stress–strain diagram; the engineering stress at which necking begins in a uniaxial tension test.

## QUESTIONS AND PROBLEMS

2.1. A 0.505 in. diameter by 2″ gauge length standard tensile test bar of $30 \times 10^6$ psi modulus is loaded to an elastic strain of 0.002. What force is required to produce this strain?

2.2. An ASTM E8-90 standard tensile bar of a copper alloy was observed to fracture at a force of 12000 lb. The true stress at fracture was 80,000 psi. What was the diameter of the bar at the point of fracture? What was the engineering and true strains to fracture? What was the ductility of this alloy? What was the true fracture stress in MPa?

2.3. An ASTM E8-90 standard tensile bar of steel was loaded to 12,000 lb.

The ultimate tensile strength was not attained. After the load was released, it was found that the gauge length had increased to 2.046 in. The modulus of this steel is $30 \times 10^6$ psi and the yield strength was observed to be $30 \times 10^3$ psi.

(a) What was the plastic strain after load was released?

(b) What was the total length of the bar before the load was released?

(c) Graph the stress–strain curve assuming a linear relation between stress and strain in the plastic region.

2.4. List five mechanical properties and five physical properties of materials that are found in most handbooks.

2.5. A specific pure metal has a shear modulus of $28 \times 10^3$ MPa and a Poisson's ratio of 0.3 What is Young's modulus of elasticity?

2.6. The yield strength of steel is defined by ASTM E8-90 as that stress necessary to achieve a 0.002 plastic strain. Why would a different offset be preferred for some materials?

2.7. Verify by solving the following problem that the reduction in area prior to fracture is a better measure of ductility than that obtained by a change in length. Consider a 2-in.-gauge-length tensile bar that was elongated such that the center section necked and thereby increased in length from 0.5 in. to 1.0 in. The remaining 1.5-in. gauge section elongated uniformly by 10%. Now consider a 10-in.-gauge-length specimen of the same material in which again the center 0.5-in.-original-length section necked and thereby increased in length to 1.0 in. while the remaining 9.5-in. length of gauge section elongated uniformly by 10%. Compute the engineering strain to fracture (the ductility) for both specimens.

2.8. What is the one bad and the one good feature of the Rockwell hardness test? What are the advantages of the diamond pyramid scale?

2.9. Approximately how much greater is the strain rate in a Charpy impact test compared to a standard tensile test?

2.10. What is a typical value for the impact energy absorbed in fracture for a specimen of a ceramic material such as alumina ($Al_2O_3$)? What is it for a mild steel such as a 1020 steel alloy?

2.11. What does the Charpy impact test really tell us with respect to the design of a steel ship or bridge?

2.12. What is the chief advantage of the fracture mechanics approach to fracture toughness over the Charpy impact test?

2.13. A steel alloy has an ultimate tensile strength of 1500 MPa. What would be a good guess for the alternating stress that it could withstand for $10^7$ cycles?

2.14. What is the difference between a creep test and a creep-rupture test?

2.15. What region of the creep curve is used for design purposes?

2.16. Explain the difference between electrical resistance and resistivity.

2.17. A 0.020 cm diameter copper wire has a resistance of 5 $\Omega$ at 20°C. What is the length of this wire if its resistivity at the same temperature is $1.67 \times 10^{-6}$ ohm-cm?

2.18. What would be the resistance of the copper wire in Problem 2.17 if it were heated to 100°C and its temperature coefficient of resistance is 0.004 per °C? What would be the new length at 100°C?

## SUGGESTED READING

Boyer, H. E., Ed., *Atlas of Creep and Stress–Rupture Curves*, ASM International, Metals Park, Ohio, 1988.

Boyer, H. E., Ed., *Atlas of Stress–Strain Curves*, ASM International, Metals Park, Ohio, 1986.

Courtney, T. H., *Mechanical Behavior of Materials*, McGraw-Hill, New York, 1990.

Hertzberg, R. W., *Deformation and Fracture Mechanics of Engineering Materials*, Wiley, New York, 1989.

Keithley Instruments, Inc., *Low Level Measurements*, 1984.

# 3

# Atomic Structure and Atomic Bonds

## 3.1  STRUCTURE OF THE ATOM

Since atomic structure is covered in introductory chemistry courses, the student should be aware by now that the atom consists of a nucleus containing neutrons and protons around which electrons orbit, the orbits being more or less confined to certain radii. (We will not be concerned here with the subatomic particles such as quarks or neutrinos.) The smaller the diameter of the orbit, the greater the attractive force between the electron and the nucleus and the greater the absolute binding energy (BE). Binding energies, by convention, are considered to be negative, so we are speaking of a large negative binding energy for the innermost orbit. The electrons in the outer orbits are bound less tightly, and the outermost electrons can to some extent be considered to be loosely bound and not necessarily residing in well-defined orbits. These are the valence electrons and are the ones that are involved in the bonding together of the atoms, and hence strongly affect all material properties: physical, mechanical, and chemical. The nature of this bond is what determines whether the substance is a metal, in which the bond is often between like atoms, or a ceramic or a polymer material, where the bond is between dissimilar atoms. Before describing these various bond types, we review the electron configuration and the related periodicity of the elements.

### 3.1.1  Atomic Mass Unit and Atomic Numbers

The mass and the charge on each of the atomic particles of interest are listed as follows:

|  | Mass (g) | Charge (C) |
|---|---|---|
| Proton | $1.673 \times 10^{-24}$ | $+1.602 \times 10^{-19}$ |
| Neutron | $1.675 \times 10^{-24}$ | $0$ |
| Electron | $9.109 \times 10^{-28}$ | $-1.602 \times 10^{-19}$ |

The atomic mass of an element is the sum of the masses of all protons, neutrons, and electrons. Whereas the number of electrons or protons in a stable atom is fixed for each element, the number of neutrons can vary somewhat among the atoms of a given element. Thus the atoms of some elements have two or more different atomic masses. These different atoms are called *isotopes*. Iron has been found to contain atoms made of 28, 31, and 32 neutrons. Natural carbon contains approximately 1% of carbon 13, along with 98.9% of carbon 12. The numbers 13 and 12 are the *atomic weights* of the isotopes. The atomic weight of natural carbon is 12.011 *atomic mass units (amu),* where the weight of 1 *amu* has been defined as the weight of one-twelfth of the mass of one atom of the common isotope of carbon 12, which is $1.66 \times 10^{-24}$ g. The *atomic weight* of an element, however, is a much larger number and is the weight in grams of 1 *mole* of atoms (i.e., $6.023 \times 10^{23}$ atoms) (Avogadro's number). This weight is the gram-atomic weight. In the case of molecules it is the weight of $6.023 \times 10^{23}$ molecules or the gram-molecular weight. The atomic weight of an element is usually presented in most periodic charts below the atom symbol, as shown in Figure 3.1.

The *atomic number* of an atom, which is a whole number and usually listed above the atomic symbol in the periodic charts, represents the number of protons in the nucleus. A neutral atom will have this same number of electrons revolving around the nucleus. If the atom has more or fewer electrons than those in the neutral state, it is said to be "ionized" and we now call the resulting atom an *ion*. If it contains more electrons than the neutral atom, it is a negative ion called an *anion,* and if fewer electrons orbit the nucleus, it becomes a positive ion or *cation.*

*Example 3.1.*  Calculate the weight in grams of a sodium atom (a) from the information listed in Figure 3.1, and (b) by the weights of each particle, and note the difference.

Metals | Nonmetals

| Iᵃ | IIᵃ | IIIᵇ | IVᵇ | Vᵇ | VIᵇ | VIIᵇ | VIII | VIII | VIII | Iᵇ | IIᵇ | IIIᵃ | IVᵃ | Vᵃ | VIᵃ | VIIᵃ | 0 | Orbit |
|---|---|---|---|---|---|---|---|---|---|---|---|---|---|---|---|---|---|---|
| 1 +1 -1<br>H<br>1.0079<br>1 | | | | | | | | | | | | | | | | | 2 0<br>He<br>4.00260<br>2 | K |
| 3 +1<br>Li<br>6.939<br>2-1 | 4 +2<br>Be<br>9.0122<br>2-2 | | | | | | | | | | | 5 +3<br>B<br>10.81<br>2-3 | 6 +2 +4 -4<br>C<br>12.011<br>2-4 | 7 +1 +2 +3 +4 +5 -1 -2 -3<br>N<br>14.0067<br>2-5 | 8 -2<br>O<br>15.9994<br>2-6 | 9 -1<br>F<br>18.998403<br>2-7 | 10 0<br>Ne<br>20.179<br>2-8 | K-L |
| 11 +1<br>Na<br>22.9898<br>2-8-1 | 12 +2<br>Mg<br>24.312<br>2-8-2 | | | | | | | | | | | 13 +3<br>Al<br>26.98154<br>2-8-3 | 14 +2 +4 -4<br>Si<br>28.08<br>2-8-4 | 15 +3 +5 -3<br>P<br>30.97376<br>2-8-5 | 16 +4 +6 -2<br>S<br>32.06<br>2-8-6 | 17 +1 +5 +7 -1<br>Cl<br>35.453<br>2-8-7 | 18 0<br>Ar<br>39.948<br>2-8-8 | K-L-M |
| 19 +1<br>K<br>39.09<br>8-8-1 | 20 +2<br>Ca<br>40.08<br>8-8-2 | 21 +3<br>Sc<br>44.9559<br>8-9-2 | 22 +2 +3 +4<br>Ti<br>47.9<br>8-10-2 | 23 +2 +3 +4 +5<br>V<br>50.941<br>8-11-2 | 24 +2 +3 +6<br>Cr<br>51.996<br>8-13-1 | 25 +2 +3 +4 +7<br>Mn<br>54.9380<br>8-13-2 | 26 +2 +3<br>Fe<br>55.847<br>8-14-2 | 27 +2 +3<br>Co<br>58.9332<br>8-15-2 | 28 +2 +3<br>Ni<br>58.71<br>8-16-2 | 29 +1 +2<br>Cu<br>63.54<br>8-18-1 | 30 +2<br>Zn<br>65.38<br>8-18-2 | 31 +3<br>Ga<br>69.72<br>8-18-3 | 32 +2 +4<br>Ge<br>72.59<br>8-18-4 | 33 +3 +5 -3<br>As<br>74.9216<br>8-18-5 | 34 +4 +6 -2<br>Se<br>78.96<br>8-18-5 | 35 +1 +5 -1<br>Br<br>79.904<br>8-18-7 | 36 0<br>Kr<br>83.80<br>8-18-8 | -L-M-N |
| 37 +1<br>Rb<br>85.467<br>-18-8-1 | 38 +2<br>Sr<br>87.62<br>18-8-2 | 39 +3<br>Y<br>88.9059<br>18-9-2 | 40 +4<br>Zr<br>91.22<br>-18-10-2 | 41 +3 +5<br>Nb<br>92.9064<br>-18-12-1 | 42 +6<br>Mo<br>95.94<br>-18-13-1 | 43 +4 +6 +7<br>Tc<br>98.9062<br>-18-13-2 | 44 +3<br>Ru<br>101.07<br>-18-15-1 | 45 +3<br>Rh<br>102.905<br>-18-16-1 | 46 +2 +4<br>Pd<br>106.4<br>-18-18-0 | 47 +1<br>Ag<br>107.868<br>18-18-1 | 48 +2<br>Cd<br>112.40<br>-18-18-2 | 49 +3<br>In<br>114.82<br>-18-18-3 | 50 +2 +4<br>Sn<br>118.69<br>-18-18-4 | 51 +3 +5 -1<br>Sb<br>121.75<br>-18-18-5 | 52 +4 +6 -2<br>Te<br>127.60<br>-18-18-6 | 53 +1 +5 +7 -3<br>I<br>126.9045<br>-18-18-7 | 54 0<br>Xe<br>131.30<br>-18-18-8 | -M-N-O |
| 55 +1<br>Cs<br>132.9054<br>-18-8-1 | 56 +2<br>Ba<br>137.3<br>-18-8-2 | 57* +3<br>La<br>138.9055<br>-18-9-2 | 72 +4<br>Hf<br>178.49<br>-32-10-2 | 73 +5<br>Ta<br>180.948<br>-32-11-2 | 74 +6<br>W<br>183.85<br>-32-12-2 | 75 +4 +6 +7<br>Re<br>186.207<br>-32-13-2 | 76 +3 +4<br>Os<br>190.2<br>-32-14-2 | 77 +3 +4<br>Ir<br>192.9<br>-32-15-2 | 78 +2 +4<br>Pt<br>195.09<br>-32-16-2 | 79 +1 +3<br>Au<br>196.9665<br>-32-18-1 | 80 +1 +2<br>Hg<br>200.59<br>-32-18-2 | 81 +1 +3<br>Tl<br>204.37<br>-32-19-3 | 82 +2 +4<br>Pb<br>207.19<br>-32-18-4 | 83 +3 +5<br>Bi<br>208.980<br>-32-18-5 | 84 +2 +4<br>Po<br>(209)<br>-32-18-6 | 85<br>At<br>(210)<br>-32-18-7 | 86 0<br>Rn<br>(222)<br>-32-18-8 | N-O-P |
| 87 +1<br>Fr<br>(223)<br>-18-8-1 | 88 +2<br>Ra<br>226.0254<br>-18-8-2 | 89** +3<br>Ac<br>(227)<br>-18-9-2 | 104 +4<br>Rf<br>(261)<br>-32-10-2 | 105<br>Ha<br>(262)<br>-32-11-2 | 106<br>(263)<br>-32-12-2 | | | | | | | | | | | | | O-P-Q |

Key to chart: Atomic Number → 50; Oxidation States → +2 +4; Symbol → Sn; Atomic Weight → 118.69; Electron Configuration → -18-18-4

Transition Elements: groups IIIᵇ through Iᵇ

| | | | | | | | | | | | | | | | |
|---|---|---|---|---|---|---|---|---|---|---|---|---|---|---|---|
| *Lanthanides | 58 +3 +4<br>Ce<br>140.12<br>-20-8-2 | 59 +3<br>Pr<br>140.9077<br>-21-8-2 | 60 +3<br>Nd<br>144.24<br>-22-8-2 | 61 +3<br>Pm<br>147<br>-23-8-2 | 62 +2 +3<br>Sm<br>150.4<br>-24-8-2 | 63 +2 +3<br>Eu<br>151.96<br>-25-8-2 | 64 +3<br>Gd<br>157.25<br>-25-9-2 | 65 +3<br>Tb<br>158.925<br>-27-8-2 | 66 +3<br>Dy<br>162.50<br>-28-8-2 | 67 +3<br>Ho<br>164.9304<br>-29-8-2 | 68 +3<br>Er<br>167.26<br>-30-8-2 | 69 +3<br>Tm<br>168.9342<br>-31-8-2 | 70 +2 +3<br>Yb<br>173.04<br>-32-8-2 | 71 +3<br>Lu<br>174.967<br>-32-9-2 | -N-O-P |
| **Actinides | 90 +4<br>Th<br>232.038<br>-18-10-2 | 91 +5 +4<br>Pa<br>231.0359<br>-20-9-2 | 92 +3 +4 +5 +6<br>U<br>238.029<br>-21-9-2 | 93 +3 +4 +5 +6<br>Np<br>237.0482<br>-22-9-2 | 94 +3 +4 +5 +6<br>Pu<br>239.052<br>-24-8-2 | 95 +3 +4 +5 +6<br>Am<br>(243)<br>-25-8-2 | 96 +3<br>Cm<br>(247)<br>-25-9-2 | 97 +3 +4<br>Bk<br>(247)<br>-27-8-2 | 98 +3<br>Cf<br>(251)<br>-28-8-2 | 99 +3<br>Es<br>(254)<br>-29-8-2 | 100 +3<br>Fm<br>(257)<br>-30-8-2 | 101 +2 +3<br>Md<br>(258)<br>-31-8-2 | 102 +2 +3<br>No<br>(259)<br>-32-8-2 | 103 +3<br>Lr<br>(260)<br>-32-9-2 | -O-P-Q |

**Figure 3.1** Periodic table of the elements. (From *Metals Handbook,* Desk Ed., ASM International, Metals Park, Ohio, 1985, p. 1–43.)

Solution.   (a) The gram-molecular weight of sodium is found from Figure 3.1 to be 22.9898 g/g-mol.

$$\frac{22.9898 \text{ g/mol}}{6.02 \times 10^{23} \text{ atoms/mol}} = 3.819 \times 10^{-23} \text{ g}$$

(b) Taking an average value of $1.674 \times 10^{-24}$ g for the 11 neutrons and 11 protons, we have

$$22 \times 1.674 \times 10^{-24} \text{ g} = 36.828 \times 10^{-24} \text{ g for protons}$$

and for the electrons

$$11 \times 9.109 \times 10^{-28} \text{ g} = 0.010 \times 10^{-24} \text{ g}$$

$$36.838 \times 10^{-24} \text{g} \quad \text{total}$$

The difference in the two values is a result of the various isotopes that are present in natural sodium.

### 3.1.2  Electron Configuration

The simplest atom to describe is that of hydrogen, since it has only one electron revolving around a proton. Niels Bohr described the hydrogen atom in this way in 1913. Bohr assumed that the electron obeyed the classical laws of mechanics and that the centripetal Coulombic attractive force between the electron and proton was balanced by the centrifugal force of the revolving electron. Bohr further postulated that only those orbits were allowed for which the angular momentum was an integral multiple of $h/2\pi$, that is,

$$mvr = \frac{nh}{2\pi} \tag{3.1}$$

where $m$ = mass electron
$v$ = velocity of electron
$r$ = radius of orbit
$n$ = an integer representing the number of the orbit [for stable hydrogen, $n = 1$ ($n$ is known as the principal quantum number)]
$h$ = Planck's constant

The quanta concept and Planck's constant were frequently used during this period in an attempt to explain particle behavior. This constant arose during Planck's description of the energy levels of vibrating atoms. Planck hypothesized in 1901 that atoms could vibrate only at fixed frequencies $v$, and that their energy at each frequency was equal to $hv$. $hv$ was called a *quanta* of energy, much like an energy packet. Shortly thereafter, Einstein in 1905 explained the photoelectric effect using the quanta concept. Electrons had been observed to be emitted from a metal when light of a certain frequency impinged on its surface. It was first observed by Hertz in 1887 when polished zinc was illuminated by ultraviolet light. The ejected electron had an energy equal to $h\nu$ minus some fixed energy term, called the *work function,* which allowed the electron to escape the surface. Whereas Planck had quantized atomic motion, Einstein had quantized electron energy.

All of these developments led to the Bohr model of the atom and enabled the calculations of the radii of the orbits and of the related energies of the electrons when they resided in the various orbits. We refer to these as energy levels. All atoms in the fundamental or ground states have electrons occupying the most stable energy-level configuration, usually that of the lowest energy. There actually exists an infinite number of allowable energy levels, but in hydrogen only one level will be occupied since there is only one electron available. If hydrogen gas is subjected to a spark discharge, the electrons can be excited from the innermost stable orbit of $n = 1$ to higher obits (e.g., where $n = 2, 3, 4, \ldots$). Equating the Coulombic attraction to the centrifugal force and solving for $r$, the orbit radius yields

$$r = \frac{n^2 h^2 \epsilon_0}{\pi m e^2} \tag{3.2}$$

where $\epsilon_0$, the permittivity of vacuum, is $8.54 \times 10^{-12}$ F $m^{-1}$. Substituting the appropriate numbers we find that for $n = 1$, $r = 0.53$ nm (nm $= 10^{-9}$ m).

An expression for the electron energy levels can be obtained by summing the values for the potential energy of the electron (which is dependent on the attractive force of the electron to the nucleus and is set at zero for $r = \infty$) and the kinetic energy, $\frac{1}{2}mv^2$. We will not bother with the math here but just state the resulting energy equation, which is

$$E = \frac{m e^4}{8 n^2 h^2 \epsilon_0^2} \tag{3.3}$$

Substituting again for the constants and using $n = 1$ yields

$$E = -\frac{13.6}{n^2} \quad \text{electrons volts} \tag{3.4}$$

The electron energy-level diagram for a hydrogen atom can now be constructed. First, it should be mentioned that from about 1885 to 1910 a number of observations of the radiation emitted from excited hydrogen were reported. The excited incandescent gas emitted a spectrum of radiation composed of lines at definite wavelengths. These characteristic spectra were named after those who discovered them and are so noted in Figure 3.2 along with the computed energy levels for various values of $n$. The frequencies of radiation emitted corresponded to transitions of electrons from one energy level to another. The Lyman spectra, for example, could be explained on the basis that electrons,

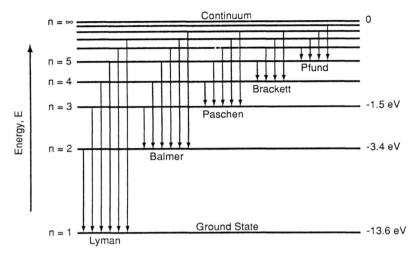

**Figure 3.2**   Energy-level diagram determined from the spectra of radiation emitted from excited hydrogen.

which had been excited to levels corresponding to $n$ values of 2, 3, 4, 5, and 6, emitted radiation of a characteristic frequency for each of these $n$ values when these excited electrons fell to the innermost stable orbit of $n = 1$. The frequency of the spectral lines arise from a transition from level 2 to 1 is given by

$$\nu = \frac{\Delta E}{h} = \frac{1}{h}(E_{n1} - E_{n2}) \tag{3.5}$$

The energy to remove the electron completely is the ionization energy, and for the hydrogen atom it is 13.6 eV.

*Example 3.2.*   Compute the frequency of the radiation for a Balmer series radiation where an excited electron falls from orbit $n = 3$ to the orbit for $n = 2$.

Solution.   First we must compute the energy change involved in the transition.

$$\Delta E = E_{n3} - E_{n2} = 13.6 \left[ \frac{1}{3^2} - \frac{1}{2^2} \right] = 1.89 \text{eV}$$

$$\nu = \frac{\Delta E}{h}$$

From Appendix A, $h = 6.63 \times 10^{-34}$ J $\cdot$ s and 1 eV $= 1.60 \times 10^{-19}$ J.

$$\nu = \frac{1.89 \text{ eV} \times 1.60 \times 10^{-19} \text{J/eV}}{6.63 \times 10^{-34} \text{J} \cdot \text{s}} = 4.55 \times 10^{14} \text{s}^{-1} = 4.55 \times 10^{14} \text{Hz}$$

The agreement between calculated and observed frequencies was excellent and represented a great triumph for the Bohr theory of the hydrogen atom and lent some solid support to the admittedly ad hoc hypothesis on which it was based. Subsequently, spectra were observed from other excited univalent atoms (e.g., Na, K, Li) and were explained on the basis that the closed inner shells of electrons provided a screen around the nucleus such that the nuclei acted as an effective positive charge. Approximate agreement with experiment could be obtained simply by dividing equation (3.2) by $Z$, and multiplying equation (3.3) by $Z^2$, where $Z$ = atomic number of the element.

Bohr developed his theory for the hydrogen atom, in which the electron was suspected to revolve around the proton in a spherical orbit. Around 1920, Sommerfeld introduced two other quantum numbers, and this atomic model is sometimes referred to as the Bohr–Sommerfeld model. The second quantum number, $l$, was introduced to explain the fine structure of the spectral lines, which showed that each main spectral line in atoms of atomic number greater than hydrogen actually consisted of several closely spaced lines when viewed with a high-resolution spectroscope. $l$ was thought to be related to the ratio of the major axis to the minor axis of the orbit, since the presence of the inner electrons would cause the orbit to be nonspherical. Thus $l$ was called the azimuthal quantum number and was related to the angular momentum when the effects of all electrons were considered. We still refer to the $l$ states by the letters $s$, $p$, $d$, and $f$, which were assigned based on the following old spectral line descriptions:

$l = 0$   sharp line or $s$ state
$l = 1$   principal or $p$ state
$l = 2$   diffuse or $d$ state
$l = 3$   fundamental or $f$ state

The third quantum number, $m_l$, also attributed to Sommerfeld, was introduced to account for the splitting of the lines in the presence of a magnetic field. It is called the *magnetic quantum number* and specifies the spatial orientation of a single atomic orbital. It has little effect on the energy.

Neither the Bohr nor Sommerfeld models could explain the very fine lines found in the atomic spectra. Goudsmit, in 1925, attributed these lines to the spin angular momentum of the electron about its own axis. The spin quantum number has only a minor effect on the energy level, but if two electrons are present in the same energy level, they must have opposite spins.

In the late 1920s these energy levels were computed in a more sophisticated manner using the theory that has come to be known as *wave* or *quantum mechanics*. Schrödinger developed his well-known wave equation about this time and from this equation the energy levels could be determined without the classical mechanical model assumed by Bohr and without making assumptions pertaining to the quantum numbers. The quantum numbers and electron energy levels emerged from the solution of the Schrödinger partial differential equation under certain boundary conditions. For the material covered in this book the quantum numbers can still be viewed as in the Bohr–Sommerfeld model. The energy levels and electron positions will be presented in that light. In fact, solutions to the wave equation for high atomic number elements require so many simplifying assumptions that the four quantum numbers of the Bohr–Sommerfeld model are not only sufficient, but probably are the best method available for specifying the energy levels of the electron in a complex atom.

We should be aware of the chief difference between the two approaches, however. The solutions to the Schrödinger wave equation tell us that the electron, which is treated as behaving like a wave, can have only discrete wavelengths and hence discrete energies. Furthermore, its precise position cannot be determined, but only the probability that it is located in the vicinity of the orbit portrayed by the Bohr model. We now tend to think more in terms of an electron cloud or charge density as depicted in Figure 3.3. Here the orbit of the electron for $n = 1$ for the hydrogen atom is shown as a precise circle, whereas the shaded region represents the electron charge cloud as per quantum mechanical calculations. Either way we can now proceed to discuss the electron configuration of the atom, in which the allowed energy states can be described in a shorthand notation.

The principal quantum number $n$ is described by integers which can take on all integral values from 1 to $\infty$. It represents the main shell or orbit, where there is a high probability of finding electrons. The $l$ quantum number serves to specify the subenergy levels that exist within the main shell and can be viewed as more closely defining the position of the electron. However, only the allowable states for which electrons are available will be filled. In the absence of any excitation the filled levels will always be those nearest the nucleus. The $l$ number can have values of $l = 0, 1, 2, \ldots, n$. The letters $s$, $p$, $d$, and $f$ are

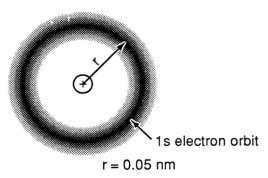

1s electron orbit

r = 0.05 nm

**Figure 3.3**   Electron charge cloud surrounding the 1$s$ orbit for the hydrogen atom.

used to define these sublevels in the shorthand notation. The $m_1$ quantum number has permissible values of "$-1$" to $+$"1," including zero. When $l = 1$, the permissible values for $m_1$ are "$-1$," "0," and "$+1$," and this pattern continues. So, in general, there are $2 l + 1$ allowable values for $m_1$. In terms of the $s$, $p$, $d$, and $f$ notations it means that there exist one $s$ orbital, three $p$ orbitals, and seven $f$ orbitals for each allowed $s$, $p$, $d$, and $f$ subenergy level. Finally, the $m_s$ quantum number is pictured as defining the spin of the electron and by some will be used as a clockwise or counterclockwise spin, and by others as a parallel or antiparallel spin. It is not really important. The only point that the $m_s$ number is trying to make is that there exist two allowable substates for each value of $m_s$. But if two electrons occupy a specific allowable level, they must be of different spins.

Periodic Table

The periodic table of Figure 3.1 lists all the elements from hydrogen, of atomic number 1, to lawrencium, of atomic number 103. There is a very important principle that determines how the allowable energy levels for each atom in the periodic table are filled with electrons. It is the *exclusion principle* first enunciated by Wolfgang Pauli, and it states that "in a single atom no two electrons can have the same set of four quantum numbers." This principle is more often stated in the form "only two electrons can occupy a state specified by the quantum numbers $n$, $l$, and $m_1$, and they must be of opposite spin." The exclusion principle plus the four quantum numbers and their restrictions will now be used to describe how the energy states are filled with electrons as we proceed from hydrogen through the periodic table. Remember that the lower-energy levels are always filled first for the most stable state of the element.

The lowest-energy state is that for which the principal quantum number $n$ is 1 and the azimuthal quantum number $l$ is zero. This is the $1s$ state. The hydrogen atom has only one electron, so it goes into that state. Helium, of atomic number 2, has both of its electrons in the $1s$ state, and this state is then filled since it can contain only two electrons. In other words, it has only one orbital of two electrons. When the $s$ state is filled, there exists a spherical symmetry of electron charge, which makes this element very stable. As we proceed to lithium of atomic number 3, the $1s$ level is full and the $2s$ level contains one electron. Since the most stable state contains two electrons, lithium is very reactive and would like to lose or gain an electron. From beryllium (atomic number 2) to lawrencium (103), the $2s$ level will always be filled. However, the elements from boron (atomic number 5), through fluorine (atomic number 9), have unfilled $p$ shells. The $p$ orbitals can hold six electrons, two in each of the three directions along the three coordinate axes, as depicted in Figure 3.4. These elements would like to fill their orbitals, so they will also react with other elements in an attempt to fill the $p$ levels. The outermost $p$ orbitals of

**Figure 3.4** Electron probability distribution for the *p* orbitals. The highest probabilities are along the three orthogonal axes, each containing two electrons of opposite spin. When the *p* orbitals are filled, there exists a nearly spherical symmetry of charge. (From D. W. Richerson, *Modern Ceramic Engineering*, Marcel Dekker, New York, 1982, p. 4.)

neon, atomic number 10, are filled along all three axes, and as such the charge density again approaches spherical symmetry. Thus neon is inert. Similarly, argon is inert because the 3*p* subshell is the outermost shell and is filled with electrons and also has a spherical charge shape. By now we can see a pattern emerging that is summarized in Table 3.1. The *d* level can hold 10 electrons and when it is filled we again have spherical symmetry. The same concept applies to the 14 electrons in the filled 4*f* shell. Any time that the *s, p, d,* and *f* subshells are filled, spherical symmetry of the cloud of electron charge exists and the corresponding elements are inert. This principle becomes more evident in Table 3.2, where the electron filling pattern is shown for all elements, and

**Table 3.1** Order in Which Stable Energy Levels in an Atom Are Filled with Electrons

| Main shell[a] | Subshell | | | | Total capacity |
|---|---|---|---|---|---|
| | *s* | *p* | *d* | *f* | |
| (*K*) *n* = 1 | 2 | | | | 2 |
| (*L*) *n* = 2 | 2 | 6 | | | 8 |
| (*M*) *n* = 3 | 2 | 6 | 10 | | 18 |
| (*N*) *n* = 4 | 2 | 6 | 10 | 14 | 32 |

[a] The letters *K, L, M, N* were assigned to the main shells to represent spectral lines observed in x-ray emissions.

**Table 3.2**  Electron Structure of the Elements

| SHELL SUBSHELL | $K(n=1)$ s | $L(n=2)$ s | p | $M(n=3)$ s | p | d | $N(n=4)$ s | p | d | f | $O(n=5)$ s | p | d | f | $P(n=6)$ s | p | d | $Q(n=7)$ s |
|---|---|---|---|---|---|---|---|---|---|---|---|---|---|---|---|---|---|---|
| CAPACITY | 2 | 2 | 6 | 2 | 6 | 10 | 2 | 6 | 10 | 14 | 2 | 6 | 10 | 14 | 2 | 6 | 10 | 2 |
| 1 H | 1 | | | | | | | | | | | | | | | | | |
| 2 He | 2 | | | | | | | | | | | | | | | | | |
| 3 Li | 2 | 1 | | | | | | | | | | | | | | | | |
| 4 Be | 2 | 2 | | | | | | | | | | | | | | | | |
| 5 B | 2 | 2 | 1 | | | | | | | | | | | | | | | |
| 6 C | 2 | 2 | 2 | | | | | | | | | | | | | | | |
| 7 N | 2 | 2 | 3 | | | | | | | | | | | | | | | |
| 8 O | 2 | 2 | 4 | | | | | | | | | | | | | | | |
| 9 F | 2 | 2 | 5 | | | | | | | | | | | | | | | |
| 10 Ne | 2 | 2 | 6 | | | | | | | | | | | | | | | |
| 11 Na | Filled (2; 2 − 6) | | | 1 | | | | | | | | | | | | | | |
| 12 Mg | | | | 2 | | | | | | | | | | | | | | |
| 13 Al | | | | 2 | 1 | | | | | | | | | | | | | |
| 14 Si | | | | 2 | 2 | | | | | | | | | | | | | |
| 15 P | | | | 2 | 3 | | | | | | | | | | | | | |
| 16 S | | | | 2 | 4 | | | | | | | | | | | | | |
| 17 Cl | | | | 2 | 5 | | | | | | | | | | | | | |
| 18 A | | | | 2 | 6 | | | | | | | | | | | | | |
| 19 K | Filled (2; 2 − 6; 2 − 6) | | | | | | 1 | | | | | | | | | | | |
| 20 Ca | | | | | | | 2 | | | | | | | | | | | |
| 21 Sc | | | | | | 1 | 2 | | | | | | | | | | | |
| 22 Ti | | | | | | 2 | 2 | | | | | | | | | | | |
| 23 V | | | | | | 3 | 2 | | | | | | | | | | | |
| 24 Cr | | | | | | 5 | 1 | | | | | | | | | | | |
| 25 Mn | | | | | | 5 | 2 | | | | | | | | | | | |
| 26 Fe | | | | | | 6 | 2 | | | | | | | | | | | |
| 27 Co | | | | | | 7 | 2 | | | | | | | | | | | |
| 28 Ni | | | | | | 8 | 2 | | | | | | | | | | | |
| 29 Cu | Filled (2; 2 − 6; 2 − 6 − 10) | | | | | | 1 | | | | | | | | | | | |
| 30 Zn | | | | | | | 2 | | | | | | | | | | | |
| 31 Ga | | | | | | | 2 | 1 | | | | | | | | | | |
| 32 Ge | | | | | | | 2 | 2 | | | | | | | | | | |
| 33 As | | | | | | | 2 | 3 | | | | | | | | | | |
| 34 Se | | | | | | | 2 | 4 | | | | | | | | | | |
| 35 Br | | | | | | | 2 | 5 | | | | | | | | | | |
| 36 Kr | | | | | | | 2 | 6 | | | | | | | | | | |
| 37 Rb | Filled (2; 2 − 6; 2 − 6 − 10; 2 − 6) | | | | | | | | | | 1 | | | | | | | |
| 38 Sr | | | | | | | | | | | 2 | | | | | | | |
| 39 T | | | | | | | | | 1 | | 2 | | | | | | | |
| 40 Zr | | | | | | | | | 2 | | 2 | | | | | | | |
| 41 Nb | | | | | | | | | 4 | | 1 | | | | | | | |
| 42 Mo | | | | | | | | | 5 | | 1 | | | | | | | |
| 43 Tc | | | | | | | | | 6 | | 1 | | | | | | | |
| 44 Ru | | | | | | | | | 7 | | 1 | | | | | | | |
| 45 Rh | | | | | | | | | 8 | | 1 | | | | | | | |
| 46 Pd | | | | | | | | | 10 | | | | | | | | | |
| 47 Ag | Filled (2; 2 − 6; 2 − 6 − 10; 2 − 6 − 10) | | | | | | | | | | 1 | | | | | | | |
| 48 Cd | | | | | | | | | | | 2 | | | | | | | |
| 49 In | | | | | | | | | | | 2 | 1 | | | | | | |
| 50 Sn | | | | | | | | | | | 2 | 2 | | | | | | |
| 51 Sb | | | | | | | | | | | 2 | 3 | | | | | | |
| 52 Te | | | | | | | | | | | 2 | 4 | | | | | | |
| 53 I | | | | | | | | | | | 2 | 5 | | | | | | |
| 54 Xe | | | | | | | | | | | 2 | 6 | | | | | | |

Transition elements (Manganides) — rows 19–28 (N subshell)

Transition elements (Technetides) — rows 37–46 (O subshell)

Empty — O subshell, rows 47–54

(*continued*)

**Table 3.2**  Continued

| SHELL SUBSHELL | | K(n=1) s | L(n=2) s p | M(n=3) s p d | N(n=4) s p d f | O(n=5) s p d f | P(n=6) s p d | Q(n=7) s |
|---|---|---|---|---|---|---|---|---|
| CAPACITY | | 2 | 2  6 | 2  6  10 | 2  6  10  14 | 2  6  10  14 | 2  6  10 | 2 |
| 55 | Cs | | | | | 2  6 | 1 | |
| 56 | Ba | | | | | 2  6 | 2 | |
| 57 | La | | | | | 2  6  1 | 2 | |
| 58 | Ce | | | |     2 | 2  6 | 2 | |
| 59 | Pr | | | |     3 | 2  6 | 2 | |
| 60 | Nd | | | |     4 | 2  6 | 2 | |
| 61 | Pm | | Filled | (same as above) |     5 | 2  6 | 2 | |
| 62 | Sm | | | |     6 | 2  6 | 2 | |
| 63 | Eu | | | |     7 | 2  6 | 2 | |
| 64 | Gd | | | |     7 | 2  6  1 | 2 | |
| 65 | Tb | | | |     9 | 2  6 | 2 | |
| 66 | Dy | | | | 10 | 2  6 | 2 | |
| 67 | Ho | | | | 11 | 2  6 | 2 | |
| 68 | Er | | | | 12 | 2  6 | 2 | |
| 69 | Tm | | | | 13 | 2  6 | 2 | |
| 70 | Yb | | | | 14 | 2  6 | 2 | |
| 71 | Lu | | | | 14 | 2  6  1 | 2 | |
| 72 | Hf | | | | |     2 | 2 | |
| 73 | Ta | | | | |     3 | 2 | |
| 74 | W | | Filled | | |     4 | 2 | |
| 75 | Re | | (2; 2 − 6; 2 − 6 − 10; 2 − 6 − 10 − 14; 2 − 6) | | |     5 | 2  Transition | |
| 76 | Os | | | | |     6 | 2  elements | |
| 77 | Ir | | | | |     7 | 2  (Rhenides) | |
| 78 | Pt | | | | |     8 | 2 | |
| 79 | Au | | | | | | 1 | |
| 80 | Hg | | | | | | 2 | |
| 81 | Tl | | | | | | 2  1 | |
| 82 | Pb | | Filled | | | | 2  2 | |
| 83 | Bi | | (2; 2 − 6; 2 − 6 − 10; 2 − 6 − 10 − 14; 2 − 6 − 10) | | | Empty | 2  3 | |
| 84 | Po | | | | | | 2  4 | |
| 85 | At | | | | | | 2  5 | |
| 86 | Rn | | | | | | 2  6 | |
| 87 | Fr | | Filled | | | | 2  6 | 1 |
| 88 | Ra | | (same as above) | | | | 2  6 | 2 |
| 89 | Ac | | | | | | 2  6  1 | 2 |
| 90 | Th | | | | | | 2  6  2 | 2 |
| 91 | Pa | | | | |     2 | 2  6  1 | 2 |
| 92 | U | | | | |     3 | 2  6  1 | 2 |
| 93 | Np | | Filled | | |     5 | 2  6 | 2 |
| 94 | Pu | | (same as above) | | |     6 | 2  6 | 2 |
| 95 | Am | | | | |     7 | 2  6 | 2 |
| 96 | Cm | | | | |     8 | 2  6  1 | 2 |
| 97 | Bk | | | | |     9 | 2  6 | 2 |
| 98 | Cf | | | | | 10 | 2  6 | 2 |

*Source*: Richards.

we can see that the inert elements Ne, Ar, Kr, Xe, and Rn have filled outer subshells as well as filled inner main shells. The overall filling pattern is called the *electron configuration* of the elements and in shorthand notation can be written as follows:

$$1s^2 \quad 2s^2 2p^6 \quad 3s^2 3p^6 3d^{10} \quad 4s^2 4p^6 4d^{10} 4f^{14} \quad 5s^2 5p^6 5d^{10} 5f^{14} \quad 6s^2 6p^6 6d^{10} 6f^{14}$$

In writing the electron configuration of any element, the principal quantum number $n$ is written first, followed by the letter or letters for any $l$ subshells, with these letters being written with a superscript number indicating the number of electrons in each subshell.

There are some rather obvious inconsistencies in the way the shells are filled that must be addressed. These discrepancies are first noticed in the elements of atomic numbers 19 through 28, and then again in the 37 through 45 group, and in many of the rare earth elements and others of high atomic number. These exceptions can be presented, and perhaps better explained, by referring to Figure 3.5, where the energy levels are plotted as a function of the atomic number Z. At a certain atomic number, in the first case that for potassium, the $3d$ energy level is lower than that of the $4s$ orbital. This difference is predicted from the quantum-mechanical picture but not from the classical Bohr–Sommerfeld orbitals. The cause of the lowering of the $3d$ level can be attributed to the shielding effect of the inner filled shells. The $4s$ orbital can be viewed as penetrating in toward the nucleus, and thus the $4s$ electrons are more tightly bound to it because of the high positive charge on the nucleus. Elements that portray this characteristic are known as *transition elements*. One of the major consequences of this filling sequence is that there now exists an imbalance of spins for the electrons in the unfilled $d$ shell, which has a profound effect on the magnetic behavior of some of these elements.

*Example 3.3.* (a) Aluminum has an atomic number of 13. Write its electron configuration without resorting to Table 3.2. What is its valency?

(b) Iron is a transitional element with an atomic number of 26. Its electron configuration is $1s^2 2s^2 2p^6 3s^2 3p^6 3d^6 4s^2$. Cobalt is also a transition element and has an atomic number of 27. Write its configuration without using Table 3.2.

Solution. (a) $1s^2 2s^2 2p^6 3s^2 3p^1$. Aluminum generally behaves as a valency of 3 (i.e., both the $s$ and $p$ electrons become involved in bonding, even though the $s$ level is filled). Given the atomic number the student should be able to write the electron configuration of any element other than the transition elements.

(b) $1s^2 2s^2 2p^6 3s^2 3p^6 3d^7 4s^2$. In the transition elements, usually the partially filled subshell, in this case the $d$ shell, will fill up with the addition of one electron with each increase in atomic number. There are some exceptions. What element is the one exception in this transition series?

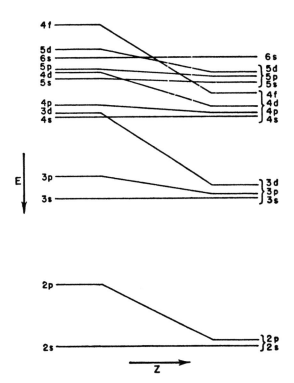

**Figure 3.5**  Dependence of energies of electron orbitals on the nuclear charge Z. (After W. J. Moore, *Physical Chemistry,* Prentice-Hall, Englewood Cliffs, N.J., 1955, p. 291.)

A final look at the periodic table is now in order. The elements that are in the same vertical row have similar outer-shell configurations and similar properties. The horizontal rows all begin with a very active, or as some prefer, reactive, metal on the left, and end with an inert element on the right. The transition elements are concentrated in the center. The active elements in the vertical rows on the left tend to form stable compounds with those elements on the right, excluding the inert elements. The term *electropositive* and *electronegative* are used to describe the elements that tend to form these compounds, with the *electropositive* elements on the left being metallic in nature and giving up electrons in forming these compounds, while the *electronegative* elements on the right receive the electrons. The most electronegative elements are in groups 6A and 7A of the periodic table. *Electronegativity* may be defined as the tendency of the atom to attract electrons. Pauling derived a somewhat empirical but useful way to express the relative electronegativity of elements as follows:

$$X_A - X_B = \left[ \frac{BE(A \text{ to } B) - \frac{1}{2}[BE(A \text{ to } A] + BE(B \text{ to } B)}{R} \right]^{\frac{1}{2}} \tag{3.6}$$

where $R$ is 96.5 kJ (a constant to convert kJ to electronegativity units), BE is the binding energy, and $X_A$ and $X_B$ are the electronegativities of elements A and B, respectively. By examining a number of compounds and by somewhat arbitrarily assigning a 4.0 electronegativity number to fluorine, it was found that the electronegativities varied from 0.9 for K to the 4.0 of F (see Table 3.3). All of this information is pertinent to or discussion of atomic bonding in the following section.

## 3.2  BOND TYPES

Chemical bonding occurs when dissimilar atoms are brought together because, by this marriage, they can satisfy their needs in terms of filled outer shells; that is, they do it because they want to. In a technical and less colloquial way of expressing this concept, they bond together because they can reduce the total energy of the system. Chemical bonds are generally classified as strong and weak bonds, while the latter have been described in the general category of *van der Waals* or *polarization* bonds.

**Table 3.3**  Electronegativities of the Elements

| H |  |  |  |  |  |  |  |  |  |  |  |  |  |  |  |  |
|---|---|---|---|---|---|---|---|---|---|---|---|---|---|---|---|---|
| 2.1 |  |  |  |  |  |  |  |  |  |  |  |  |  |  |  |  |  |
| Li | Be |  |  |  |  |  |  |  |  |  |  | B | C | N | O | F |
| 1.0 | 1.5 |  |  |  |  |  |  |  |  |  |  | 2.0 | 2.5 | 3.0 | 3.5 | 4.0 |
| Na | Mg |  |  |  |  |  |  |  |  |  |  | Al | Si | P | S | Cl |
| 0.9 | 1.2 |  |  |  |  |  |  |  |  |  |  | 1.5 | 1.8 | 2.1 | 2.5 | 3.0 |
| K | Ca | Sc | Ti | V | Cr | Mn | Fe | Co | Ni | Cu | Zn | Ga | Ge | As | Se | Br |
| 0.8 | 1.0 | 1.3 | 1.5 | 1.6 | 1.6 | 1.5 | 1.8 | 1.8 | 1.8 | 1.9 | 1.5 | 1.6 | 1.8 | 2.0 | 2.4 | 2.8 |
| Rb | Sr | Y | Zr | Nb | Mo | Tc | Ru | Rh | Pd | Ag | Cd | In | Sn | Sb | Te | l |
| 0.8 | 1.0 | 1.2 | 1.4 | 1.6 | 1.8 | 1.9 | 2.2 | 2.2 | 2.2 | 1.9 | 1.7 | 1.7 | 1.8 | 1.9 | 2.1 | 2.5 |
| Cs | Ba | La | Hf | Ta | W | Re | Os | Ir | Pt | Au | Hg | Tl | Pb | Bi | Po | At |
| 0.7 | 0.9 | 1.1 | 1.3 | 1.5 | 1.7 | 1.9 | 2.2 | 2.2 | 2.2 | 2.4 | 1.9 | 1.8 | 1.8 | 1.9 | 2.0 | 2.2 |

*Source*: Adapted from Linus Pauling, *The Nature of the Chemical Bond*, Cornell Press, Ithaca, N.Y., 1960.

### 3.2.1 Covalent Bonds

Covalent bonds arise when two atoms share electrons in an attempt to fulfill their needs in the sense of completely filling the $s$, $p$, $d$, and $f$ suborbitals. In our earlier discussion it was emphasized that the most stable states were those when the $s$ orbitals contained two electrons, the $p$ orbitals six electrons, the $d$ orbitals 10 electrons, and the $f$ orbitals 14 electrons. The simplest case of the covalent bond is that of two hydrogen atoms coming together to form a combined $s$ orbital that contains two electrons, one being contributed by each hydrogen atom. This is the classic example in chemistry texts, where the electron-dot notation is used to show the bond as follows (the two dots represent the shared electrons):

H:H

Perhaps a better way to visualize this bond is by illustrating the overlap of the two $1s$ orbitals, as depicted in Figure 3.6. Yet another way to describe this bond is that the electrons resonate between the two atoms and essentially trick the atoms into thinking that each atom always has two electrons in its $1s$ shell. Whatever the model, experiments show that the reduction in energy is about 4.5 eV when the hydrogen molecule is formed. This also means that 4.5 eV must be supplied to separate them (i.e., to dissociate the molecule). In higher-atomic-number common gases, such as nitrogen, oxygen, and fluorine, the atoms attain a stable shell of eight electrons by sharing outer-shell electrons. In nitrogen this is accomplished by the sharing of three $2p$ electrons of one atom

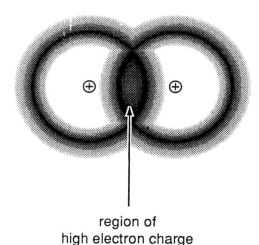

region of
high electron charge

**Figure 3.6**   Hydrogen molecule bond.

with three $2p$ electrons from another nitrogen atom. Similarly, two oxygen atoms share two $2p$ electrons, and fluorine atoms share one $2p$ electron from each atom, as shown by dot notation in Figure 3.7.

Hybridization Bonding Process

The covalent bond is fairly common in organic molecules and occurs in many cases by the "hybridization" of the $s$ and $p$ orbitals. The $s$ shell is more or less spherical in shape, whereas the $p$ shell is more like a dumbbell in shape (Figure 3.4). When they hybridize a different shape of charge density occurs. Let's examine pure carbon first. The electrons involved in the hybridization are the $2s^2 2p^2$ electrons, which are changed to $2s^1 2p^3$ states by excitation of a $2s$ electron to the $2p$ level. In pure carbon we now have four equivalent $sp^3$ hybrid orbitals, as depicted in Figure 3.8. Now each of these four orbitals shares an electron with identical orbitals from four other carbon atoms to achieve an eight-electron outer shell. Carbon in this form is crystalline (diamond structure), as shown in Figure 3.9. Even though energy is required for an $s$ electron to be excited to the $p$ level, there is a net decrease in energy when the carbon atoms are bonded together. This is an extremely strong bond and is the reason that diamond is so hard, which is reflected in its high bond energy and high modulus of elasticity. In this type of bond the electrons resonate between the atoms and have a very high probability of being in the dark regions in Figure 3.9. Covalent bonds of this type are very directional and result in a directional-

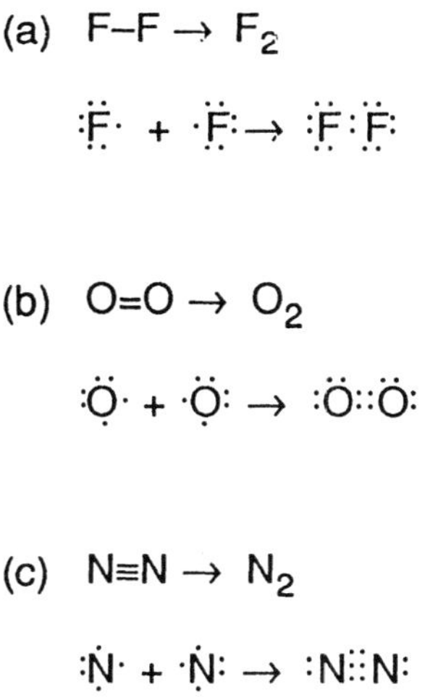

(a)  F–F → $F_2$

  $\ddot{F}\cdot$ + $\cdot\ddot{F}\colon$ → $\ddot{F}\colon\ddot{F}\colon$

(b)  O=O → $O_2$

  $\ddot{O}\cdot$ + $\cdot\ddot{O}\colon$ → $\ddot{O}\colon\colon\ddot{O}\colon$

(c)  N≡N → $N_2$

  $\ddot{N}\cdot$ + $\cdot\dot{N}\colon$ → $\colon N\colon\colon N\colon$

**Figure 3.7** Covalent bonds in some gaseous molecules.

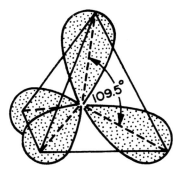

**Figure 3.8**   Four $sp^3$ hybrid orbitals for the carbon atom.

ity of properties; the strength of diamond, for example, will be dependent on the crystallographic direction in which it is measured.

When carbon atoms combine to form graphite, a different type of hybridization occurs. Graphite has a hexagonal layered structure (Figure 3.10). Here three electrons hybridize in bonds at 120° to each other, leaving one electron only weakly bonded to the others. This electron accounts for the high electrical and thermal conductivity of graphite compared to diamond. The layers are bonded together by weak secondary bonds, which allows them to slide past each other easily, thereby producing the lubricating properties of graphite.

Now let's examine the orbitals in methane ($CH_4$) in terms of the hybridization concept. Again a $2s$ electron of carbon is excited to a $2p$ state, but now the covalent bond is completed by overlapping the $s$ orbital of a hydrogen atom and sharing this electron with one electron from the four available from the

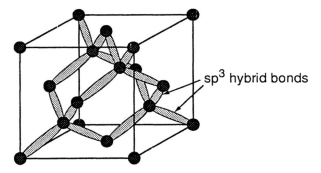

**Figure 3.9**   Diamond crystal structure formed from four $sp^3$ orbitals that combine with four $sp^3$ orbitals from four other carbon atoms.

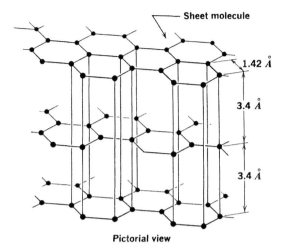

**Figure 3.10** Sheet structure of graphite. (From C. W. Richards, *Engineering Materials Science,* Brooks/Cole, Monterey, Calif., 1961, p. 45.)

carbon atom. This arrangement is depicted in Figure 3.11. Again notice the directionality of the bond, which is a characteristic of the covalent bond. Covalent bonds are prevalent in polymer molecules.

### 3.2.2 Ionic Bonds

Ionic bonds are formed by a transfer of electrons from one atom to another, in contrast to the sharing found in the covalent bond. Another distinct difference

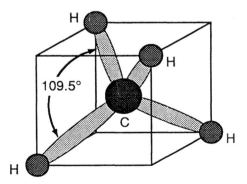

**Figure 3.11** Methane molecule formed from four $sp^3$ carbon orbitals sharing electrons with four hydrogen atoms.

between the two is the lack of directionality in the ionic bond that was so prevalent in the covalent bond. Let's examine the bond in NaCl. Sodium of atomic number 11 has an electron configuration of $1s^2 2s^2 2p^6 3s^1$, while chlorine has 17 electrons in its configuration (i.e., $1s^2 2s^2 2p^6 3s^2 3p^5$). Chlorine completes its third shell of eight electrons by receiving the $3s^1$ electron from the sodium when the two atoms are brought together. For the moment let's assume that this bond is one of 100% ionic character. Then there would be a total charge transfer resulting in a sodium ion with $+1$ charge ($+1.602 \times 10^{-19}$ C) and a chlorine ion with an equal quantity of negative charge as depicted in Figure 3.12. The Coulombic attractive force between the two ions is proportional to the product of the charges divided by the equilibrium interatomic distance, $r_0$, squared, that is,

$$F_{\text{attractive}} = \alpha \frac{Z_1 Z_2 e^2}{r_0^2} \tag{3.7}$$

where $\alpha$ is the proportionality constant. The repulsive force, due primarily to the inner electron shells of the two ions, is of the form

$$F_{\text{repulsive}} = -\frac{nb}{r_0^{n+1}} \tag{3.8}$$

where $n$ and $b$ are constants. At the equilibrium interatomic distance these forces will be equal in magnitude and opposite in sign, resulting in a net zero force.

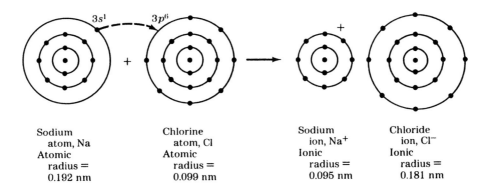

| Sodium atom, Na Atomic radius = 0.192 nm | Chlorine atom, Cl Atomic radius = 0.099 nm | Sodium ion, Na$^+$ Ionic radius = 0.095 nm | Chloride ion, Cl$^-$ Ionic radius = 0.181 nm |

**Figure 3.12**   Formation of a sodium chloride ion pair by transfer of negative charge from a $3s^1$ sodium orbital to a $3p$ chlorine orbital. (From W. F. Smith, *Principles of Materials Science and Engineering*, McGraw-Hill, New York, 1986, p. 36.)

### 3.2.3 Covalent Versus Ionic Bonds

In the discussion of the NaCl bond above it was assumed that the bond was 100% ionic (i.e., that total charge transfer had taken place). In discussing the covalent bonds it was implied that there existed a 100% covalent bond (i.e., complete sharing of electrons or complete resonance of the electron between the two atoms). In most compounds, particularly in the inorganic compounds found in ceramic materials (the halides are classified here as a type of ceramic compound), we find that the bond is neither pure covalent nor pure ionic. We speak of the degree of covalency or ionicity. Pauling, using his electronegativity system [equation (3.6)], suggested that the degree of ionicity could be approximated by the following:

$$\% \text{ ionic character} = [1 - \exp(-¼)(X_A - X_B)^2] \times 100\% \qquad (3.9)$$

Applying this equation, and using the values of $X_A$ and $X_B$ from Table 3.3 for some common compounds, we obtain the following:

| Compound | Ionic character (%) |
|----------|---------------------|
| LiF | 98 |
| KF | 98 |
| MgO | 70 |
| NaCl | 59 |
| SiC | 12 |

### 3.2.4 Metallic Bond

In the metallic bond generally we are talking about the bonding together of like metallic atoms. In metallic materials each atom gives up its valence electrons to form a cloud of negative charge in which the positive ions that remain behind are immersed, somewhat as depicted in Figure 3.13. The forces that hold the ions together are somewhat akin to the ionic bond in that there exists an electrostatic attraction between the negative charge cloud and the positive ions, with the repulsive forces being primarily those due to the inner electron shells. As the two atoms approach each other in the formation of the metallic bond, the negatively charged inner shell of one atom comes closer to the other atoms' negative shell than do the positive nuclei, which on an atomic scale remain a sizable distance apart. But one could also say that there exists some similarity to the covalent bond in that the valence electrons resonate among all atoms. The metallic bond is not nearly as directional as the covalent bond; however, metal properties do vary somewhat with crystallographic direction. The

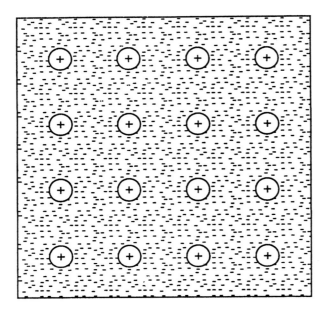

**Figure 3.13**   Metallic bond formed by the attraction of the positive nuclei to the valence electron space charge.

valence electrons are free to roam throughout the body of the material. They are constantly in rapid motion at temperatures above 0 K and in essence resonate among all atoms. These "free electrons" account for the high electrical and thermal conductivity of metals and to some extent their plasticity. Alloys of two or more metals in solid solution, a concept that we explore later, will also have a metallic bond, the bond strength of the unlike atom pairs being somewhat different, but not drastically so, than that of the like atoms.

Intermetallic compounds, such as the nickel and titanium aluminides ($Ni_3Al$, $NiAl$, $Ti_3Al$, and $TiAl$), which are currently being seriously studied for high-temperature applications, can be considered to be of a mixed bonding nature with some degree of ionicity and covalency mixed with the metallic character. The greater the difference in electronegativity of the elements, the larger the ionic character. The titanium aluminides would be expected to have more ionic character than the nickel aluminides, which have have only about 2.2% ionic character, based on Pauling's approximation. These compounds are brittle, somewhat like the ionic and covalent compounds, but their electrical and thermal conductivities are more like those of metals. Electrical conductivities are on the order of $10^6 \ (\Omega \cdot m)^{-1}$.

## 3.2.5 Weak Polarized Secondary Bonds

J. D. van der Waals proposed in 1873 a new thermodynamic equation of state of a real gas which took in account the weak attractive forces between gas molecules. These weak forces are a result of both permanent and fluctuating dipoles. On approaching each other, even the inert gases experience this attraction. A dipole moment exists when there is a nonsymmetrical charge distribution, and the moment is simply the magnitude of the charge multiplied by the distance of separation. Even though the average electrostatic field strength of an inert gas atom is zero, it is not zero within the filled electron shells if the number of electrons on one side is greater than on the other side. This fluctuating dipole creates attraction sufficient to cause liquefaction at low temperatures where thermal agitation is minimal. Neon at atmospheric pressure becomes a liquid at 27 K and a solid at 24 K. Helium becomes a liquid at 4.2 K under atmospheric pressure. The bond strengths in liquid inert elements vary from about 2 to 8 kJ/mol (0.5 to 2 kcal/mol).

There are other types of weak physical forces which are stronger than the *van der Waals* forces but yet very much weaker than the strong ionic, covalent, and metallic bonds. These bonds arise due to a permanent nonsymmetrical charge distribution in the molecule (i.e., they have permanent dipoles). Water is the most common example. In this molecule the hydrogen atoms are bonded to the oxygen atom at an angle that places the positive hydrogen atoms on one side of the molecule (Figure 3.14). This causes the positive side of the

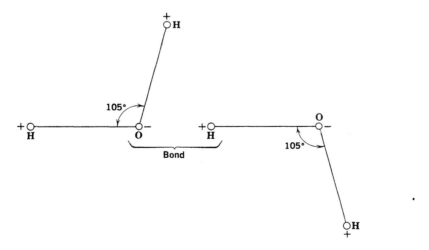

**Figure 3.14** Hydrogen bond between two water molecules. (From C. W. Richards, *Engineering Materials Science*, Brooks/Cole, Monterey, Calif., 1961, p. 26.)

molecule to be attracted to the negative side of a neighboring molecule. As water vapors are cooled the attractive forces are sufficient to cause the vapor to condense to a liquid at 273 K (100°C). At any given temperature and volume the number of bonds broken per unit time equals the number re-forming so that an equilibrium vapor pressure is set up. This bond is called the *hydrogen* bond and has an energy of about 29 kJ/mol (7 kcal/mol), compared to the strong bonds, which are in the range 50 to 1000 kJ/mol (12.5 to 250 kcal/mol). The hydrogen bond is a prevalent bond between polymer molecules and biological molecules such as DNA. These electrostatic charges on the molecule are quite small and on the order of one-sixth of the charge on the electron.

### 3.2.6  Comparative Bond Strengths

Bond strengths have been measured and expressed in a number of ways. The properties such as modulus of elasticity, heats of vaporization and fusion, and melting points are all indicative of bond strengths, and the higher these values, the stronger the bond. Lattice energies for crystalline solids can be both calculated and measured experimentally. These energies will differ somewhat from those of individual bonds, however, because more than two atoms of the crystalline lattice are involved in the bonding process. This will become more apparent after we have studied crystalline structures. The enthalpies of atomization, which is basically the same as the heats of vaporization, for metallic, ionic, and covalently bonded solids are listed in Table 3.4. For the ionic-covalent combination, the degree of ionicity or covalency is also listed. These are the energies to dissociate the solids into the gaseous state and are approximately equal to the binding energies. In Table 3.5 the strengths of several of

**Table 3.4**  Materials with Strong Binding Energies

| Substance | Bond | Enthalpies of atomization (kJ/m) |
|---|---|---|
| LiF | 98% Ionic | 849 |
| MgO | 70% Ionic | 1000 |
| NaCl | 59% Ionic | 640 |
| Al | Metallic | 324 |
| Cu | Metallic | 339 |
| W | Metallic | 849 |
| Ti | Metallic | 473 |
| Si | 100% Covalent | 450 |
| C (diamond) | 100% Covalent | 713 |
| SiC | 88% Covalent | 1230 |

**Table 3.5**  Materials with Weak Binding Energies

| Substance | Bond type between molecules | Enthalpies of sublimation at melting point (kJ/mol) |
|---|---|---|
| Argon | Van der Waals | 7.5 |
| Oxygen | Van der Waals | 7.5 |
| Water | Hydrogen | 51 |
| Ammonia | Hydrogen | 35 |

some weak bonds are listed for comparative purposes. These bond strengths are indicated by their heat of sublimation at their respective melting temperatures.

The melting temperatures of metals, and to a lesser extent ceramics, have been used for some time as a relative measure of bond strength, but it is just that, only a relative measure. The term *homologous temperature* means equal fractions of the melting point of solids in kelvin. For example, if we want to compare the creep behavior (discussed in Chapter 2) of two metals, we should compare them at equal fractions of their melting points in kelvin. Lead (Pb) has a melting point of 327.4°C which is 600 K. Aluminum has a melting point of 660°C or 933 K. Generally, creep becomes significant in the vicinity of 0.5 of the melting point in kelvin. This means that aluminum will creep at 467 K or at 194°C, while lead will creep at 300 K or 27°C, which is about room temperature. The homologous temperature comes in handy for such qualitative comparisons, but it is not a good measure of bond strength. In Table 3.6 we compare the heats of fusion with those of evaporation for both aluminum and water. Note how small the heats of fusion are compared to those of evaporation, the latter being a good measure of bond strength. In the melting process, bonds are constantly being rearranged and re-formed rather than all being broken at the same time. Much less energy is required than that needed to free the

**Table 3.6**  Heats of Fusion Compared to Bond Energies

| Substance | Heat of fusion (kcal/mol) | Heat of evaporation (kcal/mol) |
|---|---|---|
| Water ($H_2O$) | 1.4 | 10.7 |
| Aluminum | 2.55 | 67.6 |

atom completely from its environment of other atoms. Nevertheless, the melting temperature in kelvin, symbol $T_m$, will be used in future discussions when comparing temperature-dependent properties.

## 3.3 INTERATOMIC DISTANCES

The equilibrium spacings between atoms is determined by a balance between attractive and repulsive forces no matter what type of bond is involved. The bonding force between atoms may be represented approximately by the general equation

$$\frac{A}{r^M} - \frac{B}{r^N} \qquad (N > M) \tag{3.10}$$

where $r$ is the center-to-center spacing between atoms and $A$, $B$, $M$, and $N$ are constants. The first term represents the attractive force and the second term the repulsive force. When these to are equal, $F(r) = 0$, and the atoms are said to be at their equilibrium spacing $r_0$ for the temperature and pressure to which the atoms are being subjected. Any change in these two variables will alter the spacing accordingly.

Let's examine how these forces vary with distance by referring to the graph of Figure 3.15. The shape of the attractive and repulsive curves depend on the values of the constants. We will assume that one atom is fixed at the position

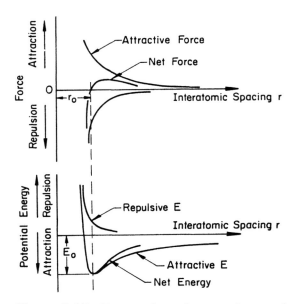

**Figure 3.15**   Forces and energies versus interatomic spacing.

$r = 0$ and that another atom is made to approach it from some large distance, infinity if you wish. As the atoms approach they are drawn together by the attractive force, which outweighs the repulsive force. If they are made to approach farther than the equilibrium spacing, the repulsive force predominates and tries to push them back toward their equilibrium position. Note that the repulsive force curve has a much steeper slope than that of the attractive force curve. The inner filled electron shells of each atom provide a very strong repulsive force between the atoms as they approach each other. It is more difficult to push atoms together than to pull them apart. Hence the second part of equation (3.10) must increase more rapidly for diminishing values of $r$ than does the first term, and $N$ is necessarily greater than $M$. The shape of the attractive force curve is the same as that for the attraction between any unlike charges. According to Coulomb's law, this force is inversely proportional to the square of the distance between the charges. Thus $M$ has the value of 2 for the strong interatomic bonds. But $M$ may be as high as 7 for the weak intermolecular bonds. The value for $N$ is more difficult to establish and is a function of the type of bond for both strong and weak bonds. Experimentally, it has been found that $N$ varies from about 7 to 12 for strong bonds, with metallic bonds generally falling in the lower part of this range and the ionic and covalent bonds in the upper part.

*Example 3.4.* Determine the equilibrium interatomic spacing $r_0$ of two atoms in terms of the constants in equation (3.10). At $r = r_0$,

Solution.

$$F = 0$$

$$\frac{A}{r_0^M} = \frac{B}{r_0^N}$$

$$r_0^{N-M} = \frac{B}{A}$$

$$r_0 = \left( \frac{B}{A} \right)^{1/N-M}$$

The $A$ in equation (3.10) can be approximated from the ionic charge on each atom; that is, it becomes the product of $Z_1$ and $Z_2 e^2$, where $Z_1$ and $Z_2$ are the atomic numbers of the two atoms. The repulsive force has the form

$$F_{\text{repulsive}} = -\frac{nb}{r^{n+1}} \qquad (3.11)$$

where $n \approx 9$ for ionic solids. The net resulting force now becomes

$$F_{\text{net}} = \frac{Z_1 Z_2 e^2}{4\Pi\epsilon_0 r_0^2} - \frac{nb}{a^{n+1}} \qquad (3.12)$$

where $\epsilon_0$ is the permittivity. Note that the attractive force is proportional to $r^{-2}$ while the repulsive force is proportional to $r^{-10}$. Thus the attractive forces predominate at large atom separations, while the repulsive forces take over at small interatomic spacings.

The equilibrium interatomic distances for most materials is on the order of a few angstroms ($\text{Å}$) ($1\,\text{Å} = 10^{-8}$ cm $= 0.1$ nm). In the net force curve of Figure 3.15, the slope at $r = r_0$ is related to the modulus of elasticity. The modulus is determined from a straight-line relationship (Figure 2.3). In most elastic deformations the atoms move less than 10% of their equilibrium spacing. For this reason we can make the statement that the elastic moduli in tension and compression are essentially the same.

### 3.3.1 Bond Energy Versus Interatomic Distance

In most cases it is better to deal with energies than with forces. The potential energy between the two atoms, which is the energy of interest here, is found by integrating the force equation as follows:

$$U(r) = \int F(r)dr = \int \left[ \frac{A}{r^M} - \frac{B}{r^N} \right] dr$$

$$U(r) = -\frac{A}{M-1}\frac{1}{r^{M-1}} + \frac{B}{N-1}\frac{1}{r^{N-1}} + C$$

Setting $U(r) = 0$ at $r = \infty$ and solving for $C$ yields

$$U(r) = -\frac{a}{r^m} + \frac{b}{r^n} \qquad n > m$$

where $a$ and $b$ are new constants related to $A$ and $B$, $m = M$, and $n = N - 1$. The equilibrium spacing is that of lowest energy, which also occurs when $F(r) = 0$, as can be illustrated by taking the derivative of the energy equation; thus

$$U(r) = \int F(r)\,dr$$

$$\frac{dU(r)}{dr} = F(r)$$

Both the force and the minimum in the energy curve are zero at the same value of $r$ (i.e., for $r = r_0$) (Figure 3.15).

### DEFINITIONS

*Anion:* negatively charged nonmetallic ion that results from an atom gaining an electron.

*Atomic mass unit (amu)*: measure of atomic mass; one-twelfth of the mass of a carbon 12 atom.

*Atomic number (Z)*: number of protons in the nucleus of an atom.

*Atomic weight*: mass in grams of a mole of atoms of an element, including all naturally occurring isotopes.

*Avogadro's number*: number of atoms or molecules in a mole of the substance. It has the numerical value of $6.023 \times 10^{23}$.

*Azimuthal quantum number*: quantum number $l$, which is related to the ellipticity of an orbiting electron.

*Cation*: positively charged metallic ion that results from a loss of an electron from an atom.

*Coulombic force*: force between charged particles that varies inversely with the square of the distance between the particles.

*Electronegative*: for an atom, the tendency to accept electrons when forming a compound.

*Electronegativity*: measure of the tendency of an atom to attract electrons.

*Electron volt*: unit of energy that is equivalent to that acquired by an electron when it moves through an electric potential of 1V.

*Electropositive*: for an atom, the tendency to lose electrons when forming a compound.

*Heat (enthalpy) of dissociation*: energy required to separate atoms within a molecule.

*Heat (enthalpy) of evaporation*: energy required to cause atoms or molecules to go from the liquid to the vapor (gaseous) state.

*Heat (enthalpy) of fusion*: energy required to melt a solid substance.

*Hybrid bond*: bond formed when the $s$ and $p$ electrons combine and act as if they were in equivalent energy levels.

*Intermetalic compound*: compound formed between two or more metallic atoms.

*Ion*: atom that has either an excess or a deficiency of electrons with respect to the number of protons in the nucleus.

*Mole*: quantity of a substance corresponding to $6.023 \times 10^{23}$ atoms in the case of an element, or molecules in the case of compounds.

*Pauli exclusion principle*: principle that only two electrons can occupy a given energy level and they must be of opposite spin.

*Photoelectric effect*: emission of electrons from a metallic surface when the surface is exposed to radiation in the visible, ultraviolet, or near-visible range of frequencies.

*Planck's constant (h)*: universal constant that has a value of $6.63 \times 10^{-34}$ J · s.

*Quantum numbers*: set of four numbers that defines allowable electron energy states. Three of these numbers describe the size, shape, and orientation of the electrons' probability density; the fourth defines the electron spin.

*Schrödinger wave equation*:   equation used in wave mechanics which can be used to compute electron energy levels and binding energies.

*Transition elements*:   elements whose *s* energy level is less than the 3*d* in one series of elements. In the higher-atomic-number elements the *s* levels are less than the *f* levels.

*Valence electrons*:   electrons in the outermost shells that become involved in the bonding process.

*van der Waals forces*:   weak forces between atoms or molecules that arise from dipoles, permanent or induced.

## QUESTIONS AND PROBLEMS

3.1. Write the electron configuration of copper (atomic number 29) without referring to Table 3.2.

3.2. List some intermetallic compounds that are formed from the transition metals of atomic numbers 22 through 28. Many inorganic compounds are listed in the *Handbook of Chemistry and Physics* (CRC Press, Boca Raton, Fla.) Explain their valencies in terms of their electron configurations (i.e., how many 3*s* and 3*d* electrons are involved).

3.3. Compute the weight of an iron atom based on the average weight of all isotopes.

3.4. How many protons and electrons are there in an atom of zinc?

3.5. Compute the approximate radius of the innermost electron orbit of a magnesium atom.

3.6. What is the electron energy level of the electron in Problem 3.5?

3.7. Compute the frequency of radiation emitted from excited hydrogen when an electron falls from an orbit of $n = 3$ to one of $n = 1$.

3.8. Compute the percent ionic character for the intermetallic compound NiAl and for the inorganic compounds $TiO_2$ and $Al_2O_3$.

3.9. What types of bonding would you expect in (a) sodium; (b) CaO; (c) $CaF_2$; (d) $Fe_2O_3$?

3.10. Which of the following metals would you expect to have the smallest bonding energy in the pure metallic state of each? Pb, Al, Cu. Explain how you arrived at your selection.

3.11. List four properties that are indicative of bond strength.

3.12. How might the energy required to separate two atoms to an infinite spacing be determined from the force versus atom separation distance (Figure 3.15)?

3.13. How many bonds are formed in the semiconducting element germanium? How many electrons are involved in the bonding process? What type of bond exists?

3.14. Compute in coulombs the maximum negative charge on a chlorine ion, a bismuth ion, and a silicon ion.

3.15. How many atoms are there in 1 g of nickel? In 1 lb?

3.16. What are the allowed values of each of the four quantum numbers?

3.17. Calculate the energy in joules and electron volts for a photon of 500 nm wavelength.

3.18. What is the percent by weight of nickel in the $Ti_3Al$ intermetallic compound?

3.19. What is valency of Ti in the $Ti_3Al$ compound?

3.20. Why are intermetallic compounds being considered for use as structural components in high-temperature applications?

3.21. Make a cardboard model of a tetrahedron.

3.22. Cartridge brass is an alloy consisting of 70 wt % of copper and 30 wt % of zinc. What is the atom percent of each element in this alloy?

## SUGGESTED READING

Miller, F. M., *Chemistry: Structure and Dynamics,* McGraw-Hill, New York, 1984.
Smith, W. F., *Principles of Materials Science and Engineering,* 2nd ed., McGraw-Hill, New York, 1990.

# 4

# Structure of Materials

## 4.1 INTRODUCTION: CRYSTALLINE VERSUS AMORPHOUS

Now that we have an understanding of atomic structure and how atoms bond together, let's proceed to a consideration of larger structures that consist of aggregates of many atoms. To a certain extent, engineering materials can be separated into two broad classes, called *amorphous* and *crystalline*. In Chapter 1 engineering materials were classified as belonging to one of four groups: metals, ceramics, polymers, and composites. (We intentionally omitted fabrics and wood, although wood is a type of composite.) These are the more traditional classes of engineering materials. Which of these four traditional classes fall into the broad classification of *amorphous,* and which fall into the *crystalline* category? Actually, these four major groups can fall into either the *amorphous* or *crystalline* groups, depending on how they are processed.

At this point it may be prudent to introduce the concept of a *phase* in order to better understand the classes of matter. We sometimes erroneously speak of crystalline and amorphous phases when referring to gases, liquids, and solids, which are the three states of matter. It is thus tempting to call these states of matter the three phases of matter. But this is not quite correct, as will become apparent in the following discussion.

A *phase* is defined as a structurally homogeneous region of matter throughout which the physical and chemical properties are uniform. Phases are separated by phase boundaries and each of the phases are distinguishable,

*103*

although high-magnification microscopes may be required for this distinction process when the quantity of a phase is minute. Phases are also thought of in theory as being mechanically separable, but in reality the mechanical separation process is almost impossible when the quantity of a phase, particularly in the solid state, is sufficiently small to require microscopic detection procedures. A gas or a mixture of gases is a phase of matter. Within a system that contains only gases there can be only one phase, which is a random distribution of molecules. In a gas phase there is no particular pattern or arrangement of molecules. It is a *disordered* phase. Disordered phases are said to be *amorphous*. In gases the words *state* and *phase* are interchangeable. A system containing two immiscible liquids, water and oil for example, is in a liquid state but consists of two phases. A system of two miscible liquids, such as water and alcohol, is a single-phase system. Liquids are disordered amorphous phases. Liquids do have some degree of local or *short-range order,* particularly liquid metals and liquid ceramics. In these liquids the arrangement of atoms over a few hundreds of atom spacings may be in a somewhat regular pattern, but these atoms are in a state of flux and constantly interchanging positions. There is no *long-range order.* In a liquid the cohesive forces are sufficiently strong to lead to a condensed state, but not strong enough to prevent a considerable translation energy of the individual molecules or atoms. There is a sharp change in properties in going from a solid to a liquid, the sharpness of the melting point being one of these properties.

Solids can exist in either ordered or disordered states. Quartz crystals, of the type used for frequency control in radio circuits, and grains of sand are composed of silicon and oxygen atoms in a definite precise arrangement of atoms. The atoms are present in the proportion of two oxygen atoms for each silicon atom, resulting in the chemical formula $SiO_2$ (silica), but there is not a distinct molecule of $SiO_2$. Each atom is bonded to a larger number of atoms which are arranged in a pattern that repeats itself throughout the three dimensions of the quartz crystal or the grain of sand. In such bodies of matter the atoms are in an *ordered* configuration and the material is said to be *crystalline*. In the example above, only one solid phase is present and there are no phase boundaries other than the surface–air boundary. Each particle of sand and each quartz crystal is a single crystal and also a single phase. Single-phase crystalline bodies usually consist of more than one grain or crystal and are termed to be *polycrystalline* structures. Crystalline bodies often contain more than one crystalline phase. A good example is plain carbon steel. It consists of regions of iron, in which the iron atoms occupy precise positions in a specific identifiable repeatable pattern or arrangement of atoms, and other regions of the compound iron carbide, which has the chemical formula $Fe_3C$ and has an entirely different pattern of atoms than that of the iron. These phases can be distinguished in the microscope (Figure 4.1). The iron phase usually consists of

**Figure 4.1.** Two-phase mixture of iron and iron carbide, the latter being the dark platelets. When the two phases are arranged in such a lamellae structure, the resulting microstructure is called *pearlite*.

many grains. Similarly, a particle of the iron carbide phase may also consist of several grains, depending on the processing history. Thus this material is both polyphase and polycrystalline, the phases being separated by phase boundaries, and within each phase the crystals are separated by grain boundaries. Iron itself, as well as a number of other elements and crystalline compounds, can change their solid-state atom arrangement when heated through certain temperatures called *phase transition temperatures*. Such materials are *polymorphic* (i.e., capable of existing in more than one solid phase). A crystalline phase can thus be viewed as an orderly arrangement of atoms.

Returning to silica for the moment, if it is heated to 1710°C, it melts and loses its ordered arrangement in becoming a liquid. If it is cooled from this amorphous state to the solid state, it may or may not become crystalline, depending on the rate of cooling through the liquid–solid phase transition. If a molten pot of silica is removed from the furnace and poured into a shallow mold or floated on molten metal, a process used in the manufacture of plate glass, the cooling rate on the order of several degrees per minute is too fast for the oyxgen and silicon atoms to take up the precise positions necessary to form crystals or grains. Thus glass is often called a *supercooled liquid,* because it is in a disordered amorphous state. It is a solid, however, and can sustain a stress, something a liquid cannot do. Thus solids can exist in either amorphous or crystalline structures, something gases and liquids cannot do. Furthermore, crystalline solids can exist in more than one type of atom arrangement.

Metals, until recently, were thought to exist in only the solid state in a crystalline arrangement of atoms, never in an amorphous state. A few decades ago it was found that liquid metals, when cooled through their solid-to-solid transition at cooling rates on the order of 1 million degrees per second could become a glasslike amorphous solid. These structures are now referred to as *metallic glasses* and have been found to possess some interesting and practical physical and mechanical properties.

Polymers can also exist in the solid state in either amorphous or crystalline atom arrangements. Unlike ceramics and metals, however, it depends more on molecular structure than on cooling rate. Some long-chain polymer molecules can align themselves in a more-or-less repeatable pattern (Figure 4.2) that are sufficiently ordered to be called crystalline. They can diffract x-rays just as metallic and ceramic crystals do, even though they do not have the degree of preciseness in their pattern of atom arrangement as do metals and ceramics (x-ray diffraction, a subject to be covered later, is a characteristic of, and a test for, crystallinity). Note in Figure 4.2 that only in certain regions are the molecules aligned. Seldom if ever do all of the molecules align in a polymer body. Ninety percent crystallinity is considered high and unusual. But many polymers are entirely amorphous. All of the thermoset polymers, although they have a three-dimensional network arrangement of atoms, are amorphous, since there is no repeat pattern in their structures.

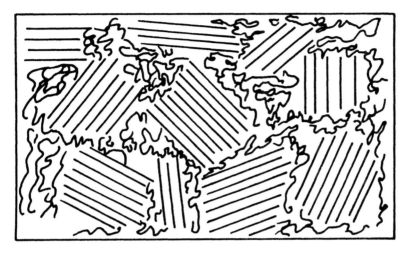

**Figure 4.2.** Partial crystallization in polymers.

## 4.2 SOLIDIFICATION OF METALS AND CERAMICS

How do crystalline solid structures form from a molten amorphous mass of material? Actually, we have already stated that amorphous solids result if the molten material is cooled faster than some specified rate, being very fast for metals and some ceramics but a normal cooling rate for glasses. The thermodynamics and atom arrangements for the formation of a metallic glass is quite involved and still under study. Suffice it to say that these structures are metastable. They are not at their lowest energy state, and as for all metastable structures, they would like to go to a lower state of energy, in this case to a crystalline structure. When amorphous silica glass structures go to the stable crystalline phase the process is called *devitrification.* The word *vitreous* means "glasslike," or "relating to glass." Thus devitrification means to do away with the amorphous glass structure. *The word "glass" implies an amorphous structure.* At room temperature the rates of devitrification are extremely slow, as witnessed by the fact that prehistoric glass artifacts remain as glasses today. Some glasses will slowly devitrify at ambient temperature. The rate of devitrification increases with increasing temperature and attains its maximum rate at about $0.7T_m$, where $T_m$ is the melting point in kelvin.

When metals cool from the molten state, except for the extremely fast solidification that produces metallic glasses, they form many small crystals called *grains.* Let's assume for the moment that the molten metal is very pure and is suspended away from any container walls. This suspended state can be achieved by *floating-zone melting,* a process that we will describe later. In such a state there are no foreign particles to promote nucleation of a solid phase. In the absence of such nucleation sites, solid nuclei then form when the temperature of the molten mass is cooled just below the melting–solidification point. Some undercooling below the freezing point is necessary to overcome the energy required to form a liquid–solid interface. When cooled very slowly, this undercooling is only a matter of a few degrees and is of no significant practical importance. Solidification begins with the formation of a few nuclei of solid crystals, as depicted in Figure 4.3a. In this nucleus a few hundred or so atoms have taken up the arrangement we will eventually find in the final solid. But in the absence of foreign nucleation sites, these nuclei form by random fluctuation of the atoms. Eventually, a cluster of a few hundred to a few thousand atoms will be large enough to become a stable nucleus. Remember: They want to take on this arrangement because at temperatures below the freezing point, the crys-

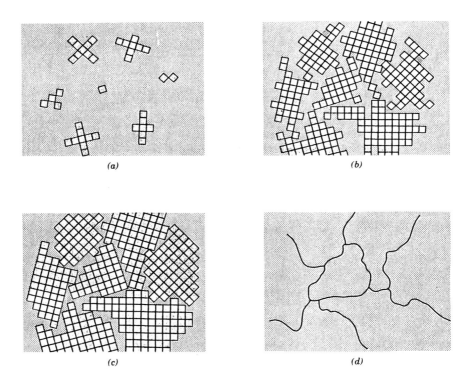

**Figure 4.3.** The solidification process begins with the formation of a few crystalline nuclei that grow until they meet at a grain boundary. Solidification of crystalline substances is thus a process of "nucleation and growth." (Adapted from W. Rosenhain, *An Introduction to the Study of Physical Metallurgy,* 2nd ed., Constable and Company, Ltd., London, 1915.)

talline-ordered arrangement is a lower-energy state and hence a more stable atom configuration. At or near the solidification temperature many nuclei form and grow until these solid particles impinge on one another, eventually consuming all of the liquid. Each nucleus eventually becomes a grain and the region of impingement a grain boundary. Since the nuclei formed at random, they will be oriented differently with respect to each other. Thus the atoms in each final grain will be oriented differently from that in a neighboring grain and within the solid as a whole. The spatial pattern of the atoms (i.e., the crystal structure) within each grain will be the same since they are all of the same solid phase. It is just that the *building blocks* (i.e., the grains) are sort of cocked with respect to each other. This solidification sequence can be followed in Figure 4.3.

Seldom do we have the ideal solidification conditions discussed above. Metals are generally melted in a pot called a crucible and poured into a mold to form an ingot for further fabrication to size, or poured into a mold designed for the final cast shape. The cross section of a typical cast ingot of brass that has been polished and chemically etched to reveal the grains is shown in Figure 4.4. Here the steel mold walls have acted as nucleation sites. The solid crystals started forming at the wall and grew into the molten mass as heat was extracted from the mold walls. This type of crystal growth produces long columnar grains. Finally, at the last stage of solidification, some equiaxed grains form in the center from the random nucleation of grains. The crater in the top of the ingot is a result of shrinkage, since the solid density is greater than the liquid. The shrinkage cavity will always be at the last region to solidify. Sometimes, due to the complicated heat extraction process, this shrinkage cavity may occur in the interior regions of a large casting and not be visible to the eye. This is an undesirable casting defect. When the ingot is further fabricated, particularly at room temperatures, the cavity may form a crack. This is one reason that ingot breakdown is done while the metal is hot, thereby affording the possibility of welding together the surfaces of the shrinkage cavity.

The solidification of molten ceramics, except for the silicates that form glasses, is the same in principle as that for metals. Commercial ceramic shapes, however, are usually formed by the pressing together of powders rather than going through the liquid-to-solid transition.

## 4.3 CRYSTALLOGRAPHY

Crystallography is concerned with the various types of atom arrangements that are possible in crystalline solids. Here we are talking about a single phase since each solid phase in a crystalline body has its own specific arrangement of atoms. In the mid-nineteenth century August Bravais, a French crystallographer, showed that there could be only 14 different and distinct ways in

**Figure 4.4.** Cast ingot of brass that was solidified in a steel mold. Note the long columnar grains that grew inwardly from the mold wall. (Courtesy of J. Harruff.)

which *points* could be arranged geometrically. These are called the *Bravais lattices* and are depicted in Figure 4.5. Note that the word *points* is emphasized. Since Bravais was a crystallographer, no doubt he was trying to correlate the internal arrangement of atoms with the observed symmetry of crystalline faces found in naturally occurring minerals. X-ray diffraction techniques were not yet available, but certain internal atom arrangements had been predicted.

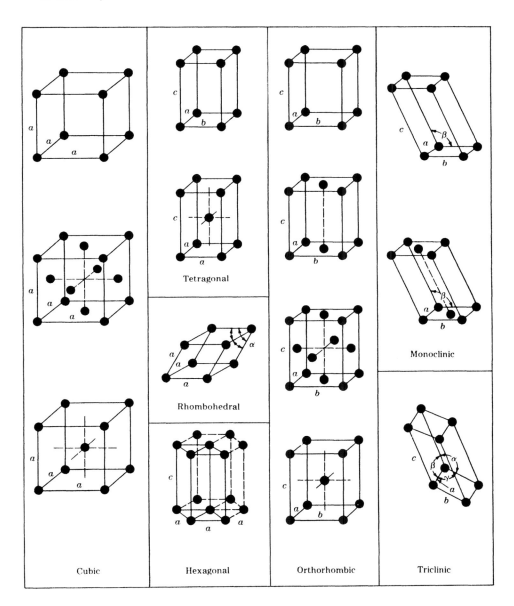

**Figure 4.5.** The 14 Bravais lattices and the seven crystal systems. (After W. G. Moffatt, G. W. Pearsell, and J. Wulff, *The Structure and Properties of Materials,* Wiley, New York, 1964, p. 47.)

Nevertheless, the outcome of Bravais's work was an arrangement of points, not atoms. The fact that there can only be 14 distinct ways to arrange these points was a problem in geometry, not materials. Now we do find that in some materials, particularly pure metals, it is convenient to mentally put atoms at the points when discussing crystal structures. For crystalline compounds it may be better to have a point represent a cluster of atoms whose positions are defined in terms of their distance from the lattice points.

The Bravais lattices can be broken down into seven crystal systems, also noted in Figure 4.5. It is found that all naturally occurring minerals fall into one of these crystal systems. Each system can be described in terms of the relative spacing between the points and the angular relationship of the coordinate axes used to outline the points. These relationships are summarized in Table 4.1.

### 4.3.1 The Unit Cell and Miller Indices

For the purposes of this book the *unit cell* will be defined as the smallest repeating unit in the crystalline lattice pattern. The lattice pattern extends through the entire grain or crystal. By defining the unit cell in this way, the Bravais lattices depicted in Figure 4.5 become unit cells. The true crys-

**Table 4.1**  Angular and Distance Relationships in the Seven Crystal Systems

| System | Axes and interaxial angles | |
|---|---|---|
| Triclinic | Three axes not at right angles, of any lengths | |
| | $a \neq b \neq c$ | $\alpha \neq \beta \neq \gamma \neq 90°$ |
| Monoclinic | Three axes, one pair not at right angles, of any lengths | |
| | $a \neq b \neq c$ | $\alpha = \gamma = 90° \neq \beta$ |
| Orthorhombic (rhombic) | Three axes, at right angles, all unequal | |
| | $a \neq b \neq c$ | $\alpha = \beta = \gamma = 90°$ |
| Tetragonal | Three axes at right angles, two equal | |
| | $a = b \neq c$ | $\alpha = \beta = \gamma = 90°$ |
| Cubic | Three axes at right angles, all equal | |
| | $a = b = c$ | $\alpha = \beta = \gamma = 90°$ |
| Hexagonal | Three axes coplanar at 120°, equal | |
| | Fourth axis at right angles to these | |
| | $a_1 = b_2 = a_2 \neq c$ | |
| | ($or\ a_1 = b \neq c$) | $\alpha = \beta = 90°, \gamma = 120°$ |
| Rhombohedral (trigonal) | Three axes equally inclined, not at right angles; all equal | |
| | $a = b = c$ | $\alpha = \beta = \gamma \neq 90°$ |

tallographer would call these *structure cells,* but that definition is not needed for the principles of crystallography that will be applied in this book. The distance that a unit cell must be translated in order to repeat itself is called the *lattice parameter.* In the cubic and rhombohedral systems, there is only one lattice parameter since $a = b = c$. In the tetragonal and hexagonal cells there are two lattice parameters and three in the orthorhombic, monoclinic, and triclinic cells. The unit cell is used (1) to define the atom positions in the lattice, (2) to enable us to identify crystal planes and directions, and (3) to describe the atomic packing density of a particular crystalline material. The need for such information arises because the properties of crystals vary with direction in the crystal. Materials in which the properties do not vary with direction are *isotropic* materials, and those that vary with direction are *anisotropic* materials. In crystals the properties are a function of atom spacing and the directionality of bonding. In addition, the planar atomic packing density determines which planes will be involved in plastic deformation of crystalline materials.

Miller Indices in the Cubic Unit Cell

To define crystal planes and directions an identification system proposed by W. H. Miller in 1839 is used. Let's examine how this method is applied to the description of a *crystallographic direction. First it should be realized that in any Bravais lattice every point has identical surroundings.* This will not be evident unless one visualizes several unit cells stacked together and project the lattice along the three Cartesian coordinate system $x$, $y$, $z$ axes. Since each point has identical surroundings, any point can be selected as the origin. Once the origin and Cartesian $xyz$ axes have been specified, the crystallographic direction of a line from the origin of the unit cell to any point whose vector coordinates are $u$, $v$, $w$ is denoted as $[u\,v\,w]$. The coordinates of the endpoints of the direction vector are the same as the components of the vector resolved along the three coordinate axes. To avoid fractions, the components are reduced to the smallest integers. The line must pass through the designated origin. Since all parallel lines must have the same direction, and hence the same indices, one must move a line that does not pass through the origin to a position such that it does pass through the origin. But it must be parallel to the position of the original line. Some simply crystallographic directions in a cubic unit cell are shown along with their Miller indices in Figure 4.6. Negative point coordinates are indicated with a bar above the number. By convention, specific direction indices are bracketed by the [ · ] configuration. Most often we are concerned with directions of a given type. For example, the edges of a cube all have a one and two zeros in their Miller indices. All of the cube edges are called *one hundred directions* and are bracketed with the < · > notation. All of the <100> directions are crystallographically equivalent and are sometimes called a family of directions. Any member of the family has the same atom spacing along its

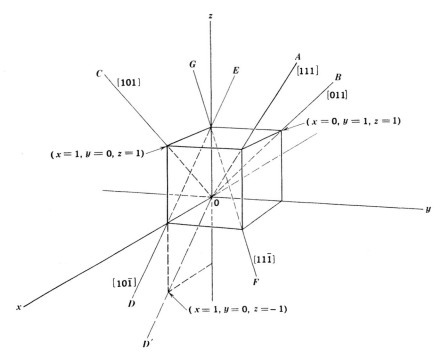

**Figure 4.6.**   Some simple crystallographic directions and their corresponding Miller indices.

directions. The cube edge family consists of the following $<100>$ specific directions:

[100], [010], [001], [$\bar{1}$00], [0$\bar{1}$0], [00$\bar{1}$]

The members of a family always comprise the set consisting of the possible ways of arranging the vector coordinates of the family. For the cube edge there are six members of the family. Other family directions commonly used in the cube system are the cube body diagonals, designated as $<111>$, and the cube face diagonals, designated as $<110>$.

*Example 4.1.*   Write the Miller indices of the specific directions comprising the equivalent directions in the $<110>$ and $<111>$ families.

Solution.   The possible combinations for the numbers 1, 1, 0 as per Miller indices are

[110], [101], [011], [$\bar{1}$10], [$\bar{1}$01], [0$\bar{1}$1], [1$\bar{1}$0], [10$\bar{1}$], [01$\bar{1}$]

Although there are six cube faces with two face diagonals each, there are only nine specific directions since parallel directions in parallel faces have the same indices.

The possible combinations for the <111> family are

$$[111], [\bar{1}11], [1\bar{1}1], [11\bar{1}], [\bar{1}\bar{1}1], [1\bar{1}\bar{1}], [\bar{1}1\bar{1}], [1\bar{1}\bar{1}]$$

There are eight members of this family.

*Example 4.2.* Determine the Miller indices for the directions *A* and *B* shown below.

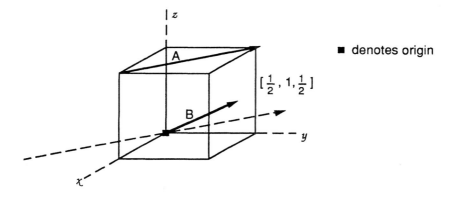

■ denotes origin

$[\frac{1}{2}, 1, \frac{1}{2}]$

Solution. The endpoint coordinates of vector *A* are 0, 1, 1. But the vector does not pass through the origin, so we move it parallel to itself until it does pass through the origin. The new position is denoted by the dashed line. Its endpoint coordinates are now $-1$, 1, 0 and its indices are written [$\bar{1}$10].

The vector *B* passes through the origin and ends in the center of a cube face at the position ½, 1, ½, which when reduced to the smallest integers become a [121] direction.

The identification of *crystallographic planes* in a cubic unit cell per the Miller method utilizes the points where the planes intercept the three Cartesian axes. There are certain rules that must be obeyed and these are listed in sequence as follows:

1. Locate the points where the plane intersects the three axes. The plane cannot pass through the origin. If it does, another plane must be constructed parallel to the original plane or the origin must be moved. Usually, it is more convenient to do the latter.

2. Take the reciprocals of the intercepts. Now it becomes apparent why the plane could not pass through the origin since the reciprocal of zero is

infinity. However, a plane that parallels an axis may be considered to have an infinite intercept and therefore a zero index when the reciprocal is taken.

3. Clear fractions but do not reduce to lowest integers.
4. Enclose the numbers in parentheses ( · ). Again, negative numbers are denoted with a bar over the number.

Just as in directions, we have families of planes and these are enclosed in braces, thus:

$$(100), (010), (001), (\bar{1}00), (0\bar{1}0), (00\bar{1}) = \{100\}$$

These are the cube faces. The planes that bisect the cube are of the $\{110\}$ family. The $\{111\}$ family of planes form an octahedron and are sometimes called the octahedral planes. Figure 4.7 illustrates the indices for some common planes.

There are a few characteristics of the Miller indices for both planes and directions that should be noted. Planes and their negatives are identical. This is not the case for directions. A [100] direction points opposite to a [$\bar{1}$00] direction. Planes and their multiples are not identical. The (100) plane has an atom arrangement different than does the (200) plane. They have a different atom density. But again this is not the case for directions. In directions a [220] line is just an extension of the [110] line.

*Example 4.3.* Write the members of the family of planes for the $\{110\}$ and $\{111\}$ families.

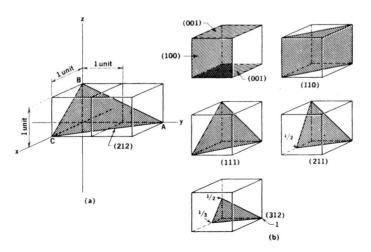

**Figure 4.7.** Miller indices of some common crystallographic planes.

Solution. Again the solution constitutes the number of ways that we can write these indices, including negative numbers. For the {110} family this becomes

$$(110), (101), (011), (\bar{1}10), (\bar{1}01), (0\bar{1}1), (1\bar{1}0), (10\bar{1}), (01\bar{1}) = \{110\}$$

For the {111} family we have

$$(111), (\bar{1}11), (\bar{1}\bar{1}1), (1\bar{1}1), (\bar{1}1\bar{1}), (\bar{1}\bar{1}\bar{1}), (1\bar{1}\bar{1}), (11\bar{1}) = \{111\}$$

*Example 4.4.* What are the Miller indices of the shaded planes *A* and *B* shown below?

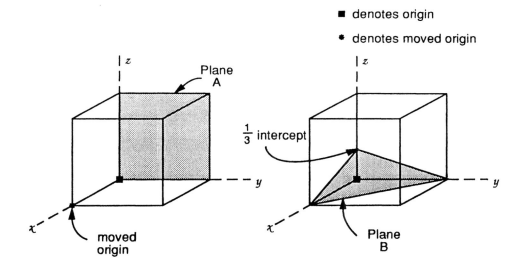

Solution. Plane *A* lies in the *yz* plane and passes through the origin. We thus move the origin the distance of the cube edge along the positive *x* axis. The intercepts are now $-1$, $\infty$, $\infty$. The reciprocals are $-1$, 0, 0, so the indices of the plane now become $(\bar{1}00)$.

Plane *B* intercepts the axes at 1, 1, ⅓. The reciprocals are 1, 1, 3 and the indices are written as (113).

### Miller Indices in the Hexagonal Unit Cell

From a close examination of all of the unit cells of Figure 4.5 it should become apparent that the same approach that was used for the cubic unit cell can be applied to all of the others except the hexagonal lattice. In the hexagonal cell we use four axes rather than three. Three of the axes lie in what is called the basal plane and are positioned 120° apart as depicted in Figure 4.8. The fourth

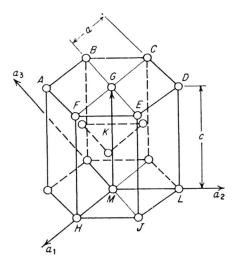

Basal plane (0001) − *ABCDEF*
Prism plane (10$\bar{1}$0) − *FEJH*
Pyramidal planes
   Type I, Order 1 (10$\bar{1}$1) − *GHJ*
   Type I, Order 2 (10$\bar{1}$2) − *KJH*
   Type II, Order 1 (11$\bar{2}$1) − *GHL*
   Type II, Order 2 (11$\bar{2}$2) − *KHL*
Digonal axis [11$\bar{2}$0] − *FGC*

**Figure 4.8.**   Important planes of the hexagonal lattice. (From G. E. Dieter, *Mechanical Metallurgy*, 2nd ed., McGraw-Hill, New York, 1976, p. 109.)

axis, called the $c$ axis, is perpendicular to the basal plane. In this unit cell the length of the $a_1$, $a_2$, and $a_3$ axes are all the same, including both positive and negative directions. There are thus two lattice parameters, namely $a$ and $c$, with $c > a$.

In hexagonal systems the intercepts of the planes on all four axes are found, their reciprocals taken, and fractions cleared, as was the case for the cubic unit cells. This leads to four intercepts, which are called the $(h, k, i, l)$ indices. Because of the redundancy of the $a_3$ axes and the geometry of the lattice, the first three integers, corresponding to the $a_1$, $a_2$, and $a_3$ intercepts, are related by the equation $h + k = -i$. Therefore, the four-digit indices can be reduced to the three-digit notation. In most technical papers the four-number system, known as *Miller–Bravais indices*, are used. The important planes for the hexagonal system are shown in Figure 4.8.

The Miller indices for crystallographic directions in the hexagonal lattice involve translations along the three positive and negative $a$ directions and also

translations along the *c* axis. In introductory-level materials courses it is seldom, if ever, necessary to specify a direction other than along these four axes.
The indices for these directions are shown in Figure 4.9 for both the three- and
four-digit index notation.

## 4.4  SOME COMMON MATERIALS STRUCTURES

Normally, one would examine the structures of engineering materials by
describing the structures separately of metals, ceramics, semiconductors, polymers, and composites. We depart from this approach to some extent and separate the structures into the broader classes of crystalline, amorphous, and composite forms. The semiconducting elements, for example silicon and germanium, are crystalline materials. The term *semiconductor* was not a technical
household word until the 1950s and 1960s. Prior to that time they, along with
boron, arsenic, antimony, and tellurium, were often called *metalloids,* since
many of their properties fell between those of metals and nonmetals. But they
are crystalline, and along with metals, ceramics, and some polymers, comprise
the crystalline category. Composites will be placed in a separate category.
Many of today's composites consist of crystalline and noncrystalline structures
intermixed. But they can also consist of two or more crystalline or two or more

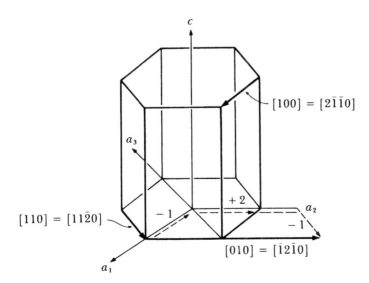

**Figure 4.9.**  Important crystallographic directions for the hexagonal lattice. (From
D. E. Askeland, *The Science and Engineering of Materials,* PWS-Kent, Boston, 1989,
p. 69.)

amorphous phases. We can also expect this class of engineered materials to expand rapidly and eventually to comprise structures that are nonexistent at the time of this writing.

## 4.4.1 Crystalline Structures

The simplest crystal structures to describe are those composed of all like atoms (i.e., the elements). At room temperature and atmospheric pressure these structures include all of the metallic and semiconducting elements, and the elements such as boron and carbon, which perhaps are better classified as insulators, although carbon in graphite form is a conductor of heat and electricity. Of the 90 or so elements found in nature, about 75 are crystalline at ambient conditions, and about 65 of these have electrical conductivities $[>1(\Omega \cdot m)^{-1}]$ sufficient to be considered as metals. Also, about 60 of the crystalline elements are found to be either the cubic or hexagonal close-packed (HCP) structures. These are the crystal structures that we now examine in some detail. In describing these structures, the hard-sphere model will be used, in which the atoms are considered to be spherical and each sphere consisting of an ion core. This is in contrast to the Bravais unit cells of Figure 4.5, where the atoms were essentially points in an open-lattice arrangement. This open lattice was used so that the geometrical arrangement of points could easily be viewed and comprehended. The more realistic picture is the hard-sphere model where atoms touch in some directions and nearly touch in all directions when they are at their equilibrium positions. The surfaces then represent those of the outermost filled shell of electrons even though the mass of the atom is in the nucleus, which comprises an insignificant part of the sphere. We say that the spheres are *hard* because they are almost incompressible, due to the high repulsive forces exerted between atoms by the filled shells of electrons. The items of interest in discussing these structures are their atom density, called the *atomic packing factor* (APF), atom spacing (lattice parameter $= a$), and atom radius $R$.

Cubic Structures: Elements

In Figure 4.5 we saw that the cubic crystal system consisted of simple cubic (SC), face-centered cubic (FCC), and body-centered cubic (BCC) structures. No elements of the simple cubic structure exist, so our discussion will be confined to the BCC and FCC forms.

The BCC unit cell is simply a cube with atoms at each corner plus one atom in the center of the cube. The open model showing the atom sites is depicted in Figure 4.10a and the isolated hard model in Figure 4.10b. Note that in the latter model, the spheres touch only in the <111>-type crystallographic directions (i.e., along the cube diagonals). The center atom touches all eight corner

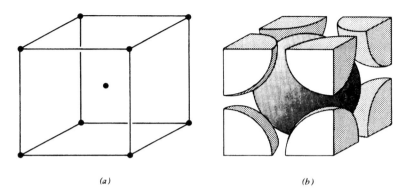

**Figure 4.10.** BCC hard-sphere and atom-site unit cell models. (From K. M. Ralls, T. H. Courtney, and J. Wulff, *Introduction to Materials Science and Engineering,* Wiley, New York, 1976, p. 157.)

atoms. It has eight nearest neighbors and is said to have a *coordination number* (CN) of eight.

Remember that all atoms have the same surroundings, so we can choose our axes so that the corner atom becomes the center atom. The CN is the same for all atoms in the BCC structure. Figure 4.11 shows the relationship between the atom radius $R$ and the lattice parameter $a$, from which the APF can be calculated. But first we need to know the number of atoms per unit cell. Note in the isolated cell of Figure 4.10b that the cube corners consist of a one-eighth section of a sphere. Each corner atom is shared by eight other adjacent unit cells. Thus we can say that we have eight corner atoms of one-eighth each plus one full center atom, resulting in a total of two atoms per unit cell. The volume of the unit cell is simply $a^3$. The APF can be expressed as

$$\text{APF} = \frac{\text{volume of atoms in cell}}{\text{volume of cell}}$$

Now using the relationship between $a$ and $R$ of Figure 4.11, that is,

$$a = \frac{4R}{\sqrt{3}}$$

and the total volume of the atoms, which is $= 4/3\pi R^3 \times 2$, the APF becomes

$$\text{APF} = \frac{2(4/3\pi R^3)}{(4R/\sqrt{3})^3} = 0.68$$

The atomic packing factor is independent of atom size.

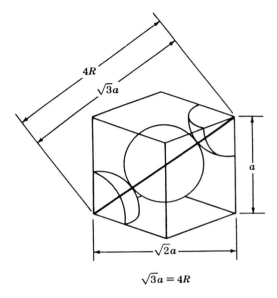

**Figure 4.11.** Relationship of the lattice parameter $a$ and the atom radius $R$ in the BCC structure.

The more common elements that are found to exist at room temperature and atmospheric pressure in the BCC structure include the alkali metals Li, Na, and K, and the transition metals V, Cr, Nb, Mo, Ta, W, and Fe (low-temperature form of iron called *alpha* iron).

*Example 4.5.* In the BCC structure, what is the planar atom density in the most densely packed plane of alpha iron?

Solution. By inspection it is observed that the most densely packed plane is a $<110>$ type, that is, the plane that bisects the cube and contains the center atom.

$$\text{Planar density} = \frac{\text{equivalent no. of atoms whose centers are intersected}}{\text{area of plane}}$$

Atoms intersected $= 1$ center atom $+ 4 \times \frac{1}{4}$ atoms

Plane area $= \sqrt{2}a \times a = \sqrt{2}a^2$

$$\text{Density} = \frac{2 \text{ atoms}}{\sqrt{2}a^2}$$

For alpha iron,

$$\text{density} = \frac{1}{\sqrt{2}(0.287 \text{ nm}^2)} = \frac{17.2 \text{ atoms}}{\text{nm}^2}$$

The hard-sphere isolated unit cell model and the atom site model of the FCC lattice is shown in Figure 4.12. In this cell the atoms are positioned at each corner of the cube and in the center of each face. They touch along a face diagonal (i.e., a $<110>$ crystallographic direction). The number of atoms per unit cell in this case can be computed as follows:

8 corner atoms at 1/8 each = 1 atom

6 face atoms at 1/2 each = 3 atoms

total atoms per unit cell = 4 atoms

From Figure 4.13 the relationship between $a$ and $R$ becomes

$$a = \frac{4R}{\sqrt{2}}$$

and the APF is computed to be

$$APF = \frac{4 \times 4/3\pi R^3 \text{ (volume of atoms)}}{(4R/\sqrt{2})^3 \text{ (volume of cell)}} = 0.74$$

The common elements that are found in the FCC system are Al, Ca, Ni, Pd, Pt, Cu, Ag, Au, Pb, and the high-temperature form of iron called *gamma* iron. In comparing the BCC and FCC lattices there are two points that should be remembered and which we will use in later discussions. The first is the packing density. In the BCC lattice 68% of the space is occupied, while in the FCC lattice a higher density of 74% is found. The BCC is a more open structure. The FCC density is the same as that which one could obtain if pool balls were stacked in layers on top of one another, with the second layer being placed such that each sphere rests in a cup formed by a hole between the spheres in the first layer. The second layer must be rotated 120° with respect to the first layer about an axis perpendicular to the layers to accomplish this fit. Now let's place a third layer on top of the second layer. If we rotate the third layer 120°

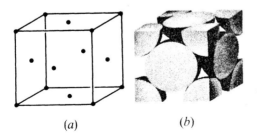

(a)        (b)

**Figure 4.12.** FCC hard-sphere and atom-site unit cell models. (From K. M. Ralls, T. H. Courtney, and J. Wulff, *Introduction to Materials Science and Engineering*, Wiley, New York, 1976, p. 154.)

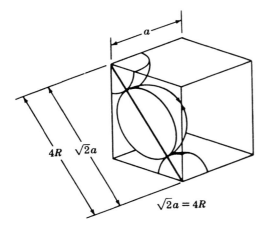

**Figure 4.13.** Relationship of the lattice parameter $a$ and the atom radius $R$ in the FCC lattice.

with respect to the second layer, we again find cups formed by a hole between the spheres of the second layer. When a fourth layer is placed on the third and rotated another 120°, the atoms of the fourth layer fall into the cups on the third layer. Note that 3 times 120° is equal to 360°, so we have come full circle, such that the fourth layer is identical to the first. This layered structure has an ABCABC stacking sequence. This atom arrangement is shown schematically in Figure 4.14. In the first layer of the closest packing of spheres, the atom centers take on a hexagonal arrangement as denoted by the hexagonal dashed line configuration. The pockets in this first layer are shown by the lines at 120° with respect to each other. To see the subsequent layers we need to examine an open structure as shown on the top left in Figure 4.15. With a little imagination the rotation of the subsequent layers can be visualized. What is not so obvious is that these atom layers are all on a [111] set of parallel planes. This is shown on the bottom left of the figure in an open schematic picture, but to really see that the {111} planes are planes of close-packed spheres, an actual model including several unit cells is required.

The other point of interest in comparing the BCC and FCC unit cells is the location of the interstitial sites. These sites are the holes that exist in the hard-sphere models. They are the sites where smaller atoms can reside in alloys and compounds. These sites are illustrated in our discussion of cubic compound structures.

The diamond cubic lattice was used in Chapter 3 to illustrate the strong directional covalent bonding of carbon atoms (Figure 3.9). In this hybridized

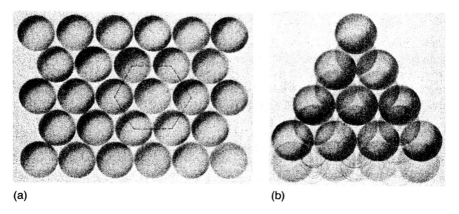

**Figure 4.14.** ABCABC stacking sequence for the FCC close-packed lattice.

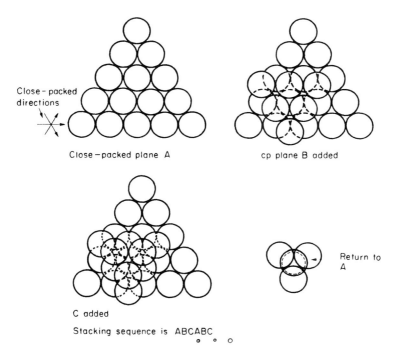

**Figure 4.15.** ABCABC stacking sequence for the FCC lattice. (From M. F. Ashby and D. R. H. Jones, *Engineering Materials*, Pergamon Press, Elmsford, N.Y., 1980, p. 44.)

structure each carbon atom bonds to four other carbon atoms. The structure consists of atoms at all eight corners and in all six faces, plus atoms in interior tetrahedral positions. If we examine a single carbon atom bonded to its four nearest neighbors, a tetrahedron results (Figure 4.16). Four of these tetrahedral groups combine to form the large cube, which is a variation of an FCC structure. There is space in the large cube for eight tetrahedrons, but only four are used. Silicon, germanium, and tin (below 13°C) have diamond cubic lattices.

Cubic Structures: Compounds

There are many compounds that have a cubic structure, but most are not SC, BCC, or FCC structures but some variation of these unit cells. A picture of the *interstitial* positions in the cubic structures (Figure 4.17) will assist in explaining the cubic compound structures. The black dots of Figure 4.17 represent the position of the larger holes in the three cubic structures. In metals these holes will often contain the small impurity atoms, such as boron, oxygen, nitrogen, carbon, and hydrogen, sometimes added intentionally and other times added inadvertently during processing. In compounds these sites will contain one of the atoms of the compound. Not all sites are necessarily filled. The largest SC interstitial site is the center of the cube. An atom in that site will have eight nearest neighbors of the other atom of the compound (i.e., its coordination number is eight). *In compounds the coordination number is always the number of nearest neighbors of unlike atoms involved in the bonding.* Atoms in the octahedral and tetrahedral sites will have coordination numbers of 6 and 4, respectively. Octahedral sites in the BCC unit cells are located in the faces of the cube, and in FCC they occur at the center of the cube and at the midpoint of the cube edges.

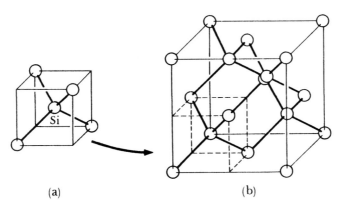

(a)                                          (b)

**Figure 4.16.**   Diamond cubic structure, which is present in silicon, germanium, and many compound semiconductors. (From D. E. Askeland, *The Science and Engineering of Materials,* PWS-Kent, Boston, 1985, p. 82.)

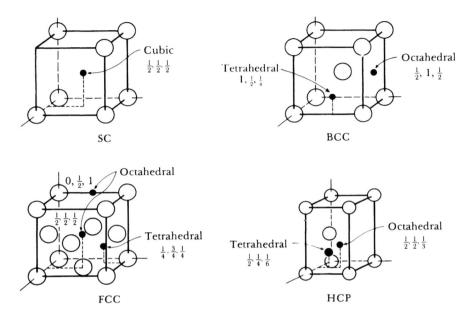

**Figure 4.17.** Interstitial sites in the cubic and HCP unit cells. (Adapted from D. E. Askeland, *The Science and Engineering of Materials,* 2nd ed., PWS-Kent, Boston, 1989, p. 73.)

The type of structure that will form between unlike atoms is dependent on the magnitude of the electrical charge on each atom (i.e., their valency) and on the relative sizes of the cations and anions. The electrical charge must be neutral. Assume a +2 charge on the anion, as in calcium. Then, for the needed neutrality to occur, there must be two negative charges somewhere nearby. In the case of the compound calcium fluoride, one calcium ion can balance the charge on two fluoride ions, giving the stoichiometric formula of $CaF_2$. But remember that more is involved in the bonding process than just three ions. A molecule of $CaF_2$ does not really exist as such in the solid state. The bonding will occur between nearest-neighbor atoms, which in cubic structures will usually be four, six, or eight atoms. There is also a weaker bond to the next shell of next-nearest neighbors.

The second factor, the relative size of cation and anion, determines the coordination number. Since it gives up electrons, the cation is usually smaller than the anion, which receives the electrons, although the charge transfer is never complete. The coordination numbers for ranges of cation–anion ratios $(r_c/r_a)$ are listed in Table 4.2 and the ionic radii for several ions in Table 4.3. The radii are dependent on the ion surroundings and will be somewhat different

**Table 4.2**  Coordination Numbers for Various Cation/Anion Ratios

| Coordination number | Cation/anion radius ratio |
|---|---|
| 2 | <0.155 |
| 3 | 0.155–0.225 |
| 4 | 0.225–0.414 |
| 6 | 0.414–0.732 |
| 8 | 0.732–1.0 |

for different coordination numbers. The most stable compound occurs when the anions surrounding a cation are in contact. In general, the larger the cation/anion ratio, the larger the CN. The following cubic structures can be viewed in Figure 4.18.

*Cesium Chloride Structure.* The cesium chloride structure is a simple cube, with the cube center interstitial site filled with the chlorine anion. The $Cl^-$ is surrounded by the eight corner cesium cations. The $(r_{Cs}/r_{Cl})$ ratio is 0.92; hence according to Table 4.2, a coordination number of 8 is required. One may be tempted to call this a BCC structure, *but the unit cell is always determined by like atoms.* Several ionic compounds have the CsCl structure, but these are not important ceramic materials. On the other hand, there are several important intermetallic compounds, including AlNi, which have this structure.

*NaCl Structure.* The $r_{Na}/r_{Cl}$ ratio is 0.56; thus a coordination number of six is required. The structure is one of two interpenetrating FCC lattices, with

**Table 4.3**  Ionic Radii of Selected Elements

| Element | Ionic radius (nm) |
|---|---|
| $Al^{3+}$ | 0.051 |
| $Ba^{2+}$ | 0.136 |
| $O^{2-}$ | 0.140 |
| $F^{-1}$ | 0.133 |
| $Na^{+1}$ | 0.10 |
| $Mg^{2+}$ | 0.072 |
| $S^{2-}$ | 0.184 |
| $Cl^{-1}$ | 0.181 |

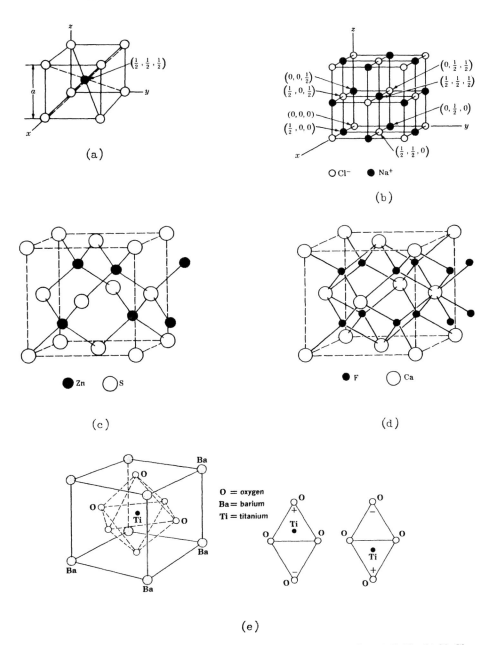

**Figure 4.18.** Common cubic structures in ceramic compounds: (a) CsCl; (b) NaCl; (c) ZnS; (d) CaF₂; (e) *BaTiO₃*.

each set of like ions forming an FCC structure. The CN is 6. MgO, CaO, FeO, and MnS are some of the compounds possessing this structure.

*Zinc Blende Structure.* Zinc blende is the mineralogical term for zinc sulfide. It was one of the first structures of this type to be studied; hence thereafter, all structures of this type have been called zinc blende. There are many important compounds possessing this atom arrangement. The unit cell has all atoms of one kind at the corners and faces of the cell, with the atoms of the other kind occupying four of the eight interior tetrahedral positions. Each zinc atom is bonded to four sulfur atoms, for a CN of 4. The cation-to-anion ratio is 0.4. If we compare this structure with the diamond cubic structure (Figure 4.16), we see identical structures, with the exception that in the diamond cubic structure of carbon, silicon, and germanium, the tetrahedron consisted entirely of like atoms. But just as the diamond cubic structure is characteristic of elemental semiconductors, so is the zinc blende structure common to many compound semiconductors. The semiconducting compounds are called 3–5 (or III–V) and 2–6 (II–VI) compounds, denoting the group on the periodic chart from whence the elements came. Gallium arsenide (GaAs), gallium phosphide (GaP), and indium antimonide (InSb) are typical group 3–5 semiconducting compounds, and zinc sulfide (ZnS), zinc selenide (ZnSe), and cadmium telluride (CdTe) are prominent group 2–6 compounds. In diamond the carbon acts as a four-valency ion covalently bonding to four other carbon atoms to satisfy the need for two electrons in each bond. The group 3–5 and 2–6 atoms combine to give a net four-valency, so they act in combination, very similar to carbon, silicon, and germanium.

Silicon carbide also has the diamond cubic structure and is high on the list of ceramics that are being considered for high-temperature structural applications. Both silicon and carbon are in group 4 of the periodic chart, and in the compound each silicon is bonded to four carbon atoms, and vice versa. It is just like removing one-half of the carbon atoms in the diamond cubic structure and replacing them with silicon atoms.

All of the cubic compounds above are sometimes referred to as AX compounds because the unlike atoms are present in equal numbers. We will now consider an $A_mX_p$ type, where either $m$ or $p$ is not equal to 1.

*Fluorite Structure.* Calcium fluorite ($CaF_2$) is the popular compound in the $A_mX_p$ class. One positive two ion, $Ca^{2+}$, is combined with two negative fluoride ions, resulting in charge neutrality. The $r_{Ca}/r_F$ radius ratio is 0.8, which requires a coordination number of 8 as in Table 4.2. Calcium ions occupy the center of the cube with fluoride ions at the corners. Since there are only half as many $Ca^{2+}$ ions as there are $F^-$ ions, only half of the cube center positions are occupied. $UO_2$ and $ThO_2$ have the fluorite structure.

*Perovskite Structure.* An important cubic crystal structure is the perovskite type. Some of these structures possess some interesting and practical thermomechanical applications. In addition, some of the recently developed high $T_c$ oxide superconductors have this structure. The perovskite structure represents an $A_m B_n X_p$ type in which three different types of ions are present. Barium titanate ($BaTiO_3$) is one of the more widely studied perovskite compounds. As can be seen from Figure 4.18, the $Ba^{2+}$ ions are positioned at the corners of the cube, a single $Ti^{4+}$ at the cube center, and six $O^{2-}$ ions in the center of the cube faces. $BaTiO_3$ is a very slightly distorted cubic structure at room temperature. The $Ti^{4+}$ and $O^{2-}$ atoms are shifted with respect to each other by about 0.012 nm, resulting in a permanent dipole. Mechanical distortion of the unit cell produces an electrical signal, and similarly, a signal can change the dimensions of the cell. Materials exhibiting this phenomenon are called *piezoelectrics*, literally "pressure electric," and are used as electromechanical transducers. Other commercial perovskite structures include solid solutions of lead zirconate ($PbZrO_3$) with lead titanate ($PbTiO_3$).

*Spinel Structures.* Several oxides have the general formula $AB_2O_4$, where the metal ions $A^{2+}$ and $B^{3+}$ combine with $O^{2-}$. The $MgAl_2O_4$ is usually the standard bearer for this group, which are known as *spinels*. The oxygen ions form a FCC lattice, with the A and B ions occupying tetrahedral and octahedral sites. All interstitial sites are not filled, nor do the B and A atoms always seek the same interstitial site.

### Hexagonal Close-Packed Structures: Elements

The cubic elemental structures are predominantly metals at ambient conditions. The same holds for the HCP structures. There are 24 HCP metals, which include 12 rare earth metals. The HCP hard-sphere model and atom site unit cell are shown in Figure 4.19. This structure represents the closest packing of spheres, just as did the FCC structure. The coordination number can be better visualized using the hard-sphere model. The center atom in the top layer of atoms, which is the focus of interest for the CN determination, is surrounded and touched by six other atoms. Now this top center atom sits in a hole or pocket formed by the three atoms in the second or middle layer. Thus it is in contact with three atoms in this layer. Now if we visually add a fourth layer of atoms on top of the first layer, where our center atom under discussion lies, this center atom will be in contact with three atoms in the just-added fourth layer. Thus this center atom is in contact with 12 other atoms, which are its nearest neighbors. The HCP lattice then has a coordination number of 12. These close-packed layers of atoms discussed above are the basal $\{1000\}$-type planes. They have the same closest packing of spheres that we had in the $\{111\}$ planes of the FCC lattice. Let's place our rack of pool balls on a flat surface as we did for the FCC lattice in Figure 4.14. The second layer is twisted around a normal to the basal plane 120° with respect to the first layer such that the atoms in the second layer fit into the pockets or holes of the first layer. But when the third layer is added, it is not twisted another 120° with respect to the

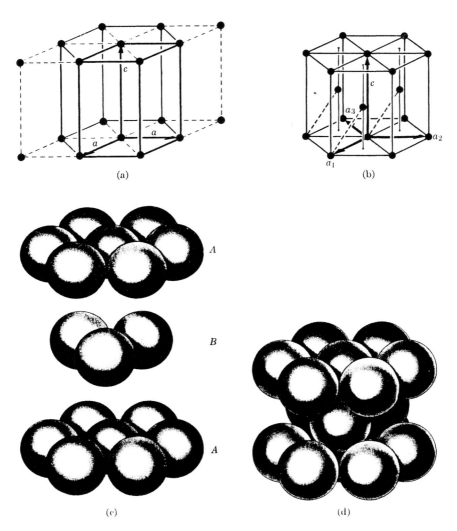

(a)

(b)

(c)

(d)

**Figure 4.19.** HCP hard-sphere and atom-site unit cell models. (From A. G. Guy, *Elements of Physical Metallurgy*, Addison-Wesley, Reading, Mass., 1959, p. 90.)

first layer as in the FCC structure. Rather, the third layer is now rotated 120° in the opposite direction to what the second layer was rotated. Now the third layer atoms rest in the pockets of the second layer, but they are also aligned directly over the first layer of atoms. This arrangement is referred to as the ABABAB stacking sequence, in contrast to the ABCABC sequence of the FCC structure. Both structures have the closest packing of spheres arrangement, but

**Table 4.4** *c/a* Ratios for Some HCP Metals

| Metal | *c/a* Ratio |
|-------|-------------|
| Cadmium | 1.886 |
| Zinc | 1.856 |
| Magnesium | 1.624 |
| Titanium | 1.587 |
| Zirconium | 1.593 |
| Beryllium | 1.567 |

whereas the parallel close-packed planes were the {111} planes in the FCC lattice, they are the {1000} planes of the HCP lattice, and as pointed out above they have different stacking sequences. Both lattices, since they are arranged as the closest packing of spheres, have the same atomic packing factor.

The ratio of the lattice parameters (i.e., *c/a*) for the ideal closest packing of spheres in the HCP cell is 1.633. Not all HCP metals have this ratio (Table 4.4). In Mg, Co, Ti, and Be the *c/a* ratio is less than ideal, meaning that these structures are slightly compressed, while the others, whose ratios are larger than 1.633, are slightly extended in the *c* direction. Thus the HCP metals are not really arranged in the closest packing of spheres form but are sufficiently close that we generally refer to them as close-packed structures.

*Example 4.6* Compute the APF for the ideal HCP structure.

Solution. The number of atoms per unit cell must first be determined. The best way to visualize this is to break up the unit cell into three body-centered tetragonal (BCT) cells. In Figure 4.8 the letters FGDE on top of the cell, and the letters HMLJ on the bottom, represent one such BCT cell. The BCC cell has two atoms per unit cell, and by the same argument the BCT also has the same number per BCT cell. But there are three BCT cells in the HCP unit cell, so the total number of atoms per HCP cell is six.

The volume of the unit cell is the product of the base area times the height. The base area is depicted below.

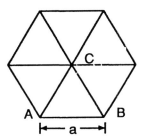

Area of base $= 6 \times$ area of equilateral triangle $ABC$

$$= 6 \left[ \frac{a^2}{2} \sin 60° \right] = 3a^2 \sin 60°$$

Cell volume $= (3a^2 \sin 60°)c = (3a^2 \sin 60°)1.633a$

From the equilateral triangle, $a = 2R$,

$$\text{APF} = \frac{\text{atom vol.}}{\text{cell vol.}} = \frac{6(4/3\pi R^3)}{[3(2R^2) \sin 60°](1.633 \times 2R)} = 0.74$$

Thus the APF for FCC = APF for HCP.

Hexagonal Close-Packed Structures: Compounds

Just as for cubic structures we must examine the interstitial sites in the HCP lattice. The tetrahedral and octahedral sites are shown in Figure 4.17 for the bottom half of the HCP cell. There will be identical sites in the top half. As one may envision the HCP, compound lattices can be very complex. Let's examine one of the more common such lattices.

*Corundum Structure.* Corundum is the name for the most common form of $Al_2O_3$. It is often expressed as the $A_2X_3$ structure, which also include the compound $Cr_2O_3$. $Al_2O_3$ is the most widely used compound in ceramic materials of the nonclay variety. It is used for spark plugs, other electrical insulators, and as an abrasion-resistant material. Remember: It was number 8 on the Mohs scale of hardness, right beneath diamond. In these structures (i.e., $A_2X_3$), the $O^{2-}$ ions form a hexagonal pattern while the metal$^{3+}$ ions are in the octahedral sites (Figure 4.20). Only two-thirds of these sites are filled.

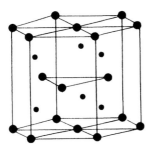

● $O^{2-}$ atoms
• Octahedral sites
$\frac{2}{3}$ occupied by $Al^{3+}$

**Figure 4.20.** $Al_2O_3$ (corundum) structure.

In all of this section on crystalline structures we have attempted to highlight some compounds as being typical of certain structures. There is one class of materials, the silicates, that we have omitted. The silicate structures are based on an $SiO_4^{4-}$ tetrahedral arrangement of silicon and oxygen. These tetrahedra can bond to metals and to themselves and take on many forms, such as chains, sheets, three-dimensional networks, and so on. These materials can be explained better in terms of these structural forms rather than unit cells. Accordingly, we will delay the discussion of the silicate structures, including glasses, until the ceramics processing section.

## Intermetallic Compounds

Intermetallic compounds such as $Ni_3Al$ and $Ti_3Al$ are being studied extensively for structural engineering applications, especially at high temperatures. They have good oxidation resistance and strength at temperatures exceeding 1100°C (2000°F). But they have the same objectionable feature characterized by ceramics (i.e., they are brittle). They are very crystalline. They are characterized by a high degree of order, just like that in ceramic compounds. $Ni_3Al$ is an ordered FCC structure, while $Ti_3Al$ has an ordered hexagonal atom arrangement. Ordering, in the sense used here, means that the titanium atoms, in the case of $Ti_3Al$, and nickel atoms in $Ni_3Al$, occupy specific positions in the unit cell, while the aluminum atoms occupy the remaining specific positions. The bonding of atoms in these compounds is a mixture of metallic, ionic, and covalent, but they are relatively good conductors of heat and electricity, and in this sense are more like metals than are the other compounds, such as the oxides, nitrides, borides, and carbides.

## Crystalline Polymers

In an earlier section (Figure 4.2), it was pointed out that crystallinity in polymers took the form of some type of alignment of the long-chain molecules. Only in the long-chain molecules is crystallinity found, and then never in 100% of the volume. The three-dimensional network polymers are never crystalline, even though on paper the structures appear to be more crystalline than the long-chain molecules. There is sufficient crystallinity in the long-chain crystalline polymers to permit the diffraction of x-rays, but not in the three-dimensional network molecules. The latter do not have a repeat three-dimensional arrangement.

The crystals in crystalline polymers are referred to as molecular crystals. Perhaps this can be better understood if we examine first a nonpolymer molecular crystal. The structure of solid iodine, a molecular crystal, is shown in Figure 4.21. A molecule of iodine $(I_2)$, which consists of two iodine atoms bonded together by van der Waals forces, is located at each lattice point. Remember: In the Bravais lattice, a point does not have to represent an atom. It

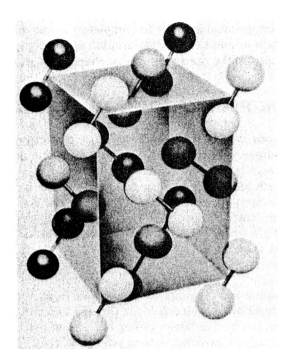

**Figure 4.21.**   Molecular crystal structure of iodine. (From W. G. Moffatt, G. W. Pearsall, and J. Wulff, *The Structure and Properties of Materials,* Wiley, New York, 1964, p. 171.)

can represent some type of atom cluster, which is the case here. The structure is orthorhombic and the molecules are bonded together by a weak secondary bond.

Now let's examine a polymer crystal. The polyethylene crystal, depicted in Figure 4.22, does have a unit cell when the long-chain molecules align. The cell is orthorhombic with 90° angles between the axes. But the unit cell points consist of a cluster of atoms (i.e., one carbon and two hydrogens), not the molecule $C_2H_4$ (i.e., two carbons and four hydrogens). But the lattice spacing can be determined and density calculations made just as in the more perfect crystalline structures found in metals and ceramics.

## 4.4.2   X-ray Diffraction

We have frequently mentioned that the diffraction of x-rays is a test for crystallinity. In the following a brief description of the diffraction process is given

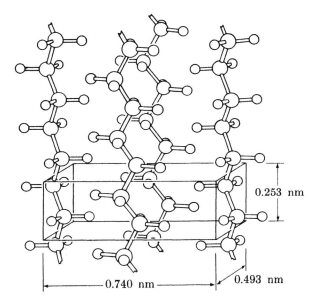

0.253 nm

0.740 nm

0.493 nm

**Figure 4.22.** Molecular crystal structure of polyethylene. (From M. Gordon, *High Polymers,* Lliffe, and Addison-Wesley, Reading, Mass., 1964; after C. W. Bunn, *Chemical Crystallography,* University Press, Oxford, 1945.)

so that the reader will feel more at ease when such a statement is made. There are many good texts on x-ray diffraction (see the Suggested Reading) and it is usually a separate and complete course in materials engineering curriculum. Courses at the graduate level are really required to master the applications of x-ray diffraction in the determination of crystal structure and in the evaluation of the degree of crystallinity in a material.

Bragg's Law

X-rays are a form of electromagentic radiation that is emitted when a beam of electrons strike a material with sufficient energy to cause an electron from an inner shell (e.g., the $K$ shell) to be excited to a higher state of energy. Then when an electron falls back into the empty space in the $K$ shell, x-rays are emitted of a wavelength determined by the difference in energy of these two states (i.e., $E_2 - E_1 = h\nu$). The wave length of the radiation is on the order of the atom spacing in crystalline solids.

When a beam of x-rays impinges on a solid it interacts with the atoms such that the atoms reradiate x-rays of the same wavelength as that of the impinging beam. From Figure 4-23a it can be seen that all of the waves of the incoming

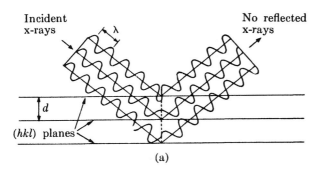

(a)

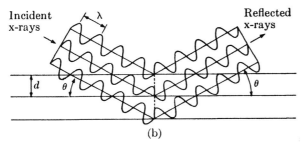

(b)

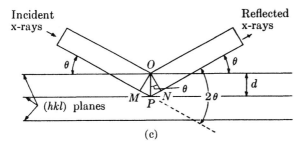

(c)

**Figure 4.23.** Bragg's law for x-ray diffraction. (From A. G. Guy, *Elements of Physical Metallurgy*, Addison-Wesley, Reading, Mass., 1959, p. 90.)

beam are in phase, but not for the reflected beam, because the rays making up the beam travel different path lengths. But when the difference in path lengths of the incident rays (Figure 4.23b) that impinge on parallel planes are an integral number of wavelengths, the reflected rays are in phase, and by being such, they reinforce each other. This difference in path length is the distance *MP* plus *PN* of Figure 4.23c. Thus for this reinforcement to occur we must have the condition where the distances $MP + PN = n\lambda$, where $\lambda$ is the wavelength and $n = 1, 2, 3....$ Now *MP* and *PN* are equal to $d \sin \theta$, where $d$ is the interplanar spacing and $\theta$ is the angle of incidence and reflection, both being the same. Therefore, the condition for a reflected beam to be reflected is

$$n\lambda = 2d \sin \theta \qquad (4.1)$$

We have used the term *reflection* here since this phenomenon is analogous to reflected light. But Bragg's law is the condition for reinforcement, which is called x-ray *diffraction*.

Most x-ray diffraction equipment consists of an x-ray source (the x-ray tube) and a detector, often a Geiger counter. The x-ray source is fixed while the detector moves on a circular path around the sample (Figure 4.24). If the sample is a single crystal, the crystal must be placed at a specific angle $\theta$ for each specific set of parallel crystallographic planes in order for Bragg's law to be satisfied and for reinforcement of the reflected x-rays to occur. If the sample is composed of powder or filings, as in the case of a metal, many particles will be oriented such that Bragg's law is satisfied for a number of positions on the circle. In the latter case a trace showing the diffraction peaks will be obtained (Figure 4.25). The intensity of the diffracted beam is measured as a function of $\theta$. Each peak represents diffraction from a specific set of parallel planes with specific $(h, k, l)$ indices and whose interplanar separation are such that at a certain angle of x-ray impingement $\theta$ reinforcement will occur. From the measurement of the angle $\theta$, the interplanar and interatomic distances can be computed. By comparison with peaks of known materials the trace of an unknown material serves to identify this material. Such diffraction data have been catalogued and now are also available on computer data bases, which makes the

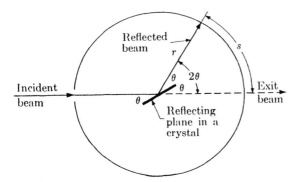

**Figure 4.24.** Relation between the x-ray source (the incident beam), the sample location, the diffraction angle $\theta$, and the detector position $s$.

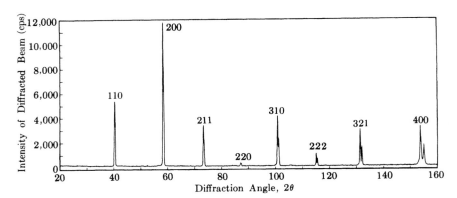

**Figure 4.25.** X-ray diffraction trace resulting from the diffraction of x-rays from randomly oriented crystals in a powder sample of tungsten. To satisfy Bragg's law, the crystals must be oriented such that for given values of the interplanar spacing *d* and the angle of x-ray incidence, reinforcement will incur. In powder samples many particles will be so oriented.

material identification a much simpler task. A material that does not yield constructive reinforcement (i.e., peaks are not observed) is an amorphous material.

### 4.4.3 Amorphous Structures

In Chapter 1, polymers were separated into three groups: the long-chain thermoplastic molecules, the three-dimensional network thermoset molecules, and elastomers. All of these structures consist of very large molecules. Their molecular weights are in thousands of amu. Elastomers are long-chain molecules that are cross-linked by other atoms or molecules, sometimes to the extent of forming structures that approach in appearance that of the three-dimensional network thermosets. We hold the discussion of elastomers for a later section. For now let's briefly examine the structure of the thermoplastic and thermosetting polymers.

Long-Chain Molecules

Most long-chain molecules are made by a process called a *chain polymerization* reaction, which is also known as the *addition* process. The classic example of this type of reaction is the polymerization of ethylene gas of formula $C_2H_4$ to form a solid polyethylene material that is frequently used for household functions such as garbage bags, sandwich bags, and a host of other similar applications. The word *polymer* means "many mers." The process begins with a monomer of ethylene gas in which the carbon atoms are joined by double covalent bonds, and which can be viewed by the electron dot structure, or more often with a solid line for the double bond, as shown below.

**H   H**
· · · ·
**· C :: C ·**
· ·   · ·
**H   H**

**Monomers**

**H H**
| |
**C = C**
| |
**H H**

**H   H**
| |
**—C—C—**
| |
**H   H**

**Mer**

Each bond has two electrons, which satisfies the need for the *s* and *p* levels to be filled. Through the use of heat, pressure, and a catalyst, the double bonds, which are said to be *unsaturated,* are broken to form single bonds. The resulting structure, called a *mer,* is now free to react with other mers and thereby form the long-chain molecule depicted in Figure 4.26. The catalyst, or initiator, used for polyethylene polymerization is a peroxide, usually $H_2O_2$, although organic peroxides have also been used. The covalent bonds between the oxygen atoms in the peroxide are broken as well as the carbon double bond in the monomer, leaving an OH group which is attached to one end of the chain. The other end of the chain is where propagation takes place by the addition of mers until a long chain containing on the order of perhaps 10,000 mers are formed. Termination of the reaction occurs by two chains combining or by OH groups becoming attached to both ends of the chain. To some extent the length of the chain can be controlled by the quantity of the initiator. Nevertheless, there will

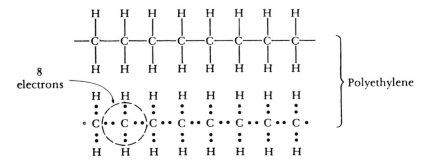

**Figure 4.26.** Polymerization by the addition mechanism of ethylene gas to yield solid polyethylene.

be a wide distribution of molecular weights in the final polyethylene mass since termination does not occur at a specific length. The number of mers per molecule is called the *degree of polymerization* (DP) and is computed by dividing the average molecular weight by the mer weight:

$$DP = \frac{\text{average molecular weight}}{\text{mer weight}} \tag{4.2}$$

The reaction is frequently written as

$$
\begin{array}{ccc}
\text{H} \ \ \text{H} & & \left[ \begin{array}{cc} \text{H} \ \ \text{H} \\ | \ \ \ | \\ \text{C} - \text{C} \\ | \ \ \ | \\ \text{H} \ \ \text{H} \end{array} \right]_n
\end{array}
$$

$$
\begin{array}{c}
\text{H} \quad \text{H} \\
| \quad \ \ | \\
\text{C} = \text{C} \ \longrightarrow \\
| \quad \ \ | \\
\text{H} \quad \text{H}
\end{array}
$$

where *n* represents the number of mers. The average molecular weight is obtained from the mass of the starting materials.

*Example 4.7.* What is the degree of polymerization in a reaction that begins with 10,000 g of ethylene gas and 1 g of $H_2O_2$ assuming that all $H_2O_2$ is consumed? (The amount consumed can be determined by chemical analysis.)

Solution. First the molecular weights of the starting materials are computed.

Mol. Wt. $C_2H_4 = 2(12) + 4(2) = 28$ g/gmol

Mol. Wt. $H_2O_2 = 2(16) + 2(1) = 34$ g/gmol

The number of molecules and mers can now be computed.

$$\frac{10,000 \text{ g } C_2H_4}{28 \text{ g/gmol}} \times 6.02 \times 10^{23} \frac{\text{molecules}}{\text{gmol}} = 2.15 \times 10^{26} \text{ mers } C_2H_4$$

Assuming that 1 molecule of $H_2O_2$ is consumed per molecule of polyethylene, we have

$$DP = \frac{2.15 \times 10^{26} \text{ mers } C_2H_4}{1.77 \times 10^{22} \text{ molecules } H_2O_2} = 12,147 \frac{\text{mers}}{\text{chain}}$$

In Figure 4.2 the structure of a crystalline polymer was shown to consist of the long chains aligning with each other in some type of ordered arrangement. In the amorphous structure the long-chain molecules are twisted and kinked, forming entanglements that could be described as being somewhat like a bowl of spaghetti if it were magnified a few thousand times. (At room temperature, polymers with very short chains with molecular weights around 100 g will exist as liquids or gases.) The schematic representation of a single polymer chain molecule that has numerous random kinks and coils is shown in Figure 4.27.

Glass Transition Temperature

The structure of polymers is very much a function of temperature. Whereas metals and ceramics experience a sharp change in the degree of order, and correspondingly in density, at their melting–solidification points, the same is not necessarily true for polymers and glasses. In the liquid state the long-chain molecules are in a sort of continuous state of rearranging themselves. These twisted and kinked entangle structures make it difficult to form unit cells and chain alignment. It is not surprising that such liquids can be solidified without forming any repeatable crystalline pattern.

We will now follow the solidification of a liquid polymer by referring to the change in volume (density could also be used) as a function of temperature, as illustrated in Figure 4.28. As the temperature is lowered from the liquid state, thermal agitation lessens and there is a decrease in volume. For those polymers that retain their amorphous structure into the solid state there is continual increase in density and decrease in volume in the entanglements rather than an abrupt change as when crystals are formed. In the liquid state there is a certain amount of *free space* among the molecules that gradually decreases in amount as thermal agitation lessens. As the molecules become more tightly packed, they form a rigid solid at the solidification temperature, but there still exists some free space and room for tighter packing. Thus as the temperature is further reduced, the packing continues to become more dense as a gradual

**Figure 4.27.** Schematic representation of a single polymer chain molecule showing numerous random kinks and coils.

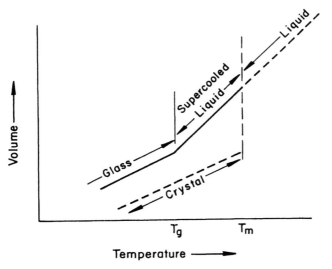

**Figure 4.28.** Glass transition and melting temperatures for amorphous and crystalline polymers.

reduction in volume occurs. In this state the material is said to be a *supercooled* liquid even though it has a rigid solid form. The structure in this region is still amorphous (i.e., glasslike). The graph is a straight line down to the point marked $T_g$, the *glass transition temperature*. At $T_g$ the molecules become so tightly packed that further reduction in temperature causes only a slight change in volume, due to a reduced amplitude of atom vibration, similar to the thermal contraction experienced in metals. Below $T_g$ the polymer is still amorphous in structure, but it now is very brittle and nonformable. The solid-state forming region lies between $T_g$ and $T_m$. $T_g$ is thus just as important, or perhaps more so, than $T_m$. Now some polymers form a high degree of crystallinity, the degree being dependent on the chain structure and the cooling rate. The volume–temperature relationship is also included in Figure 4.28 for crystalline polymers. This plot is like that found in metals and ceramics.

Network Polymers

A stepwise polymerization process, sometimes called the *condensation* method, can produce either long-chain molecules or the three-dimensional network type. Network polymers (and elastomers) are noncrystalline, essentially due to the random three-dimensional covalent bonding that prevents any rearrangements necessary for long-range ordering. In the condensation method of forming network polymers, two monomers are reacted to form a third molecule plus a by-product, usually water. The reacting monomers have more than one reactive functional group. When the functionality is greater than 2, network polymers can form. They can connect to three or more adjacent molecules. An example of the network polymers is found in the results of the reaction of gaseous formaldehyde ($CH_2O$) and the low-melting solid phenol ($C_6H_5OH$) to form phenol–formaldehyde (Bakelite). The reaction is shown in Figure 4.29a and the resulting network structure in Figure 4.29b. The network polymers do not soften when heated, as do the thermoplastic polymers. To mold phenol-formaldehyde, it is necessary to start with a partially polymerized material and allow the polymerization reaction to become complete while it is formed under pressure and with heat—hence the name *thermosets.* The polymers cannot be reheated and re-formed. They are hard and brittle and decompose on heating.

Glasses

It is of interest, and perhaps somewhat amusing, to review the various definitions of glass. As a student, you are probably of the age that in recent years you have been assigned the task of washing dishes after a holiday or special occasion dinner function and you may have heard the expression—"be careful with those glasses, that's my best crystal." It may not relieve you of your assignment, but at least you might impress the assignor if you returned with a comment of—"but those glasses are not crystalline." Webster defines glass as *an amorphous transparent or translucent substance consisting of a mixture of silicates, or sometimes borates or phosphates, formed by fusion of silicates, or oxides of boron or phosphorus, into a mass that cools to a rigid condition without crystallization.* This definition would be somewhat simpler if we omitted the terms *borates* and *phosphates,* which are not necessary to form glasses and are not even present in most glasses. In previous sections we have used such terms as *glasslike* and *metallic glasses,* until we are conditioned to accepting the definition of glass as being anything noncrystalline. This is perhaps going to the other extreme, but in technological circles the term *glass,* which for many centuries really applied to silica glasses, has been used, erroneously perhaps, to describe noncrystalline polymers and metals.

The main ingredient in glasses is the $SiO_4$ tetrahedron (Figure 4.30), which is the basic building block of all silicates, a subject we discuss at greater length in Chapter 7. But for now, let's discuss only the arrangement of these $SiO_4$ tetrahedra in crystalline and in amorphous structures. Silica ($SiO_2$) is polymorphic [i.e., the crystal structure changes with temperature (pressure has some effect)]. In fact, silica has been found in at least seven crystal structures. Quartz, the predominant constituent of most beach sands, consists of the arrangement of these $SiO_4$ tetrahedra in a complex pattern that achieves overall charge neutrality in the stoichiometric compound $SiO_2$. Each $SiO_4$ tetrahedron has a charge of $-4$. Silicon has a $+4$ charge, and each of the four oxygen

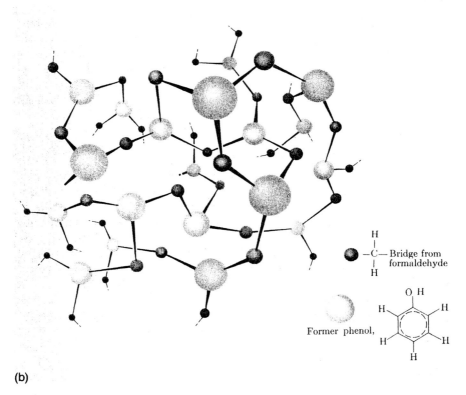

(a)

(b)

**Figure 4.29.** (a) Reaction of formaldehyde ($CH_2O$) with phenol ($C_6H_5OH$) to form phenolformaldehyde. (b) Resulting network polymer (Bakelite). (From L. H. Van Vlack, *Elements of Materials Science Engineering,* 3rd ed., Addison-Wesley, Reading, Mass., 1975, p. 231.)

**Figure 4.30.** SiO₄ tetrahedron. (From L. H. Van Vlack, *Materials for Engineering,* Addison-Wesley, Reading, Mass., 1982, p. 295.)

atoms has a $-2$ charge, resulting in a net $-4$ charge for the tetrahedron (which is usually written as $SiO_4^{4-}$). These tetrahedra are joined through their four oxygen atoms to form quartz via a double three-dimensional helical structure (Figure 4.31). $SiO_2$ in sand form can be melted (1710°C), and if cooled extremely slowly, a large quartz single crystal can be grown. These are used commercially, in small cut slices, for frequency control in radio circuits. But if cooled normally (e.g., turn off the power to the molten pot and let ambient air convection currents cool the molten mass), the chances are that a crystalline structure will not be formed. In ordinary plate glass, crystallization is further

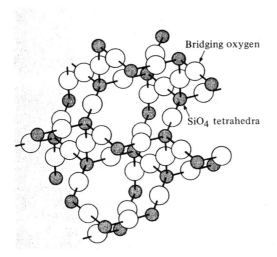

**Figure 4.31.** Quartz structure. (From L. H. Van Vlack, *Materials for Engineering,* Addison-Wesley, Reading, Mass., 1982, p. 296.)

discouraged by the addition of sodium or potassium atoms, usually in the form of their oxides, which prevent the tetrahedra from connecting at all points, such as they do in forming a quartz crystal. The structure of these *network-modified* glasses are compared to that of the quartz in Figure 4.32.

The volume–temperature relationship for glasses is schematically the same as that shown for polymers in Figure 4.28. Since glass has been manufactured a lot longer than polymers, the glass temperature was used in glass technology long before that in polymers. The polymers chemists found that it was convenient and also revealing to describe amorphous polymers in the same manner in which, for many years, glasses had been described.

Liquid Crystals

The term *liquid crystals* refers to a phase that lies between the rigidly ordered solid phase, where the mobility of individual molecules is restricted, and the isotropic phase where molecular mobility and a general lack of order exists. This phase has sometimes been called the fourth state of matter. About 5% of all organic molecules exhibit liquid-crystal behavior. These molecules fall into

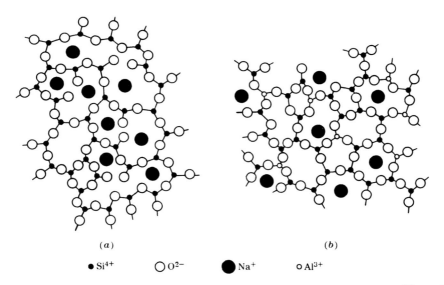

(*a*)                                                                 (*b*)

● $Si^{4+}$        ◯ $O^{2-}$        ⚫ $Na^+$        ○ $Al^{3+}$

**Figure 4.32.** Network-modified glass: (a) soda-line glass uses Na modifiers; (b) alumina–silica glass uses Al modifiers. In (a) and (b) the Na atoms are not part of the network, but in (b) the Al ions are part of the network. (From O. H. Wyatt and D. Dew-Hughes, *Metals, Ceramics and Polymers,* Cambridge University Press, Cambridge, 1974, p. 263.)

one of the three classes: nematic, chloristic, or semeitic. For a complete explanation of these classes, see the Suggested Reading. The nematic liquid group are the ones used for electronic display devices such as in digital watches and hand-held calculator displays. A thin layer (about 20 $\mu m$ in thickness) of the liquid-crystal substance is held as a dielectric between two plates that contain a thin conductive layer of tin or indium–tin oxide. The liquid crystal has a strong negative dielectric anisotropy and can form regular patterns or domains at low voltages. A type of ordering occurs that allows numbers and patterns to be displayed.

## 4.4.4 Composites

Engineering composite materials are the result of combining two or more materials from the metals, ceramics, or polymer groups in the micro to macro size range, most often favoring the latter. We are speaking here of situations where the individual constituents can be viewed by eye or at least by low-power microscopes. Their structures can take on many forms, which we discuss at more length in Chapter 9. Fiberglass, used to reinforce polymer matrices, is a classic example. Concrete is really a composite composed of rocks, sand, gravel, and mortar. When we add reinforcing steel rods it really becomes a multicomponent composite. Cemented carbides, used for cutting tools and wear-resistant applications, are composed of hard carbide particles *cemented* together by a metal, often cobalt. Honeycomb structures, used mostly in aircraft applications, consist of aluminum cells held in place by aluminum sheets. These composites provide a stiff, rigid, strong, and lightweight material. It should be apparent by now that there is no easy way to portray composite materials. They are products of engineering and imagination that will continue to amaze us for years to come.

## DEFINITIONS

*Addition method*:   polymerization process that involves adding mers together to form long-chain polymers.
*Amorphous*:   state of matter that is without long-range order.
*Anisotropic*:   state of a material in which the properties are dependent on the direction in the body of the material in which they are measured.
*Atomic packing factor*:   density of atom packing in crystalline structures.
*Basal plane*:   plane of closest packing of atoms in the hexagonal close-packed structure. This plane is of the {1000} type.
*Bravais lattices*:   arrangement of points in space that is used to describe crystal structures. There can be only 14 such arrangements.

*Condensation method*:   process for forming large polymer molecules whereby two monomers are reacted to form a third molecule plus a by-product, usually water or alcohol.

*Crystalline*:   state of matter in which there is a high degree of long-range order with respect to atomic or molecular positions.

*Disorder*:   lack of a repeating arrangement of atoms or molecules.

*Glass (transition) temperature*:   temperature in the cooling of glasses and polymers at which the rate of volume contraction is markedly reduced, further reduction being by thermal contraction only.

*Intermetallic compound*:   compounds formed between two metals in which the bonding is of mixed type but often more metallic than ionic or covalent, and of which there is a fixed stoichiometric composition.

*Isotropic*:   state of matter whereby the properties of a body are the same in all directions of measurement through that body.

*Lattice*:   construction of points that extends throughout a single-phase region of a crystalline body and is used to describe atom positions.

*Lattice parameter*:   distance a unit cell must be moved in a lattice to find itself in exactly the same atom arrangement in which it existed prior to the move.

*Mer*:   smallest repeating unit of a polymer that has the same chemical formula as the monomer used to make the chain.

*Metallic glass*:   amorphous metal formed by very rapid solidification.

*Monomer*:   molecules used to form polymers.

*Network polymers*:   three-dimensional array of atoms in an amorphous thermoset polymer.

*Nuclei*:   plural of *nucleus*.

*Nucleus*:   in the sense used in this chapter, it is a small group of atoms that have become arranged in the way that all the atoms will be when the phase transition is complete.

*Phase*:   region of a substance that is structurally homogeneous and in which the properties are uniform throughout. Phases will be separated by phase boundaries.

*Phase transition*:   rearrangement of atoms or molecules when a new phase is formed from the old.

*Polymerization*:   process of forming very large molecules from single molecules called monomers.

*Polymorphic*:   characteristic that allows a material to have an entirely different arrangement of atoms by altering the temperature and/or pressure.

*Supercooling*:   type of cooling that occurs when a material is cooled through a temperature whereby the atoms should rearrange themselves to form a new phase but such rearrangement does not occur. Such conditions and phases are metastable or unstable.

*Undercooling*: slight cooling below a phase transition temperature that is necessary to cause the new phase to nucleate and grow.

*Unit cell*: smallest repeat unit of a crystalline lattice structure.

*X-rays*: electromagnetic radiation that is emitted when an electron falls from one orbit to an orbit closer to the nucleus.

*X-ray diffraction*: constructive interference that occurs when x-rays are reflected from parallel planes of atoms in a crystal.

## QUESTIONS AND PROBLEMS

4.1. Explain the difference in the phrases *states of matter* and *phases of matter*.

4.2. Give an example of a multiphase alloy and the names of the two phases.

4.3. Oil and water are two immiscible liquids. In nonhomogenized milk, cream rises to the top. Are these immiscible liquids?

4.4. Why can molten silica be cooled at rather slow rates and retain an amorphous structure, yet metals must be solidified at a rate of the order of $10^6$ degrees/s to obtain a noncrystalline structure? Do amorphous polymers normally result if they are cooled at rates similar to the glasses? Explain.

4.5. Explain the term *devitrification*.

4.6. Why do the atoms (or molecules) of any molten substance want to form an ordered arrangement on solidification?

4.7. What causes shrinkage porosity in a cast metal ingot?

4.8. Aluminum has a FCC structure. The distance of closest approach of the atoms (the distance between nuclei in the direction of closest packing) is 2.862 Å (28.62 nm). What is the lattice parameter of aluminum? What is the interplanar spacing for the (100) planes?

4.9. Make the sketches showing the planes represented by the following Miller indices.
    (a) (010), (101), (011)
    (b) (111), (1$\bar{1}$1), (11$\bar{1}$), (1$\bar{1}\bar{1}$)
    (c) (221), (123), (124)
    (d) (10$\bar{1}$2), (10$\bar{1}$0) hexagonal indices

4.10. What are the family indices for the planes forming the sides of the unit cell in the HCP structure?

4.11. Would you expect the properties of a small-grain-size polycrystalline metal to be isotropic or anisotropic? Explain.

4.12. What are the Miller indices for the directions indicated by vectors *A*, *B*, and *C* in the following sketch?

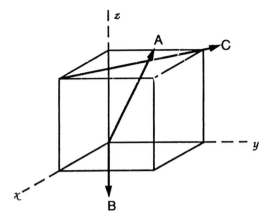

4.13. Show that the ideal closest packing of spheres in the HCP structure occurs for a *c/a* ratio of 1.633.

4.14. Selenium and tellurium have been called *metalloids*. What other elements fit this description?

4.15. What is the diameter of the largest atom that will just fit into the largest interstitial site for BCC iron? The lattice parameter at room temperature of BCC iron is 2.8664 Å.

4.16. Calculate the planar density of atoms for the $\{111\}$ and the $\{100\}$ planes of BCC iron.

4.17. BCC iron changes to FCC iron when it is heated to 912°C. What is the term that describes materials that can change their crystal forms? (Many materials possess this characteristic. The metals that do so can be found in the tables of the *Metals Handbook,* Desk Ed., ASM International, Metals Park, Ohio, 1985, pp. 1–44 through 1–48. Also note the wealth of information on crystallography, physical properties, and so on, contained in these tables.)

4.18. Copper, an FCC structure, has a lattice parameter of 3.615 Å and an atomic weight of 63.54 g/mol. Compute its density in $g/cm^3$.

4.19. Describe the *structure* (not crystal structure) of two common composite materials other than those listed in this chapter.

4.20. Four spheres of diameter *d* are arranged in a triangle, touching each other. Show that the diameter of the largest sphere that will fit into the voids between the spheres is $0.155d$.

4.21. What are the Miller indices for the following planes?

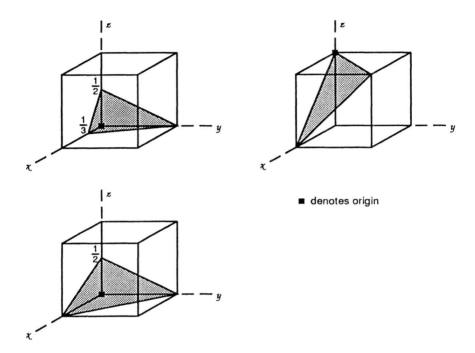

■ denotes origin

4.22.    Explain the use of parentheses, braces, and brackets as used in Miller indice notations.

## SUGGESTED READING

Callister, W. D., Jr., *Materials Science and Engineering,* Wiley, New York, 1990.
Cullity, B. D., *Elements of X-ray Diffraction,* 2nd ed. Addison-Wesley, Reading, Mass., 1978.
Guy, A. G., *Elements of Physical Metallurgy,* Addison-Wesley, Reading, Mass., 1959.
Ralls, K. M., Courtney, T. H., and Wulff, J., *Introduction to Materials Science and Engineering,* Wiley, New York, 1976.
Smith, W. F., *Principles of Materials Science and Engineering,* 2nd ed., McGraw-Hill, New York, 1990.

# 5

# Imperfections, Diffusion, and Plastic Deformation in Crystalline Materials

## 5.1 INTRODUCTION

At first glance the title of this chapter appears to encompass three separate subjects, and in many books they are treated in separate chapters. But we will find that the mechanisms of diffusion and plastic deformation in crystalline bodies are dependent on the existence of crystalline imperfections on the atomic size level, and accordingly, I have elected to tie them together in one chapter, hopefully to present a more understandable picture to the reader.

Perhaps this chapter's title should have been "Material Imperfections," since the scope of this book includes both crystalline and amorphous materials. However, imperfections at the atomic level in amorphous polymers and amorphous ceramics (glasses) do not play a major role in their properties. Imperfections on the order of a micrometer ($10^{-6}$ m) and larger in glasses and crystalline ceramics do affect their mechanical strength and fracture toughness, but this subject is treated in Chapter 7.

Defects in crystalline materials at the atomic level have a profound effect on their mechanical properties and also on some physical properties (e.g., their electrical conductivity). The presence of linear configurations of dislocated atoms (dislocations) in metals account for their plasticity and formability. Vacancies, unoccupied lattice sites, make it easier for atoms to migrate through crystals, a process called *diffusion,* which is essential to phase transformations and recrystallization of metals. Foreign atoms can have both beneficial and harmful effects on crystalline bodies. Crystalline imperfections play a major role in the electrical behavior of semiconductors and certain electronic ceramics.

It is convenient to classify these small crystalline imperfections in terms of point, linear, and planar defects. It should be realized that the term *defect* does not necessarily mean a deleterious item. Many imperfections and defects are introduced intentionally for their beneficial effects.

## 5.2  POINT DEFECTS

Point defects consist of either a vacant lattice site or a foreign atom. The foreign atom can occupy a site normally occupied by an atom of the main body of material of interest, or it can be shoved in between or among the *host* atoms. The terms *host, parent,* and *solvent* have essentially the same meaning and are interchangeable. They are the major constituent. The term *solvent* as used in the solid state may be a little confusing, so let's elaborate on that concept a bit at this point. The term "solid solvent" arises simply because in solids we can have *solid solutions,* somewhat comparable in concept to single-phase miscible-liquid solutions, or single-phase liquid solutions that contain dissolved solid solutes, such as sugar dissolved in coffee. Liquid solutions, being a single phase, are uniform in composition throughout this phase. But we can dissolve only so much solid sugar in liquid coffee. That amount of sugar that can just be dissolved and leave no solid residue at the bottom of the cup is called the *solubility limit* of sugar for that temperature of the solvent coffee. If we exceed this quantity of sugar, we will now have a two-phase system consisting of a liquid solution containing dissolved sugar (no longer a solid), and a second phase, which is the undissolved solid sugar. Similarly, we can have a solid solution of a solvent containing foreign atoms called the *solute.* The foreign atoms can occupy a host atom site or reside in an interstitial site between or among the host atoms. But there is no liquid in this solution. When the added foreign atom content exceeds the solubility limit, a solid precipitate forms, usually a compound. So we now have a system consisting of a solid solution plus a solid precipitate, a two-phase system, very similar in concept to the liquid solution containing a solid precipitate. The solid solution, being a single phase, is uniform in composition and crystal structure throughout this phase.

### 5.2.1  Vacancies

A vacant lattice site is termed, as you might expect, a *vacancy.* Why do they exist? Would not a perfect crystal, with no vacant lattice sites, be at a lower-energy state and thus more stable than one with unoccupied sites? Why don't the vacancies, which could be considered as accidents of growth of the solid phase from the liquid during solidification, just ooze out (diffuse) to the surface and escape the crystal, which really should not want this defect anyway? The answer is related to the *entropy* of the system, and if this word is not yet a part of your technical vocabulary, do not be too concerned. Entropy is a measure of randomness. Gases have a high entropy because there is no order. Perfect crystals have a low entropy, because everything is organized. I suppose that organized persons have less entropy than some of the rest of us, but that would be carrying the concept to an extreme. Entropy, I believe, applies only to particles, not to personal habits.

The lowest-energy state that we have been discussing up to now dealt with potential energy [i.e., that related to position (although in electron energy levels, the kinetic energy of the electron was a factor)]. In chemical thermodynamics, the lowest-energy state is the one with the lowest free energy. The free energy, *F*, can be stated mathematically as

$$F = E - TS \tag{5.1}$$

where $E$ = internal potential energy
$T$ = temperature in kelvin
$S$ = entropy

Thus the lowest free-energy state is one that contains some entropy or disorder. This is merely a long-winded discourse to try to convince you that crystals really like to have some vacancies, and the higher the temperature, the larger the number of vacancies that will exist in equilibrium with the atoms that occupy the normal sites.

## 5.2.2 Foreign Atoms

### Substitutional Solute Atoms

A substitutional solute atom is just that—a replacement or substitute of a host atom by a foreign atom. It is just as if we pulled an atom out of a pure crystal and inserted another atom in its place. An example of the replacement of an aluminum atom by a copper atom in the FCC lattice is illustrated in Figure 5.1. Of course, it does not happen in such a simple mechanical fashion. Aluminum is melted and a small quantity of solid copper added to either the solid charge or to the molten aluminum, into which it dissolves to form a liquid solution. But when the liquid is solidified, the aluminum solvent forms its own FCC lattice, in which in every so many unit cells of aluminum, depending on the quantity of copper added, we will find a copper atom in place of an aluminum atom. This is a "solid solution" of copper in aluminum.

Since the copper atom is of a different size and valency from the aluminum atom, it does not fit perfectly into the site provided for it, and as such, introduces a certain amount of strain into the aluminum lattice. There is a limit to how much strain—or in other words, how many copper solute atoms—the host

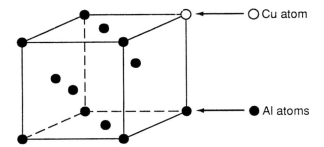

**Figure 5.1.**   Substitutional solid solution of aluminum in copper.

aluminum lattice will tolerate (just as in our own lives, sometimes there may be too many guests for comfort). So the host lattice rejects the guest atoms but they do not leave the system. The foreign atom combines with a host atom, which is also pulled from its lattice site, to form a distinct second phase, frequently a compound. In our present example, a $CuAl_2$ compound is formed, consisting of thousands to millions of copper and aluminum atoms arranged in a tetragonal structure, quite distinct from the FCC host lattice. This compound will exist in small regions of the system, not all in one place. Each compound particle will be surrounded by the solid solution phase. The $CuAl_2$ phase is distinct and readily visible via microscopic examination, but it coexists with the FCC solid solution phase. This compound forms when the solid solubility limit, discussed previously, is exceeded. This solubility limit increases with temperature (Figure 5.2), since the atom spacing and vacancy concentration increases, thus allowing more room for more foreign atoms on the host sites. Also, the modulus of elasticity decreases with increasing temperature (Figure 2.4), and this in itself causes less strain per solute atom. The less the strain per atom, the more solute atoms that host will tolerate.

Interstitial Solutes

Foreign atoms that are very small, such as hydrogen, carbon, nitrogen, boron, and sometimes oxygen, can often fit between (or among) the host atoms. They occupy the interstitial sites discussed in Chapter 4 and are actually squeezed into these interstitial sites, because, except for hydrogen, the interstitial atoms are larger than the size of the available interstice in the host lattice. This results in considerable strain in the host lattice, even more than that for substitutional solute atoms. The solubility is thus more restricted than that for the substitutional situation. But again the solubility increases with temperature, and when the solubility limit is reached at any given temperature, a second phase compound is formed. A good practical example of this situation is that of intersti-

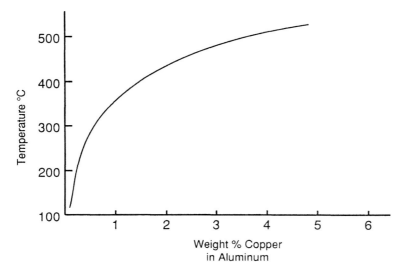

**Figure 5.2.** The solid solubility limit for copper in aluminum increases with increasing temperature.

tial carbon atoms in the BCC iron lattice. Figure 5.3 shows the solubility limit curve as a function of temperature along with the interstitial solid solution sites. In BCC iron the tetrahedral sites are the larger of the two and can accommodate an atom of 0.36 Å radius (i.e., $0.29R$, where $R$ is the radius of the host atom). The carbon atom is about 0.7 Å in radius and obviously introduces some considerable strain into the surrounding lattice. Actually, the carbon atom seems to prefer the smaller octahedral site because of the directionality of the elastic moduli in the BCC lattice. In either case the strain is large enough to limit the maximum solid solubility of carbon in BCC (alpha) iron to about 0.02 wt %. The compound, $Fe_3C$, forms when the solid solubility is exceeded. Again, this compound, which has an orthorhombic structure, is a distinct second phase and not a part of the parent iron lattice. We will find later that the strength of iron–carbon alloys, called steels, can be changed by a factor of 3 or more just by changing the size, shape, and distribution of these carbide particles, without changing the amount of carbon. This strength increase is accomplished through heat-treating processes, which we explore in detail in Chapter 6.

The defects discussed above are the basic point defects that are found in metals. But identical ones exist in ceramics and semiconductors. Solute atoms and vacancies have an even more profound effect on the electrical conductivity of semiconductors than they do on either the mechanical or physical properties

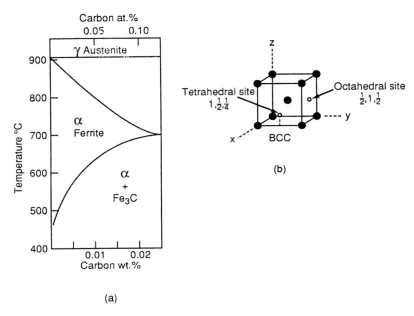

(a)

**Figure 5.3.** (a) Solid solubility limit as a function of temperature for carbon in BCC iron; (b) the BCC crystal structure, showing the interstitial sites (interstices).

of metals. Silicon and gallium arsenide, and all semiconductors for that matter, have been intentionally injected, by one means or another, with minute quantities ($<0.01$ at %) of foreign solute atoms, often called *dopants.* We say that the host material has been doped with foreign atoms. Most of these foreign atom dopants reside as substitutional solute in sites normally occupied by the host atom. Some electronic ceramics, such as the oxide superconductors, ferri- and ferroelectrics, and dielectrics materials, are also affected by minute quantities of solute atoms, sometimes added intentionally to achieve the desired properties.

### 5.2.3 Thermal Effects: Diffusion

At any temperature above 0 K, the atoms vibrate about their equilibrium position, although not necessarily in a symmetrical fashion. Atom vibrations have been loosely described as crystalline imperfections, because they do affect some properties, most notably the electrical conductivity of metals. These atom vibrations, together with the vacancies and the solvent–solute atom arrangements, constitute the ingredients necessary for solute atom diffusion to occur. Accordingly, we explore this subject now so that we will be prepared for the subjects of phase transformations, second-phase precipitation, and semiconductor doping, all introduced in later sections.

Substitutional Solute Diffusion

The driving force for the diffusion process is the concentration gradient $dc/dx$. To a first approximation the diffusion process is strictly statistical in nature (i.e., assuming that there is no chemical attraction between the unlike atoms). Cu–Ni substitutional alloys approach this condition, and furthermore, there is no solubility limit (because both atoms are of nearly the same size and the same crystal structure). We can put as many nickel atoms into copper as we wish, and vice versa, without forming a compound or second phase of any type. At nickel atom concentrations below 50% of the total, the nickel is called the solute, and similarly, when the nickel becomes the majority atom in number, copper is the solute. Both pure nickel and pure copper have FCC structures, so the alloy consists of some sites occupied by nickel and the remaining sites by copper atoms. Incidentally, when the nickel atoms are in the majority, the alloys are called Monels, a trademark of the International Nickel Corp. They are widely used in marine environments because of their good corrosion resistance to seawater. When copper is the majority element, they are called Cupro-Nickels, which are not significant alloys among the 1000 or so most widely used alloys.

Let's consider now a bar of copper welded along a straight-line interface to a bar of nickel. This arrangement is called a diffusion couple in studies of diffusion, as shown in Figure 5.4. There exists a certain probability that after a period of time any given atom of nickel will reside on the copper side of the interface in a copper lattice position, and similarly, for a copper atom to have

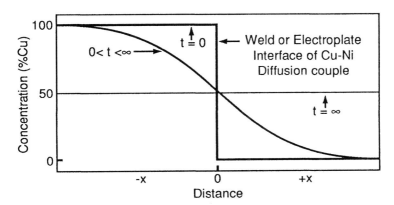

**Figure 5.4.** Copper–nickel diffusion couple showing the copper concentration at time = 0, time = some finite value, and time = ∞.

crossed into the nickel section of the diffusion couple. The atom makes its journey across the interface and into the other metal via the atom–vacancy interchange mechanism. The vacancies are constantly moving around, and when they appear adjacent to a solute atom, an interchange occurs. The higher the temperature, the higher the amplitude of atom vibration, the larger the concentration of vacancies, and the greater the probability of the copper atom being on the nickel side of the interface. In fact, this probability and the corresponding diffusion rate increase exponentially with temperature. The atoms can migrate by switching places with other atoms or by a coordinated movement of several atoms (ring mechanism), but the atom–vacancy interchange requires the least energy and accordingly is the most prominent substitutional solute diffusion mechanism. A certain activation energy is required for the atoms to move from one stable position to another.

Returning to the statistical nature of diffusion, since the atoms are moving randomly in all directions, and since there is a 100% concentration of copper atoms just one interatomic spacing to the left of the interface in Figure 5.4, statistically speaking there must be a net flow of copper atoms to the right, and similarly, a net flow of nickel atoms to the left. This is expressed mathematically by *Fick's first law* of diffusion:

$$J = D\frac{dc}{dx} \tag{5.2}$$

where $J$ = flux of atoms crossing any particular plane per unit time
$\quad\quad D$ = proportionality constant called the diffusivity
$\quad\quad dc/dx$ = concentration gradient at this particular plane

We can solve for the units of $D$ as follows:

$$J\,(\text{atoms/cm}^2/\text{s}) = D\,(\text{atoms/cm}^3/\text{cm})$$

Therefore, the units of $D$ are $\text{cm}^2/\text{s}$.

There is also a *Fick's second law* of diffusion, written as

$$\frac{dC_x}{dx} = \frac{d}{dx}\left(\frac{D\,dC_x}{dx}\right) \tag{5.3}$$

The second law takes into account the fact that $D$ changes with concentration—thus the subscript $x$ on $C$. This concentration effect is based on the fact that diffusion is not strictly a statistical process; as discussed above, each atom type behaves differently in the presence of unlike atoms.

Note that in Figure 5.4, the concentration as a function of distance $x$ is shown for $t = 0$, $t$ = some finite time, the curve shape being a function of temperature and time, and for $t$ = infinity. Theoretically, since these elements are mutually soluble in all concentrations, the concentration throughout the bar will eventually be a uniform 50% Ni–50% Cu alloy, although at room temperature the time could be millions of years to establish this uniform concentration. At temperatures near the melting point of copper, it may require only a few hours to attain uniformity. The diffusivity $D$ varies exponentially with temperature according to the Arrhenius equation,

$$D = D_0 e^{-Q/RT} \tag{5.4}$$

where $D_0$ = a constant of the same units as $D$, but independent of temperature.
   $Q$ = activation energy for the process, usually expressed in cal/mol
   $R$ = gas constant, approximately 2 cal/mol/K
   $T$ = absolute temperature in K

Diffusivity measurements are performed by materials specialists, but basically they involve measurement of the concentration of the diffusing species as a function of time $t$ and distance $x$ at a constant temperature $T$, and then computing the values of $D$ from Fick's law. For those systems where radioactive isotopes are available, the diffusivity can be measured quite accurately. In fact, the self-diffusivity (e.g., Cu in Cu) cannot be measured without radioactive isotopes. Taking the natural logarithms of equation (5.4) gives

$$\ln D = \ln D_0 e^{-Q/RT} \tag{5.5}$$

This is the equation of a straight line. Thus by plotting $\ln D$ versus $1/T$, as shown in Figure 5.5, the intercept on the $y$ axis becomes $\ln D$ and the slope equal to $-Q/R$. Sometimes diffusion data will be presented in this fashion, as shown in Figure 5.6, but more often in handbooks the constant $D_0$ and $Q$ are given and $D$ must be computed for the temperature of interest. One of the most complete sources of $D_0$ and $Q$ values for metals is that found in Tables 13.1 to 13.4 of *Smithells' Metals Reference Book*, 6th ed. (E. A. Brandes, Ed., Butterworth, Stoneham, Mass., 1983). Some selected values of these constants for some common metal systems are listed in Table 5.1.

   *Example 5.1.* From the tables in *Smithells*, the values of the constants for copper atoms diffusing into nickel at a concentration of 45.4% Cu–54.6% nickel are given as $D_0 = 2.3$ cm²/s and $Q = 60.3$ kcal/mol. Compute the diffusivity $D$ for copper in nickel at this concentration (values vary with concentration—see *Smithells*) at temperatures of (a) 0 and (b) 1000°C.
   Solution.   (a) 0°C = 273 K

$$D = 2.3 \text{ cm}^2/\text{s}^{-1} \exp \left[ -\frac{60{,}300}{2(273)} \right]$$

$$= \frac{2.3}{\exp(110.4)}$$

$$= \frac{2.3}{9.2} \times 10^{-47} \text{ cm}^2/\text{s}$$

$$= 2.5 \times 10^{-48} \text{ cm}^2/\text{s}$$

It is dubious if we would live long enough to measure the penetration of copper in nickel at room temperature.

(b) $1000°C = 1273$ K

$$D = 2.3 \text{ cm}^2/\text{s} \exp\left[ -\frac{60,300}{2(1273)} \right]$$

$$= \frac{2.3}{\exp(23.7)}$$

$$= 1.96 \times 10^{-10} \text{ cm}^2/\text{s}$$

When one wants to determine the concentration of solute at a given distance $x$ and for time $t$, knowing the diffusivity and initial concentrations, solutions to the partial differential equation (5.3) must be found. Frank's *The Mathematics of Diffusion* (see the Suggested Reading) lists solutions for most common

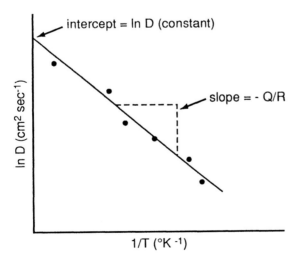

**Figure 5.5.**  Schematic of the ln $D$ versus $1/T$ relationship, showing how the activation energy $Q$ and constant $D_0$ are obtained.

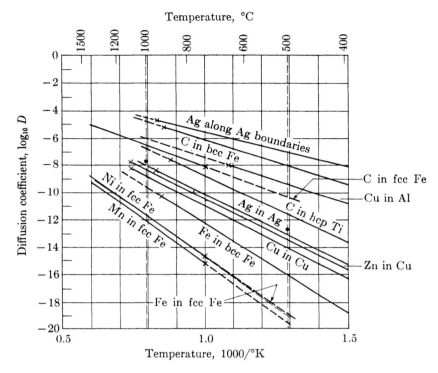

**Figure 5.6.** Diffusion coefficients versus temperature. (From L. H. Van Vlack, *Elements of Materials Science and Engineering*, 3rd ed., Addison-Wesley, Reading, Mass., 1975, p. 127.)

geometries of diffusion couples (e.g., gases diffusing through cylinder walls, surface carbuerization of steel, and many other situations encountered frequently in industry). For example, the diffusion of carbon into iron for case hardening is given by the equation

$$C_s - \frac{C_x}{C_s} - C_0 = \text{erf}\left[\frac{x}{2\sqrt{Dt}}\right] \qquad (5.6)$$

where $C_s$ = surface concentration of carbon
$\quad C_x$ = concentration of carbon at distance $x$
$\quad C_0$ = initial concentration of carbon in iron
$\quad D$ = diffusivity of carbon
$\quad t$ = time

**Table 5.1**   Diffusion Data for Selected Metals

| Solvent | Solute | $D_0$ (cm$^2$/s) | $Q$ (kcal/mol) |
|---|---|---|---|
| Cu | Cu (self-diffusion) | 0.78 | 49.3 |
| Cu | Zn | 0.73 | 47.5 |
| Al | Cu | 0.65 | 32.3 |
| Al | Mg | 0.06 | 27.4 |
| Cu | Ni | 3.8 | 56.8 |
| Ni | Cu | 0.57 | 61.7 |
| Fe(a) (350–850°C) | C | $6.2 \times 10^{-3}$ | 19.2 |
| Fe(g) (900–1600°C) | C | 0.1 | 32.4 |

*Source*: After E. A. Brandes, *Smithells' Metals Reference Book,* 6th ed., Butterworth, Stoneham, Mass., 1983).

Error function (erf) values can be found in most volumes of standard math tables (e.g., in books that list trigonometric values). The same equation would apply, for example, to a chromium-plated steel. In this case the penetration of chromium into steel at room temperature would be negligible, but if an errant worker happened to heat this component to several hundred degrees Celsius, the chromium atoms could diffuse into the steel to the extent that undesirable brittle phases would be formed.

Interstitial Solute Diffusion

Interstitial atoms diffuse through the parent material by jumping around from one interstitial site to another. Since these atoms are relatively small, the diffusivities are much larger than that for substitutional solute, often of the order of $10^5$ times greater (Table 5.1 and Figure 5.6). Similarly, the activation energies for interstitial diffusion are much less than those for substitutional diffusion. The substitutional and interstitial mechanisms, along with their required activation energies, are shown schematically in Figure 5.7. All of the same diffusion equations apply for interstitial atoms as were presented above for the substitutional solute diffusion situations. Self-diffusion in pure metals occurs by the same vacancy–atom interchange. This would occur, for example, when annealing a pure metal, and in some cases significant diffusion can occur at room temperature.

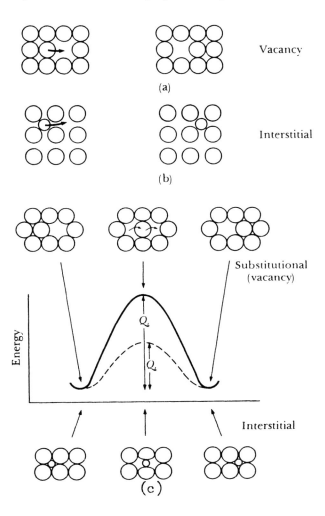

**Figure 5.7.** Diffusion mechanisms: (a) vacancy–atom interchange; (b) interstitial diffusion; (c) corresponding activation energies. (Adapted from D. R. Askeland, *The Science and Engineering of Materials*, 2nd ed., PWS-Kent, Boston, 1989, pp. 121, 123.)

*Example 5.2.* A 1010 steel (0.10 wt % C) is being gas carburized at 927°C. Calculate the time it takes for the carbon content to achieve a 0.40 wt % C content at a depth of 1/16 in. (1.5 mm). Assume that the surface carbon concentration is 0.90 wt %.

From the handbook we find that the diffusivity of carbon in BCC iron (which would be about the same for a 1010 steel) is $1.3 \times 10^{-7}$ cm²/s at 927°C.

Solution.

$$C_s - \frac{C_x}{C_s} - C_0 = \text{erf} \left( \frac{x}{2\sqrt{Dt}} \right)$$

$$C_s = 0.9\% \qquad C_0 = 0.1\% \qquad C_x = 0.4\%$$

Substitution of these values gives

$$\frac{0.9 - 0.4}{0.9 - 0.1} = \text{erf} \left( \frac{0.015}{2\sqrt{1.3 \times 10^{-7} \, \text{cm}^2/\text{s}}} \right)$$

$$0.625 = \text{erf} \left( \frac{210}{\sqrt{t}} \right)$$

Let $z = (210/\sqrt{t})$. Thus $0.625 = \text{erf} \, z$. We need a number for $z$ whose error function (erf) $= 0.625$. From handbook tables this number is found to be $0.623 = z$. Therefore, $0.623 = 210/\sqrt{t}$.

$$\sqrt{t} = \frac{210}{0.623} = 337$$

$$t = 337^2 = 113,569 \, \text{s} = 31.5 \, \text{h}$$

There is an approximation formula for computing time when $D$ and $x$ are known, or $x$ when $t$ and $D$ are known. This approximation is

$$x = \sqrt{Dt} \qquad \text{or} \qquad t = \frac{x^2}{D} \qquad\qquad (5.7)$$

What does $x$ mean in this case? It is often said that it represents the depth to which appreciable diffusion occurs. Actually, we can be a little more precise than this. The $x$ in equation (5.7) represents the distance at which the concentration becomes one-half of the difference between the initial and final values.

*Example 5.3.* Assume that we would like to have a significant carbon content (say 0.4 wt % C) at a depth of 1.5 mm (0.15 cm). Approximate the diffusion time required at 927°C. ($D$ can be obtained from Example 5.2.)

Solution.

$$t = (0.15)^2 \, \text{cm}^2/1.3 \times 10^{-7} \, \text{cm/s} = 1.7 \times 10^5 \, \text{s} = 48.1 \, \text{h}$$

*Example 5.4.* Let's now apply this approximation to Example 5.1 and estimate the time required for significant penetration of copper into nickel at (a) 0
and (b) 1000°C to a depth of 0.01 cm (0.004 in.).
  Solution.   (a) At 0°C (273 K),

$$t = \frac{x^2}{D} = \frac{10^{-4}}{2.5 \times 10^{-48}} = 10^{40}\,h$$

which confirms our earlier suspicion. Even Methuselah would not see
significant penetration of copper into nickel at 0°C in his life span.
  (b) At 1000°C (1273 K),

$$t = \frac{x^2}{D} = \frac{10^{-4}}{1.96 \times 10^{-10}} = 5.1 \times 10^6\,s = 142\,h$$

## 5.3  LINE DEFECTS: DISLOCATIONS

Dislocations are the only line defects that exist in crystalline solids. In the strict
geometrical sense they are really cylindrical defects of about five atom spacings in diameter. They thread their way through the crystal in all sorts of directions, not usually as straight lines. But they are one of the more important crystalline imperfections, even though we have known about their existence for
only a little more than 50 years. They were predicted in a series of separate
papers by E. Orowan, M. Polyani, and G. Taylor in 1934, to explain the
mechanism of plastic flow in metallic crystals. But we now know that their
presence in semiconductors can be very deleterious to the performance of such
devices. Extreme care is taken in the growth of single semiconductor crystals
in order to minimize the presence of dislocations. Claims have been made that
silicon crystals of 12.7 cm (5 in.) diameter have been grown essentially
dislocation-free. In typical as-grown metallic single crystals, their concentration is on the order of $10^6$ per square centimeter. This means that if a cross-
sectional thin slice of 1 cm$^2$ area was cut from a metal crystal, then 1 million
dislocations would intersect this surface. This has been verified experimentally
by using special etching techniques that cause a pit to form at the point where
each dislocation line emerges from the surface (Figure 5.8). These pits can be
seen through an optical microscope at magnifications on the order of 100 to
1000 times. Since dislocations are usually not straight-line defects, a better way
of expressing their density is in terms of the centimeters of length of dislocation lines per cubic centimeter of volume of the material. This definition gives
the same units as the etch pit technique (i.e., number/cm$^2$). These dislocation
lines in the volume of the material can be viewed in a thin section of a crystal
by transmission electron microscopy (Figure 5.9). But it is not practical to
measure their length per unit volume in this way. The dark lines that we see

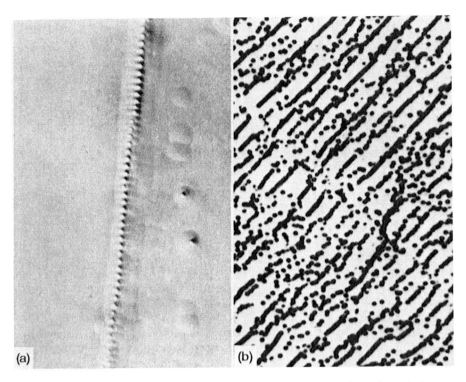

**Figure 5.8.** Dislocation etch pits in germanium. (a) In a small angle grain boundary; (b) Dislocation etch pits in subboundaries produced by polygonization; (c) Dislocation etch pits aligned on traces of slip planes in germanium that has been deformed at a low temperature. (From ASM Metals Handbook, Vol. 8, 8th ed., 1973, p. 148.)

are a result of the strain fields surrounding the dislocation line. The dislocations form during solidification and recrystallization. They can also be generated in significant quantities during plastic deformation. An annealed metal that contains a dislocation density of $10^6$ per $cm^2$ will contain about $10^{12}$ per $cm^2$ after the metal has been severely deformed (e.g., by 80% reduction in area). There are two different configurations of atoms in dislocations, called *edge* and *screw* types.

### 5.3.1 Edge Dislocation

The structure of the edge dislocation is depicted in Figure 5.10. It can be viewed as if a half-plane of atoms have been removed from the lattice and then the neighboring atoms collapsed to fill the planar void. One might think that this would be a planar defect, but not so. Look at the atom positions. At points

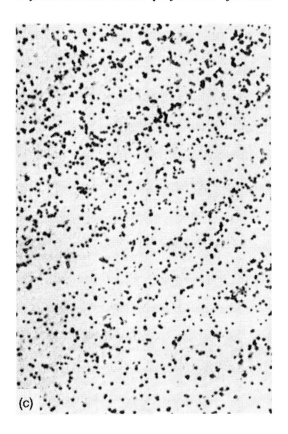

(c)

only a few atom spacings away from the missing half-plane of atoms, all of the other atoms are in near-perfect registry, all around the dislocation core. If one mentally projects a distance of about 10 atom distances, the perfect registry can be visualized. The disregistry resulted from the collapse of the planes.

The configuration shown here is by convention a positive edge dislocation. But for every positive edge dislocation there exists a negative edge dislocation where the missing half-plane of atoms have been removed from the part above the plane denoted by the letters "SP" (Figure 5.11). We chose "SP" since we will discover later that this is the slip plane on which plastic deformation occurs. The dislocation lines here are perpendicular to the plane of the paper, but lie in the slip plane. Not all dislocations lie in slip planes, but at this point we are concerned only with those that do since these are the ones that permit our understanding of plastic flow in crystalline bodies.

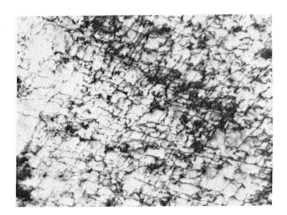

**Figure 5.9.** Uniform dislocation structure in iron foil deformed 14% at −19°C as viewed in a transmission electron microscope. ×40,000. (From *Metals Handbook,* Vol. 8, 8th ed., ASM International, Metals Park, Ohio, 1973, p. 219.)

### 5.3.2 Screw Dislocations

In the screw dislocation there is no missing half-plane of atoms. It can be viewed as if a narrow ribbon of material in two parallel planes had been twisted with respect to each other (Figure 5.12). All along this ribbon the atoms are out of registry, again resulting in a cylinder about five atoms in

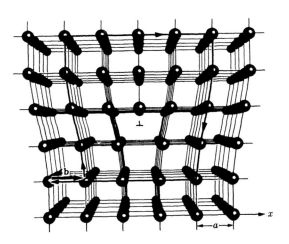

**Figure 5.10.** Structure of an edge dislocation. (From A. G. Guy, *Elements of Physical Metallurgy,* Addison-Wesley, Reading, Mass., 1959, p. 110.)

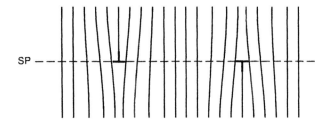

**Figure 5.11.** Negative and positive edge dislocations on the same slip plane. The dislocation lines are in a direction perpendicular to the plane of the page.

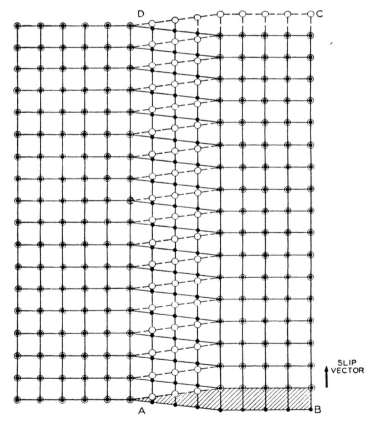

**Figure 5.12.** Arrangement of atoms around a screw dislocation. The plane of the figure is parallel to the slip plane. (From W. T. Read, *Dislocations in Crystals*, McGraw-Hill, New York, 1953, p. 17.)

diameter of dislocated atoms. This cylinder threads its way through the crystal, and just like the edge type, the screw dislocation is considered to be a line defect.

### 5.3.3 Dislocation Loops

Dislocations exist in the crystal in irregular-shaped loops and wander through the crystal in many directions, as one might guess from the way the dislocation lines appeared as if they were sort of entwined and entangled groups when viewed in the electron microscope (Figure 5.9). Their atomic configuration in the form of an enclosed loop on a slip plane can be viewed in Figure 5.13. Here we are showing only one-fourth of the loop, but the other three-fourths would have similar atom position irregularities. The point that one should obtain from this picture is that dislocations exist of pure edge or pure screw orientation only at certain points along the line, as denoted in Figure 5.13. The remaining portions of the loops are mixed dislocations; that is, the atom arrangements can be considered as being formed from components of both

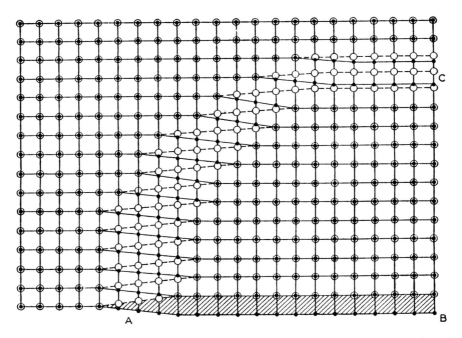

**Figure 5.13.**   Arrangement of atoms in a curved dislocation loop (only one-fourth shown). Position C is of edge orientation, and position A is of screw orientation. (From W. T. Read, *Dislocations in Crystals*, McGraw-Hill, New York, 1953, p. 18.)

edge and screw orientation. These atom positions in the mixed part of the dislocation line are difficult to visualize but their actual positions are not important to our understanding of the dislocation mechanism of plastic deformation of crystalline solids.

### 5.3.4 Diffusion Along Dislocations

As stated previously, diffusion takes place by atom–vacancy interchanges in the cases of substitutional solute and self-diffusion, and by "interstice hopping" for interstitial solute migration. A dislocation core is almost like a line of vacancies along which atoms can diffuse in a very rapid fashion. Diffusivities of solute along these linear defects are on the order of $10^5$ to $10^7$ times that in ordinary lattices, where diffusion takes place via the atom–vacancy interchange mechanism. One could expect diffusivities at 400 to 500°C along dislocations to be comparable to that at around 700°C via the common lattice atom–vacancy interchange mechanism. Of course, the volume area of the total dislocations present is a much smaller volume than the lattice proper, so the total diffusant of solute is quite small. However, in semiconductors, where one would like to have a uniform front of diffusing atoms, the dislocations cause long spikes and nonuniform distribution of dopant atoms. Such a nonuniform distribution of dopant atoms interfere with the operation of semiconductor devices—hence the need to have a very low dislocation count in semiconductor chips.

At temperatures below where normal diffusion takes place, there also exists a tendency for dislocations to attract solute atoms. A solute atom that is larger than the solvent creates a region of compressive strain. Beneath the slip plane of a positive edge dislocation (Figure 5.10) a tensile strain exists. Thus the larger solute atom is attracted to this region of tensile strain to reduce the overall total strain. This binding of solute atoms to the dislocations restricts their motion. The dislocation is said to be locked in place and a stress higher than that to move unlocked dislocations is required to release the dislocation from the solute. But once the dislocation is released, it can move at a lower stress. We then have a yield point, very common for carbon atoms in iron, where there exists an upper yield stress followed by a more normal lower yield stress to move the dislocations after they have become unlocked (Figure 2.9a).

### 5.4 PLASTIC DEFORMATION OF SINGLE CRYSTALS

Many years before dislocations were conceived, it was thought that the plastic deformation of crystalline materials occurred by the sliding of parallel planes of atoms over one another, much as a card or a thin packet of cards could be extended beyond the other cards of a deck of playing cards, somewhat like the schematic of Figure 5.14. The experimental evidence suggested such a

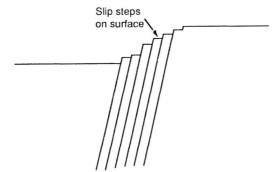

**Figure 5.14.** Schematic of slip planes protruding from the crystal after slip has occurred.

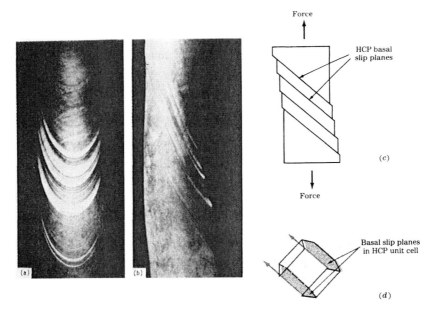

**Figure 5.15.** Slip bands on the surface of a plastically deformed zinc crystal. (Courtesy of Prof. Earl Parker, University of California, Berkeley.)

mechanism because carefully controlled experiments on stressed crystals did show surface markings, as if a thin section of the crystal protruded from the remaining portion (Figure 5.15). These markings are even more evident where slip has been caused to occur by a hardness impression on a polished metal surface (the dark region of Figure 5.16 is the impression). These markings are called *slip lines*.

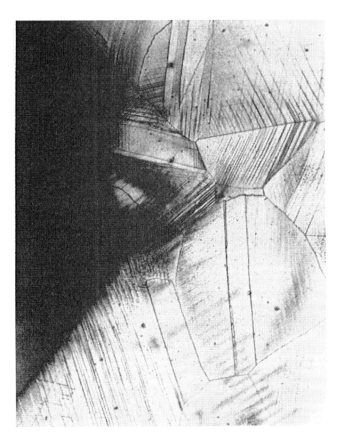

**Figure 5.16.** Slips lines produced by a hardness indentation on a polished and etched surface of an Inconel 600 specimen. × 80. Slip lines in annealing twins are at an angle to each other that demonstrates the mirror image of atoms in an annealing twin. Note the intersecting slip lines near the dark edge of the hardness indentation.

### 5.4.1 Metal Crystals

From the markings of the type shown in Figure 5.15 and 5.16, and in conjunction with x-ray diffraction crystal orientation determinations, it was found that in metal crystals the slip planes consisted of the most closely packed planes, and that the slip direction was also that of the highest linear atom density. The combination of a slip plane and a slip direction within that plane comprise a *slip system.* The primary slip systems, the most densely packed planes and directions, are the ones requiring the lowest applied shear stress for slip to occur. In the FCC system these are the {111}-type planes, of which there are four distinct planes. Each of these four slip planes contains three <110>-type slip directions, as depicted in Figure 5.17. Thus there are 12 primary slip systems in the FCC structures. A similar analysis shows that 12 primary slip systems exist in the BCC structures, consisting of six {110}-type planes, each containing two <111> slip directions. But there are also 36 other BCC slip systems with reasonably high atomic packing densities, and these secondary slip systems require only about 10% more shear stress for slip to occur than do the primary systems. Thus we often say that the BCC structures have 48 slip systems. Slip in the hexagonal system occurs most easily on the {1000} basal planes, each of which contain three <1120> slip directions, for a total of three slip systems. Some hexagonal metals (e.g., magnesium) are brittle, due in large part to having only three slip systems. It can be shown (von Mises criterion) that five independent slip systems are required to deform a crystalline

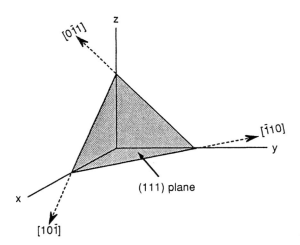

**Figure 5.17.** The {111} slip plane in an FCC crystal shows the three <110> slip directions that lie in this slip plane. The combination represents three slip systems.

body to any arbitrary shape. Some slip systems and the shear stresses required to cause slip are listed in Table 5.2. This stress to cause slip is called the *critical resolved shear stress* (CRSS). It is primarily a function of the atom bond strength and the atom spacing in the crystal.

Any force applied to a body can be resolved into a force normal to a plane and into shear components along a plane. The method of resolving a longitudinal force on a cylindrical body is depicted in Figure 5.18. The resolved shear stress, $\tau_r$, along a slip plane (shaded plane in Figure 5.18) and in a slip direction resulting from a stress $s$ along the bar is given by

$$\tau_r = \frac{F_r}{A} \tag{5.8}$$

where $F_r$ is the resolved force along the slip plane resulting from the applied force $F$ along the bar axis, and $A$ is the cross-sectional area of the slip plane. In the figure, the angle $\phi$ is that between the normal to the slip plane and the direction of $F$, and the angle $\lambda$ is that between the slip direction and the direction of the applied force $F$. Now

$$F_r = F \cos \lambda$$

$$A = \frac{A_0}{\cos \phi}$$

$$s = \text{normal stress} = \frac{F}{A_0}$$

Therefore,

$$\tau_r = \frac{F \cos \phi}{A_0/\cos \phi} = s \cos \phi \cos \lambda \tag{5.9}$$

**Table 5.2** Critical Resolved Shear Stress for Selected Metals[a]

| Metal | Crystal structure | Purity (%) | Slip plane | Slip direction | CRSS [MPa (psi)] |
|-------|-------------------|------------|------------|----------------|------------------|
| Al | FCC | 99.994 | {111} | <110> | 0.7 (100) |
| Cu | FCC | 99.98 | {111} | <110> | 0.95 (138) |
| Ni | FCC | 99.98 | {111} | <110> | 5.0 (725) |
| Mg | HCP | 99.99 | {1000} | <11$\bar{2}$0> | 0.7 (100) |
| Zn | HCP | 99.999 | {1000} | <11$\bar{2}$0> | 0.25 (36) |
| Fe | BCC | 99.96 | {110} | <111> | 20.0 (2900) |
| MO | BCC | — | {110} | <111> | 49.0 (7100) |

[a] The values are averaged from a variety of sources. Also note that the BCC values are much higher than those for the FCC and HCP close-packed structures.

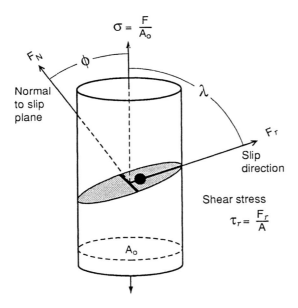

**Figure 5.18.** Computation of the resolved shear stress on plane $A$ (shaded) due to an axial load $F$.

Actually, the resolved shear stresses can be obtained along any plane from any direction and magnitude of force $F$. The material does not even have to be crystalline. The reader may have encountered such an operation in a statics course. Here we choose the slip plane and direction because we are interested in the resolved shear stress of the slip system, since when this stress attains the critical value (CRSS), slip occurs.

Dislocation Mechanism of Slip

When the stress to cause slip is computed based on the bond strength of the atoms and the assumption that all bonds are broken simultaneously, CRSS values on the order of $10^6$ psi (6900 MPa) are obtained. Note in Table 5.2 that the CRSS values are less than 1% of this computed value. This is the reason dislocations were conceived (i.e., to explain this discrepancy). The values in Table 5.2 vary with purity and defect content, and thus the values from different sources will be somewhat different.

Plastic deformation of metallic crystals takes place by the movement of dislocations. Substantial plastic deformation is observed at temperatures as low as $0.2T_m$, where $T_m$ is the melting point in kelvin. The stress required for slip generally increases with decreasing temperature (Figure 5.19), because the

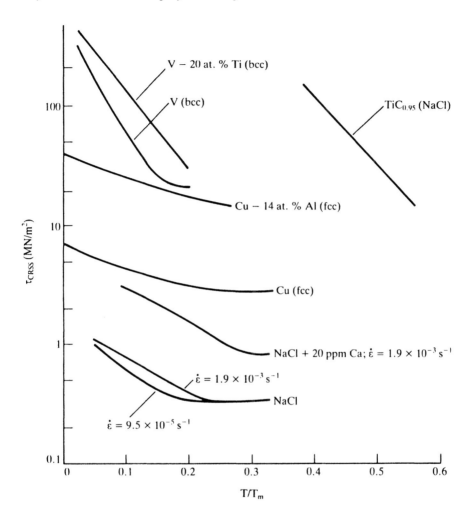

**Figure 5.19.** Variation of the CRSS value with temperature for a variety of materials. Note that the temperature is expressed in terms of the fraction of the melting point in kelvin. In this way the variation of the CRSS with the types of bonds can be compared. Note that the BCC metals have higher values of the CRSS and also a greater temperature dependency. (From T. H. Courtney, *Mechanical Behavior of Materials*, McGraw-Hill, New York, 1990, p. 141.)

modulus (i.e., the bond strength) increases with decreasing temperature. When a dislocation moves, it causes plastic deformation to occur by breaking bonds along a single row of atoms simultaneously instead of breaking all bonds simultaneously across a large plane of atoms. Furthermore, the atoms in a row along a dislocation are not directly above or below another row of atoms, but in between two rows on the plane below it, such that the stress to break these bonds, to a first approximation, is zero. This will become apparent by observing the movement of an edge dislocation in a crystal by a sequence of steps as pictured in Figure 5.20. Just a small applied stress will move the dislocation because the atoms in the row marked A, really want to line up with those in row B, since this is the stable lattice configuration. So the atoms in row B really attract the atoms in row A. But because of the atom arrangement in a dislocation the row of atoms go on to the position past row B as depicted in the second step of the sequence of Figure 5.20. The atoms really only move one interatomic distance from a point of low energy over an energy hump to another position of low energy. In these low-energy positions, called *saddle points*, the atoms are attracted equally by the rows in front and back of row A. So the dislocation continues to move from saddle point to saddle point under a very low stress, much like a wave moves through the water. The wave moves to the shore, but the molecules of water move only a part of a wave length. The stress to move a dislocation $\tau_p$ (the Periels–Nabarro stress), is given by

$$\tau_p = \frac{2G}{1 = \mu} \exp\left[\frac{2\mu a}{b(1 - \mu)}\right] \tag{5.10}$$

where $G$ = shear modulus
   $\mu$ = Poissons ratio
   $a$ = distance between atom planes
   $b$ = distance between atoms in the slip direction

Since the modulus increases with decreasing temperature, the stress to move dislocations increases with decreasing temperature, although other factors are also involved, but to a lesser degree. But the atoms, after the dislocation has moved through the lattice, are again in perfect registry, except that the last row extends beyond the bottom row one interatomic distance. This is the mechanism of plastic deformation. You might wonder how a metallic crystal could be twisted, bent, elongated, compressed, cupped, and whatever other deformation that is required to achieve a certain desired shape simply via a slip-plane effect. But there are trillions of dislocations moving in many directions, because there are many slip systems (i.e., 12 combinations of the type {111} <110> in the FCC structure) and many more in the BCC structure. Each of these dislocations produce an offset of one interatomic distance in its direction of movement so that we have many offsets in many directions. In plastic deformation, we do

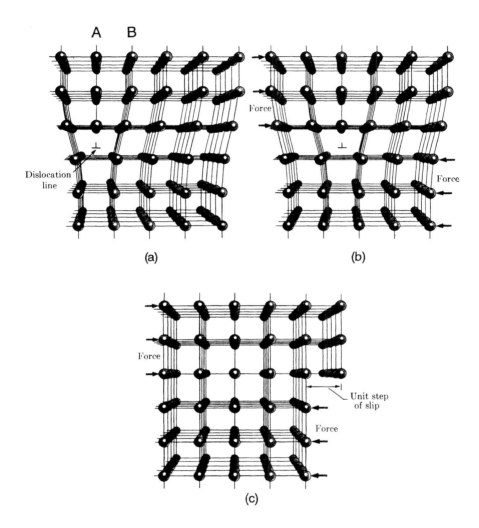

**Figure 5.20.** Motion of an edge dislocation and the resulting unit slip. (From W. T. Read, *Dislocations in Crystals,* McGraw-Hill, New York, 1953, p. 109.)

not really squash the crystals into an amorphous state. The crystals are squashed into whatever shape we desire, but the material internally remains as an orderly array of atoms. Plastic deformation does not destroy crystallinity. Screw dislocations also move out of the crystal and produce an offset as shown in Figure 5.21. Actually, they both produce an offset in the same direction, but

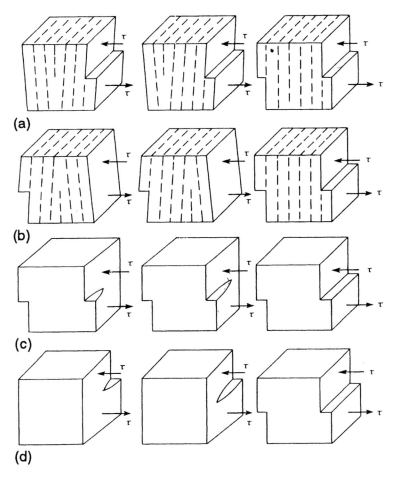

**Figure 5.21.** Ways that the four basic orientations of a dislocation move under the same applied stress to produce unit single slip. (From R. E. Reed-Hill, and R. Abbaschian, *Physical Metallurgy Principles,* 3rd ed., PWS-Kent, Boston, 1992, p. 154.)

the dislocation lines of the screw and edge move perpendicular to each other. Remember that in a large crystal, the dislocation line is in the form of a loop, consisting of edge, screw, and mixed components. Under an applied stress the loop expands and moves completely out of the crystal, producing an offset as depicted in Figure 5.22.

The student may not be able to visualize the above without a crystalline model or an exceptional amount of concentration. For the nonmaterials

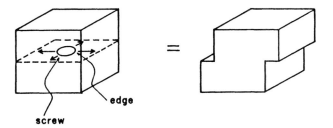

**Figure 5.22.** Slip produced by the expansion of a dislocation loop.

engineer it is not important to understand the mechanism, but instead, just have a little faith that it does happen in this fashion. This mechanism has been verified experimentally, and for all but the doubting "Thomases" this should suffice. The more curious-minded students can check the references listed at the end of this chapter for more detailed presentations.

Stress–Strain Curves of Metal Crystals: Strain Hardening

Stress–strain behavior of single crystals have little practical value but they have been examined extensively in research labs in an effort to better understand the more complex behavior of polycrystalline bodies. In portraying the stress–strain curves for single crystals (Figure 5.23), it is more informative to use the resolved shear stress along the slip plane rather than the normal applied stress as we did in Figure 2.5. We will also use the amount of slip plane displacement (shear glide) in computing the shear strain. But just as the figure for polycrystalline metals showed strain-hardening effects, so do single-crystal bodies. In all cases the shear stress increases with strain as one would expect if strain hardening is occurring during the slip process. The hexagonal metal crystals (Mg, Zn, Cd) in Figure 5.23 show very little strain hardening, and this is because the crystals have been oriented so that slip is occurring on only one set of parallel planes (i.e., one slip system). Remember, hexagonal crystals have only three slip systems, so it is easy to orient the crystal so that the resolved shear stress on only one slip system exceeds the CRSS.

For the FCC crystals there are 12 primary systems possible for slip, so perhaps four or maybe even six or more systems will be oriented for slip at the stress applied. These slip planes intersect, as can be viewed by the slip traces in Figure 5.16. Without going into the specific details, suffice to say that intersecting slips is one of the prime sources of strain hardening. It requires a larger stress to move dislocations when they must pass through other dislocations. We will return to this subject later when polycrystalline strain hardening is discussed.

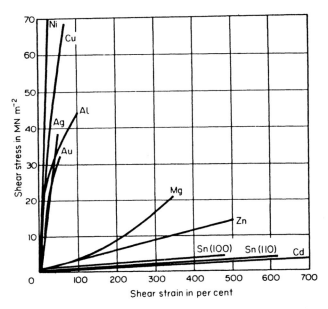

**Figure 5.23.** Shear stress versus shear strain for different metal crystals. After E. Schmid and W. Boas, *Kristallplastizitat,* Springer-Verlag, Berlin, 1935.)

### 5.4.2  Ceramic Crystals

Dislocation motion and strain hardening in ceramic crystals occur pretty much in the same way that they do in metallic crystals. Both have slip systems and intersecting slip planes for most crystal orientations. But there are some very important differences between dislocation motion in the two materials. Slip in some directions in ceramic crystals bring ions of like charge into close proximity, and because of their electrostatic repulsion, considerable stress may be required for substantial dislocation motion to occur. As a consequence of this possibility, the direction of easiest slip often will be such that ions of like sign will not become juxtaposed. In NaCl and MgO, for example, slip occurs in the <110> direction. Second, there are fewer slip systems in most ceramic crystals because of their complex structures, many being of a hexagonal type. Some of the simpler ionic crystals, NaCl, LiF, MgO, and $CaF_2$, for example, do exhibit extensive dislocation motion and concomitant plasticity at room temperature under slow strain rates, and the absence of notches. Figure 5.24 shows the amount of slip strain that was obtained as a function of temperature, as well as the fact that creep can occur in these crystals. Finally, the bond strength of many ceramics are higher than that found in most metallic crystals, although

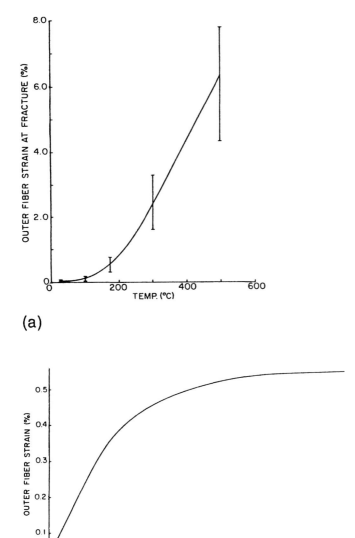

(a)

(b)

**Figure 5.24.** (a) Outer fiber strain in bending as a function of temperature of $CaF_2$ single crystals; (b) creep in bending at 175°C for a single crystal of $CaF_2$. (From R. Burn and G. T. Murray, *J. Am. Ceram. Soc.*, Vol. 45, No. 5, 1962, p. 251.)

there is some overlap. As a consequence, ceramics are often stronger and show only limited slip prior to fracture. The stress for plastic flow is very close to that for fracture, hence little ductility is obtained. There have been some examples of extensive ductility in AgCl at room temperature and in MgO at elevated temperatures (see the paper by Stokes in the book edited by Liebowitz cited in the Selected Reading), but these are exceptions. The fracture stress of ceramics can also be markedly reduced by the presence of surface scratches, notches, and the like. We say that such materials are "notch sensitive." Several references of plastic flow in ceramic crystals are listed at the end of this chapter.

## 5.5  PLANAR DEFECTS

Grain boundaries and phase boundaries are planar defects. Grain boundaries are formed as crystals meet during solidification. Each crystal has a different orientation of its atomic planes, even though each crystal has the same crystal structure. But the crystals are somewhat "cocked" with respect to each other, and the atoms in the region where the crystals meet during the final periods of solidification do not belong to either crystal (Figure 4.3). Our best estimates indicate that these boundaries are regions of disorder about five atom spacings in width, which cannot really be described in a crystallographic sense. They could be visualized as consisting of vacancies, interstitial solvent atoms, and perhaps even as dislocations, but we will treat these planar defects as disordered or amorphous solids. But there is some rather strong binding in this disordered solid structure boundary, more than sufficient to keep the crystalline body intact. We do know that at high temperatures, about $0.7~T_m$, the grain boundary regions are weaker than the interior crystalline part of the body, even to the extent that the grains slide past one another at relatively low stresses. This is one of the mechanisms of elevated temperature creep discussed in Chapter 2 (but not the only creep mechanism). But at lower temperatures, the grain boundaries contribute a significant strengthening effect to the polycrystalline body, due primarily to a higher dislocation density in the grain boundary regions that occurs during plastic deformation. Diffusion also takes place at a much faster rate along grain boundaries, much as it does in the disordered array of atoms in the core of a dislocation.

Phase boundaries (i.e., the regions where two phases meet) are also considered to be planar-type crystalline defects. But here the adjoining regions of the two phases consist of two entirely different crystal structures and as such have less disorder at their interfaces. The intermediate atoms pretty much know to which phase they belong. Thus the boundaries are very thin and not prone to being amorphous, nor do they tend to slide past one another at elevated temperatures. Diffusion along these boundaries is more like that in the lattice proper. Often the boundaries are between a ductile and a brittle intermetallic compound phase. As such, they are often points where, under stress, cracks may sometimes initiate and even propagate to failure. But more likely, when the brittle intermetallic compounds are small compared to the total volume, the more ductile "matrix" phase plastically deforms under stress and the matrix phase flows around these precipitate particles. However, the particles do provide a strengthening effect, because they do impede dislocation motion. And anything that retards dislocation motion strengthens the material as a whole. We explore this concept in greater depth when we examine the strengthening effect in two-phase alloys.

## 5.6 PLASTIC DEFORMATION AND FRACTURE OF POLYCRYSTALLINE SOLIDS

### 5.6.1 Deformation of Metals

Polycrystalline metals can consist of many grains (crystals) of the same phase, or of crystals of two or more phases, which we will call polyphase polycrystalline bodies. The latter will be discussed later when we consider the several strengthening mechanisms involved in two- or even three-phase solids. One such mechanism was just mentioned in Section 5.5. Also, we need a better understanding of the heat-treating methods used to control phase quantities, particle size, and distribution. What we will consider now is the strengthening effect of grain boundaries in single-phase polycrystalline solids. This encompasses the concept of "strain hardening" that we discussed early for single crystals. But the presence of grain boundaries enhances the strain-hardening effect. Whereas strain hardening in single crystals can be pictured pretty much as that due to dislocations intersections and some entanglements, which restrict further dislocation motion, the presence of grain boundaries, alters this picture to some extent.

When we twist, bend, push, pull, cup, or beat a polycrystalline metallic sheet to a desired shape, for example as in forming an automobile fender, the metal becomes stronger. This is somewhat contrary to our intuition, since we usually think that such brutal treatments should weaken a body, and of course it does so for human bodies, probably the source of our intuition. Let's see how this strengthening by deformation occurs in polycrystalline metals. Referring to Figure 5.25, a dislocation line segment is pinned at two points. This pinning could occur in a number of ways, one of the most common being that the dislocation at some point goes in a direction that moves it out of the slip plane, as illustrated in Figure 5.25. Nothing says that it has to stay in the slip plane. Remember that the dislocation line can run more or less in random directions throughout the crystal. As a shear stress is applied along the slip plane, the dislocation segment lying therein is bowed outward by virtue of the pinning of

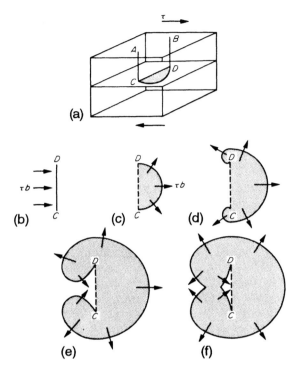

**Figure 5.25.** Operation of a Frank-Read source. (From R. W. K. Honeycombe, *The Plastic Deformation of Metals,* 2nd ed., ASM International, Metals Park, Ohio, 1984, p. 44.)

the ends of the segment. As the stress continues to be applied, the dislocation line bows completely around, to the point where the two segments connect, thereby forming a dislocation loop (Frank–Read mill). As the stress continues, more loops are formed, and the mill continues to pump out dislocation loops. That is why it is called a mill—it manufactures dislocations by the trillions. In polycrystalline bodies, grain boundaries restrict dislocation motion. As a result, the stress to maintain flow increases to the point that secondary slip systems begin to operate and the amount of intersecting slip increases substantially. Figure 5.16 shows this increase of intersecting slip planes near the grain boundaries. As plastic flow continues, dislocation entanglements and concomitant strain hardening increases much more rapidly in the vicinity of the grain boundary. This effect has been verified by hardness measurements (Figure 5.26) and electron microscopic observations. Hence polycrystalline materials are much stronger than single crystals, and small grain metals are much stronger than

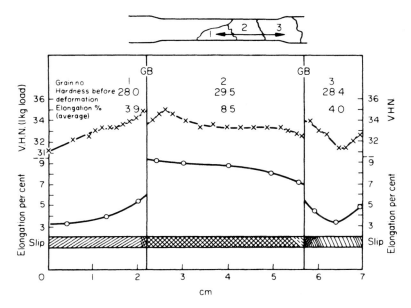

**Figure 5.26.** Inhomogeneity of plastic deformation near the grain boundaries in aluminum as revealed by hardness measurements. A higher dislocation density is formed near the boundaries during deformation. (From R. W. K. Honeycombe, *The Plastic Deformation of Metals*, 2nd ed., ASM International, Metals Park, Ohio, 1984, p. 224.)

large grain metals (at temperatures $<0.5T_m$). Fine-grain-size strengthening of metals is a strengthening mechanism that has been used for many years, long before dislocations had been visualized. In Figure 5.27 the stress–strain curves of single and polycrystalline bodies are compared. (Crystal 5 was oriented in a way that involved extensive dislocation locking effects.)

Dislocation densities increase with strain by virtue of the Frank–Read dislocation mills. Experimentally, it is found that the shear stress to maintain plastic flow increases as the square root of the dislocation density, and as previously stated, the dislocation density increases more rapidly with increasing strain in the vicinity of the grain boundary.

Strain hardening of polycrystalline metals can easily be demonstrated in a simple laboratory experiment by measuring the increase in hardness of a rolled metal strip as a function of the decrease in thickness after it is passed through a rolling mill. Such a curve for brass strip is shown in Figure 5.28. Strain hardening is a common method of strengthening metals. But strain hardening and grain size strengthening are interrelated. There is also a grain size yield

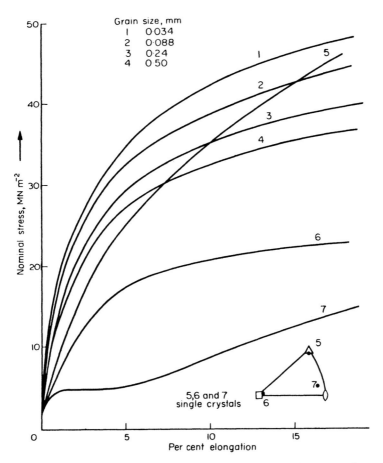

**Figure 5.27.** Effect of grain size on the stress–strain curves for pure aluminum. (From R. W. K. Honeycombe, *The Plastic Deformation of Metals,* 2nd ed., ASM International, Metals Park, Ohio, 1984, p. 238.)

strength effect independent of plastic deformation, since of course at the point of yielding, plastic deformation has just begun. The increase in yield strength, $S_y$, via a reduction in grain size has been found to follow the equation

$$S_y = S_0 + K_y d^{-0.5} \tag{5.11}$$

where $d$ is the average grain diameter and $S_0$ and $K_y$ are material constants.

It should be mentioned at this time that there is a prescribed ASTM method of measuring grain size. Let $n$ represent the grain size number and $N$ the average

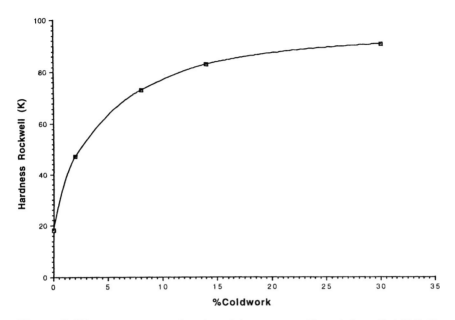

**Figure 5.28.** Hardness as a function of the percent cold work for rolled 70% Cu–30% Zn brass.

number of grains per square inch as measured on a polished and etched surface at a magnification of 100 times. These parameters are related through the equation

$$N = 2^{n-1} \qquad (5.12)$$

A grain size number of 7 is considered to be a fine-grain-size metal and one usually sought to obtain grain size strengthening effects. The ASTM has also prepared grain size charts with grain size numbers from $n = 1$ to 10, a very convenient way to estimate grain size.

### 5.6.2 Annealing of Cold-Deformed Metals

In earlier statements we have used the term *anneal* or *annealing,* which implied the state or process of softening of a metal when it is heated. Upon solidification, the metal is in a relatively soft state (i.e., with respect to cold-deformed metals). The hardness in the cast state will depend on the atom bond strength of the metal in question, the grain size (which is frequently large in the cast state), and the absence or presence of second-phase particles, and if present, their quantity, shape, size, and distribution. We will now concern ourselves with the annealing of "cold"-deformed single-phase metals. The term *cold* is relative. We will define it for the moment as that temperature where the dislocation density has increased several orders of magnitude during the deformation process. *Annealing* is the process of heating the metal to restore it to its prior soft state by essentially removing those dislocations that were introduced during the cold deformation process. Each dislocation is surrounded by a strain field as a result of their atoms being out of registry. So when we remove dislocations we have removed the internal strain that was introduced during the strain-hardening process. At the same time during plastic deformation the grain shape is altered in order to permit the body to conform to the shape dictated by the forces applied. In the case of the rolling deformation process, the more or less equiaxed grains of the annealed, hot-worked, or cast state become squashed and elongated (Figure 5.29a). The process of annealing, while actually a continuum of events with no clear boundaries, can be divided into three stages for the purpose of discussion. These stages are referred to as recovery, recrystallization, and grain growth.

(a)

**Figure 5.29.** Cold-worked and recrystallized structures of brass: (a) cold rolled 32% R.A.; (b) small grain size just at the end of the recrystallization process; (c) large grains after the grain growth stage. ×200. (Courtesy of John Harruf, Cal Poly.)

(b)

(c)

1. *Recovery.* In this first stage of annealing the elastic stresses in the metal grains are reduced by the movement of dislocations into less-stressed arrangements, and the mutual canceling of some positive and negative dislocations. With low levels of cold work this is all that occurs, there being insufficient strain energy left to power subsequent annealing stages. There is no visible change in the microstructure, and only a slight decrease in hardness and strength, but the metal becomes much more resistant to stress corrosion cracking. An anneal in this range is often referred to as a *stress relief* anneal, since internal residual stresses produced during cold working are relieved. Stress relief anneals also minimize distortion that sometimes accompanies the machining of cold worked metals.

2. *Recrystallization.* The nucleation and growth of new, essentially dislocation-free metal grains is the characteristic of the recrystallization stage of annealing. The nucleation of new grains result from random shifts in atomic position due to thermal vibrations. The "incubation period" required for the nuclei to form and grow to perceptible size, as well as the time required for their subsequent growth as new grains, is reduced by higher annealing temperatures and greater amounts of cold work. Higher temperatures provide more random shifts per unit time, while the high amounts of strain energy from cold working encourages atoms to shift into the less stressed positions of the new grains. The release of strain energy by these shifts is what powers grain growth during the recrystallization stage. These new grains grow by consuming the cold-worked grains and are easily seen in the microstructure. The hardness and strength of the metal drops off sharply as inverse functions of the percentage of metal that has recrystallized, the new grains being quite soft and weak. This stage is considered to end with the absorption of the last bit of cold-worked grain structure. At this point, the microstructure consists of relatively small, uniformly sized, equiaxed grains, and has the smallest uniform grain size of the annealing process. Figure 5.29b shows very small grains that indicate that the recrystallization has just been completed. This is a desirable microstructure since the small grains give a strengthening effect, but not nearly as much strengthening as can be obtained by the strain hardening produced during the cold-working process. The latter condition may be undesirable in some applications, due to the greater potential for corrosion plus the reduction in ductility and formability. Again we see that the application dictates the required processing method for the material.

3. *Grain growth.* Some of the grains formed during recrystallization will have lattice orientations and boundaries more favorable for growth than others, and these will continue to grow at the expense of their neighbors, often to sizes visible to the unaided eye. The energy for this growth is provided by the reduction in the grain boundary surface-to-volume ratio (i.e., in the number of relatively high stress and high energy grain boundaries). The hardness does not

drop much in this stage, the grains merely becoming larger, not significantly freer from dislocations nor softer. The recrystallization process increases exponentially with temperature simply because the diffusion process increases exponentially with increasing temperature. Grain growth is also a diffusion process and the recrystallized grains grow quite rapidly with increasing temperature. The larger grains "swallow" the smaller grains. With sufficiently long anneals one can obtain a single crystal (single grain) via the grain growth process. The large grains resulting from the grain growth stage are shown in Figure 5.29c.

In Figure 5.28 it was shown that brass was hardened from $R_k$ 18 to $R_k$ 90 as a result of a reduction of area by rolling of 32% at room temperature. We called this cold rolling earlier and defined it as that where the dislocation density increased during deformation. There is nothing wrong with this definition, but *cold working* is more commonly defined as that conducted at any temperature below the recrystallization temperature, and *hot working* as that performed above the recrystallization temperature. Cold working produces strain hardening, while hot working does not. Above the recrystallization temperature the dislocation density does not increase, because as fast as they are introduced, they are removed by the annealing process that is being carried on at the same time as that of the hot deformation.

The recrystallization temperature depends on the diffusion rate, which in turn depends on the bond strength. In general, the recrystallization temperature will be between 0.35 and $0.5T_m$. The hardness of the cold-rolled brass, as depicted in Figure 5.28, was reduced in hardness by annealing, as shown in Figure 5.30. As the hardness decreases during the recrystallization process the ductility increases. Intermediate anneals are frequently needed during the processing of an ingot to small thicknesses or small diameters in order to prevent crack formation. After a reduction of about 25 to 50% in area, depending on the metal, the cold-deformed metal is usually annealed to restore the ductility required for further cold deformation. In many cases the metal is given a rather large hot reduction in order to break up the undesirable cast structure and to minimize the amount of cold reduction required. Almost all final working to finished size is done cold, however, because of the greater sizing precision obtained. There are some exceptions to this sequence. Aluminum extrusions (e.g., window frames and shapes difficult to draw) are extruded directly to final size. Aluminum is often extruded warm (i.e., just below the recrystallization temperature). Other metals may also be extruded directly to final size where precision sizing is not required.

The three stages of the annealing process are evident in Figure 5.30. There is no precise boundary between these three stages. Furthermore, they are dependent on factors other than temperature, albeit temperature is by far the most prominent factor. Since these changes occur by a diffusion process, the

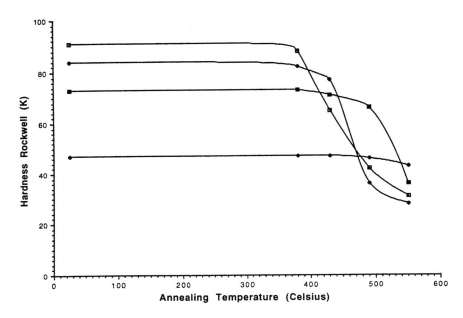

**Figure 5.30.** Reduction of hardness of cold worked brass during the annealing process.

rate varies exponentially with temperature but only linearly with time. But it is also dependent on the amount of prior cold work and the prior grain size. The recrystallization temperature of the curves of Figure 5.30 will be displaced to the left on the temperature axis as the amount of cold work increases prior to the annealing process. The higher the degree of cold work (the top curve), the greater the amount of strain energy stored, which is really the driving force for the recrystallization process, and the faster will be the diffusion process. Smaller grain size prior to the cold deformation also speeds up the recrystallization process via the higher dislocation densities in the vicinity of the grain boundaries. New grains are often nucleated during recrystallization at the points where three grain boundaries come together (triple point). During deformation these regions achieve a very high vacancy and dislocation concentration, and coupled with the extra grain boundary area available, make perfect sites for new strain-free nuclei to form.

### 5.6.3 Fracture of Metals

The subject of fracture in itself could easily comprise a complete volume. But we will not deal in any great depth with fracture mechanisms but concentrate on the appearance and prevention of fracture.

Ductile Fracture

When sufficiently plastically deformed, all ductile metals will fracture and thereby terminate the plastic deformation stage. Ductile fractures are easily identified by visual observation, as evidenced by the presence of the large amounts of plastic deformation that preceded fracture. In Figure 2.15 we saw how the fracture surface of a fractured charpy bar appeared in the low-magnification optical microscope and also at higher magnifications in the scanning electron microscope. The ductile electron microscopic fracture appearance of two tensile fractures is shown in Figure 5.31. It is sort of a tearing action, often the cracks being nucleated at grain boundary points or second-phase particles. The difference in void sizes in these two specimens can in part be attributed to a larger grain size and larger precipitate particles in the specimen fractograph shown in Figure 5.31b. In a ductile tensile test, microcracks initiate in the necked region (Figure 5.32) and sort of link up to form larger cracks or voids (Figure 5.33) that propagate to failure in a sort of sawtooth fashion. But it is these small microcracks that give rise to the dimpled appearance, called *microvoid coalescence,* that we see in the electron microscope fractographs. Ductile fracture is also preceded by extensive plastic flow. As the stress to maintain plastic flow increases, it eventually reaches the fracture stress level, indicated in Figure 5.34 by a horizontal line. This fracture stress is dependent more on atom bond strength and is really dependent on strain hardening only in the sense that the flow stress eventually approaches the bond strength. Hence we are permitted to represent the fracture stress as a constant for a given temperature and material. (Remember that the true stress is very high in the necked region of a tensile specimen.) One can better visualize this strain-hardening process by taking a soft paper clip and continually bending it by hand in a to-and-fro motion. It becomes more difficult to bend with each cycle due to strain hardening, but after about 10 cycles or so it breaks. The strain hardening has attained the fracture stress level for this material. This is not a fatigue process, because fatigue occurs well below the macroplastic yield point. Ductile fractures usually occur by "overload" (i.e., the metal has simply been stressed beyond its design limit). These fractures are often due to errors in stress analysis or improper loading, but can also be blamed on an unwise selection of materials for the intended application.

Brittle Fracture

Brittle metallic failures are of much more concern. They often occur catastrophically and without warning. Since the failure of many cargo "liberty ships" in World War II, the publications on this subject probably number in the hundreds to thousands range. In terms of appearance the fracture surface has a distinct granular appearance visually because each grain has fractured (*cleaved* is actually a better word) along a specific crystallographic plane cutting right

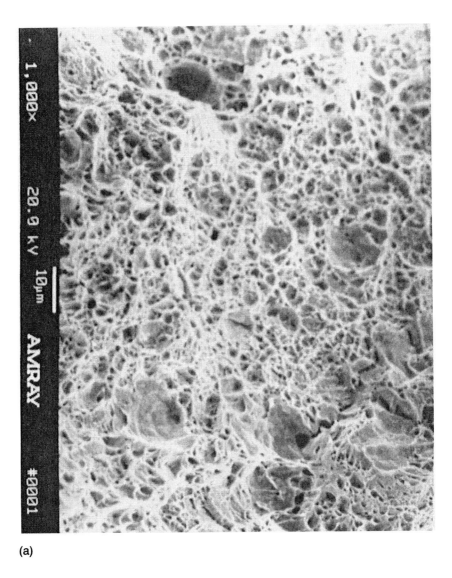

(a)

**Figure 5.31.** Ductile fractures of PH 13-8 Mo stainless tensile specimens. (From G. T. Murray et al., *Corrosion*, Vol. 40, No. 4, 1984, p. 148.)

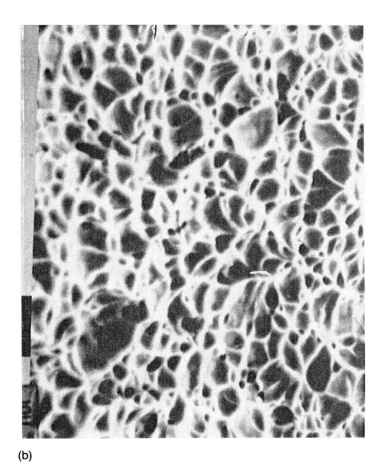

(b)

through the grain. At higher magnifications in the electron microscope, we can easily distinguish cleavage steps, where the crack has propagated along a different plane of the small parallel set of potential cleavage planes (Figure 5.35).

There is no reason to confuse ductile with brittle failure if the fracture surface is examined with the scanning electron microscope. The Charpy test (Figure 2.15) is most often used to determine at what temperature a material becomes brittle. However, the fracture mechanics approach is becoming more frequently employed, because this method yields numbers for fracture toughness that can be used in design. The B-1 bomber was the first U.S. military air-

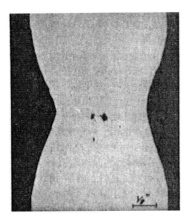

**Figure 5.32.** Voids formed in the necked region of a tensile specimen. (From R. E. Reed-Hill and R. Abbaschian, *Physical Metallurgy Principles,* 3rd ed., PWS-Kent, Boston, 1992, p. 801.)

craft employing fracture mechanics in the detail design, development, and testing of airframe structural components.

Brittle fracture is favored by high strain rates, low temperatures, and stress concentrations. The Charpy test employs all of these conditions. Metallurgical factors such as grain size, unwanted impurities and phases, and processing defects are also prominent causes of brittle fracture. Thus quality control is of utmost importance.

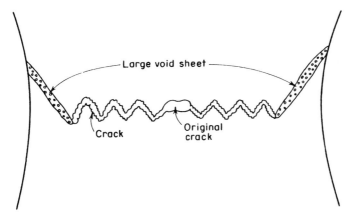

**Figure 5.33.** Void linkups to form large void sheets that show in the SEM fractograph as microvoid coalescence. (From R. E. Reed-Hill and R. Abbaschian, *Physical Metallurgy Principles,* 3rd ed., PWS-Kent, Boston, 1992, p. 803.)

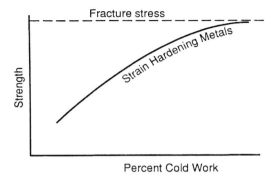

**Figure 5.34.** Strength increase during strain hardening eventually attains the fracture stress of the metal.

**Figure 5.35.** SEM fractograph of a brittle fracture in a PH stainless steel. Note cleavage steps caused by the crack propagating on different planes. (Courtesy of John Lasiter, Cal Poly.)

Brittle fractures can also be caused by certain environments, especially hydrogen. But these subjects are covered in Chapter 11.

### 5.6.4 Plastic Deformation and Fracture of Ceramic Materials

Ceramic materials are of a brittle nature. Granted that there has been some success in obtaining plastic deformation in single crystals of some ceramic materials, notably the halides and MgO (which have cubic structures), but there has been really no breakthroughs in producing ductile ceramics for commercial structural applications. In fully dense form polycrystalline MgO necks down to a point fracture at about $0.75T_m$. This is a rare exception. Even in single-crystal ceramic materials that have showed some considerable plasticity at room temperature and slightly above, it was found that in polycrystalline form at the elevated temperatures necessary for substantial dislocation movement and the associated plastic flow, grain boundary sliding became prominent and resulted in crack formation at the grain boundaries.

Considerable effort has been devoted to the development of ceramic components for high-temperature use in aircraft gas turbine engines, and even in the lower-temperature applications in automobile engines. At the time of this writing, it appears that ceramic piston rods and engine block liners could be available by the year 2000. For the aircraft industry it is believed that ceramic composite parts may be the first high-temperature materials available for this application. Because of its better fracture toughness, silicon carbide is a likely candidate material here.

There is a current flurry of activity in the category of improving the fracture toughness of ceramic materials by processes labeled as *transformation toughening, microcrack toughening,* and *deflection toughening.* Most of these ceramics are of the $ZrO_2$ or $ZrO_2-Y_2O_3$ systems. Fracture toughness values of 20 Mpa $\sqrt{m}$ have been reported.

Why are ceramics so brittle? First, dislocation motion is restricted by high bond strengths and complex crystal structures. Second, because of their low fracture toughness, once a crack forms it can readily propagate to catastrophic failure. Finally, grain boundaries contribute to their brittleness, both at room and elevated temperatures. Nevertheless, the rewards are so great that I think that it is inevitable that ceramics are going to play an important role in high-temperature structural engine components in the not too distant future.

### DEFINITIONS

*Activation energy:* energy necessary to overcome a resistance to a chemical reaction, such as the diffusion of an atom from one position of stability to another stable position.

*Annealing*:   generic term to denote the heating of a piece to cause a certain process to occur (e.g., recrystallization).

*Brittle fracture*:   cleavage type of crystalline fracture where a crack propagates with little macroscopic or microscopic plastic deformation. In environmental and amorphous material fractures, which are also brittle, a slightly different definition will be used.

*Cold deformation (working)*:   processing conducted below the recrystallization temperature of metals.

*Concentration gradient*:   slope of the concentration–distance curve. The driving force for the diffusion process.

*Critical resolved shear stress*:   stress necessary to cause planar slip (dislocation motion).

*Diffusion*:   migration of atoms through a lattice.

*Diffusion coefficient*:   proportionality constant between the diffusion flux and the concentration gradient in Fick's first law.

*Dislocation*:   linear crystalline defect consisting of a disregistry of atoms. They consist of "edge," "screw," and "mixed" components. Dislocation motion causes plastic deformation.

*Doping*:   intentional addition of certain foreign elements to semiconductors to increase their electrical conductivity.

*Driving force*:   impetus or power behind a reaction.

*Ductile fracture*:   that occurring in conjunction with substantial plastic deformation. Microvoid coalescence will be observed in SEM electron fractographs.

*Entropy*:   measure of randomness.

*Fracture toughness*:   resistance of a material to crack propagation.

*Free energy*:   minimum energy state at equilibrium. The energy available to power a reaction to attain equilibrium.

*Glassy structure*:   amorphous structure usually applied to describe glass ceramics, most polymers, and very rapidly cooled metals.

*Grain*:   individual crystal in a polycrystalline body.

*Grain boundary*:   planar defect that separates grains.

*Grain growth*:   increase in grain size that occurs during annealing at temperatures substantially above the recrystallization temperature or for longer times at the recrystallization temperature.

*Grain size number*:   ASTM method for expressing grain size.

*Host atom*:   major element in a polyatomic structure. The parent atom. In a solid solution, the solvent atom.

*Hot working*:   processing conducted above the recrystallization temperature where strain hardening is absent.

*Interstitial solid solution*:   solid solution where small atoms occupy interstitial positions between or among the solvent or host atoms.

*Interstice*:   interstitial position in a lattice.

*Planar defects*:   disregistry of atoms occurring between grains (crystals) or phases.

*Point defects*:   lattice vacancies or foreign atoms in the host lattice.

*Precipitate*:   in the solid state it is a phase, usually a compound, that forms when the solubility limit has been exceeded. In liquid solutions it is some type of solid phase.

*Primary slip system*:   slip system with the lowest value of the CRSS required for slip to occur (see Slip system).

*Recovery*:   first stage of the annealing of a cold-deformed metal where realignment and dislocation disentanglements occur accompanied by a slight hardness decrease and an improvement in corrosion resistance.

*Recrystallization*:   second stage of the annealing process of a cold-deformed metal where new grains are formed as the internal strain is removed. A substantial decrease in hardness occurs.

*Resolved shear stress*:   stress parallel to any plane resulting from any arbitrary force applied to a body.

*Secondary slip system*:   slip system that has a higher CRSS than the primary slip system.

*Slip lines*:   lines that appear on a crystal surface resulting from the slipping of one or more planes over other parallel planes. What we see are clusters of slipped planes.

*Slip system*:   combination of a slip plane and a slip direction that lies within the slip plane.

*Solubility limit*:   maximum amount of solute that can be dissolved in the solvent at any given temperature.

*Solvent*:   major constituent of a system.

*Substitutional solid solution*:   solution where the solute atom occupies a host atom site.

*Vacancy*:   vacant lattice site that is normally occupied by a host atom.

*Von Mises' principle*:   principle deduced by von Mises in 1928 which showed that five independent strain components were required to cause any arbitrary change in the shape of a body. For crystalline bodies this means that five independent slip systems are required.

## QUESTIONS AND PROBLEMS

5.1.   Why do imperfections of atomic size have little or no effect on the mechanical properties of amorphous materials?

5.2.   Distinguish among the following:
host atom; parent atom; solvent atom.

5.3.   Vacant lattice sites exist after solidification of a crystalline body. Are these vacancies accidents of growth of the solid crystal? Explain.

5.4. Explain how crystalline grain boundaries are formed during solidification of a molten metal. Sketches will be helpful.

5.5. Compute the maximum size (assume that it is spherical) of an interstitial site in an aluminum lattice. The lattice parameter of aluminum is $4.05 \times 10^{-8}$ cm. Would you expect an oxygen or a carbon atom to be able to fit into this site?

5.6. Using the data from Table 5.1, compute the diffusivity for copper in aluminum at 300°C.

5.7. What is the solubility limit for copper in aluminum at 300°C? (Fig. 5.2.) What compound is formed when this solubility limit is exceeded?

5.8. What is the crystal structure of $Fe_3C$? What is the solubility limit of carbon in iron at 727°C?

5.9. Will carbon diffuse faster in nickel than it does in copper? Explain.

5.10. Estimate the distance that carbon will significantly diffuse into iron after 100 h at 900°C? At 700°C?

5.11. Find the diffusivities of carbon in BCC iron at 200, 500, and 700°C. Plot $\ln D$ versus $1/Q$ and determine the values of the activation energy and the constant $D_0$.

5.12. Explain the difference between the edge and the screw dislocation. Does this difference have any significant effect on plastic flow in metals? Explain.

5.13. What is a typical dislocation density of an annealed metal? What would you expect would be the maximum approximate density that one could obtain in a metal? How could this density be achieved?

5.14. By what factor would the flow stress increase for a dislocation density increase of four times?

5.15. A cylindrical single-crystal bar is being pulled in a direction coinciding with the bar axis. The normal to the slip plane of this bar is 15° from the direction of the bar axis. The slip direction in this plane is 35° with respect to the bar axis. If the critical resolved shear stress of this material is 500 psi, what force must be applied to cause plastic slip to occur?

5.16. The BCC structure has 12 slip systems of the {111} <110> type. What are the secondary slip systems?

5.17. List the typical values for the CRSS values of FCC, BCC, and HCP crystals. Generally, the CRSS values of the BCC crystals are higher than those of the others. There are two reasons for this. Name at least one reason.

5.18. List two reasons why ceramic crystals are less ductile than metallic crystals.

5.19. How does the grain boundary affect the strength of metals?
   (a) At temperatures of $0.2T_m$
   (b) At temperatures of $0.7T_m$

5.20. Does the grain boundary strengthen or weaken ceramic materials?
     (a)   At temperatures of $0.2T_m$
     (b)   At temperatures of $0.8T_m$
5.21. Diffusion is (faster, slower) along grain boundaries than through the lattice proper.
5.22. Discuss and explain the mechanism of the increase in dislocation density via plastic deformation.
5.23. If 100 grains per square inch were observed at $100\times$ magnification, what would be the ASTM grain size number?
5.24. If the grain size number decreases by a factor of 4, what would be the factor of increase in the yield strength?
5.25. Sketch the curve of the decrease in hardness as a cold-worked material is annealed at increasingly higher temperatures, and describe what is occurring at the atomic level in the various stages of this curve.

## SUGGESTED READING

American Society for Metals, *Diffusion,* ASM International, Metals Park, Ohio, 1972.

Frank, J., *The Mathematics of Diffusion,* Clarendon Press, Oxford, 1975.

Honeycombe, R. W. K., *The Plastic Deformation of Metals,* 2nd ed., ASM International, Metals Park, Ohio, 1984.

Liebowitz, H., ed., *Fracture,* Academic Press, New York, 1972. See the paper, "Microstructure Aspects of Fracture in Ceramics," by R. J. Stokes, p. 157.

Reed-Hill, R. E., *Physical Metallurgy Principles,* 2nd ed., D. Van Nostrand, Princeton, N.J., 1973. See Chapter 4, "Dislocations and Slip Phenomena," and pp. 768–793 on fracture.

# 6

# Processing and Properties of Metallic Materials

## 6.1  INTRODUCTION

Despite all the hoopla one may hear about polymers, composites, and ceramics, it is much too early to predict the demise of metals. Figure 1.1 of Chapter 1 implies a leveling off of metal usage. But the increase in the population alone will probably offset the reduction in metal consumption by their replacement with some of the newer materials. We will continue to use metals in very substantial quantities for many years to come. Progress is continually being made in property improvement. The microalloyed steels provide a superb example of a case where a marked advance in the strength of structural steels has been made. In the 1930s we were using structural steels for bridges, large buildings, load-bearing automotive components, and the like of yield strengths the order of 207 MPa (30 ksi). Today, the controlled rolled microalloyed steels attain yield strengths on the order of 550 MPa (80 ksi). The development of aluminum–lithium alloys is an example of a significant increase in strength-to-weight ratio for structural aircraft components. Titanium alloys are coming into the picture, and with a 50% or so price reduction, via increased usage and processing advancements, it may yet become a leading nonferrous metal. Therefore, it is important that we continue to study metals, to the same extent that we devote to the other engineering materials. Otherwise, we may awake some day to find that the physical metallurgist has gone the way of the dinosaur, unless metals are given proper emphasis in our materials engineering curricula.

## 6.2  PHASE EQUILIBRIA

From our discussions in earlier chapters, a *phase* and the many possible atom arrangements within phases should now be a familiar concept. We have also discussed the coexistence of phases in both equilibrium and nonequilibrium systems. These examples included the solubility limit curves of sugar–coffee and of the solid solution–compound systems. The coexistence of phases in equilibrium occurred when the solubility limit was exceeded (Figures 5.2 and 5.3). The solubility limit curves are used to construct *phase diagrams* or, as they are frequently called, *equilibrium diagrams. Equilibrium* means that there is no macroscopic change with time in a phase or system of phases. In the water–water vapor enclosed system, for example, molecules continue to leave the liquid and enter the vapor phase, and vice versa. At equilibrium the forward and reverse reactions are occurring at the same rate, as indicated by the equivalent length of the arrows in the following:

$$H_2O_{liquid} \rightleftarrows H_2O_{vapor} \tag{6.1}$$

*A phase diagram is simply a map, with temperature plotted on the ordinate and the concentration of a given element on the abscissa.* At any point on this map, unless the point lies exactly on a solubility limit line (which is rare, because theoretically this line has an infinitely small width), the phase diagram tells us what phases exist for that particular temperature and for that particular overall elemental composition of the system. For example, 40% Zn–60% Cu represents a composition point on the abscissa of the Cu–Zn phase diagram, a portion of which is shown in Figure 6.1 for illustration purposes. If we proceed vertically upward along the 40% Zn composition line until we reach the 600°C horizontal temperature line, we have arrived at the point of interest on this map (marked A) for our present discussion. Now at this point on the map, the map tells us that at this temperature and alloy composition, two phases exist in this 40% Zn–60% Cu system. By convention the phase on the left is called the $\alpha$ (alpha) phase and the one on the right the $\beta$ (beta) phase. Do not take this convention too seriously, because it may sometimes lead to confusion. We could turn the diagram around and have zero percent copper on the left side. Then the solid solution of copper in zinc would be the alpha phase. Most handbooks, however, list their diagrams in alphabetical order (i.e., the Cu–Zn diagram would appear with Cu on the left). The Zn–Cu diagram, which is the same diagram but with Zn on the left, never appears in most handbooks. When we have phases that change with temperature, the lower-temperature phase is the $\alpha$ phase and the higher-temperature one is the $\beta$ phase. There are exceptions to this rule, the most common being in the C–Fe system (which, by the way, is plotted with Fe on the left, contrary to our first convention), where the low-temperature phase is called the $\alpha$ phase and the high-temperature phase, the $\gamma$ (gamma) phase. The $\beta$ phase does not exist in the Fe–C system. At one time a

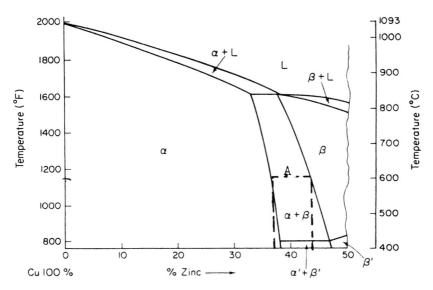

**Figure 6.1** Cu–Zn phase diagram to 50% Zn. The phase compositions at our point of interest, marked "A," is determined by the vertical lines constructed from the tie lines that intersect the $\alpha$ and $\beta$ phase boundaries.

$\beta$-crystalline form was thought to exist but was later found to be the same structure as the $\alpha$ iron, but was nonmagnetic.

But we can learn a lot more from this map (phase diagram). If we construct a horizontal line (a tie line) at the 600°C temperature, through our point of interest until we reach the phase boundary lines, we can obtain the composition of each phase (i.e., the maximum and minimum solubility limits of zinc in both the $\alpha$ and $\beta$ phases respectively). Remember, the overall system composition is 40% Zn–60% Cu, but the composition of the two phases differ from this overall composition. Now if we drop vertical lines from each point of intersection of the phase boundaries with the horizontal tie line, we can read the composition of each phase, that of the alpha being 63.1 Cu–36.9 Zn and the beta 55.9 Cu–44.1 Zn. From this horizontal line and the compositions of each phase, we can also compute the quantities of each phase within the system. We will hold this computation for the Cu–Ni system, which is covered in the following section.

So this phase diagram, equilibrium diagram, or temperature–composition map, or whatever you call it, yields a considerable amount of information. It is an indispensable source of information for the physical metallurgist. From such diagrams, which have been compiled in many handbooks (see the Suggested

Reading), the metallurgist cannot only predict the properties of alloys but can invent new alloys. The properties of multiphase alloys are controlled primarily by the quantity, size, shape, and distribution of the secondary-phase particles. The nonmaterials engineers will not be inventing alloys, but an understanding of some simple diagrams, plus the more complex Fe–C (steel diagram), will provide a better comprehension of the materials handbooks, vendor literature, and some technical articles that deal with the properties and selection of materials. For example, if you have purchased golf clubs recently, the metal club heads were probably made from 17-4 stainless steel. Why? Just be patient and we will find out why, and a lot more about the materials selection process. And just for fun we will also include the popular graphite–boron composite golf club shaft (which is already getting competition from the new titanium shafts).

The advent of the reflecting light metallurgical microscope, credited to H. C. Sorby around 1870, changed metallurgy from an art to a science. With this microscope phases could be seen, and along with the emerging methods of constituent analysis, phase equilibria could be understood and explained. Sorby suggested that the microscopic examination of railroad rails might be a worthwhile project, and in 1886 he published a paper describing a "pearl-like" constituent in the rail steel, now called *pearlite* (see Figure 4.1). Prior to this time the treatment and development of steels was done on a trial-and-error basis. The authorities were called blacksmiths. The first bound edition of the ASM *Metals Handbook* appeared in 1948 and contained about 130 alloy phase diagrams. The latest binary alloy phase diagram book (ASM International, 1991) contains almost 5000 such diagrams, including revisions of earlier publications.

Before proceeding to examine some typical alloy phase diagrams, it should be pointed out that most of our discussion will deal with binary phase diagrams, that is, those consisting of only two elements. Many of our alloys consist of two major elements plus small additions of other elements for purposes other than altering the phase quantities. Such alloys can be adequately treated with a binary phase diagram. But there are other alloys that contain three major components, and these must be studied by using the more complex ternary phase diagrams. Pressure is a third parameter in the phase equilibria determination, but most diagrams assume atmospheric pressure. Pressure is important in a few situations (e.g., in the formation of diamond as shown in Figure 6.2). Here composition is not a factor since pure carbon is utilized. We will include in our discussions the following types of phase diagrams: (1) complete solid solubility alloys: Cu–Ni, (2) eutectic-type alloys: Ag–Cu and Au–Ni, (3) precipitation hardenable alloys: Al–Cu, (4) a system with many intermediate phases, (5) eutectoid type: iron–carbon, and (6) a ternary system: Fe–Cr–Ni.

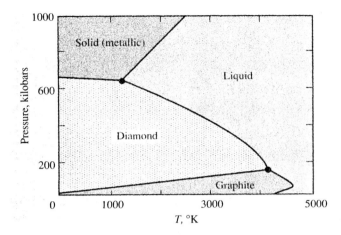

**Figure 6.2** Pressure-temperature phase diagram for carbon. (From C. G. Suits, *Am. Sci.*, Vol. 52, 1964, p. 395.)

## 6.3  SOME COMMON PHASE DIAGRAMS AND ALLOYS

### 6.3.1  Complete Solid Solubility Systems

Cu–Ni System

There are not many common alloy systems with complete solid solubility at all temperatures and compositions in the solid state. In such systems the elements must be of similar size and possess in the pure form the same crystal structure. The Cu–Ni system is the classic example and this phase diagram is depicted in Figure 6.3. There is only one solid solution phase in the solid state and it is of the substitutional type. You can call it a solid solution of nickel in copper (usually done up to 50% nickel) or a solid solution of copper in nickel, when nickel is the major element. Both copper and nickel have FCC structures, so an atom of one element is replaced by one of the other element. Note that 100% copper represents the zero percent nickel composition point on the left side of the diagram and 100% nickel–0% copper is placed on the right side according to our alphabetical way of arranging elements at the terminus of a phase diagram. In this diagram the composition is plotted in weight percent nickel. In some cases it may be of interest to know the composition in atomic percent (many handbooks show both). A 50% Cu–50% Ni weight percent alloy means that if the total quantity was 100 g, the alloy would contain 50 g of Cu and 50 g of Ni. A 50% Cu–50% Ni atomic percent alloy contains equal numbers of Cu and Ni atoms. A mole of Cu atoms weighs 63.54 g compared to 58.71 g per mole of nickel atoms. Thus the atomic percentages and weight percentages are similar.

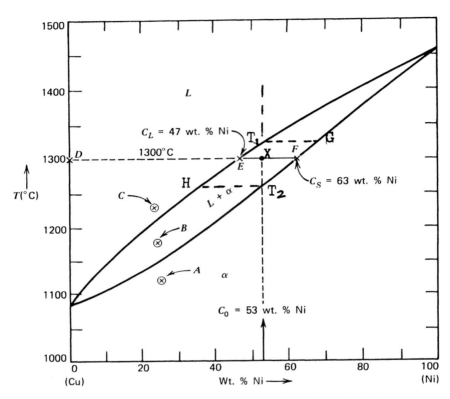

**Figure 6.3** Cu–Ni phase diagram. (Adapted from K. M. Ralls, T. H. Courtney, and J. Wulff, *Introduction to Materials Science and Engineering,* Wiley, New York, 1976, p. 317.)

A 50–50 wt % alloy converts to an approximate 52% Cu–48% Ni on an atom percent basis.

For comparison purposes let's choose a point on our Cu–Ni map at 1300°C and 53 wt % Ni, which we will label by the letter X. The diagram tells us that at this point we are in a two-phase region consisting of a liquid solution and a solid solution. The horizontal 1300°C tie line that intersects the liquid-phase boundary, called the *liquidus,* at point E tells us that the liquid phase has a 53 Cu–47 Ni wt % composition. Similarly, the intersection at the *solidus* phase boundary at point F yields a 37 Cu–63 Ni wt % solid solution phase. Only a liquid solution exists above the liquidus and only a solid solution is present below the solidus line. In both the liquid and solid solution phases the composition is uniform throughout.

*Example 6.1.* A 100-g lot consisting of 47 g of Cu and 53 g of nickel was melted and cooled to 1300°C. How many grams of copper are there in (a) the liquid phase and (b) the solid phase?

Solution. (a) First we must determine the quantity of each phase. This is done by what is often called the inverse lever rule. Let the total length of the tie line represent the total quantity of material. Then the ratio of the segment of line from our point of interest *A* to the solidus intersection point *F* to the total tie-line length, gives us the fraction of the liquid phase, that is,

$$\frac{F - X}{F - E} = \text{liquid fraction of total}$$

Similarly, the ratio

$$\frac{X - E}{F - E} = \text{solid fraction of total}$$

We could measure these line lengths very accurately with a vernier caliper, but since we usually do not have this gauge readily at hand, we can accomplish the same result by reading off the compositions at these three points and let them represent our line segment lengths, thus

$$\frac{F - X}{F - E} = \frac{63 - 53}{63 - 47} = \frac{10}{16} = 0.625 \text{ liquid} = 62.5 \text{ g liquid}$$

Note that the liquid phase contains 53 wt % Cu. Thus $0.53 \times 62.5$ g of liquid = 33.125 g of Cu in the liquid phase.

(b) Now we will do the same for the solid phase. The ratio of the length of the line segment from our point of interest *A* to the liquidus intersection at point *E* to the total tie line length gives us the fraction of solid.

$$\frac{X - E}{F - E} = \frac{53 - 47}{63 - 47} = \frac{6}{16} = 0.375 \text{ solid} = 37.5 \text{ g solid}$$

The solid phase contains 37 wt % Cu; thus $0.37 \times 37.5$ g of solid = 13.875 g of Cu in the solid phase. *Note*: The grams of liquid and grams of solid = 100 g. The grams of Cu in liquid and grams of Cu in solid = 47 g, which is the same quantity of Cu that we added to the 53 g of Ni to melt our alloy. We neither create nor destroy material by phase diagrams (or by any other trick). There are three other points on the diagram, labeled *A*, *B*, and *C*. If the temperature and composition correspond to point *A*, the alloy is a single-phase solid solution; at *B*, a two-phase mixture of solid and liquid solutions; and at point *C*, a single-phase liquid solution.

There are many ways of determining phase diagrams that are beyond the scope of this book. One method is by the use of *cooling curves*. We will illustrate these curves for the Cu–Ni system, not for the purpose of becoming experts in phase diagram determination, but to better understand the solidified microstructure, which in turn affects the properties of the alloy.

To obtain equilibrium, theoretically we must cool at an infinitely or impractically slow rate; that is, our system must be at equilibrium at each point on the temperature–time cooling rate curve. In liquids the diffusion rate is sufficiently fast that equilibrium can be attained or at least approached in a few minutes. In solids it requires a much longer time and usually, cooling curves are not used to detect solid/solid phase transformations. Nevertheless, for purposes of illustration we will assume that the cooling curves in Figure 6.4 were obtained under equilibrium conditions. In this figure our cooling curves are depicted for 100% Cu, 47% Cu–53% Ni (by weight), and 100% Ni. Note that the curves have a horizontal constant-temperature section at the melting point of the pure metals. This results from the latent heat of fusion (i.e., the heat required to melt the material, which is also equal to that released during freezing of the metal) being released as solidification takes place and which exactly balances the rate of heat removal. Under equilibrium conditions this heat release is sufficient to maintain a constant temperature of the liquid–solid mixture until solidification is complete. There are two phases, liquid and solid, coexisting during solidification. Now let's consider the alloy cooling curve and at the same time refer to the phase diagram of Figure 6.3. Solidification begins at

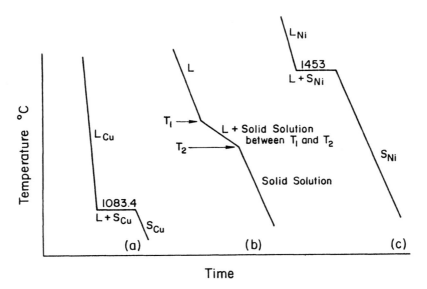

**Figure 6.4**  Cooling curves for (a) 100% Cu; (b) Cu 47%–Ni 53% alloy; (c) 100% Ni.

point $T_1$ on the phase diagram and is completed at point $T_2$. But there is no horizontal segment on the cooling curve. Some heat of fusion is being released, which changes the slope of the cooling-rate curve, as the temperature is reduced through the two-phase liquid–solid region of Figure 6.3, but this heat of fusion is not sufficient to maintain a constant temperature. At any point along the vertical line representing the composition Cu 47–Ni 53 we can con-struct a horizontal tie line and determine the quantities and compositions of the two phases just as we did in Example 6.1. The compositions change during cooling, that of the liquid phase following the liquidus curve and that of the solid phase following the solidus curve. By connecting the solidification points of the pure metals with point $E$ we can construct the liquidus line, and with point $F$, the solidus line. The phase diagram has now been completed with only three cooling curves. For greater accuracy, we should use more alloy composi-tions.

Now from a practical point of view we seldom cool under equilibrium or even near-equilibrium conditions. The commercial Cu–Ni alloys (called Monels, an International Nickel trademark) are melted and poured into a mold to produce an ingot that is subsequently fabricated to certain desired forms via the usual rolling or drawing processes. When an ingot is cast, the composition will not be uniform throughout the ingot. Referring again to the 47 wt % Cu–53 wt % Ni alloy of Figure 6.3, when an ingot of this molten alloy is cast, the part to solidify first, that at the mold walls, will have more nickel and less copper than the overall composition (see point $G$). This part of the ingot will be nickel rich. As solidification proceeds toward the ingot center the alloy com-position changes until at the ingot center the last to solidify (point $H$) will be copper rich compared to the overall starting composition of 47% Cu–53% Ni.

Usually, columnar grains grow out from the wall of a mold toward the ingot center. In Figure 4.4 we showed such a grain structure resulting from the pour-ing of a 70% Cu–30% Zn alloy into a steel mold. These grains form by nuclea-tion and growth. The material near the grain boundaries, the last to solidify, will be copper rich. Thus we have segregation throughout a grain as well as an ingot. In many metals *dendrites* are formed during solidification. During the period of columnar growth the solid–liquid interface is not planar, but develops protuberances along the grain boundaries because of compositional changes along the growing front. These shapes are called *dendrites*. A typical dendritic structure in cast brass is shown in Figure 6.5. These segregated cast structures are undesirable. The composition gradients can be smoothed out by heating the ingot at high temperatures for long periods of time (from Friday to Monday has been a convenient time in many plants). This treatment is called a *homo-genization anneal* by the metallurgist and a *soaking operation* by the plant fore-man. Hot working serves to break up the columnar grains and dendritic struc-tures, which otherwise would cause anisotropic properties in the ingot. But

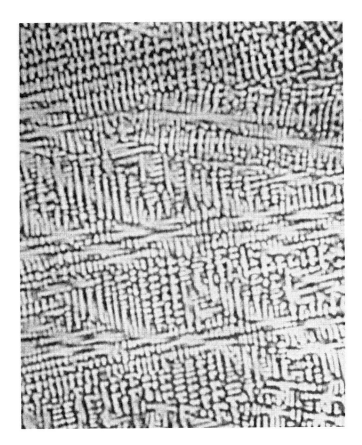

**Figure 6.5**   Dendritic solidification structure in Cu 70%–Zn 30% brass. (Courtesy of J. Haruff, Cal-Poly.)

homogenization prior to hot working is necessary to reduce the alloy segregation to an acceptable level.

Most Monels contain small amounts of other elements. K-Monel contains about 3.0% Al for precipitation hardening effects. Monels to be used in as-cast shapes contain small amounts of silicon to promote fluidity, which helps fill the intricate molds designed for final cast shapes. Monels have good corrosion resistance and strength properties (Table 6.1). They are used in chemical industries, marine applications, laundry, and pharmaceutical industries.

Other alloy systems that show complete solid solubility include Ge–Si, Ir–Pt, Mo–W, Mo–V, Ni–Rh, Se–Te, and V–W. Two elements that are com-

**Table 6.1** Tensile Strength, Yield Strength, and Ductility of Selected Nonferrous Alloys

| Alloy | Wt % | Tensile Strength MPa (ksi) | Yield Strength MPa (ksi) | Ductility Elongation | Conditioning[a] | Applications |
|---|---|---|---|---|---|---|
| 400 Monel | 64 Cu, 31 Ni, 2.5 Fe | 550 (80) | 240 (35) | 40 | A | Marine engineering, chemical and hydrocarbon processing equipment, pumps, heat exchangers, valves |
| K-500 Monel | 64 Cu, 31 Ni, 2.5 Al, 2.0 Fe, 1.5 Mn | 1100 (160) | 790 (115) | 20 | PH | Pump shafts, marine propeller, shafts, oil well tools |
| 1050 Aluminum | 99.5 Al min. | 76 (11) | 28 (4) | 39 | A | Chemical and brewing industries, food containers |
|  |  | 159 (23) | 145 (21) | 7 | 75% CW | Chemical and brewing industries, food containers |
| 2024 Aluminum | 93 Al, 4.4 Cu, 2.5 Mg | 476 (69) | 393 (57) | 10 | PH T6 | Aircraft structures, hardware, truck wheels |
| 7075 Aluminum | 90 Al, 5.5 Zn, 1.5 Cu, 2.5 Mg | 531 (77) | 462 (67) | 7 | PH T6 | Aircraft, forged structural parts |
| C10100 Copper | 99.95Cu | 220 (32) | 69 (10) | 45 | A | General stamped and drawn products, busbars, lead-in wires, cables, tubes |
|  |  | 380 (55) | 345 (50) | 4 | CW 60% | General stamped and drawn products, busbars, lead-in wires cables, tubes |
| C26000 | 70 Cu, 30 Zn | 300 (44) | 75 (11) | 68 | A | Radiator cores, socket brass shells, cartridge cases, springs, plumbing fittings, stampings |
|  | 70 Cu, 30 Zn | 700 (100) | 450 (65) | 5 | CW 50% | Radiator cores, socket brass shells, cartridge cases, springs, plumbing fittings |
| C17200 Cu-Be | 97 Cu, 2.0 Be | 1240 (180) | 1070 (155) | 7 | PH | Flexible metal hose, springs, bellows, valves |
| C61000 Al bronze | 92 Cu, 8 Al | 480 (70) | 205 (30) | 65 | A | Bolts, shafts, tie rods, pump parts |
| AZ80A Magnesium | 90 Mg, 8.5 Al, 0.5 Zn | 380 (550) | — | 7 | PH T5 | Extruded and forged products, aircraft, automotive, electrical |
| Ti-6Al-4V | 90 Ti, 6 Al, 4 V | 1178 (156) | 977 (142) | 16 | Aged | Jet engine fan disks, heat exchangers, nuts, bolts, aircraft structural parts |
| Ti 1023 | 85 Ti, 10 V, 2 Fe, 3 Al | 1242 (180) | 1173 (180) | 10 | Aged | High-strength air frame components, high toughness |

[a] PH, precipitation hardened; CW, cold worked; A, annealed.

pletely soluble in each other in the solid state are very similar in chemical behavior, atom size, and, of course, must be of the same crystal structure.

### 6.3.2  Eutectic Systems

The term *eutectic* is derived from the Greek word *eutektos,* which translates to *easily melted.* For simplicity we divide the eutectic diagrams into the *simple eutectic* systems and those with some solid solubility. The eutectic composition is the lowest melting point of all the possible compositions of alloys in these binary systems. The reason that this composition is the lowest melting point of all compositions can be deduced from the schematic of Figure 6.6. The solid curve on the left represents the lowering of the melting point of metal A by the addition of metal B, and vice versa for A added to B. These curves intersect at a point called the *eutectic point.* The continuation of the two curves, represented by the dashed lines, no longer represent melting points. At all temperatures below the point of intersection the system is in the solid state. Thus the *eutectic point* is the lowest melting point in this alloy system, and since it is a point on our phase diagram map it is also called the *eutectic temperature* and the *eutectic composition,* all being the same point. Since everything below this temperature is a mixture of two solid phases, we can just construct a horizontal line from 100% A to 100% B passing through the eutectic point, and presto, the phase diagram is complete.

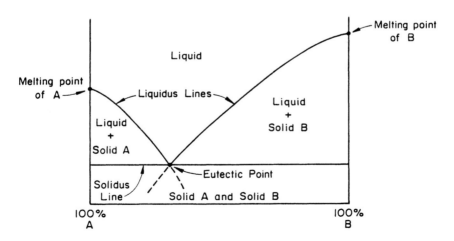

**Figure 6.6**  Schematic of an eutectic phase diagram.

Simple Eutectic Systems

The Au–Si system (Figure 6.7) is one of the very few simple eutectic systems (i.e., alloys with no solid solubility in each other). The solid solubility of Si in Au is virtually zero, and vice versa. The Au–Si system is also of considerable practical importance in electronic devices because layers of silicon and gold come into direct contact with each other, and at temperatures above the eutectic temperature (363 °C) a liquid phase will form.

The melting point of most metals will be reduced by the addition of a second metallic element. Referring to Figure 6.7, we see that the melting point of gold is reduced from 1064 °C to 363 °C by the addition of 19 at % of silicon. Similarly, the melting point of silicon is reduced from 1412 °C to 363 °C by the addition of 81 at % gold.

Cooling curves for the eutectic composition of the Au–Ni alloy and also for the pure metals are constructed along with their corresponding schematic microstructures in Figure 6.8. The cooling curve for the eutectic composition has a horizontal section at the eutectic temperature. Just as for pure metals, the eutectic composition solidifies and melts at a constant temperature. Only pure metals and alloys of eutectic composition have this particular feature. The heat released during solidification is sufficient to maintain a constant temperature

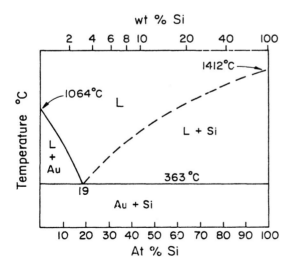

**Figure 6.7** Au–Si phase diagram (eutectic). (Adapted from R. P. Anantatmula et al., *J. Electron. Mater.*, Vol. 4, No. 3, 1975, p. 445.)

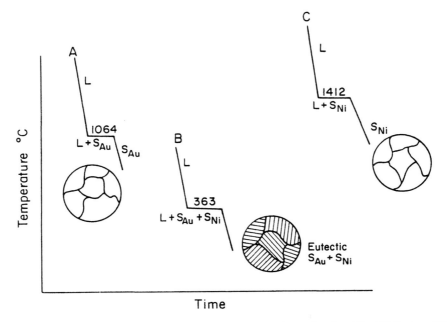

**Figure 6.8**  Cooling curves and microstructure schematics for (A) 100% Au; (B) eutectic composition; (C) 100% Si.

until solidification has been completed. But unlike the pure metals, where during solidification the pure solid and pure liquid are in equilibrium during solidification, for the eutectic composition we have two solid phases and one liquid phase (in the present case solid Au, solid Si, and a single-phase liquid solution of the two elements), all in equilibrium at the eutectic temperature. This condition can be expressed in the form of the *eutectic reaction* chemical equation as follows:

$$\text{liquid solution of A and B} \rightleftarrows \text{solid A } + \text{ solid B} \tag{6.2}$$

and in the present case of the Au–Si alloy, the reaction during solidification is

$$\text{liquid solution of Au and Si} \longrightarrow \text{solid Au } + \text{ solid Si}$$

Thus we have three phases existing together during solidification at the eutectic point. If we held the system at this temperature, there would be a reverse reaction arrow, just as we had in equation (6.2). The eutectic reaction is called an *invariant reaction* since it occurs under equilibrium conditions at a fixed temperature and alloy composition that cannot be varied.

When a eutectic composition solidifies at a relatively slow cooling rate the solid phases are arranged in a parallel platelet form as depicted in Figure 6.8B, and is known as the *eutectic microstructure.* In the Au–Si system, if a composition to the left of the eutectic composition (e.g., one of 10 at % Si) is slowly cooled from the molten (all liquid) state, pure Au will precipitate out of the liquid when the temperature drops just below the liquidus line (about 760°C). As we continue to cool, more solid Au forms and thus the liquid will contain a higher percent of Si. By constructing tie lines at successively lower temperatures, it becomes obvious that the liquid composition follows the liquidus line until it reaches the eutectic temperature, where all of the remaining liquid solidifies, just as if the system contained only liquid of eutectic composition. In fact, we could strain out all of the solid gold at the eutectic temperature and then the remaining liquid would solidify with the eutectic microstructure depicted in Figure 6.8B. But since we do not strain out the solid gold that exists just above the eutectic temperature, this solid is present and the microstructure is like that depicted in Figure 6.9, along with the corresponding cooling curve. Note that the horizontal section on the cooling curve for the 10 at % Si compo-

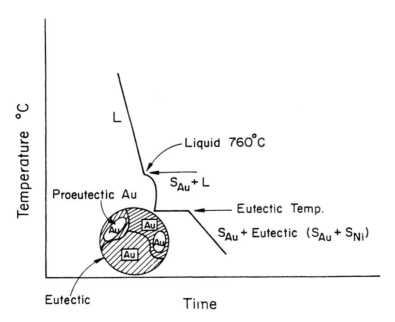

**Figure 6.9** Cooling curve and microstructure schematic of a Au–Si alloy composition which on solidification results in a combination of a proeutectic Au phase plus the Au–Si eutectic phase arrangement microstructure.

sition represents the solidification of the eutectic liquid composition—and only that composition. The nonhorizontal part of the cooling curve is that due to the precipitation of solid gold as we cool through the liquid–solid two-phase region. Again, some heat is released as the alloy is cooled through the liquid–solid region, but it is insufficient to maintain a constant temperature. The gold that precipitates out in the solid form prior to reaching the eutectic temperature is called the *proeutectic* solid (i.e., it forms prior to reaching the eutectic point).

*Example 6.2.* Convert the atomic percent of the eutectic composition to weight percent (19 at % Si–81 at % Au).

Solution. For Si we have 28.1 g/mol/6.02 × $10^{23}$ = 4.67 × $10^{-23}$ g/Si atom. For Au we have 197g/mol/6.02 × $10^{23}$ = 32.724 × $10^{-23}$ g/Au atom. Assume 100 atoms of eutectic composition.

$$19 \text{ atoms of Si} = 19 \times 4.67 \times 10^{-23} \text{ g/Si atom} = 88.73 \times 10^{-23} \text{ g of Si}$$

$$81 \text{ atoms of Au} = 81 \times 32.724 \times 10^{-23} \text{ g/Au atom} = 2650.6 \times 10^{-23} \text{ g of Au}$$

$$\text{Wt } \% \text{ Si} = \frac{88.73}{88.734 + 2650.6} \times 100 = 3.2$$

$$\text{Wt } \% \text{ Au} = \text{balance} = 96.8$$

*Example 6.3.* For a 60 at % Si–40 at % Au alloy at a temperature of 800°C, compute the atomic percent of solid Si in equilibrium with the liquid.

Solution. Construct a tie line from the liquidus phase boundary (40 at % Si) to the solid 100% Si phase boundary at the 800°C temperature.

$$\text{Total tie line length} = 100 - 40 = 60 \text{ units}$$

Since this temperature is closer to the liquidus line than the solidus (the horizontal line at 363°C), it suggests that the bulk of the material is still in liquid form. Thus

$$\frac{100 - 60}{100 - 40} \times 100 = 66.7 \text{ at } \% \text{ at liquid}$$

or

$$\frac{60 - 40}{100 - 40} \times 100 = 33.3 \text{ at } \% \text{ Si in solid form}$$

The inverse lever rule is sometimes confusing because the length of the segment of the tie line that represents the solid is the segment that intersects the liquidus line. A more logical reasoning, and an easy way to recall which segment goes with which phase, is to follow the vertical composition line downward. Shortly after it intersects the liquidus line there has not been much time for solid to form. Thus the shortest segment must represent the quantity of solid. After cooling further we approach the solidus line. Now there has been a lot of time expired and a lot of solid formed. Therefore, the shortest segment must represent the quantity of liquid remaining.

Eutectic Systems with Some Terminal Solid Solubility

To understand these phase diagrams we must first define or explain the phrase *terminal solid solubility*. The word *terminal* refers to the endpoints on the abscissa (composition axis) of the phase diagram. In Figure 5.2 the solid solubility limit of copper in aluminum was depicted showing the increase in solubility with temperature. To show all of the terminal solid solubility, this solubility line must be extended to the eutectic temperature and then back up along the solidus line until it reaches the melting point of the pure metal of interest. The Ag–Cu phase diagram is a good example of this type of eutectic system (Figure 6.10). In this diagram the liquidus line runs from the melting point of each metal to the eutectic point (779°C and 28.5 wt % Cu). But now the solidus, indicated as a heavy line, shows that the solid solubility of Cu in Ag decreases with increasing temperature at temperatures above the eutectic temperature. Thus the $\alpha$ region on the left end of the diagram consists of a solid

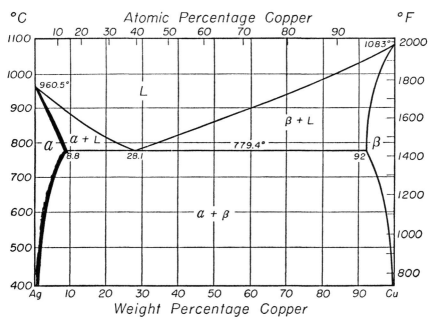

**Figure 6.10** Ag–Cu phase diagram. (From *Metals Handbook*, ASM, Metals Park, Ohio, 1948, p. 1148; by Cyril Stanley Smith.)

solution of Cu in Ag, and the $\beta$ region on the right end is a solid solution of Ag in Cu. A cooling curve of the eutectic composition (28.1 wt % Cu) would appear just as that in the simple eutectic case, and the eutectic microstructure would be the same except that the layers now would consist of alternate layers of $\alpha$ and $\beta$. But let's construct a cooling curve (Figure 6.11) for a 92.5 wt % Ag-7.5 wt % Cu. First the $\alpha$ solid solution would begin to precipitate out of the liquid solution phase as it crossed the liquidus at about 900°C. As we continue to cool through the $\alpha$ + liquid region, the liquid becomes richer in Cu. But we will never have an eutectic structure in this alloy because we pass into the all-solid $\alpha$ range prior to attaining the eutectic temperature. When we cross the $\alpha - \alpha + \beta$ line on the phase diagram, some $\beta$ will precipitate out of the $\alpha$ phase on very slow cooling. This change will never show on a typical cooling curve. They are not suitable for detecting solid–solid transformations. This line can be determined by careful examination of the microstructure for the appearance of the second phase. The microstructure would appear as that in a pure metal or a single-phase solid solution (i.e., it would show only grain boundaries) plus some $\beta$. This alloy composition is called *sterling silver* and is used for *silverware* and jewelry. Ag alone is too soft for even these relatively mildly stressed applications.

On the other hand, if we cool a 80 wt % Ag-20 wt % Cu alloy from the liquid state, we do cross the eutectic temperature. The cooling and microstructure will be much like that of Figure 6.9 except that a solid solution of proeutectic $\alpha$ silver rather than pure Au is present, and the eutectic structure will consist of alternate layers of the $\alpha$ and $\beta$ solid solution phases rather than pure metals.

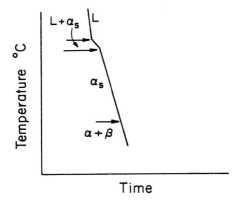

**Figure 6.11** Cooling curve for a Ag 92.5%–Cu 7.5% alloy (sterling silver). Note that there is no break in the curve at the solid $\alpha - \alpha + \beta$ phase boundary.

Other alloy systems of this type include Pb–Sn, from which many commercial solders are derived: Bi–Sn, Pb–Sb, and Cd–Zn. The Al–Cu system has an eutectic point at 548°C and at 66.8 wt % Al (Figure 6.14). The microstructure of a 70% Al–30% Cu by weight showing some proeutectic $\alpha$ in a large amount of the layered eutectic structure is shown in Figure 6.12.

### 6.3.3 Some Systems with Intermediate Phases

Most alloy systems are not as simple as those portrayed in the preceding section. A large number of phase diagrams show intermetallic compounds or

**Figure 6.12** Eutectic microstructure of an Al 70%–Si 30% by weight alloy containing some proeutectic $\alpha$ aluminum × 400. (Courtesy of J. Haruff; Cal-Poly.)

intermediate phases, which are usually positioned more or less in the central portion of the diagram. The intermetallic compound is one of fixed stoichiometry, such as $CuAl_2$ or $Fe_3C$, and will show on the phase diagram as a straight vertical line. An intermetallic compound could be considered as a type of intermediate phase, but this term is more often thought of as applying to a compound that has a range of compositions, usually fairly narrow, that is near stoichiometry (e.g., from $A_{0.9}B_{1.1}$ to $A_{1.1}B_{0.9}$). It could also be considered as a solid solution phase of a certain crystal structure that varies in percent of each atom over the range in which the phase exists. The $\beta$ phase in the Cu–Zn system of Figure 6.1 is an intermediate phase. The compound Cu–Zn falls within the $\beta$ range, but so do compositions near 50 at % Cu–50 at % Zn. This phase has the BCC structure, but many intermediate phases have complex structures. They are usually brittle and presently not of great commercial significance.

Another type of diagram is that where eutectics are formed between intermetallic compounds instead of elements. The Cr–Si phase diagram (Figure 6.13) is such an example. Four eutectics and five intermetallic compounds are found in this alloy system. This diagram is of some interest to electronic device manufacturers. Chromium is frequently deposited on silicon wafers because it acts as an adhesive layer. For example, gold does not bond well to silicon. But if a gold layer is desired, often for corrosion protection or high conductivity, chromium is first deposited, which adheres well with silicon, and then a layer of gold deposited, which adheres to the chromium. But if the device happens to experience elevated temperatures, a brittle intermetallic phase could form.

## 6.4  PRECIPITATION-HARDENING SYSTEMS: ALUMINUM ALLOYS

These are the most important group of alloys in the nonferrous category. They could be classified under the preceding section on intermediate phases, because the precipitates are usually intermetallic compounds. But because of their prominence and distinct processing methods, they warrant separate treatment. They have gained their prominence simply because the strength of the base metal, aluminum for example, can be increased manyfold by small additions of a second element that cause a compound to be formed when the solid solubility limit for this element has been exceeded. Aluminum of commercial purity in the annealed state has a tensile strength of about 28 MPa (10 ksi), and in the cold-worked state on the order of 56 MPa (20 ksi), whereas in precipitation-hardened aluminum alloys, tensile strengths on the order of 483 MPa (70 ksi) are common (see Table 6.1). This phenomenon was first recognized in aluminum alloys as early as World War I, when attention was focused on Duralium, an Alcoa heat-treatable alloy. The advent of the aircraft industry propelled aluminum to the number 1 position in the nonferrous alloy category, since

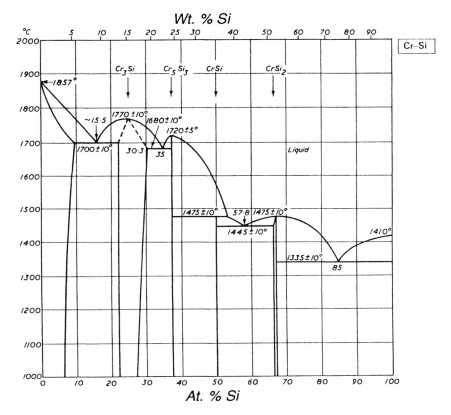

**Figure 6.13** Cr–Si phase diagram. (From E. A. Brandes, ed., *Smithells' Metals Reference Book*, Butterworth, Stoneham, Mass., 1983, p. 11–213; as taken from M. Hansen and K. Anderko.)

strong, lightweight materials were needed to replace parts made of wood and fabrics used in the first aircraft that were developed (see Chapter 1). The Al–Cu alloys were the first precipitation-hardened alloys developed, have been the most studied, and are the best understood of the precipitation-hardening alloys. We will therefore use this classic alloy system to explain the phenomenon of *precipitation hardening,* or as it was known in the 1930s and 1940s, as *age hardening.*

We described previously the solid solubility limit curve (Figure 5.2), so there is no need to repeat it here. Any phase diagram in which the solid solubility of a particular solute in a particular solvent increases with temperature is a potential precipitation-hardening alloy. But it requires more than just the pre-

cipitate formation to achieve significant hardening. The hardening comes about
in the way the precipitate forms, and in fact maximum hardening is achieved
prior to the completion of the formation and appearance of the precipitate compound. Most precipitates are intermetallic compounds, such as $CuAl_2$ in the
Al–Cu system. The complete phase diagram of the Al–Cu system can be found
in the phase diagram books listed in the Suggested Reading. We will be
interested only in the aluminum-rich end of the phase diagram, and accordingly
show that part of the diagram ranging from 100 wt % Al to about 72 wt % Cu
(28 wt % Al) in Figure 6.14. This diagram has a eutectic point at 66.8 wt % Al
and at 548°C, but this point really has no bearing on our study of precipitation
hardening in this system. The two points in the diagram that are of very much
significance is that of the maximum solid solubility of about 5.7 wt % Cu in
Al, which occurs at the eutectic temperature, and the point of appearance of the
$\theta$ phase, which has the composition of $CuAl_2$ and appears on the phase diagram
at about 54 wt % Cu–46 wt % Al. Note that in this compound there are two
atoms of aluminum for every three total atoms; thus its composition in atomic
percent is 66.7 at % Al. If we examine the diagram closely, we notice that the
$\theta$ phase is not represented by a straight vertical line on the diagram. Therefore,
it is an intermediate phase. But the region of the $\theta$ phase is very narrow in
composition and hence we will consider it as being the $CuAl_2$ stoichiometric
compound.

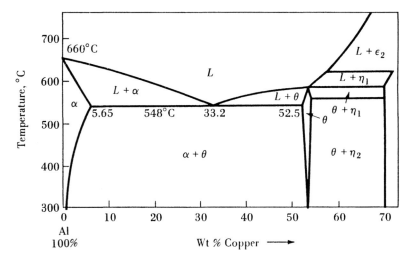

**Figure 6.14**   Al–Cu phase diagram to 72% Cu. (After K. R. Van Horn, Ed., *Aluminum,* Vol. 1, ASM International, Metals Park, Ohio, 1967, p. 372.)

One of the earliest developed Al–Cu alloys, and still probably the most widely used, is the 2024 alloy, which contains 4.4 wt % Cu, along with small amounts of Mg and Mn to lower the *solution treating temperature.* This temperature, 493°C, is used to form an all-$\alpha$ solution. When this alloy is quenched from the all $\alpha$ region, the cooling time is too short to allow the formation of the equilibrium precipitate phase $CuAl_2$. Thus the $\alpha$ phase containing 4.4 wt % Cu exists at room temperature. The equilibrium diagram says that such a phase cannot exist at such a low temperature. But the equilibrium diagram deals with equilibrium conditions. Our quenched alloy is thus a supersaturated solid solution of Cu in Al. This is a nonequilibrium metastable condition. The solid solubility of Cu in Al at room temperature is nil. The Cu wants to precipitate out in the form of $CuAl_2$ as dictated by the equilibrium phase diagram. The manner in which this happens and how the properties are affected comprises the entire subject of precipitation hardening.

The aging process begins with the formation of clusters of Cu atoms in the aluminum *matrix*. The term *matrix* as used here means the major part and the continuous phase of the system. In the solid solution state, be it an equilibrium or supersaturated solid solution, the copper solute atoms are distributed more or less randomly on the lattice positions normally occupied by aluminum atoms. But in the supersaturated state the compound $CuAl_2$, which contains 33.3 at % Cu atoms, wants to form in order to achieve equilibrium. 4.4 wt % Cu translates to about 2 at % Cu (rounded off for simplicity). This means that for every 100 atom sites in the supersaturated state, two will be occupied by Cu atoms. But to form $CuAl_2$ we need 33 sites per 100 to be occupied by Cu atoms. Thus the first stage of the aging process begins with the diffusion of Cu atoms to form regions containing approximately one-third Cu atoms. Since a diffusion process is involved, precipitation is time and temperature dependent. But in the Al–Cu system sufficient diffusion can take place at room temperature to obtain significant clustering of atoms in a period of about 5 h. This is enough to cause an increase in hardness and strength. The clusters provide a greater barrier to the passage of dislocations through the matrix than do the same number of atoms distributed randomly, the latter being the normal solid solution hardening discussed earlier. The supersaturated state is stronger than that of pure aluminum, but the cluster state is stronger yet. Remember that anything that restricts dislocation motion strengthens the material. It has been found, by use of x-ray diffraction and the electron microscope, that the Cu atoms tend to cluster on the {100} planes of the FCC aluminum lattice, such that there exists in small regions of about 8 to 10 nm diameter and about 0.6 nm thickness an arrangement of atoms in which every third {100} plane is occupied by Cu atoms. These are frequently called *GP1* zones, named after Guinier and Preston, who pioneered the concepts of the aging process in 1938.

The second stage of aging, often called the *GP2* stage, involves the formation of an intermediate lattice ($\theta''$ phase) which has a tetragonal structure. The {100} planes of the $\theta''$ phase is coherent (attached) to the {100} planes of the matrix. Thus it is still not a true $CuAl_2$ compound. But the coherency strains between the matrix and the $\theta''$ atom configuration provide a tremendous resistance to dislocation motion. The coherency strains are related to the mismatch in atom spacing between the matrix and $\theta''$. This stage coincides with maximum strengthening due to the large coherency strains. The $\theta''$ phase particles are larger than the GP1 zones by about a factor of 10 in both thickness and diameter. A sketch of the distorted lattice surrounding and including the $\theta''$ particle is depicted in Figure 6.15.

The next stage involves the formation of a $\theta'$ phase which is incoherent with the matrix; that is, it has broken away from the matrix and has a distinct tetragonal structure. The strength and hardness begin to decrease with the loss of the coherency strains.

The final stage involves the formation of the equilibrium $CuAl_2(\theta)$ phase, which is also tetragonal but of different atom spacing than that in the $\theta'$ phase. The strength decreases even more. As these particles grow in size, the alloy softens, even to the point of being in a softer condition than that of the supersaturated solid solution. This is termed the *overaging* stage. Under stress the soft aluminum flows around the large $\theta$ particles. A somewhat crude analogy, but one that may assist in understanding the aging process, is to consider marbles (as large $CuAl_2$ precipitate particles) embedded in putty (as the pure

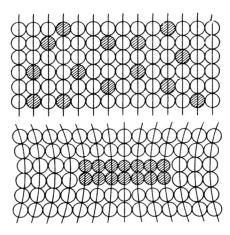

**Figure 6.15** Schematic of coherency strains created from a supersaturated solid solution (shown above) during the aging of an Al–Cu alloy. (From R. E. Reed-Hill and R. Abbaschian, *Physical Metallurgy Principles,* 3rd ed., PWS-Kent, Boston, 1992, p. 269.)

aluminum matrix). The putty under stress can flow around the marbles, which provide little resistance to plastic flow. Now if we take the same quantity of marbles and putty but grind the marbles until they consist of very small glass particles and then mix them with the putty, we find that they provide considerable resistance to plastic flow of the putty. The fine particles are analogous to the GP2 stage and $\theta'$ phase (probably more so like the latter, since there is really no coherency between the glass particles and the putty). The point to be made here is that one cannot produce a strong Al–Cu alloy by slow cooling from the liquid or $\alpha$ phases, because such cooling allows large $CuAl_2$ particles to form. To obtain the fine distribution of smaller particles, the alloy must be quenched from the $\alpha$-phase region and aged to a point that precedes the formation of large particles but instead permits the formation of a coherent intermediate precipitate with the corresponding coherency strains.

A schematic of the various stages of the aging process as revealed by hardness measurements as a function aging time is shown in Figure 6.16. Not all of these stages will be obvious at all aging temperatures. At room temperatures of around 20°C, significant hardening occurs in about 1 week, but overaging has not been observed to occur at this temperature. But if room temperature were that which we could experience in the desert (e.g., 50°C), overaging over a long period of time is possible. Hardness versus aging time for higher temperatures is shown in Figure 6.17. At temperatures on the order of 250 to 300°C the process occurs very rapidly; some stages may not be readily apparent. In essence the alloy passes through some stages too rapidly to permit detection. Also, the precipitation process is not uniform throughout the alloy. Because of

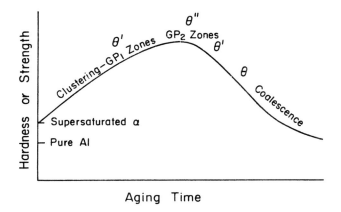

**Figure 6.16** Schematic of the various stages of the precipitation process as revealed by plotting hardness versus aging time.

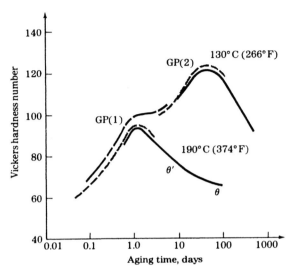

**Figure 6.17**  Hardness versus aging time in an Al–4% Cu alloy. (From J. M. Silcock, T. J. Heal, and H. K. Hardy in *Aluminum*, K. R. Van Horn, Ed., ASM International, Metals Park, Ohio, 1967, p. 123.)

faster diffusion and easier nucleation of new phases along the grain boundaries, precipitation often begins in these regions first.

Often the terms *natural* and *artificial* aging appear in the literature and textbooks, *natural* meaning aging at room temperature and *artificial* signifying aging above room temperature. These were, unfortunately, a very poor choice of words and should be avoided. First, room temperature can vary by at least 100°C [e.g., between that in the dead of winter in Alaska versus the heat of summer in the Sahara desert (or even in western U.S. deserts, Arizona for example)]. Second, there is nothing artificial about the aging or precipitation of atoms. It is a diffusion process that varies exponentially with temperature, all the way from 0 K to the solubility limit line. And remember, atoms only recognize temperature in kelvin.

Not all precipitation systems show all four stages distinctly. When the lattice parameters (interatomic distances) of the matrix and precipitate are vastly different, coherency strains may be very strong and short lived, or absent altogether. Some strengthening occurs due to small precipitates, which restrict dislocation motion. Some may call this process *dispersion* hardening rather than precipitation hardening, which involves coherency strains. (Dispersion hardening is common in superalloys and occurs by an aging heat treatment that causes extremely small particles to precipitate.) There are many aluminum alloys that are strengthened by the precipitation-hardening process. These include the Al–Mg, Al–Zn, Al–Si, Al–Mg–Si, Al–Mn, Al–Li, and Al–Ni alloys. The prime requirement is that the solid solubility increase with temperature. Alloys of the Al–Zn system are the higher-strength aluminum alloys. The Al–Li alloys were recently introduced to reduce the weight of aircraft structures. Li weighs 0.534 g/cm$^3$ at 20°C compared to 8.96 and 7.14 for copper and zinc, respectively. It has been estimated that replacing the current Al–Cu and Al–Zn alloys in a Boeing 747 aircraft with Al–Li alloys would reduce the weight of the 747 by that equivalent to one large elephant. The chief benefit, however, would be in fuel savings per passenger mile, elephants, or persons. The strengths of various Al alloys aged to peak hardness are presented in Table 6.1.

### 6.4.1 Other Nonferrous Precipitation-Hardening Alloys

Copper Base

One could search the phase diagrams and come up with a number of potential hardening systems (Problem 6.3). In the copper-based alloys, the Cu–Be alloys are the strongest of all Cu alloys. The phase diagram is shown in Figure 6.18. In these alloys the CuBe compound is the final precipitate. Commercial alloys such as C17100, C17200, and C17300 contain about 2 wt % Be, and can be aged to strengths on the order of 966 MPa (140 ksi) compared to 345 MPa (50 ksi) for cold-worked copper (see Table 6.1). Other copper-based age-hardening systems include Cu–Ti and Cu–Si. Ni–Be is also an age-hardening system that has not yet been fully exploited.

Magnesium Base

Several magnesium alloys are precipitation hardenable; the most notable is that of Mg–Al. These alloys have strengths on the order of 345 MPa (50 ksi). The portion of the Mg–Al phase diagram that pertains to the precipitation-hardening process is shown in Figure 6.19. Mg–Zn alloys are also precipitation hardenable, and many alloys contain both Al and Zn. The ASTM has a standardized numbering system for magnesium alloys, where the first two letters indicate the alloying elements of major significance, followed by two numbers containing the approximate percentages of these alloying elements. The next letter indicates the series designation for alloys developed of similar compositions. A letter and a number indicate the heat treatment. For example, AZ61A means the alloy contains about 6 wt % Al and 1 wt % Zn, and the letter A indicates that it was the first alloy of this series. The aging temperatures follow those of the aluminum alloys (i.e., T6 being the aged condition for highest strength). These alloys also have a UNS numbering system.

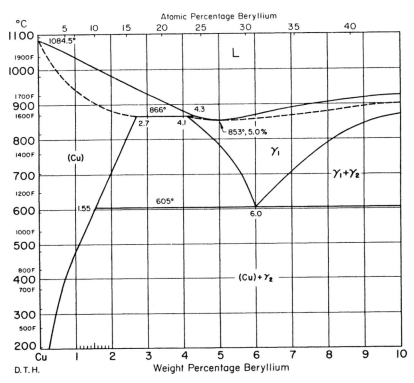

**Figure 6.18** Cu–Be phase diagram. (From *Metals Handbook,* Vol. 8, 8th ed., 1973, ASM International, Metals Park, Ohio, 1973, p. 271.)

Mg has a density of 1.74 g/cm³ compared to Al of 2.7 g/cm³. It is thus lighter than Al, but because of lower strength and ductility, plus higher cost, has not really challenged Al for aircraft usage. Nevertheless, a limited number of aircraft parts are made from magnesium alloys. These alloys are also used in some automotive parts, such as brackets and supports. Small Mg alloy parts can be found in a variety of industrial machinery. Probably the best thing that magnesium has going for itself is its abundance. Seawater contains large quantities of magnesium chloride and carnallite, a magnesium-containing sea deposit.

The properties of the more prominent precipitation-hardenable alloys are compared in Table 6.1 along with their counterpart base metal alloys in the solid-solution and cold-worked state. It is apparent that precipitation hardening is a very potent strengthening mechanism.

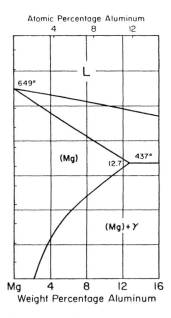

Atomic Percentage Aluminum

Weight Percentage Aluminum

**Figure 6.19** Mg–Al phase diagram. (From *Metals Handbook,* Vol. 8, 8th ed., 1973, ASM International, Metals Park, Ohio, 1973, p. 261.)

## 6.5  OTHER NONFERROUS ALLOYS

### 6.5.1  Superalloys (Nickel and Cobalt Base)

Nickel-based alloys, although not used in the quantities of the copper and aluminum alloys, still play a major role in industrial applications, particularly at elevated temperatures. *Superalloys* have been defined as those possessing good high-temperature strength and oxidation resistance. They are alloys of nickel, cobalt, and iron, which contain large amounts of chromium (e.g., 25 to 30%), for oxidation resistance. They are generally classified as iron–nickel-, nickel-, and cobalt-based alloys. For many years cobalt-based superalloys held the edge, but due to the precarious availability of cobalt, primarily from South Africa, the nickel-based superalloys have now been studied more extensively and have replaced many of the cobalt-based alloys.

The physical metallurgy of these alloys is due basically to the precipitation of a very fine distribution of small particles, primarily $Ni_3Al$ and $Ni_3Ti$, which have the generic name *gamma prime* in a gamma matrix (Figure 6.20). The nickel–iron-based alloys also have a $\gamma''$ phase in the form of a $Ni_3Nb$ com-

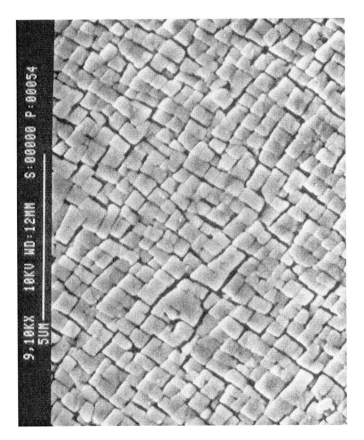

**Figure 6.20**   Gamma prime in a gamma matrix in a Mar 247 alloy. (Courtesy of T. Devaney, Cal-Poly.)

pound. The superalloys are really *dispersion-hardened* alloys because they achieve their strength by a fine dispersion of these compounds. Although these compound particles are most often obtained via an aging or precipitation heat treatment, they do not develop the coherency strains that the true precipitation hardening alloys do. These particles still resist dislocation motion and thereby strengthen the base metal. But their beauty is that these particles resist growth (coalescence) at elevated temperatures. Thus they have found their niche in the turbines and hot components of jet aircraft. Dispersion-hardening alloys do not overage as readily as precipitation-hardening alloys. This is their basic difference. Furthermore, dispersion-hardening alloys can be made by processes other than the conventional precipitation reaction. Oxide dispersion-hardening alloys are manufactured by mixing oxide powders with metallic powders by a mechanical alloying process that involves the repeated fracturing and rewelding of a mixture of powder particles in highly energetic ball mills. Nickel alloys do have a numbering system which in some ways attempts to relate the alloy numbers to the trade name. But most alloys are still known by their trade name. The letters "MA" preceding the number signify that the alloy was produced by the mechanical alloying process.

It should be pointed out here that the time to fracture at a certain stress and temperature is more important for superalloys than the decrease in strength as a function of temperature. The stress for rupture after 100 h versus temperature for superalloys is summarized in Figure 6.21. For life at 1000 and 10,000 h, see the reference for Figure 6.21. Some typical stress–rupture data are listed in Table 6.2. Results are also included in this table for the single-crystal turbine blades.

In Chapter 2 one of the mechanisms of creep and rupture discussed was that due to the grains sliding past one another at high temperatures under only moderate stresses. In the 1970s, single-crystal superalloy turbine blades were developed (Figure 6.22) to eliminate grain boundaries and their sliding. These single-crystal alloys are now used in many jet engines, especially in high-speed, high-temperature military aircraft. Ceramic and composite turbine blades are the materials predicted at this time to replace the single-crystal

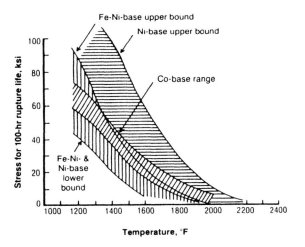

**Figure 6.21** Stress–rupture versus temperature data for a 100-h life of superalloys. (From M. J. Donachie, *Metals Handbook,* Desk ed., ASM International, Metals Park, Ohio, 1985, p. 16–5).

**Table 6.2**   Stress–Rupture Data for Selected Superalloys

| Alloy | Major element composition (wt %) | 1000-h stress–rupture | Temperature (0°C) |
|---|---|---|---|
| Inconel 718 | 52.5 Ni, 19.0 Cr, 18.5 Fe | 225 (37) | 750 |
| Inconel × 750 | 73.0 Ni, 19.0 Cr, 7.0 Fe | 125 (18) | 800 |
| Udimet 720 | 55 Ni, 17.9 Cr, 14.7 Co | 340 (49) | 800 |
| Incaloy MA 754 | 78 Ni, 20 Cr, 0.5 Ti, 1.0 Fe, 0.6 oxide | 149 (21.6) | 760 |
| Incoloy MA 956 | 74 Fe, 20 Cr, 0.5 Ti, 4.5 Al, 0.5 oxide | 63 (9.1) | 982 |
| Haynes 188 | 37 Co, 22 Cr, 22 Ni, 14 W | 165 (24) | 750 |
| Incoloy 901 | 13 Cr, 40 Ni, 45 Fe | 240 (35) | 750 |
| Mar M 246 cast | 60.0 Ni, 10.0 Co, 9 Cr, 10 W, 5.5 Al | 470 (68) | 800 |
| CMSX 2 single crystal | 64 Ni, 8 Cr, 5.0 Co, 8.0 W, 6 Ta, 5.6 Al | 325 (47) | 870 |

superalloy blades, but these materials will require considerable improvement and testing before their reliability is established.

Not all nickel-based alloys are used for high-temperature applications, the Monels and some solid-solution Inconels being the most notable exceptions. The properties of some nickel-based alloys are listed in Table 6.1. For more extensive coverage, see Table 5, pages 437–440, of Vol. 2 of the ASM *Metals Handbook,* 10th ed.

### 6.5.2   Titanium and Zirconium Alloys

Titanium was described as the *wonder metal* during its early development period of the late 1940s. It possesses a density of 4.54 g/cm$^3$, which is much lower than iron, but light enough and strong enough to be used in aircraft structural members. One of the first alloys developed, Ti–6% Al–4% V (UNS R56400), has a higher strength-to-weight ratio than that of either aluminum or steel, or of any other commercial alloy. This alloy still accounts for about 55% of titanium use. This alloy has tensile strengths on the order of 1140 MPa (160 ksi), with ductilities as high as 60% R.A.

Titanium is polymorphic. Below 883°C, it has the HCP structure called α-titanium, but transforms to the BCC beta structure when it is heated to 883°C. A pseudo-binary-phase diagram, which shows how the Ti + 6% Al alloy phases vary with tmperature and vanadium content, is shown in Figure 6.23.

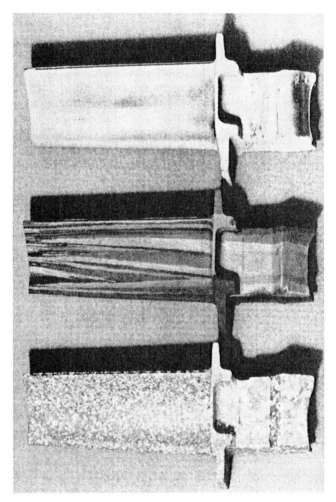

**Figure 6.22** Single-crystal superalloy turbine blade. The left blade is polycrystalline, the center one has long columnar grains, and the right one is a single crystal. (From D. H. Maxwell and T. A. Kolakowski, *Quest: New Technology at TRW*, Autumn 1980, p. 48.)

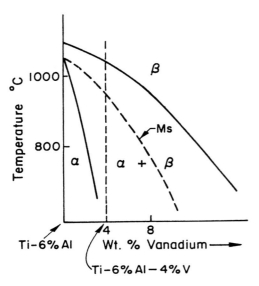

**Figure 6.23**   Pseudo-binary-phase diagram showing how the phases of the Ti-6% Al alloy varies with temperature and vanadium content.

This type of phase diagram avoids the complexity of ternary diagrams. When titanium and some of its alloys are water quenched, a type of martensite is formed. But it is not a hard martensite like that obtained in the steels, partially because it is not supersaturated with interstitial elements. The Ms line is shown in Figure 6.23 for this alloy. By a variety of heat treatments, such as quenching, aging, and so on, the quantities of the $\alpha$ and $\beta$ phases can be altered as well as the distribution of these phases. Ti-6% Al-4% V is thus known as an $\alpha$-$\beta$ alloy. The heat treatments on this and other titanium alloys are quite varied. Polymorphic alloys have this unique feature of the ability to change the quantities and distribution of the high- and low-temperature phases via the heat treatment process. This is somewhat different from the control of phase distribution by quenching and aging that was described in Section 6.4 for the precipitation of small particles. The latter dealt with a rather small quantity of the second phase (compound), whereas in the $\alpha$-$\beta$ titanium alloys we are dealing with more massive quantities of the two phases. In both cases, of course, a wide range of properties are available, dependent on the specific heat treatment selected.

   More recently, some all $\beta$, and $\beta$ plus small quantities of $\alpha$, alloys have been developed with somewhat higher strengths than the $\alpha$-$\beta$ alloys, but with less ductility. One of the more promising $\beta$ alloys is the one designated as Ti

1023, of composition 85% Ti, 10% V, 2% Fe, 3% Al. Some typical titanium alloy properties are listed in Table 6.1.

Because of its high strength-to-weight ratio, titanium has been used mostly in the aircraft industry as forged support components and as the fuselage for aircraft flying beyond Mach 1 speeds. The Concorde and some military fighter aircraft have fuselage skin temperatures that are too high for the weaker aluminum alloys.

Titanium has not quite lived up to its initial hopes and predictions. For aircraft usage it is limited to temperatures of about 500°C, due to its high reactivity with oxygen and nitrogen at higher temperatures. Unless protective coatings can be developed, titanium cannot be used in the hot regions of the jet engines. Furthermore, as these operating temperatures increase, Ti alloys will not possess the required high-temperature strength. The melting points of titanium alloys are not much greater than that of the superalloys.

While these aerospace developments were proceeding, other applications for titanium have been realized. Titanium is a passive metal. Just like aluminum and chromium, an adherent oxide film ($TiO_2$) forms a protective surface layer almost immediately upon exposure to air. It can be anodized to control the oxide thickness. It is now used for surgical implants and prosthetic devices, where good corrosion resistance, strength, and ductility are required. Chemical processing equipment, parts exposed to salt water, and similar corrosion-resistant applications have used some titanium alloys. Even in the automotive industry, some limited usage has been found in valves, springs, gears, and so on. Probably the single most significant factor limiting titanium usage is its high cost, being more than five times that of aluminum on a weight basis. This cost is slowly being reduced by increased consumption and cost-reduction processing, but barring a major breakthrough, it will probably never compete on a cost basis with aluminum and steel.

Zirconium, a sister metal to titanium, was first used extensively in the nuclear industry because of its transparency to neutrons. But zirconium also has excellent corrosion resistance in a variety of media. Accordingly, it has found to be of some use in the chemical industry. Zirconium is polymorphic, and like titanium, heat-treatable alloys have been developed with strengths as high as 587 MPa (85 ksi). It is not a light metal, however (6.5 g/cm³ density), and because of its high cost and only moderate strength is unlikely to be used in many structural applications.

### 6.5.3 Low-Melting-Point Metals

Zinc is probably better known as the galvanized coating on steel than as a structural member. Zinc castings, particularly die castings, which are of rather odd shapes and difficult or expensive to manufacture by other means, is one case where zinc is used. Die castings are usually limited to low-melting-point metals (419°C for zinc). Zinc castings are used in the automotive industry as carburetors, wiper parts, speedometer frames, and a host of similar small parts, primarily because of the ease with which it can be cast into intricate shapes. Zinc castings are also used extensively in general hardware, electronic and electrical fittings, and domestic appliances. Zinc, however, is also rolled, extruded, forged, and drawn. These wrought products include roof flashing, dry-cell battery cans, ferrules, gaskets, and solar collectors, to name a few.

Tin is used as a protective coating for steel (tinplate). It is a low-melting-point, low-strength metal. Its most popular application is as a solder when alloyed with other low-melting-point metals, such as lead, antimony, or silver. Since lead has been banned for internal household plumbing, silver is now frequently used, a 95 wt % Sn–5 wt % Ag being a popular composition.

Lead has found limited usage as lead–tin babbitt alloys (bearings). Pure lead is used widely for nuclear shielding and in Pb–$PbO_2$ batteries.

Indium, another low-melting-point metal, is used in solders. In the electronic industry it has been the most prominent solder, essentially as pure indium, since it makes low-ohmic contacts to silicon and gallium arsenide substrates. But pure indium is expensive (about $500 per pound) and is often sold by the troy ounce.

### 6.5.4  Precious Metals

Precious metals are considered to be those high-priced, fairly inert metals that include gold, platinum, palladium, and silver. Silver is relatively inexpensive compared to the others and as such is more widely used, particularly as silverware. It is less inert than the others. The precious metals over the years have been used primarily as jewelry, art objects, and dental alloys (also as corrosion-resistant platings). The electronic industry is now a large user of these metals in both thin- and thick-film form as conductors, capacitors, and protective layers.

### 6.5.5  Refractory Metals

Because of their high melting points (2468 to 3387°C), the refractory metals niobium, tantalum, molybdenum, tungsten, and rhenium have been widely studied for use in the higher-temperature regions of jet engines. Protective coatings, however, are needed since these metals are very reactive with hot gases. This course of action is still being pursued. The strengths at high temperatures for some refractory metal alloys are given in Table 6.3.

Tungsten, of course, has been used for many years as light-bulb filaments. Tantalum has good corrosion resistance and accordingly is used to some extent

**Table 6.3** High-Temperature Strength of Some Refractory Metal Alloys

| Alloy | Composition (wt %) | Temperature (°C) | Strength [MPa (ksi)] |
|---|---|---|---|
| FS 85 | Nb, 28 Ta, 10 W, 1 Zr | 1000 | 414 (60) |
| HWM | Nb, 25 Mo, 25 W, 1 Hf | 1000 | 1352 (196) |
| W–1% Re | | 1400 | 862 (125) |
| Ta–10% W | | 1000 | 304 (44) |

in chemical process equipment. But the largest use of tantalum is in electrolytic and foil capacitors. Molybdenum is an alloying element in steel (< 3%). Molybdenum-based alloys have found some applications in the missile industry as high-temperature structural parts such as nozzles, leading edges of control surfaces, support vanes, and struts. Niobium, the lowest-melting-point refractory metal, has also found some limited use in hypersonic flight vehicles. It is easier to fabricate than molybdenum and competes with it for high-temperature applications. Niobium also has a low-neutron-absorption cross section and competes to some extent with zirconium for nuclear applications. Niobium–tin and niobium–titanium are superconducting alloys at temperatures in the vicinity of 20 K. Protective coatings for niobium are further advanced than that for the other refractory metals and thus have a good possibility for more extensive high-temperature use.

## 6.6 IRON–CARBON ALLOYS: THE EUTECTOID SYSTEM

The production of iron comprises about 90% of the total world production of metals from their ores (Table 1.1 of Chapter 1). In 1988, U.S. steel production was approximately 100 million tons.

$Fe_2O_3$ is extracted from the ore and reduced by carbon in a blast furnace (at about 1600°C) to achieve a product called *pig iron*. Pig iron contains about 90% Fe, 4.4% C, with the balance being primarily Si, Mn, P, and S. This pig iron is converted to steel by the basic oxygen process (Figure 6.24). Approximately 30% steel scrap is added to the pig iron and charged into a refractory-lined converter. Oxygen is blown into the converter and thereby reduces the pig iron to a purity sufficient for steel production. Furnaces used in steelmaking include electric, induction, and consumable electrode arc melting furnaces. About 60% of today's steel ingots are continuously cast in slab form and then soaked (homogenized) prior to being hot-rolled to smaller shapes. Final sizing is usually done by cold rolling or drawing. The production of steel is a much

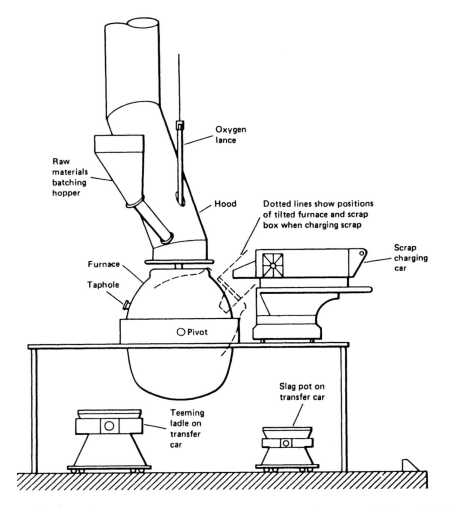

**Figure 6.24** Basic oxygen process for converting pig iron to steel (From *Metals Handbook,* Desk ed., ASM International, Metals Park, Ohio, 1985, p. 22-6.)

more complex process than that summarized above. The interested reader can find a good review of this subject in "Steel Processing Technology" by R. I. L. Guthrie et al. in Vol. 1 of the 10th ed. of the ASM *Metals Handbook,* p. 107.

In Chapter 1 we classified steels into the categories of cast iron, plain carbon steels, alloy steels containing 2 to 5 wt % of alloying elements, high-strength low-alloy steels (which include the microalloyed steels), and specialty steels. The reader is advised to review this short section in Chapter 1 prior to reading the following sections on phase diagrams and the heat treatment of steels.

## 6.6.1  Iron–Iron Carbide Phase Diagram

In the iron–carbon phase diagram we use only that part of the diagram from 100 wt % Fe to 93.3 wt % Fe–6.7 wt % C. The compound $Fe_3C$ contains 6.67 wt % C (25 at % C), so it is more convenient to use the Fe–$Fe_3C$ diagram rather than the Fe–C diagram. Actually, the difference between the two is insignificant, as can be seen in Figure 6.25, where both diagrams are displayed, the Fe–$Fe_3C$ diagram being shown as the solid lines while the Fe–C diagram is shown by the dashed lines. We also usually round off the 6.67 wt % C content of the $Fe_3C$ compound to 6.7 wt % C for convenience in our calculations. Actually, for all but the cast irons, we can use the abbreviated diagram of Figure 6.26 for the heat treatment processes to be described herein. Also note that in this diagram the 0.77 wt % C point has been rounded to 0.8 wt % C, and the 0.022 wt % C point to zero in order to facilitate some computations to be illustrated later.

## 6.6.2  Plain Carbon Steels

The plain carbon steels are those containing up to about 1 wt % carbon and to 1.5 wt % Mn. plus trace amounts of phosphorus and sulfur. Higher-Mn-content alloys are found in the 1300 series. These alloys can be explained using only the small left portion of the phase diagram in Figure 6.26. Once the steel is melted, solidified, and cooled into the all-solid-$\gamma$-phase region, we are no longer concerned with the $\delta$-Fe and associated regions near the melting point of pure Fe. Since all heat treatments involve heating the steel into the $\gamma$ region, cooling at various rates to room temperature, plus, for some steels, reheating to lower temperatures, the abbreviated diagram of Figure 6.26 will be used.

There are a number of terms, which may be called the *ites,* that will be referred to throughout this and the following sections on the heat treatment of steel. It is important to remember which of the *ites* are phases and which are microstructures, or microstructural constituents. Also remember that a crystalline solid phase is a specific arrangement of atoms (crystal structures) (e.g., FCC, BCC), which can only be *seen* by x-rays, while a *microstructure* is a certain arrangement of a phase or phases that are viewed through the microscope. These *ites* are listed below and described briefly with reference to Figure 6.26. Perhaps it will be easier to remember the *ite* phases if we use the Greek letter notation.

1.  *Ferrite.* This is a phase called the $\alpha$ phase since it is on the extreme left of the diagram. It is an interstitial solid solution of carbon in BCC iron where

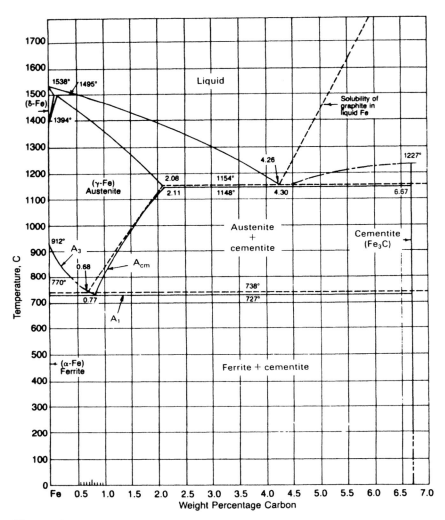

**Figure 6.25**   Fe–Fe$_3$C (solid lines) and Fe–C (dashed lines) phase diagram. (From G. Krauss, *Metals Handbook,* Desk ed., ASM International, Metals Park, Ohio, 1985, p. 28–2.)

the carbon resides in the largest interstitial sites shown in Figure 5.3. We will refer to this phase hereafter as α-iron in an attempt to reduce the number of *ites* one must remember. The α phase is a relatively weak (soft) iron since it contains a maximum of 0.022 wt % C only at 727°C (eutectoid temperature).

2.   *Austenite.* This is the γ phase, which consists of an interstitial solid solution of carbon in FCC iron, where the carbon atom resides in the largest

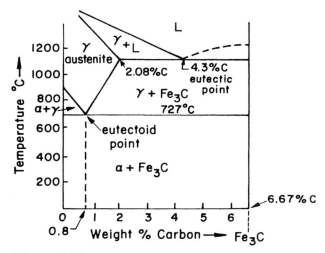

**Figure 6.26**    Abbreviated Fe–Fe₃C phase diagram to 6.7% carbon.

interstitial site of coordinates ½, ½, ½ (i.e., in the center of the cube). The FCC lattice has a larger interstitial site than does the BCC lattice and therefore can contain more carbon, up to 2.11 wt % C at 1148°C and up to 0.8 wt % C at 727°C. Under equilibrium conditions this phase cannot exist below 727°C.

3.    *Cementite.* This phase is Fe₃C, the iron carbide compound, and will be so called hereafter. It has an orthorhombic crystal structure (which you should forget) and, like intermetallic compounds, is hard and brittle (which you should remember).

4.    *Martensite.* This is a metastable phase that results from the rapid cooling (*quenching*) of a steel from the γ region to room temperature. It cannot be found on the equilibrium diagram. It has a tetragonal structure with the carbon atoms residing in a body-centered-tetragonal (BCT) crystal structure, in a similar position as is the BCC unit cell. The BCT cell is merely an elongated BCC cell, and at low carbon contents the martensite is actually a BCC structure (i.e., the elongation is barely if at all detectable). Martensite is considered to be supersaturated with carbon because it would be more stable if the carbon atoms were not present. This metastable phase is very hard and brittle, and most often is not present in a finished product. It plays the role of an intermediate step in the heat treatment process, where the objective is to obtain a more desirable microstructure, somewhat analogous to the supersaturated phase in the quenched Cu–Al alloy, although the latter is not hard and brittle. We do not have a Greek symbol for martensite, so we will use the letter *M*. The following three *ites* are microstructures rather than phases.

5.    *Pearlite.* A microstructure consisting of parallel platelets of α and Fe₃C. This product results when the γ phase of 0.8% C is rather slowly cooled

(e.g., in ambient air or in the furnace after the power has been switched off). It has a higher hardness than the $\alpha$ phase but much less than the hard martensite phase.

6. *Bainite.* A microstructure consisting of small particles of $Fe_3C$ in an $\alpha$ matrix. It is obtained by cooling the $\gamma$ phase at moderately slow rates. This microstructure has very desirable mechanical properties. Its hardness is between that of pearlite and the hard but brittle martensite, yet it has good ductility and toughness.

7. *Spheroidite.* A microstructure consisting of spherically shaped $Fe_3C$ particles in an $\alpha$ matrix. This is a very soft product, even softer than pearlite of the same carbon content.

Eutectoid Steel

The *eutectoid point* on our abbreviated Fe–$Fe_3C$ map of Figure 6.26 is the lowest temperature and composition at which the $\gamma$ phase can exist. It is that point corresponding to a temperature of 727°C and a composition of 99.2 wt % Fe-0.8 wt % C. This is the *eutectoid point* of the Fe–iron carbide system. This point resembles the eutectic point described in Section 6.3.2. When a liquid of eutectic composition is slowly cooled through the eutectic point, it is transformed into two solid phases that exist in the parallel-plate formation known as the eutectic microstructure. At the eutectoid point, however, we have the eutectoid reaction

solid A $\longrightarrow$ solid B + solid C

Students often confuse the terms *eutectic* and *eutectoid.* One memory gimmick that's worth a try is to spell the word *solid.* Notice that it contains the letters "oid," which you can associate with the eutectoid reaction, which deals with a solid-to-solid transformation, whereas the eutectic reaction begins with a liquid. At the eutectoid point, the *eutectoid reaction* takes place on cooling a 0.8 wt % C alloy composition slowly through the eutectoid temperature. At this temperature the eutectoid reaction is

$\gamma \longrightarrow \alpha + Fe_3C$

This reaction is one of a solid transforming to two different solids, and forms a parallel-plate microstructure (just like the eutectic) of the two phases called *pearlite* (Figure 6.27). This reaction takes place by nucleation and growth of the $Fe_3C$ platelets, which usually nucleate at the $\gamma$ grain boundaries (Figure 6.28) but sometimes grow out from previously nucleated colonies. The strength and hardness vary somewhat with cooling rate, the faster cooling rates (e.g., air cooling) forming thinner plates but more than are formed at a slower cooling rate. The finer pearlite is the stronger of the two. But in general the hardness of pearlite is in the vicinity of Rc 20 and has a corresponding tensile

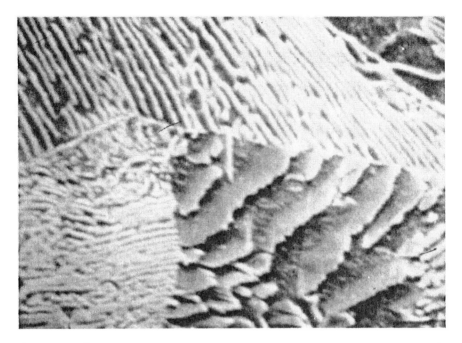

**Figure 6.27** Pearlite composed of $Fe_3C$ platelets (white) separated by ferrite platelets. Plates of $Fe_3C$ in lower right were cut almost parallel to surface. $\times 7000$. (Courtesy of Ryma Wallingford, Cal Poly.)

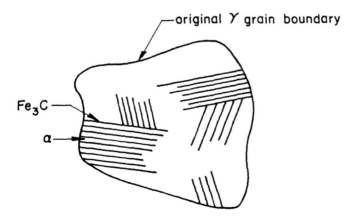

**Figure 6.28** Schematic of pearlite formation from $\gamma$-iron (austenite).

strength of about 773 MPa (112 ksi). A plain carbon steel of eutectoid composition is a 1080 steel (see Chapter 1).

*Example 6.4.* Compute the percentages on a weight basis of the $\alpha$ and $Fe_3C$ phases in a 100% pearlite microstructure. (Refer to Figure 6.26.)

Solution. Using the 0.8% C vertical line and 20°C as our point of interest, construct a tie line from the left boundary of 0% C to the right boundary of 6.7% C. We can now use our inverse lever rule as follows:

$$\% \; \alpha = \frac{6.7 - 0.8}{6.7 - 0} \times 100 = 88\%$$

$$\% \; Fe_3C = \frac{0.8 - 0}{6.7 - 0} \times 100 = 12\%$$

Pearlite will always consist of 88% $\alpha$–12% $Fe_3C$ no matter how fine or coarse the platelets and no matter in which plain carbon steel it exists.

Hypoeutectoid Steels

Steels with carbon contents below 0.8 wt % C (*hypo,* meaning *less than*) are known as hypoeutectoid steels. Let's examine the slow cooling of a 1040 steel (i.e., one that has a 99.6% Fe–0.4% C composition) from the $\gamma$-phase region (refer to Figure 6.26). When the alloy has been cooled to the line that separates the $\gamma$ from the $\alpha + \gamma$ region (approximately 800°C), the $\alpha$ phase begins to form. As cooling continues one can construct tie lines from the 100% Fe boundary (since the maximum carbon content is only 0.022% in the $\alpha$ phase, we will assume that it is zero for computational purposes) to the boundary running from pure iron to the eutectoid point, and thereby determine the composition of the remaining $\gamma$ phase at each temperature and corresponding composition along this boundary. Of course, as low-carbon $\alpha$-iron precipitates out of the $\gamma$ phase, the $\gamma$ contains more and more carbon until at 727°C and 0.8% C it has the eutectoid composition. On further cooling the remaining $\gamma$ transforms to pearlite at the eutectoid point. The microstructures obtained on cooling at various points along the 0.4% C vertical line are sketched in Figure 6.29. This 1040 slowly cooled steel will be somewhat softer than the 1080, since it contains some of the softer *proeutectoid* $\alpha$ phase. The phase that precipitates out prior to the eutectoid temperature is called a *proeutectoid* phase, just as in the case of the eutectic system the first phase precipitating out of the liquid was called the proeutectic phase. A 1040 microstructure of the type sketched after the pearlite (eutectoid) formation will have a Rc hardness less than 10 (at this Rc level it is advisable to use a different scale), or a DPH hardness number of about 190 and a tensile strength of around 518 MPa (75 ksi). It should be realized that these mechanical properties will vary with pearlite spacing and $\alpha$ grain size, but it gives one an idea of how the strength increases with the percent pearlite. Figure 6.30 shows the final 1040 slowly cooled microstructure.

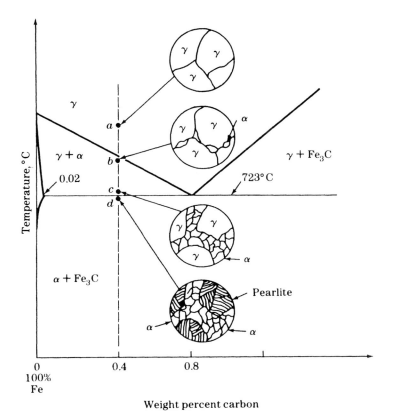

**Figure 6.29** Transformation of a 1040 (0.4% C) hypoeutectoid steel on slow cooling. (From W. F. Smith, *Principles of Materials Science and Engineering,* McGraw-Hill, New York, 1986, p. 447.)

*Example 6.5.* Compute the weight percent of the proeutectoid $\alpha$ phase and the weight percent of pearlite for a 1040 steel slowly cooled from the $\gamma$ region. Also compute the total $\alpha$.

Solution.   In this problem we must construct a tie line at a temperature just above the eutectoid temperature (e.g., 728°C). The % $\gamma$ at this point is obtained as follows:

$$\% \ \gamma = \frac{0.4 - 0}{0.8 - 0} \times 100 = 50\%$$

Obviously, the proeutectoid $\alpha$ is the balance, or 50%. This can be checked by the inverse lever rule:

**Figure 6.30** 1040 steel slowly cooled microstructure showing the proeutectoid $\alpha$ phase plus the eutectoid pearlite at room temperature. (Courtesy of W. Morris, Cal Poly.)

$$\% \text{proeutectoid } \alpha = \frac{0.8 - 0.4}{0.8 - 0} \times 100 = 50\%$$

Assume 100 g of starting $\gamma$. At 728°C only 50g of $\gamma$ remains, which transforms to pearlite via the $\gamma \rightarrow \alpha + Fe_3C$ reation. The pearlite contains 88% eutectoid $\alpha$. Thus $50 \times 0.88 = 44$ g of eutectoid $\alpha$. Total $\alpha = 50$ g of proeutectoid $\alpha + 44$ g of eutectoid $\alpha = 94$ g. Thus the total $\alpha = 94\%$. We can check this by constructing a tie line at 20°C, still using the vertical 0.4% C line as our composition of interest. Using the inverse lever rule, we have

$$\% \text{total } \alpha = \frac{6.7 - 0.4}{6.7 - 0} \times 100 = 94\%$$

Below the eutectoid temperature only the $\alpha$ and $Fe_3C$ phases exist. But the computation immediately above does not tell us how the $\alpha$ phase is distributed. To determine the distribution we must use a tie line immediately above the eutectoid temperature as we did in the first part of the computation above.

Hypereutectoid Steels

Steels of carbon content greater than the 0.8% C eutectoid composition, but less than the 2.1% C content, where the cast irons begin, are termed *hypereutectoid* steels (*hyper* meaning "more than"). In Figure 6.31 we can follow the slow cooling of a 1.2% C steel and examine the resulting microstructures. In this case, proeutectoid $Fe_3C$ first precipitates out of the $\gamma$ when the line proceeding from the eutectoid point upward to the right is crossed. Between this line and the horizontal eutectoid temperature line, the phases $\gamma$ and $Fe_3C$ exist. The proeutectoid $Fe_3C$ forms at the $\gamma$ grain boundaries. As the alloy is cooled through this two-phase region, the remaining $\gamma$ is continually reduced in carbon content until it reaches the 0.8% C eutectoid composition, at which point it transforms to pearlite. This structure is impossible to deform plastically. The $Fe_3C$ is almost as brittle as glass and cracks under the slightest shock. If a piece of this alloy in this condition were dropped from a height of several feet onto a concrete floor, chunks of pearlite and shattered particles of $Fe_3C$ would result. The actual final microstructure after slow cooling is shown in Figure 6.32.

*Example 6.6.* (a) Compute the percent proeutectoid $Fe_3C$ that would result from slow cooling a 1095 steel from the all $\gamma$ range. (b) Compute the total $Fe_3C$.

Solution. (a) Again using our inverse lever rule:

$$\% \text{ proeutectoid } Fe_3C = \frac{0.95 - 0.8}{6.7 - 0} \times 100 = 2.5\%$$

$$(b) \ \% \text{ total } Fe_3C = \frac{0.95 - 0}{6.7 - 0} \times 100 = 14.2\%$$

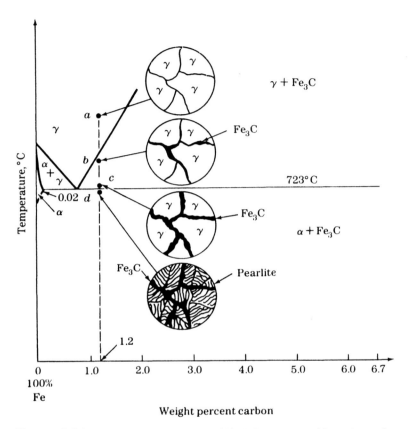

**Figure 6.31**   Transformation of a 1.2% C hypereutectoid steel on slow cooling. (From W. F. Smith, *Principles of Materials Science and Engineering,* McGraw-Hill, New York, 1986, p. 449.)

The equilibrium structures and properties of the plain carbon steels have a variety of applications. Low-carbon 1010 steel sheet would be used where good formability is required (e.g., in forming automobile fenders). Sheet steel of even lower carbon (e.g., 1005 to 1007) is used in most auto bodies. The low-carbon steels can be cold worked to increase the strength levels where both low carbon and higher strength are required. Cold-rolled sheet steel, which is subsequently tin plated and formed into cans, has a carbon content of about 0.08%. 1080 steels have been used for years for rather massive parts such as railroad rails. The hypereutectoid plain carbon steels are used in the heat-treated condition (quenched and tempered) for high-strength applications such as springs, cutting tools, and files.

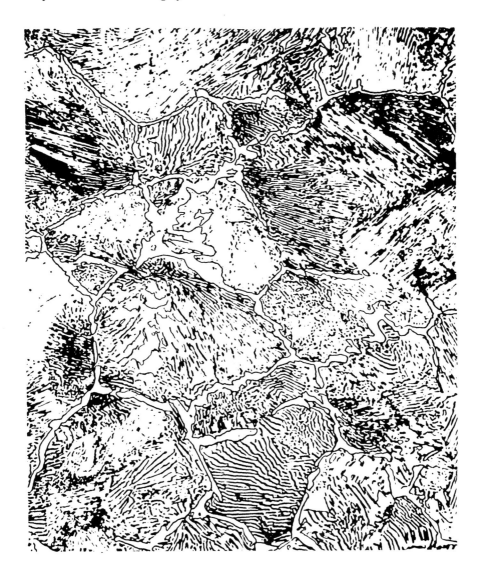

**Figure 6.32** Microstructure of a slowly cooled high-carbon (1.5% C hypereutectoid) steel. Note the white proeutectoid $Fe_3C$ surrounding the pearlite colonies. × 1000. (Courtesy of W. F. Forgeng, Cal Poly.)

## 6.7 NONEQUILIBRIUM PROCESSING OF STEELS

Although many steel forms, such as hot-rolled plate and rod in both hot- and cold-rolled conditions, cold-drawn wire, and cold-rolled sheet, are made and used from near-equilibrium microstructures, there is a large quantity of steel where the manufacturing process involves a nonequilibrium stage. This is the quenched or rapidly cooled stage of processing, where $\gamma$ is transformed to martensite, that is,

$\gamma$ (rapidly cooled) $\longrightarrow$ martensite

This is a nonreversible reaction. One cannot attain the $\gamma$ state directly from martensite upon reheating. When martensite is heated it first goes to the equilibrium forms shown on the phase diagram. Below the eutectoid temperature this reaction is

M (heated) $\longrightarrow \alpha + Fe_3C$

Even if one tried to go directly to the $\gamma$ form from M by placing the steel in a furnace at a temperature in the all-$\gamma$ region, it probably could not happen. In heating the steel to temperatures above the eutectoid temperature it must first pass through the $\alpha + Fe_3C$ region of the phase diagram, and it would start to decompose to these products before it could attain the $\gamma$ temperature range. One could try to heat very rapidly a thin wire or sheet (e.g., via a high-amperage current pulse) to the $\gamma$ temperature range, but even here the BCT martensite would probably reject its carbon and become BCC $\alpha$ iron before it would go directly to the FCC $\gamma$ phase. This argument is presented only to emphasize that certain products of reactions are controlled by the rate of heating and cooling rather than by what the equilibrium conditions predict. And now we are prepared to examine the results in detail of the nonequilibrium cooling rates used in the heat treatment of steels.

### 6.7.1 Isothermal Transformation Diagrams

The isothermal transformation diagrams (IT curves) were developed so that time could be introduced as a variable, which, as we will see, has a tremendous impact on which *ite* of the seven we noted in Section 6.6.1 results from the cooling of $\gamma$.

Let us consider a plain carbon steel of eutectoid composition (1080) that we will cool from the $\gamma$ region at different rates of heat extraction. One can obtain different cooling rates by quenching in salt water, plain water, liquid polymers, oil, air, or simply furnace cooling in which the power is turned off and the part cools over a period of hours, depending on the size and insulation of the furnace. These quenching or cooling media are listed above in descending order with respect to cooling rate. The IT curve for a 1080 steel is shown in Figure

6.33. The horizontal line represents the eutectoid temperature (727°C). For a 1080 steel nothing happens until the metal has been cooled below this temperature, at which time it wants to transform according to what the equilibrium diagram dictates, that is,

$$\gamma \rightarrow \alpha + Fe_3C$$

But as we pointed out earlier, the formation of $Fe_3C$ requires substantial diffusion of carbon atoms. This requires time, which is plotted on the abscissa of the IT curve as log time, since the cooling rates vary from seconds to hours or even days. Note that the $\gamma$ phase exists below the eutectoid transformation temperature and to the left of the first curve with respect to the time axis. This means that $\gamma$ can exist for short periods of time below where it wants to transform to the $\alpha$ and $Fe_3C$ phases. But if we cool the $\gamma$ to room temperature by quenching in salt water (which extracts heat faster than pure water), or even

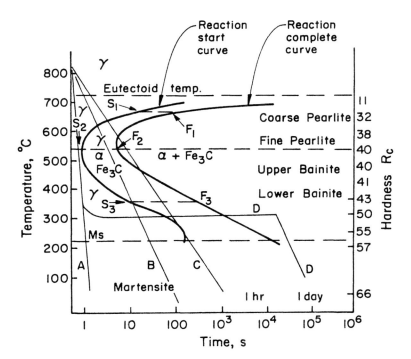

**Figure 6.33** Isothermal transformation (IT) curves for a 1080 steel, listing the microstructures obtained by a variety of cooling rates from the $\gamma$ region of the phase diagram.

in pure water, we may be able to prevent the reaction of $\gamma$ going to the equilibrium products and instead induce the reaction

$\gamma \longrightarrow$ martensite

This reaction begins at the horizontal line marked Ms for *martensite start,* being about 200°C for a 1080 steel, and is completed by the time it reaches room temperature. Martensite is a very hard ($\approx$ Rc 65) brittle phase with a distinct microstructure (Figure 6.34). At carbon contents less than 0.6%, it consists of domains of laths of slightly different orientations. At higher carbon contents it has more of a platelet appearance. The martensite will remain in this metastable state for infinite periods at room temperature. Martensite has been found in meteorites of millions to billions of years old.

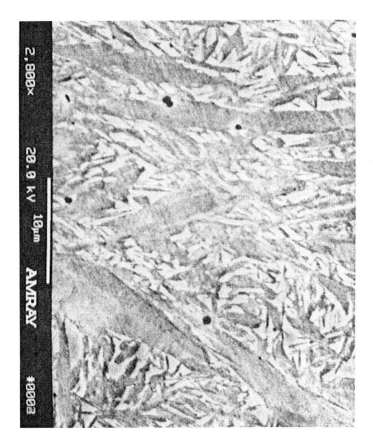

**Figure 6.34** Microstructure of martensite in a background of untransformed $\gamma$.

Now how are these points on the IT curves determined? Let's examine the points on the left curve labeled S1, S2, S3, and so on. S represents the start of the reaction

$$\gamma \longrightarrow \alpha + Fe_3C$$

So we have two possibilities, either $\gamma$ goes to M or to $\alpha + Fe_3C$. The driving force for the reaction of $\gamma$ to $\alpha + Fe_3C$ is larger the lower the $\gamma$ phase is below the eutectoid temperature. Since it wants to transform at the eutectoid temperature, obviously the more we reduce it below this temperature, the more it wants to transform. (This is the free-energy concept of Section 5.2.1.) On the other hand, the lower the temperature, the lower the diffusion rate for carbon atoms and the longer it takes for $Fe_3C$ to form. The transformation starts in the least time at the nose of the left curve (i.e., at the point marked S2). Here the driving force is moderately large and so is the diffusion rate of carbon. At higher temperatures (e.g., S1) the diffusion rate is large but the driving force small. At S3 the reverse is true.

In constructing these curves we first quench a number of samples from the $\gamma$ phase into a constant-temperature bath (often a molten salt bath) and withdraw them at different periods of time and quickly quench them into water. Then these samples are polished, etched, and examined under the microscope. Microstructures containing $Fe_3C$ appear dark, so the sample in which it is first noted is marked S1. This point represents the time it was held in the constant-temperature bath. When all of the samples have been examined for this transformation temperature, we finally arrive at a point on the time scale where the microstructure ceases to become darker, thus indicating that the reaction has finished. This point is the F1 point on the second curve (Figure 6.33). This isothermal transformation series of specimens is also conducted at other temperatures, labeled S2–F2 and S3–F3. When a sufficient number of points have been located, the curves can be constructed. The left curve represents the start of the reaction to $\alpha + Fe_3C$, referred to below as the start curve, and the right curve represents the completion of this reaction, called the finish curve. Isothermal transformation curves for the common steels can be obtained from the steel manufacturers. The actual photographs from which IT curves were obtained for a 4150 steel are shown in Figure 6.35.

Now we will superimpose on the IT curves of Figure 6.33 three lines of different slopes, which represent approximately the cooling rates for quenching in water (line A), in oil (line B), and cooling in air (line C). The actual cooling rate would not be a straight line, but such a line will suffice for explanation purposes. The line marked C, air cooling, called a *normalizing treatment,* usually produces fine pearlite. We will also assume that the sample is very thin, so that the heat is extracted at the same rate from inside this thin sample as it is from the surface. Sample A is cooled so fast (<1 s) that the $\gamma \longrightarrow \alpha + Fe_3C$

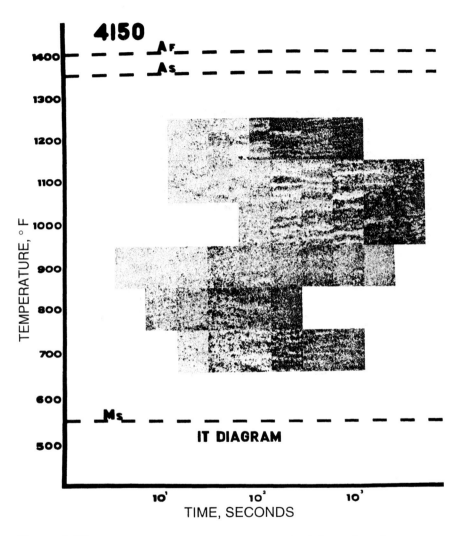

**Figure 6.35** IT diagram of a 4150 steel as derived from the microstructures obtained after quenching from successive increases in holding times at constant temperatures. (Courtesy of Robin Churchill, Cal Poly.)

reaction never begins. But the $\gamma \longrightarrow$ M reaction starts at the Ms temperature of about 220°C and is completed just above room temperature. This sample is all martensite. Martensite is a phase but is also considered to be a distinct single-phase microstructure. So in order to obtain martensite the sample must be cooled very rapidly. It must pass to the left of curve that represents the start of the $\alpha +$ Fe$_3$C reaction (i.e., it must pass the nose of the IT curve without intersecting it). If the plain carbon 1080 sample was cut from a large 1-in.-thick plate, it could not be cooled fast enough to produce martensite throughout the entire thickness. It would have a martensite surface but not so in the interior of the thick plate. We will return to this subject in the following section.

In the oil-quenched sample, line B, the heat is not removed as fast as in the water quench, thereby causing the sample to be cooled to room temperature such that it passes between the two curves. At the temperature where line B intersects the start curve, the reaction $\gamma \longrightarrow \alpha +$ Fe$_3$C begins and continues until the cooling line B crosses the start curve again at the lower temperature of about 470°C. But not all of the $\gamma$ has been decomposed. This remaining $\gamma$ will transform to martensite when the sample passes through the Ms horizontal line. The phases resulting from the oil quench are $\alpha$, Fe$_3$C, and M. But what about the microstructures in this sample? When samples are isothermally transformed at the higher temperatures, between about 723 and 550°C (roughly above the nose of the IT curves), pearlite is formed, being coarse pearlite at the higher temperatures and fine pearlite in the lower part of this region. Samples that transform at temperatures from 550°C to 250°C form the microstructure called *bainite*, named after E. C. Bain, who first explored the products of isothermal transformations in detail in the 1930s. In bainite the $\alpha$ and Fe$_3$C phases try to form a layer structure, but the transformation rates are such that complete platelets are never formed. Instead, much smaller *Fe$_3$C* particles are formed, and when viewed through the electron microscope at very high magnifications (Figure 6.36) they appear to be slanted from each other in a sort of partial herringbone arrangement. These small particles cannot always be observed at the optical microscopic magnifications to which Bain was limited during his studies. A distinction is made between upper bainite, which is formed by isothermal transformations between about 550 and 350°C, and lower bainite, which forms around 350 to 250°C. Upper bainite is characterized by larger rodlike particles which are often visible in the optical microscope at about 1000 to 1500 × magnifications, whereas lower bainite has very small Fe$_3$C particles. Both are shown in Figure 6.36. Bainite is a very desirable microstructure. Hardness values are in the range 40 to 45 Rc, with tensile strengths of 1208 MPa (175 ksi) to 1450 MPa (210 ksi) and ductilities on the order of 30% R.A. or higher.

The line C cooling rate could be obtained by allowing the samples to cool in air. It may take several minutes to reach room temperature. The sample

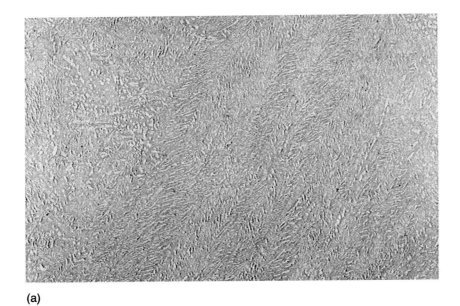

(a)

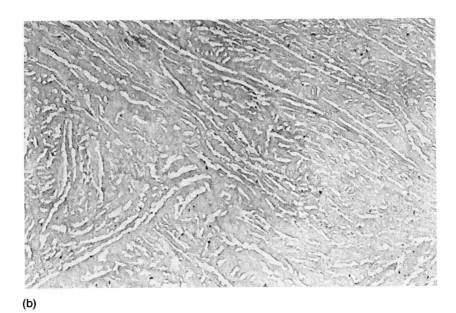

(b)

**Figure 6.36** The microstructure of (a) lower and (b) upper bainite in a eutectoid steel. Note that the Fe$_3$C particles form a herringbone arrangement as if they attempted to align as they do in pearlite. × 15,000. (Courtesy of P. S. Trozzo, U.S. Steel.)

transforms to pearlite, as you might expect, since equilibrium conditions are approached at such slow cooling rates. Line C intersects both the *start* and *finish* curves at high temperatures, where both coarse and fine pearlite is formed. It does not matter what cooling rate is used after the finish curve is crossed because all of the $\gamma$ has been transformed. One could quench the sample after crossing the finish line, and no martensite or bainite would be formed.

Since bainite is a desirable microstructure, let's find a way to obtain a 100% bainitic structure. Cooling the metal according to line D accomplishes this objective. The steel is quenched into a constant-temperature salt bath and then held just above the Ms temperature but below the nose of the curve. The piece is held at this temperature for a time sufficient to cross the finish curve. It can then be removed and quenched, air cooled, or cooled at any rate. The result will be an all-bainitic microstructure. This treatment is termed *austempering*, for reasons that will become more apparent after we have discussed *tempering*. The phases and microstructures resulting from these various quenching procedures are summarized in Table 6.4.

## 6.7.2 Continuous Cooling Curves

In the normal industrial practice, except for austempering, the steel is never transformed isothermally. Yet the curves of Figure 6.33 were obtained via the isothermal cooling of samples. Continuous cooling curves have been developed for many of the more common alloys. In Figure 6.37 a continuous cooling curve for a eutectoid 1080 steel is superimposed on the IT curves discussed above. In the continuous cooling diagram the start and finish lines are shifted to slightly longer times and slightly lower temperatures. The dashed lines represent various cooling rates and the microstructures resulting from these cooling rates are listed at the end of each dashed line. IT curves for most commercial steels are available in handbooks, many published by the steel

**Table 6.4**   Phases and Microstructures of 1080 Steel After Various Cooling Rates

| Cooling media | Phases | Microstructures | Hardness Rc |
|---|---|---|---|
| Line A: Water Quench | Martensite | Martensite | 65 |
| Line B: Oil quench | $\alpha$-$Fe_3C$, martensite | Pearlite, bainite, martensite | 40–45 |
| Line C: Furnace cool | $\alpha$-$Fe_3C$ | Pearlite | 15–20 |
| Line D: Austempered | $\alpha$-$Fe_3C$ | Bainite | 40–45 |

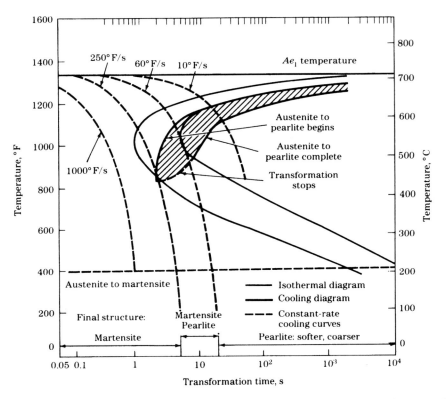

**Figure 6.37**  Continuous cooling diagram for a 1080 steel. (After R. A. Grange and J. M. Kiefer as adapted in E. C. Bain and H. W. Paxton, *Alloying Elements in Steel*, 2nd ed., ASM International, Metals Park, Ohio, 1966, p. 254.)

manufacturers. Some books will show the continuous cooling curves, although these are not as readily available as are the IT curves.

### 6.7.3  Hardenability: Low-Alloy Steels

*Hardenability* is defined as the *capability to harden* or as more often expressed as the *depth of hardening* of a particular steel. In the IT curves it was mentioned that very thin samples were used and that martensite could be obtained only to a small depth on a thick 1080 steel plate. Plain carbon steels have a low hardenability. They lack depth of hardness when quenched. Low-alloy steels were developed to overcome this problem. By the addition of small quantities (from about 0.5 to 3%) of elements, such as Mn, Ni, Cr, and Mo, one can obtain martensite at slower cooling rates, even as slow as air cooling in some steels. Thus the low-alloy steels have much better hardenability than do the plain carbon steels. Whereas the plain carbon steels can be hardened to a depth approaching 1/16 in., the low-alloy steels can be hardened to depths in excess of 1 in. The standard alloy steels and their compositions are listed in Table 6.5.

How do these small additions of alloying element achieve this feat? Let's return to our basic reactions that may occur at temperatures below the eutectoid temperature: namely,

(a)   $\gamma \longrightarrow \alpha + Fe_3C$

or

(b)   $\gamma \longrightarrow$ martensite

These are the two possible reactions that can occur on cooling steel from the $\gamma$ range. If reaction (a) can be retarded in some manner, reaction (b) can proceed. Reaction (a) requires significant diffusion of the carbon atoms. The alloying elements retard the diffusion of carbon. The carbon atoms like to stay in the vicinity of the solute atoms and do not jump around and migrate to the extent that they do in the presence of iron atoms alone. Thus the formation of $Fe_3C$ is slowed and allows reaction (b) to take over. This is a somewhat simplified explanation but will suffice for our purposes.

**Table 6.5**  Composition of Selected Alloy Steels

| Steel | C | Mn | Ni | Cr | Mo | Other |
|---|---|---|---|---|---|---|
| | | Low-carbon quenched and tempered steels | | | | |
| A514 | 0.15–0.21 | 0.80–1.10 | — | 0.50–0.80 | 0.18–0.28 | — |
| HY 80 | 0.12–0.18 | 0.10–0.40 | 2.00–3.25 | 1.00–1.80 | 0.20–0.60 | 0.25 Cu |
| | | Medium-carbon ultrahigh-strength steels | | | | |
| 4130 | 0.20–0.33 | 0.40–0.60 | — | 0.8–1.10 | 0.15–0.25 | — |
| 4340 | 0.38–0.43 | 0.60–0.80 | 1.65–2.00 | 0.70–0.90 | — | — |
| | | Carburizing bearing steels | | | | |
| 3310 | 0.08–0.13 | 0.45–0.60 | 3.25–3.75 | 1.40–1.75 | — | — |
| 5120 | 0.17–0.22 | 0.70–0.90 | 0.55 | 0.70–0.90 | — | — |
| 8620 | 0.20 | 0.80 | — | 0.50 | 0.20 | — |
| | | Heat-resistant chromium–moyldenum steels | | | | |
| UNSK12122 | 0.10–0.20 | 0.30–0.80 | — | 0.50–0.80 | 0.45–0.54 | — |
| UNSK31545 | 0.15 | 0.30–0.060 | — | 2.65–3.35 | 0.80–1.06 | — |

Hardenability Measurements: The Jominy End-Quench Test

One could obtain a good idea of the hardenability by quenching a round bar or plate, then cutting a cross-section sample some distance away from the ends of the piece and making hardness measurements across the diameter or plate thickness. Such hardness traverses that might be obtained on steels of both poor and good hardenability are depicted schematically in Figure 6.38. The best hardenability was obtained in the alloy containing 1.0% Cr plus 3% Ni. The good hardenability steel shows a nearly constant hardness traverse, decreasing only slightly in the center. It is nearly all martensite throughout its cross section. The plain carbon steel contains a martensitic shell with pearlite in the remaining cross section. The latter microstructure was, of course, a result of the decrease in cooling rate as we proceed inwardly toward the center of the bar. In low-hardenability steel, pearlite forms at rather shallow depths and throughout the central sections—thus the large dip observed in the hardness traverse.

W. E. Jominy and A. L. Boegehold in 1939 proposed a convenient test for hardenability now known as the *Jominy end-quench test,* where the bar is cooled by water quenching on one end only (Figure 6.39). Thus there exists a

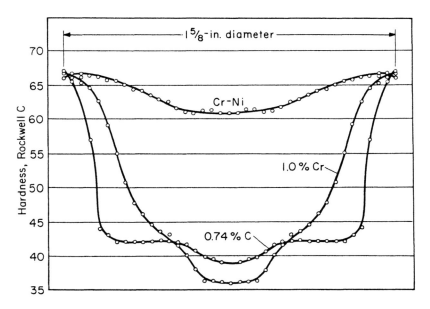

**Figure 6.38** Hardness traverses across cross sections of low- and high-hardenability steels. (From E. C. Bain and H. W. Paxton, *Alloying Elements in Steel,* 2nd ed., ASM International, Metals Park, Ohio, 1966, p. 156.)

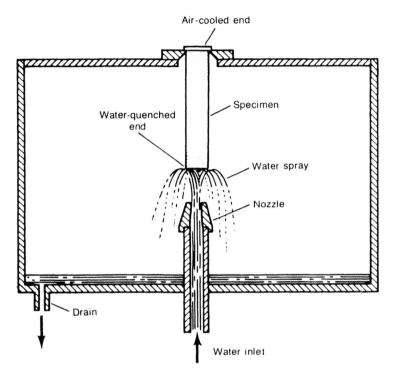

**Figure 6.39** Jominy end-quench test. (From *Metals Handbook,* Vol. 1, 10th ed., ASM International, Metals Park, Ohio, 1990, p. 466.)

spectrum of cooling rates along the bar. This can be visually observed by the change in color along the length of the bar as it is being cooled. The quenched end experiences a very high cooling rate, while at the opposite end the cooling rate approaches that of air cooling (i.e., the water has little or no effect on the extraction of heat on this end). Figure 6.40 shows the cooling rates along the specimen. Specimen dimensions and other details are given in ASTM A255 specification.

After the bar is fully cooled to room temperature, which will require about 15 minutes, it is removed from the quench apparatus, a flat is ground on one side, and the hardness measured along the bar length, the hardness spacings being closer together near the quenched end. Hardenability curves for a 1050 and a 4150 steel are shown in Figure 6.41. Some steels have the letter H at the end of the numerals (e.g., 4340H). The H specifies that the hardness values must fall within certain maximum and minimum values along the hardenability curve. The ASM *Metals Handbook,* Vol. 1, 10th ed., shows these H-band

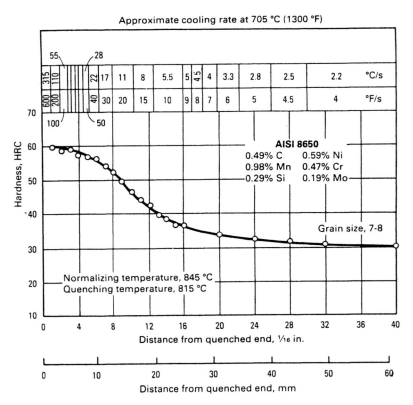

**Figure 6.40** Cooling rates along the length of a Jominy end-quench specimen. (From *Metals Handbook,* Vol. 1, 10th ed., ASM International, Metals Park, Ohio, 1990, p. 466.)

hardenability curves (pp. 485–570) for 85 carbon and low-alloy steels. A 4340 steel, for example, will have a hardness of about 45 Rc 2 in. from the quenched end. This means that a water-quenched 4-in.-diameter rod that is being cooled from all sides, would have a minimum hardness of 45 Rc if one made a hardness traverse across a cross section. Such a steel would be used for shafts. A 1090 steel, on the other hand, might be used in thin sections such as in razor blades, files, or cutlery. The plain carbon steels are less expensive. The low-alloy steels are thus used only when a great depth of hardening is required. The IT curve of a high-hardenability steel would be displaced to the right on the time axis with respect to the plain carbon steels (Figure 6.42).

Although the hardness and strength of plain carbon and low-alloy steels can be markedly altered by microstructural changes, for a given microstructure, the

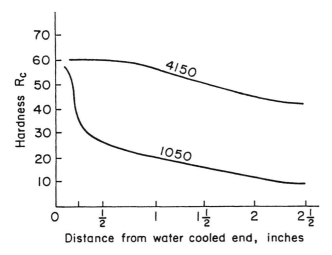

**Figure 6.41** Hardenability curve comparisons for 1050 and 4150 steels. (Adapted from E. C. Bain and H. W. Paxton, *Alloying Elements in Steel,* ASM International, Metals Park, Ohio, 1966, p. 136.)

hardness is primarily a function of carbon content (Figure 6.43). The spheroidite structure whose hardness is shown by the lower line is the softest of all steel microstructures. It is obtained by heating the steel for long periods of time at temperatures about 50 to 75°C below the eutectoid temperature. It does not matter whether the initial structure is martensite, bainite, or pearlite; its lowest-energy state is when the carbide particles are spherical. The spheroidite microstructure obtained from heating a 8620 steel for 20 h at 600°C is shown in Figure 6.44. A steel with this microstructure can easily be plastically formed to any desired shape. Often, steel is purchased in a spheroidized state, formed to the desired shape, and then heat treated to higher-strength conditions.

### 6.7.4  Tempering of Steel

*Tempering* is an annealing process that is used to transform the metastable martensite to the stable products of $\alpha + Fe_3C$. Generally, one would prefer to have a less brittle material than martensite, even at the sacrifice of hardness or strength. This is accomplished by the tempering process (the blacksmith would call this drawing the steel, as if it were drawn back to a softer state). The reaction is

$$\text{martensite (plus heat)} \longrightarrow \alpha + Fe_3C$$

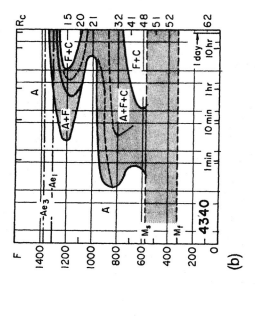

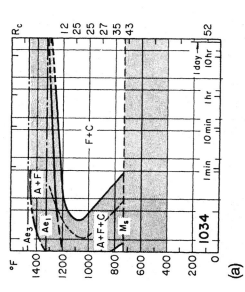

**Figure 6.42** Isothermal transformation diagrams for (a) a 1034 steel of 0.34% C and (b) a 4340 steel of 0.40% C plus 1.7% Ni, 0.8% Cr, and 0.25% Mo. Note how the Ni, Cr, and Mo alloy additions push the IT curves to the right on the time axis. (From *Metals Handbook*, Vol. 2, 8th ed., ASM International, Metals Park, Ohio, 1964, p. 38.)

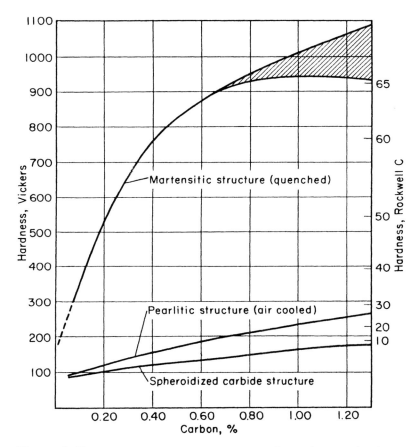

**Figure 6.43** Hardness versus carbon content for various steel microstructures. (From E. C. Bain and H. W. Paxton, *Alloying Elements in Steel,* ASM International, Metals Park, Ohio, 1966, p. 37.)

Theoretically, tempering can be done following the reaction above at any temperature below the eutectoid temperature. Usually, it is heated around 400 to 450°C for about 30 minutes to an hour, depending on the size of the piece and the alloy composition. For higher hardness, such as that needed for some tools, a tempering temperature of 200°C may be used. Figure 6.45 shows the range of hardnesses obtained on tempering as a function of carbon content and temperature. The product of tempering is called *tempered martensite*. This is another confusing term, especially to students, and a poor selection of words, because after tempering, martensite no longer exists. The product is $\alpha + Fe_3C$

**Figure 6.44**  Spheroidite microstructure in a 8620 steel obtained by heating for 20 h at a temperature of 600°C. ×2500. (Courtesy of Allen Chien, Cal Poly.)

(Figure 6.46). The Fe₃C particles are very small, similar to those in bainite. At very high magnifications in the electron microscope, the microstructures are very similar, except that in the tempered state they are more randomly distributed in orientation than they are in bainite. Also, bainite and tempered martensite can have similar hardnesses (i.e., 40 to 45 Rc) and similar strength values.

Both bainite and tempered martensite are the most popular and widely used heat-treated steel microstructures. They have good toughness, strength, formability, and machinability. Thus we have two methods of obtaining these desirable microstructures, quench and temper, or austemper to 100% bainite. The quench and temper route is more easily controlled. But often, continuous cool-

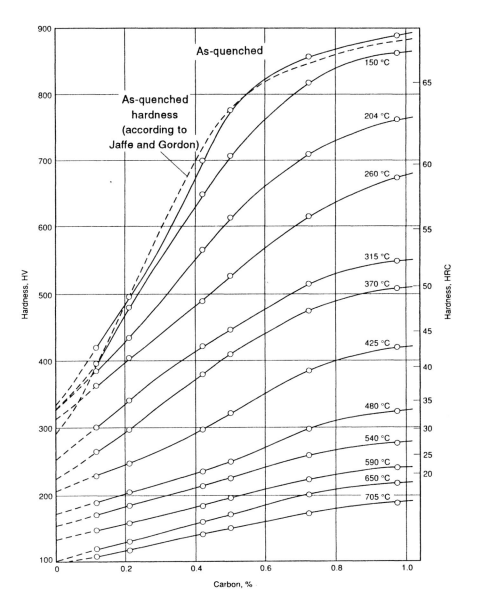

**Figure 6.45** Tempered hardness of plain carbon steels versus carbon content and tempering temperature. (From R. A. Grange, C. R. Hibral, and L. F. Porter, *Met. Trans. A.*, 1977, p. 1775.)

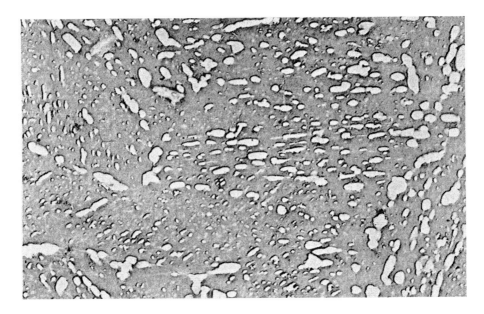

**Figure 6.46** Microstructure of tempered martensite showing the Fe₃C particles in an $\alpha$-ferrite matrix. (Courtesy of W. D. Forgeng, Cal Poly.)

ing can result in large percentages of bainite, a process that is less expensive than either austempering or the quench-and-temper route. Now we can explain the austempering term. The isothermal transformation of $\gamma \longrightarrow$ bainite, at temperatures just above the Ms temperature, is analogous to the tempering of austenite ($\gamma$) well below the eutectoid temperature.

The properties and applications of the plain carbon and low-alloy steels are quite varied. This, of course, is one of the reasons that steel use exceeds that of all of the other metals and most nonmetallic materials.

## 6.8 HIGH-STRENGTH LOW-ALLOY STEELS (HSLA): MICROALLOYED STEELS

In Chapter 1 we discussed this category of steels briefly and will repeat two important points. These steels have a lower alloy content and less strength than those of the alloy steels discussed in the preceding section. The second point is that they are primarily structural steels for rather large buildings, bridges, ships, and oil and gas pipelines. Today, they are also finding their way into the automotive applications. The microalloyed steels are a relative newcomer to this group but are rapidly replacing many of the older structural steels. The

ASM *Metals Handbook* separates these steels into four groups, two of which are heat treatable. The other two groups, which account for the major use of structural steels, fall into the categories of carbon–manganese and microalloyed steels. The typical structural steel is a carbon–manganese steel of about 0.18% carbon and 1.3% manganese and has the ASTM designation of A36. The microalloyed steels, of which the ASTM A572 group is the most common, contain small amounts of niobium, titanium, or vanadium, all of which are strong carbide formers. The strength is achieved by a fine dispersion of these carbides plus a carefully controlled hot-rolling process to produce a fine grain size. The small carbides themselves restrict grain growth. Both of these strengthening mechanisms produce structural steels of up to 550 MPa (80 ksi) yield strength, with an ultimate tensile strength in excess of 690 MPa (100 ksi). Because these steels are strengthened by the strong carbide formers, most of the carbon is removed from the ferrite solid solution. In welding processes the steel is at some point during the cooling process in the $\gamma$-phase region. At moderately fast cooling rates the undesirable brittle martensite phase could form. The microalloyed steels have little carbon available for martensite formation, another plus for these steels.

Another more recent addition to the HSLA family of steels is the dual-phase steels. They are now classified as HSLA steels, although when they were first introduced around 1980, there was some confusion concerning their classification. Their microstructure consists of regions of hard martensite in a much softer $\alpha$-ferrite phase matrix (Figure 6.47) and is achieved by quenching from the $\alpha + \gamma$ region of the iron–iron carbide phase diagram. The $\gamma$ phase transforms to martensite on quenching, while the $\alpha$ phase remains as such. Their carbon contents are about 0.1%, thus the major phase is $\alpha$ ferrite, while the martensite comprises only about 20%. They have strengths of the order of 621 MPa (90 ksi) with good formability and a very high rate of strain hardening. With carbon contents in the range 0.2 to 0.4, strengths of 1040 MPa (140 ksi) have been attained after cold forming. The uses of these steels to date have been primarily in the automotive industry.

The treatment, properties, and applications of some typical SAE steels are summarized in Table 6.6.

## 6.9  SPECIALTY STEELS

### 6.9.1  Stainless Steels

The stainless steels account for the majority of what we will call specialty steels. The stainless variety can be separated into five groups: the austenitic, which account for well over half of the use of all stainless steels; martensite; ferritic, precipitation hardening; and duplex stainless steels.

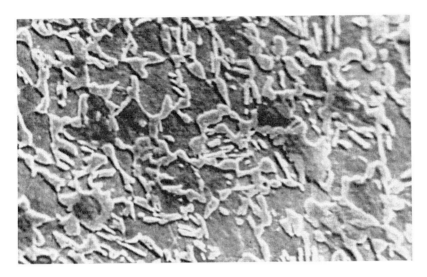

**Figure 6.47** Microstructure of a dual-phase steel quenched from 730°C which contains approximately 28% martensite (light areas). ×2000. (Courtesy of Barbara Blumenthal, Cal Poly.)

Stainless steels are stainless because of their chromium content. Chromium is a passive metal. Like aluminum and titanium, it forms a thin oxide layer, $Cr_2O_3$, on immediate exposure to air. When the chromium is in solid solution with iron or nickel, about 10.5 wt % Cr is required to make the alloy passive. The iron–chromium phase diagram is shown in Figure 6.48. The important feature is the γ (austenite) loop on the left side of the diagram. Chromium and α iron are very similar metals with respect to atom size, valency, and crystal structure. Thus the BCC iron can contain lots of BCC chromium in a substitutional solid solution. But the γ-FCC field can dissolve only about 12.7% Cr. Also, the γ region is restricted to temperatures between about 830 and 1390°C.

Ferritic Stainless Steels

Ferritic stainless steels are Fe–Cr alloys containing between 12 and 30% Cr. The upper limit is determined by the σ phase. This phase, located in the central portion of the diagram, is very brittle and undesirable and should be avoided. Ferritic steels on the high end of the Cr content scale have good corrosion resistance but relatively low strength compared to the other stainless steels. Since they do not contain nickel, they are less expensive than the austenitic. The 409 and 430 are common ferritic stainless steels.

**Table 6.6** Properties and Applications of Plain Carbon, Alloy Steels, and HSLA Steels

| Alloy type and condition | Ultimate tensile strength [MPa (ksi)] | Elongation (%) | Applications |
|---|---|---|---|
| Plain carbon steels | | | |
| Low carbon <0.2% C Hot rolled | 290 (42) to 414 (60) | 40 to 25 | Nails, stampings, cans, common sheet steel products, auto bodies, consumer goods, deep-drawn products |
| Medium carbon 0.2–0.5% Hot rolled | 414 (60) to 552 (80) | 25 to 40 | Machine parts, rivets, carbuerized gears, fasteners, cams, camshafts, forgings, general-purpose heat-treated parts |
| Medium-high carbon 0.5–0.9% C Hot rolled | 552 (80) to 828 (120) | 20 to 8 | Oil-hardening gears, set and socket screws, spring steel, ball bearings |
| High carbon 0.9–1.5% Hot rolled | 828 (120) to 966 (140) | 8 to 1 | Cutting tools, files, saws, knives, boring tools (often quenched and/or quenched and tempered) |
| Low-alloy steels | | | |
| Cr–Ni–Mo A736 A517 A542 A543 | 760 (118) to 930 (135) | 18 (min.) | Cr-Mo-Ni steels have a myriad of applications: automotive and other machinery, shafts of many types, steel, pressure vessels |
| Cr–MO A387 A832 | 380 (55) to 690 (100) | 18 (min.) | Cr-Mo steels are widely used in oil and gas industries and in nuclear power plants and fossil fuel plants |
| Ni–Mo A302 A533 A645 | 550 (80) to 860 (125) | 18 (min.) | Pressure vessels |
| HSLA and structural carbon steels | | | |
| Microalloyed A572 A588 | 480 (70) to 345 (50) min. | 18 min. | High-strength niobium–vanadium steels of good structural quality |
| Structural carbon A36 | 400 (50) min. | 20 | General structural steel |

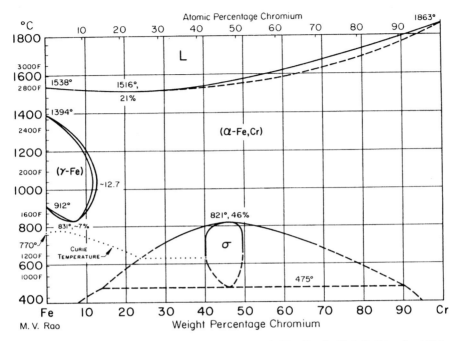

**Figure 6.48** Fe–Cr phase diagram. (From *Metals Handbook,* Vol. 8, 8th ed., ASM International, Metals Park, Ohio, 1973, p. 291.)

## Martensitic Stainless Steels

Martensitic stainless steels become martensitic just like the other steels (i.e., by quenching from the $\gamma$ region). The Fe–Cr phase diagram suggests that the maximum Cr content would be about 12.7%. But the carbon content expands the $\gamma$ region to the extent that larger chromium contents are possible. Common alloys are 410, containing 12% Cr and low carbon, and alloy 440 of 17% Cr with a high carbon content. The martensitic are the strongest of all stainless steels, having strengths to 1897 MPa (275 ksi). But at such high strength levels they lack ductility. They are used for cutlery, surgical instruments, springs, and applications that in general involve high strength and corrosion resistance.

## Austenitic Stainless Steels

These steels are a good compromise between the ferritic and martensitic stainless steels and hence their widespread usage. These steels are ternary alloys, and as such really require a ternary phase diagram for a complete understanding of the phase relations. By the addition of nickel in quantities on the order of 8%, the $\gamma$ transformation temperature, which in pure iron is 916°C, can be depressed to the extent that the $\gamma$ can be retained at room temperature. Actually, the $\gamma$ phase is metastable at room temperature, but the tendency to transform is so sluggish that for all practical purposes it can be considered to be stable. Higher nickel contents (e.g., 10%) can ensure $\gamma$ stability.

Most alloys of austenitic stainless steels contain 16 to 25% Cr, 7 to 20% Ni, and the balance Fe. The chief advantage is their greater ductility and hence formability that goes with the FCC structure. Their strengths are in between those of the ferritic and martensitic steels, but via cold working and the related strain hardening, strengths of around 759 MPa (110 ksi) with ductilities of 50% R.A. are attainable. The austenitic steels are the 300 series of steels, with the 301 and 304 series of roughly 18% Cr–8% Ni being the most widely used.

Most austenitic steels are susceptible to *sensitization* when heated in the range of about 425 to 870°C or slowly cooled through this temperature range. The chromium becomes tied up as chromium carbide and is not available to form the passive $Cr_2O_3$ surface film. In this state they are *sensitive* to corrosive media. The chromium carbide frequently forms first in or near the grain boundaries and hence corrosion begins in the grain boundary regions. Two steels have been developed to minimize sensitization, 347 containing about 0.6% niobium, and 321 containing similar amounts of titanium. These elements readily form carbides, leaving the chromium free to form $Cr_2O_3$. This is not a *cure-all* but it helps. The 321 and 347 are called *stabilized* stainless steels. The best solution to the problem is to avoid this temperature. During the welding of 321 and 347 the niobium and titanium carbides go back into solution near the weld and make this region again subject to sensitization. A 304 L steel, L meaning "low carbon," was developed for situations where welding is required.

Ternary Phase Diagrams

The austenitic stainless steel discussion is a good place to introduce the subject of ternary phase diagrams. The metallurgist often tries to circumvent these diagrams by one means or another because of their complexity. Also, in many cases in important alloy systems they may not exist. But these diagrams are very commonplace in ceramic systems. Let's examine their concept and see how it works in the iron–chromium–nickel system.

In the ternary diagram the composition of the three components are defined in the form of a triangle, as depicted in Figure 6.49. The binary alloy compositions *AB*, *BC*, and *AC* are represented on the three edges of the triangle. The temperature is a constant throughout this triangle. To find the composition of an alloy denoted by point *X* in the diagram, a line is constructed from each corner representing the pure metals to the side of the triangle opposite that corner and perpendicular to these sides. At that point where the line from corner *A*, for example, intersects line *BC*, the % *A* is zero. The total length of

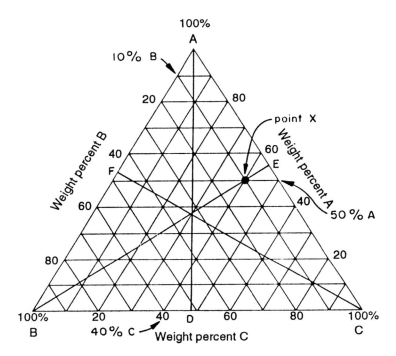

**Figure 6.49**   Schematic of a three-element ternary phase diagram.

the line from *D* to *A* represents 100% *A*. Point *X* is on an isocomposition line at 50% *A*; thus the percentage *A* in the alloy is 50%. Note that the line from the point of interest *X* to line *AC* is perpendicular to the line *AD*. Similarly, the % *B* is read from the line *BE*, which is constructed from 100% *B* perpendicular to line *AC*, and which in this case turns out to be 10% *B*. The balance must be the % *C*, which can be verified by constructing line *CF*. This is the composition for only one temperature. To know the phases that exist at any given temperature, the composition triangle must be known for that particular temperature. These are called *composition isotherms*. The isotherms for different temperatures for the Fe–Cr–Ni system are shown in Figure 6.50.

Precipitation-Hardening (PH) Stainless Steels

These steels can attain very high strength levels, in excess of 1380 MPa (200 ksi) with good ductilities (50 to 70% R.A.). They attain their high strengths by precipitation of intermetallic compounds via the same mechanism as that found in aluminum alloys. The compounds usually are formed from Fe or Ni with Ti, Al, Mo, and Cu. Typical compounds are $Ni_3Al$, $Ni_3Ti$, or $Ni_3Mo$. Chromium

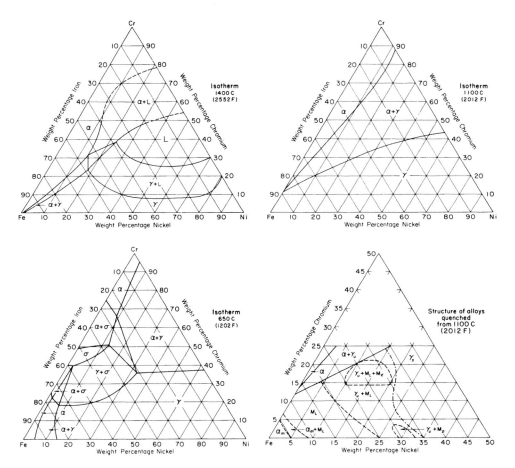

**Figure 6.50** Cr–Fe–Ni ternary phase diagram isotherms at four different temperatures. (From *Metals Handbook,* Vol. 8, 8th ed., ASM International, Metals Park, Ohio, 1973, p. 425.)

contents are in the range 13 to 17%. These steels have been around for several decades but are now being recognized as a real alternative to the other stainless steels. They have the good characteristics of the austenitic steels plus strengths approaching the martensitic steels. One of the early problems centered around forging difficulties, but these problems have been overcome to some extent. They are of two types, martensitic and semiaustenitic, the latter being able to be converted to martensite by special heat treatments. The most popular are the 13–8, 15–5, and 17–4 martensitic types, where the numbers refer to their Cr

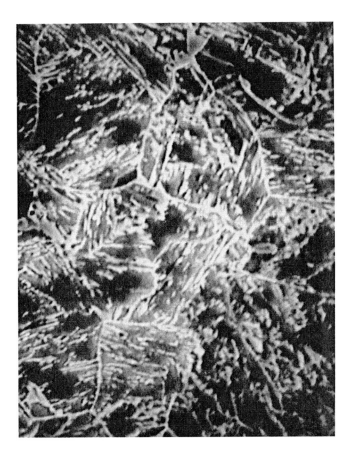

**Figure 6.51**   Microstructure of a PH 13-8 Mo stainless steel that has been solution annealed and aged for 4 h at 1050°C. (Courtesy of T. Mousel, Cal Poly.)

and Ni contents, respectively. They are susceptible to hydrogen embrittlement, but this too can be overcome to some extent by overaging to a strength level in the vicinity of 1173 (170 ksi).

These steels are finding a myriad of uses in small forged parts and even in larger support members in aircraft designs. They have been considered for landing gears. Many golf club heads are made from these steels by investment casting techniques, and the manufacturers proudly advertise these clubs as being made from 17-4 stainless steel. Golfers do not have the least idea what the numbers mean. They suspect that perhaps these clubs will reduce their handicaps from 17 to 4. The 17-4 is the most popular, although the 13-8 can

**Figure 6.52** Microstructure of Ferrallium 255 duplex stainless steel quenched from 1120°C showing islands of $\gamma$ in an $\alpha$-ferrite matrix. ×200. (Courtesy of Kerri Page, Cal Poly.)

attain higher-strength levels. The precipitate in an aged 13-8 steel is shown in Figure 6.51.

Duplex Stainless Steels

The duplex stainless steels contain both austenite ($\gamma$) and ferrite ($\alpha$) phases. These steels can be heat treated to strengths of 828 MPa (120 ksi) with 35% elongation. The chief advantage of these steels is their good corrosion resistance in salt water, making them candidates for marine applications. Chromium contents are about 25%. They have trade names such as Ferrallium 255, U50, and 7-Mo plus. They contain about 50% austenite–50% ferrite. They have a relative high solubility for molybdenum and it is the Mo content of about 2% that results in a good corrosion resistance to chloride environments. The microstructure of a duplex stainless is shown in Figure 6.52.

The properties of the various stainless steels and some of their applications are listed in Table 6.7. Many stainless steels are used at elevated temperatures, and for this reason they are sometimes classified in the category of *heat-resistant alloys*. The maximum operating temperatures of the austenitic, ferritic, and martensitic grades are listed in Table 6.8 and their creep and stress–rupture properties are shown in Figure 6.53.

### 6.9.2 Maraging Steels

The maraging steels, as a group, are the highest-strength steels currently available. Their yield strengths range from 1030 MPa (150 ksi) to 2420 MPa (350 ksi). The lower-strength maraging steels are on the same order of strength as the low-alloy high-hardenability steels such as 4130, 4340, and 8640, but the higher-strength maraging steels exceed the strength levels of all other steels.

The maraging steels are hardened by the precipitation of intermetallic compounds (e.g., $Ni_3Mo$ and $Ni_3Ti$), and in this sense their physical metallurgy is similar to that of the PH martensitic stainless steels. The latter, however, have a high chromium content, for good corrosion resistance. The maraging steels do contain sufficient carbon to form a low-carbon iron–nickel soft lathe martensite, but the intermetallic compounds really account for their strength. These steels typically contain high nickel (18%), cobalt (~9%), and molybdenum (~5%). Because of their low carbon content, their weldability is excellent, and the high alloy content provides good hardenability. They also have good fracture toughness, and except for the very high-strength alloys, adequate ductility. The properties in the aged condition of two of these steels are listed as follows:

| Grade | Ultimate tensile strength | | Yield strength | | | Fracture toughness | |
|---|---|---|---|---|---|---|---|
| | MPa | ksi | MPa | ksi | R.A. (%) | Mpa $\sqrt{m}$ | ksi $\sqrt{in}$. |
| 18 Ni (250) | 1800 | 260 | 1700 | 247 | 55 | 120 | 110 |
| 18 Ni (300) | 2050 | 297 | 2000 | 290 | 40 | 80 | 73 |

Probably their chief disadvantage is their high cost and somewhat lower ductility than that of most low-alloy steels. They have been used as missile

**Table 6.7** Room-Temperature Properties and Applications of Stainless Steels

| Type | Condition | Minimum Y.S. MPa (ksi) / U.T.S. [MPa (ksi)][a] | | R.A. % | Applications |
|---|---|---|---|---|---|
| Martensitic 410 | Quenched and tempered, 500°F | 800 (116) | 1000 (150) | 60 | High-strength and elevated-temperature applications; not suitable for cryogenic uses |
| Martensitic 431 | Quenched and tempered, 500°F | 750 (109) | 1000 (150) | 65 | Similar to 410; both have acceptable resistance to SCC at hardness below Rc 22 |
| Ferritic 430 | Annealed bar | 205 (30) | 450 (65) | 45 | Popular general purpose where welding is not required; low toughness after welding |
| Ferritic 409 | Annealed plate | 205 (30) | 380 (55) | — | General-purpose construction steels for automotive exhaust systems and light applications |
| Ferritic 405 | Annealed bar | 205 (30) | 415 (60) | 45 | Toughness satisfactory after welding |
| Austenitic 304 | Annealed bar | 205 (30) | 515 (75) | 50 | Good formability; use 304L (low carbon) for welding structures; general-purpose corrosion resistant |
| | Cold-worked bar ¼ hard | 515 (75) | 860 (125) | — | High-strength, corrosion-resistant structural parts; railway passenger cars, truck trailers, missiles, cryogenic |
| Austenitic 316 | Annealed bar | 205 (30) | 515 (75) | 5 | Mo added to increase pitting, general corrosion resistance to chloride environments |
| Austenitic 347 | Annealed bar | 205 (30) | 515 (75) | 5 | Stabilized with Nb and Ta to reduce sensitization in range 480°C (900°F) to 813°C (1500°C) |
| PH 17-4 martensitic | Quenched and aged, 1025°C | 1000 (145) | 1070 (155) | 45 | Weldable, strong, corrosion resistance, varied applications for small forged precision cast parts; aerospace, hydrogen embrittles |
| PH 13-8 Mo martensitic | Quenched and aged, 1025°C | 1210 (175) | 1275 (185) | 45 | Highest-strength stainless steels, aerospace, fasteners, cooler sections of aircraft engines, susceptible to hydrogen embrittlement |
| Duplex ferralium 225 | Annealed | 550 (80) | 760 (110) | 65 | Good corrosion resistance to chlorides, marine applications |

[a] Y.S., yield strength; U.T.S., ultimate tensile strength.

**Table 6.8**  Maximum Service Temperatures for the Major Stainless Steels

| AISI type | Maximum service temperature (°C) | |
| --- | --- | --- |
| | Continuous | Intermittent |
| | Austentic grades | |
| 201 | 815 | 845 |
| 202 | 815 | 845 |
| 301 | 840 | 900 |
| 302 | 870 | 925 |
| 304 | 870 | 925 |
| 308 | 925 | 980 |
| 309 | 980 | 1025 |
| 310 | 1035 | 1150 |
| 316 | 870 | 925 |
| 317 | 870 | 925 |
| 330 | 1035 | 1150 |
| 347 | 870 | 925 |
| | Ferritic grades | |
| 405 | 705 | 815 |
| 406 | 815 | 1035 |
| 430 | 815 | 870 |
| 442 | 980 | 1035 |
| 446 | 1095 | 1175 |
| | Martensitic grades | |
| 410 | 705 | 815 |
| 416 | 675 | 760 |
| 420 | 620 | 735 |
| 440 | 760 | 815 |

*Source*: *Metals Handbook*, Vol. 1, 10th ed., ASM International, Metals Park, Ohio, 1990, p. 878.

cases, aircraft structural forgings, springs, shafts, bolts, and punches. A cobalt-free maraging steel has been developed for the nuclear industry and a few stainless grades have been produced.

### 6.9.3  Tool Steels

Any steel used to make metals by cutting, forming, or shaping is a tool steel. For simplicity and brevity we will classify tool steels into the two broad categories of cold- and hot-worked tool steels. The ASM *Metals Handbook*, Vol. 1, 10th ed., pp. 757–792, has listings of compositions and properties in much more detail than will be presented here.

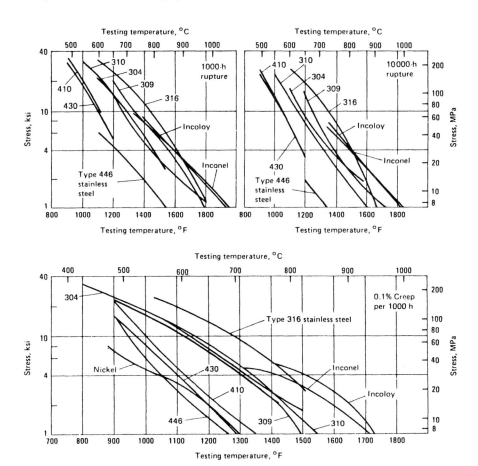

**Figure 6.53** Creep and stress–rupture comparisons for selected iron-based heat-resistant alloys. (From *Metals Handbook,* Vol. 3, 9th ed., ASM International, Metals Park, Ohio, 1980, p. 195.)

## Cold-Worked Tool Steels

Most of the cold-worked tool steels are subclassified according to the quenching media employed in the heat-treating process. *Water-hardening* tool steels known as the W type contain from about 0.6 to 1.4% carbon. They are essentially plain carbon steels that are inexpensive, have low hardenabilities, and as such must be used in thin sections. Typical applications include chisels, drills, lathe tools, and files. The *oil-hardening* (O) type have carbon contents in the range 0.9 to 1.5% with some small additions (0.5 to 1.0%) of chromium, tungsten, and manganese for hardenability. Thus they can be quenched in larger sections at slower rates and still possess a high hardness value. The quenched structure consists of martensite, bainite, some retained γ, and undissolved carbides (carbides that do not go into the γ phase during the austenizing anneal).

Generally, they are tempered slightly to reduce the brittleness of martensite. These steels are the most widely used tool steels. They have good toughness in the tempered condition and are subject to less distortion on quenching than are the W-type tool steels. They are used for shear blades and a wide variety of cutting tools. The *air hardening* (A type) contain 1 to 2% C and larger quantities of Cr, Mn, Mo, V, or Ni to achieve high hardenability. The as-quenched martensite is later tempered, but these steels tend to retain some γ. Sometimes at room temperature this retained γ transforms to martensite, which could result in dimensional changes. Applications for the A-type steels include punches, forming dies, blanking, and trimming dies.

The *shock-resistant* or S type have good toughness due to a relatively low carbon content of about 0.5%. They also contain small amounts of alloying elements, giving a medium range of hardenability. Most are oil quenched. These steels are used for repetitive wear applications, such as rivet sets, shear blades, punches, pipe cutters, and concrete drills.

There is one more cold-worked tool steel, called the D type for dies. They have high carbon (1.5 to 2.3%) and high chromium (12%) contents and were originally developed for high-temperature applications. However, they have been replaced by other high-temperature steels and are now used primarily for wear resistance, such as in long-run drawing dies, rolls, deep drawing dies, thread rolling, and slitters.

Hot-Worked Tool Steels

All hot-worked tool steels have one thing in common, and that is referred to as *secondary hardening*. This can be better explained by Figure 6.54, where the hardness versus tempering temperature is shown for molybdenum-containing hot-worked tool steels. The hot-worked steels actually show a hump in the tempering curve that is due to secondary hardening brought about by carbide precipitation, usually of Mo, W, or Cr carbides. These are more stable than Fe carbides. The hot-worked tool steels are used for extrusion and forging dies, die cast dies, and for high cutting rates of hard materials where the tool itself becomes very hot.

The hot-worked tool steels are classified as H, M, T, or W type. This sounds very confusing, but such classifications are often based on the chronological order in which the alloys were developed.

The H type are most often used as extrusion and forging dies. There are three types, based on the alloying element used to provide secondary hardening. They are listed as follows:

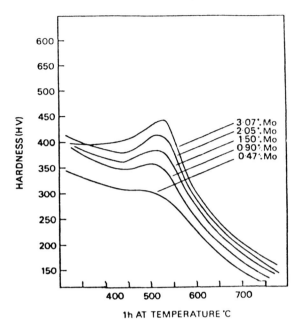

**Figure 6.54** Secondary hardening in Mo-containing tool steels. (From K. J. Irvine and F. B. Pickering, *J. Iron Steel Inst.*, Vol. 194, 1960, p. 137.)

| H 1 to H 19 | | Cr alloying element |
| H 20 to H 39 | W | Cr alloying element |
| H 40 to H 59 | Mo | Cr alloying element |

They all exhibit secondary hardening effects. The Cr steels have the higher hardenability characteristics, while the Mo-containing steels have a very marked secondary hardening peak. The W-containing steels, also used for extrusion and forging dies and mandrels, have a secondary hardening peak similar to the Mo steels but of less magnitude and at lower temperatures.

Finally, we have high-speed tool steels of the T and M types, containing tungsten and molybdenum, respectively. These are somewhat different from the H-type steels, in that the molybdenum contents are much higher than in the M-type steels of the H 40 to H 59 group of hot-working tool steels. However, in the T type the tungsten contents are only slightly higher than the tungsten contents of the H 20 to H 39 group, the latter containing from 8 to 19% W, while the T type contain around 12 to 19% W. Carbon contents are around 0.75 to 1.5 in both of these tungsten-containing tool steels. The M steels are much more popular than the T steels, primarily because of the approximately

40% lower cost of Mo compared to W. The M steels were developed after the T type, when large deposits of molybdenum were found in Colorado in the late 1930s. However, England still continued the use of the T steels during World War II and later, perhaps because of their familiarity and some better temperature control on the older T-type steels. The compositions and applications of the various types of tool steels are listed in Table 6.9.

## 6.10  CAST IRONS

Cast irons are iron–carbon alloys that contain carbon in amounts larger than the steels, varying from 2 to 4 wt % C. As such they are brittle alloys and must be cast rather than formed to their final shapes. Cast irons do not have the strength levels of most steels, except for perhaps the low-carbon steels in the annealed or hot-formed condition. Cast irons are easily melted, have a high fluidity, and thus can easily be cast to sizes and shapes that are difficult to form by other means. Their competition, from a metal-forming viewpoint, are forgings. Many products that are cast can also be forged to near-final shape. The choice must take into consideration properties and costs, the latter frequently being the determining factor. In large structures, such as machine tool bases, the cast irons have the ability to absorb vibrations more so than other metals. Also, most large structures cannot be forged. Cast irons are widely used in automotive engines, as the engine block, camshafts, and other parts, to the extent of comprising about 95% of the engine weight.

If we examine the iron–iron carbide phase diagram again (Figure 6.25) we find that the cast irons solidify with some eutectic component. Furthermore, at high carbon contents they can solidify with the Fe–carbon (graphite) structure rather than the Fe–$Fe_3C$ steel structures. The cast irons have wide variations in mechanical properties, depending on whether they form the stable austenite ($\gamma$)-graphite (C) or the metastable austenite-$Fe_3C$ eutectic. $Fe_3C$ is a metastable phase, even in low-carbon steels. But in the latter steels the tendency to decompose is virtually nil, so we treat the iron carbide as a stable phase. But this is not so in the cast irons, particularly the high-carbon alloys and those containing silicon.

There are essentially four types of cast irons: gray, white, malleable, and nodular (sometimes called ductile cast iron, even though it has limited ductility compared to other metals). Their properties are dependent on heat treatment and the quantities of carbon and silicon. Microstructures of the four types are presented in Figure 6.55.

### 6.10.1  Gray Cast Iron

In gray iron the carbon is in the form of graphite flakes. Silicon additions assist in making the $Fe_3C$ unstable. As the metal slowly cools in the mold the $Fe_3C$

**Table 6.9**  Compositions and Applications of Various Types of Tool Steels

| Type | Typical compositions | Applications |
|---|---|---|
| | Cold-worked tool steels | |
| W: Water hardening | 0.7–1.4% C | Chisels, drills, files, lathe cutting tools |
| O: Oil hardening | 0.8–1.5% C<br>0.5–1.0% Cr, W, Mn | Shear blades, wide variety of room-temperature cutting tools |
| A: Air hardening | 1–2% C, Cr, Mn, Mo, V,<br>and Ni to approx. 3% total | Punches, forming dies, blanking and trimming dies |
| S: Shock resistant | 0.5% C, 0.5–1.8% Cr,<br>0.1–1.5% Mo | Repetitive-wear applications, rivet sets, shear blades, punches |
| | Hot-worked tool steels | |
| H Type<br>H10–H19<br>H20–H39<br>H40–H59 | 0.3–0.5% C, 3–5% Cr<br>0.2–0.5% C, 8–19% W<br>0.5–0.7% C, 4–6% Mo | Shearing knives<br>Extrusion and forging dies for brass<br>Hot-worked steel tools |
| High-speed T | 0.65–1.6% C<br>0.9–5% V<br>12–19% W | High-speed tool steels, both T and M types, used for heavy-duty work calling for high hot hardness and abrasion resistance; hot forging and extrusion dies and mandrels, cutting tools |
| High-speed M | 0.7–1.5% C<br>1–6% W<br>4–11% Mo | |

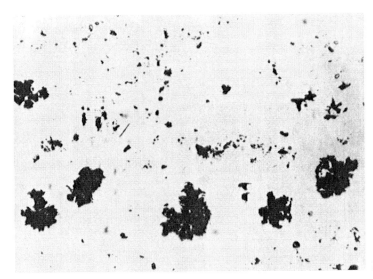

(b)

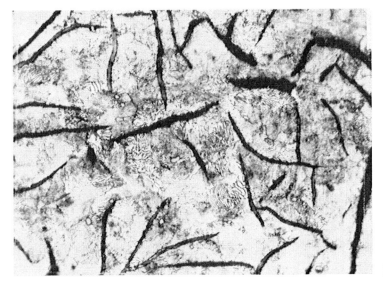

(a)

(d)

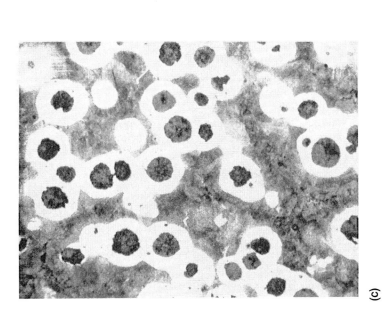

(c)

**Figure 6.55** Typical microstructures of cast irons: (a) gray 400×; (b) malleable 200×; (c) nodular 200×; (d) white, 480×. (Courtesy of John Haruff and Roxanne Hicks, Cal Poly.)

will decompose to graphite. Its name comes from the gray appearance of a fractured surface. The gray irons are noted for their good machinability and ability to absorb vibrations.

## 6.10.2  White Cast Iron

In white cast iron, the silicon and carbon contents are relatively low and the cooling rates moderately high. The carbon is thus primarily in the form of $Fe_3C$, which results in a very hard, white-looking microstructure. White cast iron is used where good wear resistance is important. Most white cast iron consists of pearlite and massive $Fe_3C$ regions. From examination of the phase diagram, an alloy containing about 3% C will solidify with some eutectic plus the $\gamma$ phase. At lower temperatures the $\gamma$ will be of the eutectoid composition and thus converts to pearlite on cooling through the eutectoid temperature. Very fast cooling can prevent pearlite from forming and this product, called *chilled cast iron,* will contain considerable amounts of martensite. Ni, Cr, and Mo additions promote the formation of martensite.

## 6.10.3  Malleable Cast Iron

This iron is made by heating white cast iron for long periods of time at about 950°C, causing the $Fe_3C$ to decompose to $\alpha$ and graphite, the graphite being in the form of small particles rather than as flakes. At the eutectoid temperature, on slow cooling, more $Fe_3C$ decomposes yielding more carbon. This carbon is called *temper carbon.* This malleable iron has slight ductility and varies considerably in properties, depending on the cooling rate from the graphitization temperature. If the alloy is fast cooled from the 950°C anneal to about 750°C, and then slowly cooled to room temperature, the $\gamma$ transforms to $\alpha + C$, with the graphite depositing on the existing temper carbon. This ferrite matrix cast iron is called ferritic malleable iron. Cooling very slowly to 870°C, followed by air cooling, allows the $\gamma$ to transform to pearlite. The microstructure then consists of temper carbon nodules in a pearlite matrix, known as pearlitic malleable iron. By cooling to around 850°C, holding for about 30 minutes to homogenize the $\gamma$, and then oil quenching to room temperature, a martensite matrix is produced. This martensite is subsequently tempered, resulting in a final microstructure consisting of carbon nodules in a tempered martensite matrix. In summary, malleable iron can consist of either a ferritic, pearlitic, or martensitic matrix containing graphite nodules, and the properties will vary accordingly.

## 6.10.4  Nodular Cast Iron

In this iron small additions of magnesium and/or cerium cause the graphite to form spherical particles instead of the flake form found in gray iron. The flake form has better machinability, but the spheroidal form yields much higher strengths and ductility. As in the case of the malleable iron the matrix can be ferritic, pearlitic, or martensitic, depending on the heat treatment process. The most common form is that shown in Figure 55c, where the graphite nodules are surrounded by white ferrite, all in a pearlite matrix.

From the brief presentation above one can see that cast iron metallurgy is just as varied and complex as that of steels. By understanding the Fe–Fe$_3$C phase diagrams and the associated heat treatments, it is relatively easy to comprehend the processing of cast irons. The nonmaterials engineer, however, is more interested in properties than in processing. The properties and some applications are summarized in Table 6.10. The nonmaterials engineer, by being familiar with the microstructures, has a better understanding of the advantages and disadvantages of the variety of cast irons available.

In recent years an austempered ductile iron has appeared on the market which can have a matrix varying from ferrite, to ferrite plus pearlite, to pearlite, to bainite, and even to martensite, depending on the austempering heat-treat cycle. Yield strengths can vary from 550 to 1300 MPa (80 to 188 ksi) and elongations from 1 to 10%. It has excellent wear properties and is used for gears, crankshafts, chain sprockets, and high-impact and high-fatigue applications.

## 6.11  FORMING AND SHAPING OF METALS

Most metal-forming operations comprise the ways in which a cast ingot can be formed into the final desired shape. In one situation these operations may consist of the manufacturing of mill products into rod, sheet, wire, and tube form, which are warehoused in various thicknesses and diameters for subsequent sale to manufacturers for their secondary processing to final size and shape. Mill manufacturing operations generally will include primary forming operations such as hot ingot breakdown by rolling, extrusion, or forging, followed by cold reduction operations that may include further rolling and drawing processes to a specific size and hardness for stocking in the warehouse. Secondary processing includes bending, spinning, shearing, deep drawing, stretch forming, and machining to final size and shape.

The forming to final shape by casting, die casting, powder pressing, and forging are operations that attempt to approach the final shape as nearly as possible in one operation. Near net shape forming has come a long way in the past decade (see the February 1990 issue of *Advanced Materials and Processes Journal* published by ASM International). This paper and the articles referenced included polymers and composites as well as metals, but we will cover only metals at this time. More recent developments and improvements in the

**Table 6.10** Types, Properties, and Applications of Cast Irons

| Type | Composition (wt %) | Yield strength [MPa (psi)] | Elongation % | Applications |
|---|---|---|---|---|
| | | Gray | | |
| UNS No. F10006 | 3.1–3.14% C, 0.6–0.9% Mn, 2.3–1.9% Si | 207 (30) | — | Cylinder blocks, pistons, brake drums, clutches |
| UNS No. F10008 | 3.0–3.3% C, 0.6–0.9% Mn, 2.2–1.8% Si | 276 (40) | — | Diesel engine castings, liners, cylinders, pistons, camshafts |
| | | Ductile (nodular) | | |
| UNS No. F32800 (ferrite matrix) | 3.5–3.8% C, 2.0–2.8% Si | 276 (40) | 18 | Shock-resistant parts, valves and pump bodies |
| UNS No. F34800 | 3.5–3.8% C, 2.0–2.8% Si | 483 (70) (mostly pearlite) | 3 | Best combination of strength and wear resistance; can be surface hardened: crank |
| | | Malleable | | |
| UNS No. F22200, ASTM A338 | 2.3–2.7% C, 1.0–1.7% Si | 224 (32) | 10 | Low-stress parts required, good machinability: steering-gear housings, brackets, carriers |
| ASTM A602 M5503 | 2.2–2.9% C, 0.9–1.9% Si | 379 (55) (tempered martensite) | 3 | For machinability and response to induction hardening |
| | | White | | |
| Class I, D | 2.5–3.6% C, 1.3% Mn, 1.0–2.2% Si, 5–7% Ni, 7–11% Cr | Hardness[a] 500–600 Brinell, tempered martensite | — | For abrasion resistance |

*Source:* Adapted from *Metals Handbook*, Vol. 1, 10th ed., ASM International, Metals Park, Ohio, 1990.

[a] Hardness only reported for white cast irons.

near net shape forming technology of metals include vacuum casting, powder metallurgy, hot isostatic pressing (HIPing), mechanical alloying followed by HIPing, forging through computer simulation, and metal injection molding.

## 6.11.1  Forming by Mechanical Means

Hundreds of processes have been developed for metalworking applications; however, these processes may be separated into a few categories listed as follows (after G. E. Dieter): (1) direct compression, (2) indirect compression, (3) tension-type processes, (4) bending processes, (5) shearing processes. These basic processes are depicted in Figure 6.56. The direct compression forces are the *rolling and forging* operations, where the metal moves at right angles to the

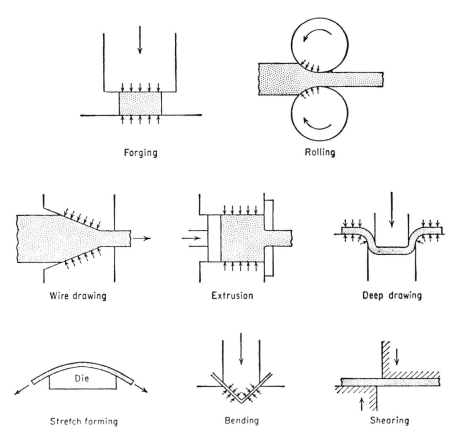

**Figure 6.56**  Mechanical metal-forming processes. (From G. E. Dieter, *Mechanical Metallurgy*, 3rd ed., McGraw-Hill, New York, 1986, p. 504.)

compression direction. Indirect compression includes *drawing* and *extrusion,* where the metal flows under the action of combined stresses that include a high compressive stress. Seamless tubes are made by extruding the metal over a mandrel.

The forging operation includes ingot breakdown as well as forging to the near final shape. In the former the ingot is often reduced to size by repeated blows from a drop hammer, whereas in the latter a single squeezing action is employed, causing the metal to be slowly squeezed by pressure into a mold of the final shape desired. Rolling includes both sheet and rod shape, the latter not being shown in Figure 6.56. In rod or round rolling the matching roll surfaces contain grooves, starting at one side of the roll with a large diameter and progressing in steps to the other side of the roll, which contains the smallest groove. The work piece is manually passed through the largest groove. The operator on the exit side turns the piece around and passes it back through the next smaller groove. Round rolling is an expensive operation (labor intensive) but is an easier way to introduce the desired amount of strain hardening than by drawing a rod through a die, or by cold extrusion. The smallest diameter produced in the round rolling operation is about 1/8 in. Further processing is done by drawing the 1/8-in. rod through successive wire drawing dies—a very fast, inexpensive, continuous operation. At what diameter does a rod become a wire? Who knows, but it could be the point at which drawing begins, and that will vary with the alloy in question.

There also exists the less frequently used *swaging* operation (Figure 6.57), which is employed for small runs to form rod and large-diameter wire. In this process the rod is usually fed through die sections that when placed together form a circular shape. There are two and four die-section swagers. The metal diameter is reduced by a hammering action created by a set of revolving hammers, which alternately strike the die sections as the hammers revolve through 360°. It is a very noisy and expensive operation. Gun barrels are made this way by predrilling a hole in the rod and hammering the metal down onto an inserted mandrel of the finished diameter size. Swaging provides a compressive stress and is useful for metals that cannot withstand the tensile force of a drawing operation. It is also useful in the laboratory since wire drawing and extrusion equipment is often difficult to fit into a small laboratory.

*Superplasticity* is a unique forming process that can be used on metals and alloys that exhibit superplastic behavior. Superplasticity occurs by grain boundary sliding and is accompanied by grain rotation. It occurs in certain metals at high temperatures (e.g., $0.7T_m$) and at low strain rates on the order of $10^{-3}$ per second. Tensile elongations as high as 4800% have been obtained in some superplastic alloys. A fine grain size, < 10 $\mu$m, is required for manifestation of this behavior. The grain shape is preserved during deformation, contrary to other mechanical forming methods. Superplastic forming of commercial

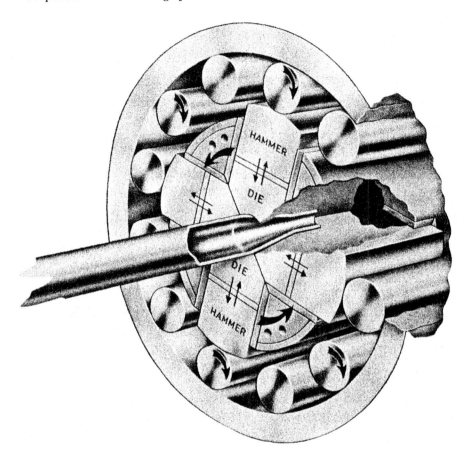

**Figure 6.57** Rotary four-die swaging machine. (Courtesy of Fenn Mfg. Co.)

aluminum-, titanium-, and nickel-based high-temperature alloys is done routinely. But not a large number of alloys exhibit this behavior. Another drawback to superplastic forming is the low strain rates required, which result in low production rates and the corresponding high costs.

*Metal injection molding* is a near net shape process that has been used for years in the polymer industry but has only recently been considered seriously for metal processing. The starting material consists of fine metal powder mixed with up to 50 vol % of a polymer binder material. After being injected into mold cavities, the binder is removed by either solvent extraction or thermal treatment. The metal powder forms are then sintered to densities of around

95%. For small parts this process competes with investment and die casting, powder metallurgy methods, and machining processes.

## 6.11.2 Forming by Casting Methods

Casting involves pouring or injection of a molten metal into a mold of the shape of the desired product. Casting methods can be broken down into *sand casting, investment casting,* and *die casting.*

Sand Casting

This casting method is the oldest of all methods and foundries have produced medium- to large-size parts by this method for many years. A two-piece mold is formed by packing sand around a pattern of the final desired shape. The second mold piece filled with packed sand fits over the first one containing the pattern. They are called cope and drag sections, respectively. Gates are provided in the cope section, into which the metal will be poured. The pattern, usually wood, is carefully removed from the drag section, and then the cope section with the gates is placed on top of the drag section that originally contained the pattern. Molten metal is then poured into the gates. The number and placement of gates depends on the size and shape of the part. Sand casting in the steel industry is used for rather large parts, such as machine bases and engine blocks. Bronze plaques and some aluminum and brass parts are also manufactured by the sand cast process.

Investment Casting

Other terms for this process include precision casting and the lost wax process. The pattern is usually made from wax or a low-melting easily formed polymer. The pattern is surrounded by plaster of paris or a similar ceramic slurry, and then the wax pattern is burned out after the ceramic hardens. Molten metal is then poured into the resulting cavity. Whereas in sand casting the pattern is reusable but the mold is expendable, in investment casting a reusable metal pattern die is used to produce expendable wax patterns, which in turn are used to produce expendable ceramic molds. These ceramic slurry molds are much faster and easier to form than are the sand molds. This sequence provides for mass production of parts. The molten metal is "thrown" into the mold by a centrifugal casting machine. In more recent years some manufacturers have used vacuum pumps to pull the molten metal into the mold. Investment casting has been used for over 50 years to produce intricate detailed shapes, such as jewelry rings, other jewelry of odd shapes, and dental alloy fixtures such as crowns, inlays and partial plates. But in more recent years it has also been applied to the manufacture of turbine blades, golf club heads, many titanium alloy shapes, and other rather high-tech parts.

Die Casting

Die casting is very similar to injection molding except that liquid metal is forced into a permanent mold, often made of steel. It is very useful for long runs of low-melting-point metals such as small parts of zinc, magnesium, and aluminum alloys. The split die is closed and the molten metal flows into a warm piston pump chamber where a plunger drives the metal into the die while the air escapes through vents. The die is then opened, the part ejected, the die then rapidly water cooled, closed (locked), and ready for the next cycle. Thousands of pieces can be made from each die before rework or replacement is required. It is a fast (usually greater than 1 part per minute), inexpensive process.

Continuous Casting

This process, which was developed 40 to 50 years ago and first employed in the copper industry, was then used primarily to produce long (several feet) ingots in order to minimize the amount of ingot breakdown and primary working in general. The molten mass of metal was continuously poured into a mold arranged such that the solidified ingot could also be withdrawn simultaneously from the other end. In the 1970s it was introduced into the steel industry and subsequently replaced some of the blooming and slabbing ingot breakdown operations. Today's continuous casters produce 150- to 320-mm (6- to 13-in.)-thick slabs in various widths, which are subsequently hot-rolled to final strip thickness.

### 6.11.3 Powder Metallurgy Methods

Powder metallurgy dates back to the years prior to World War II when the process was used on a limited scale that involved the pressing of metal powders, usually for small parts that were difficult to cast or machine, and the subsequent sintering of these powder compacts to bind the powders together. The sintering process is illustrated in Figure 6.58. Here we consider three small spherical powder balls that have been pressed with sufficient plastic deforma-

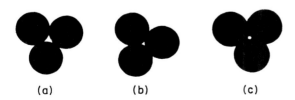

(a)          (b)          (c)

**Figure 6.58**   Three-particle sintering model.

tion that they are in contact along a short segment of their interface. When they are heated above about $0.5T_m$, diffusion of the atoms occurs sufficient to form a bridge between the particles and thereby decrease the pore size that existed among the three particles. As diffusion continues, the pore size decreases further, as shown in Figure 6.58c. Ideally, one would like to have zero porosity, but this is probably never achieved with pressing and sintering operations. Densities of the order of 95% of theoretical are considered quite good.

Some time in the 1950s, hot pressing, at least on an experimental level, began to be used. Pressing and sintering simultaneously caused much more plastic deformation to occur and thus provide a much larger interface contact area over which diffusion could take place. Densities in excess of 95% were common with the hot pressing process.

Room-temperature isostatic pressing was the next development, which, via the application of a hydrostatic pressure equally in all directions on the part, yielded a more uniform density. This process, commonly referred to as *cold isostatic pressing* (CIP), uses high fluid pressures to densify powders in rubber molds. After pressing and vacuum sintering, uniform densities in excess of 95% of theoretical are quite common.

*Hot isostatic pressing* (HIP) has been developed over the past 25 years and is now an established but expensive process used in the aerospace and defense industries, and is beginning to make inroads into the automotive, marine, medical, and electronic industries. The HIP process involves the application of high pressures via inert gases to pressure-tight containers in which the powders are enclosed. After pressing at the required high temperatures the metal container is removed. Uniform densities approaching 100% of theoretical have been achieved. This process has been used for near net shape fabrication of titanium alloys, superalloys, stainless steels, and aluminum alloys. A 16-in.-diameter HIP unit that attains temperatures to 1343°C (2450°F) and pressures to 310 MPa (45 ksi) is shown in Figure 6.59. Units to 60 in. in diameter have been constructed.

Although not envisioned originally, a rather significant beneficial aspect of the HIP process has been found to be that of densification of castings. The minimization or elimination of porosity in castings has been a long-sought goal. HIP has been described as a "defect-healing" process for investment castings of titanium- and nickel-based superalloys. HIPing of previously cold isostatic pressed parts has been found to be a good two-step process to achieve high densities and net or near net shapes. The initial CIP step produces a "green" preform shape using the inexpensive elastomeric tooling under hydrostatic pressure. The cold consolidated preform is then sintered to a density of about 95% to a near shape product. Subsequent HIPing achieves densities approaching 100%. Since the as-sintered preform has a closed porosity, the HIPing can be done without the necessity of canning the preform. This two-

**Figure 6.59** Sixteen-inch-diameter HIP Press. (Courtesy of International Materials Technology, Inc.)

step process, sometimes referred to as the CHIP process, has been used for titanium, nickel, iron, tantalum, and molybdenum alloys. Some typical parts are shown in Figure 6.60. The HIP and CHIP processes have also been extended to the forming of near net shape ceramic parts, metal matrix composites, and to high-pressure impregnation of carbon to create carbon–carbon composites.

Powder metallurgy has now entered the "engineered materials age." *Rapid solidification,* a process whereby liquid alloy particles are solidified at rates of $10^4$ to $10^6$ °C per second, has produced powders with structures not attainable by other means. These structures include metastable phases, amorphous structures, and dispersion-strengthed alloys.

Another approach that is now being used for superalloys and some aluminum alloys is that called *mechanical alloying.* This process involves the alternate shearing and cold welding together of particles of two or more species of greatly different hardness. This attrition takes place in high-intensity ball mills, the product subsequently being formed to the desired shapes by any of the pressing methods. Again, a unique nonequilibrium structure can be obtained.

The spray deposition process whereby the particles are formed and deposited in a single operation is also a powder metallurgy technique that we may see

**Figure 6.60** Number of high-temperature alloys that have been produced in a variety of shapes by the CHIP process. (Courtesy of Dynamet Technology, Inc.)

more of in the future. It offers the direct production of preform shapes along with the direct inclusion of reinforcing materials. Pushing the horizon even further are the techniques being developed to produce particles a few nanometers in size which can be pressed to 100% dense compacts with some rather interesting properties.

However, despite all the new processes developed or being explored, the press and sinter method is still the backbone of the powder metallurgy industry, primarily because of cost. The production of iron and steel powder compacts by the press and sinter route is used by the cost-conscious auto industry as well as by other industries involved in large quantities and low cost per part.

## 6.12  METAL JOINING

### 6.12.1  Welding

The American Welding Society defines a weld as a "localized coalescence of materials, wherein coalescence is produced by the heating to suitable temperatures, with or without the application of pressure and with or without the application of a filler material." Note that no mention is made of melting! Furthermore, there is no mention of the type of heat source used. The various processes include: (1) ultrasonic welding, where vibration of the part produces the atomic closeness and cleanliness requisite for bonding on the atomic scale; (2) low-temperature "forge welding" operations, where heat and pressure are superimposed to produce joining; (3) diffusion bonding, where lower pressures and higher homologous temperatures (still below melting) are used; (4) transient liquid processes; (5) resistance welding; and (6) the common arc and beam processes, where melting of the materials produces closeness and cleanliness on the atomic scale.

The most common arc welding processes are shielded metal arc welding (SMAW), gas metal arc welding (GMAW), gas tungsten arc welding (GTAW or TIG), plasma arc welding (PAW), and submerged arc welding (SAW). Typically, SMAW is manual (travel speed and wire feed controlled manually) or automatic (no manual control). The process selection depends on the material, the plate thickness, joint accessibility, and other factors. High-energy density processes or automated processes can often overcome material weldability problems, but typically increase the cost of capital equipment and joint preparation. Electron beam fusion welding is an example of such a process.

Welding is a decidedly nonequilibrium process, and produces changes not only in the fused metal, but also in the adjacent base metal, called the heat-affected zone (HAZ). The HAZ for a typical one-pass submerged arc weld in steel is depicted schematically in Figure 6.61. The rapid thermal cycle associated with the passage of the welding arc produces nine distinct regions in the vicinity of the weld, each with different microstructures and properties. Histor-

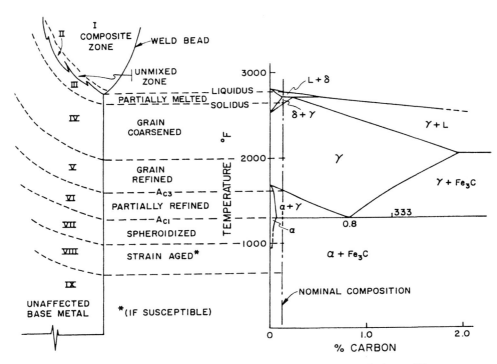

**Figure 6.61**   Schematic of the heat-affected zones (HAZ) in the fusion welding process. (Courtesy of D. Walsh, Cal Poly.)

ically, in steels, many problems have been associated with the partially melted and grain-coarsened regions, where the rapid thermal cycling may produce hard and brittle martensite.

Often, hydrogen present during the welding operation will diffuse into these areas and produce hydrogen-induced cracking. This may occur immediately postwelding, or after a delay of as long as several weeks. This insidious problem obviously poses a challenge for nondestructive evaluation. Postweld heat treatment, preheat, and welding energy input limitations (to control cooling rates) are used in an attempt to circumvent all such problems.

Figure 6.62 highlights the grain-coarsened region of a SAW weld. Figure 6.63 shows the fusion-line region of a weld in 316L stainless steel. The dendritic structure of the fusion zone is evident. Ferrite is the dark etching vermicular phase. The partially melted region is also evident as the complete grains in the HAZ are surrounded by liquated material. Full liquation of several alloy-

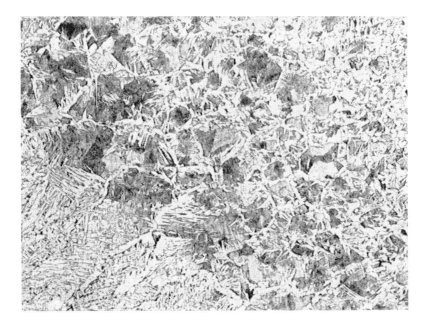

**Figure 6.62** Microstructure showing the coarse-grained region in a submerged arc weld in a 1030 steel. ×200. (Courtesy of D. Walsh, Cal Poly.)

rich inclusions is also evident. Such constitutional liquation is caused by rapid heating associated with the welding operation.

*Hardfacing* is the application of a hard, wear-resistant material to a metal surface by welding, spraying, or allied welding processes. This definition of hardfacing excludes nitriding, ion implantation, flame hardening, and similar processes. Hardfacing materials include a variety of cobalt-, nickel-, and iron-based alloys plus some ceramic materials such as carbides. Service performance requirements dictate both the hardfacing material and the process selection. *Thermal spraying* is considered here as an allied welding process since the surface is fused. As a rule, welding is preferred for hardfaced coatings.

## 6.12.2 Brazing and Soldering

Brazing and soldering involve the use of low-melting-point filler metals that are placed between two higher-melting-point metals. The latter are not fused, whereas the filler metal is melted and chemically bonds to the metals being joined. At one time it was thought that the difference between brazing and soldering was that in brazing a chemical bond was formed, whereas in soldering it

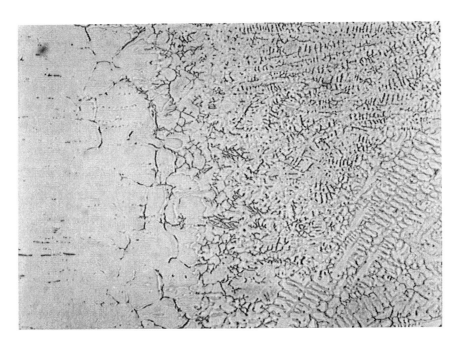

**Figure 6.63** Fusion line and dendritic structure in a 316L stainless steel weld. ×250. (Courtesy of D. Walsh, Cal Poly.)

was a more mechanical-like bond, with the solder sort of interlocking with irregularities on the metal surface. It is now realized that in both processes a chemical-like metallurgical bond is obtained, albeit to a miniscule depth. So the American Welding Society (AWS) has, somewhat arbitrarily perhaps, defined soldering as metal coalescence below 426.67°C (800°F). (Obviously, the Fahrenheit scale was used first.) The term *coalescence* could also have a variety of meanings. Perhaps the best way to describe soldering is that "the filler metal is distributed between the closely fitted surfaces by capillary action." But to add to the confusion, brazing is defined in the same way except that the temperature is above 800°F (incidentally, some sources use a temperature of 840°F). A number of definitions in metallurgy depend on the age of the definer.

Brazing a large number of units is normally performed in a furnace, while torch brazing is used for smaller production runs or repair operations on a single unit. Filler metals are usually silver alloys or Cu–Zn alloys. The silver alloys have melting temperatures in the range 593°C (1100°F) to 704°C

(1300°F), while the Cu–Zn alloys are somewhat higher, in the neighborhood of 871°C (1600°F). Fluxes, usually consisting of boric acid, borates, or fluorides, are required to decompose oxides (except in vacuum brazing operations).

Formerly, soldering alloys consisted primarily of alloys of lead and tin. But since lead has been banned for household plumbing, we now find alloys of silver ( ≈ 5%) with tin or other low-melting-point metals being used in household plumbing. The main requirements for a solder alloy system is that it have a low-melting-point eutectic composition and be relatively inexpensive. Other solders include Sn–Sb for stainless steels, Sn–Zn and Al–Zn for aluminum alloys, and indium-based solders for nonmetallic materials. Indium solders, often as pure indium, which is expensive (about $500 per pound) and has a melting point of 156.4°C (313°F), are used throughout the semiconductor industry since indium makes low-contact-resistance bonds with silicon.

## DEFINITIONS

*Age hardening*:    heat-treating process that increases hardness via precipitation of some type of phase, often a compound.

*Austempering*:    isothermal heat treatment of steel below the nose of an IT curve but above the Ms temperature that results in an all-bainitic structure.

*Bainite*:    microstructure that consists of a mixture of $\alpha$-iron with small $Fe_3C$ particles that has a hardness in the Rc range 40 to 45.

*Binary system* or *Binary phase diagram*:    alloy or phase diagram where more than 80% of the composition is made up of two elements.

*Cementite*:    $Fe_3C$ (iron carbide).

*Dendrites*:    grain structure in alloys where growth is favored in certain directions, giving a *Christmas tree*–like appearance. Snowflakes also have a dendritic appearance.

*Die casting*:    process of forcing a liquid metal into a specially shaped die.

*Dual-phase steels*:    steels composed of the $\alpha$-ferrite and martensite phases.

*Equilibrium*:    that condition whereby phases can exist in the same system with zero net change among the phases.

*Equilibrium diagram* (also phase diagram):    map that plots the conditions under which phases can exist in equilibrium, the conditions being temperature, composition, and pressure.

*Eutectic microstructure*:    microstructure consisting of parallel platelets of the $\alpha$ and $\beta$ phases that are formed when the eutectic structure solidifies.

*Eutectic point*:    lowest melting point and composition in a binary alloy.

*Eutectic reaction*:    liquid $\longrightarrow A_s + B_s$ ($s$ = solid).

*Eutectoid composition*:    point and a temperature on a phase diagram where one solid decomposes to form two different solid phases.

*Eutectoid microstructure*:   parallel platelets of two solid phases that result from the transformation that occurs on passing through the eutectoid point; this transformation is the *eutectoid reaction.*

*Extrusion*:   process of forcing a material through a die into a certain desired shape.

*Ferrite*:   α phase on the Fe–C diagram, which is a solid solution of carbon in iron (BCC).

*Forging*:   method of shaping objects by forcing them into a die or into a shape dictated by the forging machinery and mold or die shape.

*Gamma* (γ):   in steel, the solid solution phase, consisting of carbon in FCC iron.

*Gamma prime* (γ′):   phase, usually an intermetallic compound, that exists in superalloys and provides high-temperature strength.

*Gray cast iron*:   iron–carbon alloy of about 2 to 4 wt % carbon in which the carbon exists as flakes and a fractured surface has a gray appearance.

*Hardenability*:   capability to harden steel. The depth to which a steel bar can be hardened.

*Homogenization anneal*:   heat treatment, generally at high temperatures and for long periods of time, used to diminish segregation of elements. It promotes the establishment of equilibrium conditions.

*HSLA steels*:   high-strength low-alloy steels that are used primarily in large structures and in the machinery industry.

*Isothermal transformation diagram (IT)*:   diagram that shows the phases of steel (also used in other alloys) that exist during cooling as a function of time and temperature.

*Jominy end-quench test*:   method of measuring hardenability whereby a specimen is quenched from one end only.

*Liquidus*:   line or lines on a phase diagram above which the system is completely liquid.

*Malleable cast iron*:   cast iron produced by annealing white cast iron.

*Martensite*:   metastable phase formed by quenching.

*Matrix*:   in an alloy, the major and continuous phase.

*Mechanical alloying*:   process whereby alloying is achieved by repeated fragmentation and welding together of particles in a ball mill.

*Metastable*:   condition whereby a phase exists even though thermodynamic equilibrium does not exist.

*Microalloyed steels*:   type of HSLA steel that results from a small dispersion of carbides combined with a fine grain size.

*Monel*:   International Nickel Corp. trademark for a series of Cu–Ni alloys.

*Nodular cast iron*:   cast iron consisting of graphite nodules usually surrounded by ferrite and pearlite.

*Overaging*:   aging   beyond   the   point   of   maximum   strength   in   the precipitation-hardening process. The loss of coherency strains and coalescence of the precipitate occurs.

*Pearlite*: eutectoid microstructure in Fe–C alloys that is formed from alternate platelets of $\alpha$-iron (ferrite) and $Fe_3C$ (cementite).

*Phase diagram*: *see* Equilibrium diagram.

*Pig iron*: product that results from the reduction of iron oxide to iron in a blast furnace.

*Precipitation hardening*: *see* Age Hardening.

*Proeutectic phase*: solid phase that forms first in solidification of a liquid solution of two elements that form an eutectic system. It appears at temperatures below the liquidus and above the eutectic temperature.

*Proeutectoid phase*: solid phase that precipitates out of a single-phase solid solution as the solid solution phase enters a two-phase solid range (on the phase diagram) above the eutectoid temperature. Occurs only in eutectoid alloy systems.

*Refractory metals*: high-melting-point metals Re, W, Mo, Ta, and Nb.

*Segregation*: nonequilibrium separation of elements that occurs during alloy solidification.

*Solidus*: line or lines on a phase diagram below which all phases are in the solid state.

*Spheroidite*: Small $Fe_3C$ particles in ferrite that result from a *spheroidizing anneal* for long periods at temperatures just below the eutectoid temperature.

*Supersaturated solid solution*: solid solution that has solute in excess of that dictated by the phase diagram. It is a metastable phase usually formed by quenching from a temperature where the solute is not in excess of the solid solubility limit.

*Tempering*: process of annealing martensite whereby it decomposes into $\alpha + Fe_3C$.

*Terminal solid solubility*: solid solubility permitted by the equilibrium diagram near the pure base metal position on the diagram.

*Ternary phase diagram*: phase diagram that shows the phases that exist as a function of temperature and composition for a three-base-element system.

## QUESTIONS AND PROBLEMS

6.1. In Example 6.3 the liquid quantity in a composition of 60 at % Si–40 at % Au at 800°C was 66.7% of the total on an atom percent basis. Convert this to weight percent silicon precisely and also by interpolating from the weight percent figures given on the diagram.

6.2. For the composition in Problem 6.1, sketch the resulting microstructure if this alloy were slowly cooled to room temperature.

6.3. List three nonferrous and one ferrous precipitation-hardening alloys along with their strengths and ductilities in the aged condition.

6.4.  Using the Cu–Zn phase diagram of Figure 6.1, for the overall alloy composition, compute the relative quantities of the $\alpha$ and $\beta$ phases that exist at 600°C assuming equilibrium conditions.

6.5.  Why must elements that have complete solid solubility in the solid state possess the same crystal structure?

6.6.  For a 70% Cu–30% Ni alloy at equilibrium at a temperature of 1175°C, compute the amount of the solid phase. What is the composition for all phases within this alloy system at this equilibrium temperature?

6.7.  If the alloy in Problem 6.6 became a solid via being poured into a mold, sketch or describe how the composition gradient might exist across the ingot. Sketch the grain structure that would exist in this casting if it solidified slowly but yet too fast for equilibrium to be attained, and show which regions have a nickel content greater than 30% Ni.

6.8.  Why is K Monel so much stronger than other Monel alloys?

6.9.  Assume that a gold foil was placed on a silicon wafer (assume that foil is one-tenth the weight of the silicon) and heated to 365°C for a period of several hours and then suddenly cooled. Sketch the microstructure and identify the phases that would be present in the solidified compact.

6.10.  For a 60 wt % Cu–40 wt % Ag alloy that was melted and cooled to 770°C, where equilibrium was established, compute the quantity of proeutectic $\alpha$, eutectic $\alpha$, and total $\alpha$ in the resulting microstructure. Sketch this microstructure.

6.11.  Estimate the hardness that one would obtain on a 2024 Al alloy (a) quenched from the all-$\alpha$ region; (b) quenched and aged at 190°C for 24 h; and (c) furnace cooled from the all-$\alpha$ region.

6.12.  Consider pure copper and a Cu–2 wt % Be alloy. Which has the higher strength and which has the higher electrical conductivity? Which would be the better material for a power line and which would be more suitable for electrical contacts where impact is involved?

6.13.  The 2024 Al alloy is an age-hardened alloy, whereas most superalloys are considered to be dispersioned-hardened alloys. Explain the difference between these two alloy types.

6.14.  Estimate the room temperature strength-to-weight ratio for a 2024 Al alloy aged to the T6 condition, an aged Ti–6Al–4V titanium alloy, and the highest strength low alloy steel (Tables 6.1 and 6.6). Assume that the densities of the alloys are the same as that for the pure base metals, i.e., Al = 2.7, Ti = 4.5, and Fe = 7.87 g/cm$^3$.

6.15.  The refractory metals have a high density. Why are they of so much interest to the aerospace industry?

6.16.  For a 1020 steel cooled to room temperature under near-equilibrium conditions, compute the total of (a) the proeutectoid $\alpha$ phase; (b) the total $Fe_3C$ phase. Sketch the resulting microstructure and estimate the alloy's strength.

6.17. Describe two methods of obtaining a very desirable microstructure of around 40 to 45 Rc hardness in a 1080 steel.

6.18. What would be the benefits of adding small amounts (e.g., 0.5 wt %) of Cr and Mo to a 1080 steel? What would be the difference in the IT curves?

6.19. What heat treatment would you give to a 10100 steel slowly solidified alloy to make it useful? What strength would it have after your heat treatment?

6.20. Are there any advantages to oil quenching a 1080 steel versus water quenching? List the phases and the microstructures that would result in each case.

6.21. Why are microalloyed steels suitable for welding? What types of steel should not be welded?

6.22. Construct a table listing the relative advantages and disadvantages of the three stainless steels: ferritic, martensitic, and austenitic.

6.23. How is malleable cast iron heat treated to obtain its intended properties?

6.24. Where do the dual-phase steels fit into the overall HSLA category of steels in terms of strength and formability?

6.25. What is the highest-strength practical steel developed to date, and for what purpose might it be used?

6.26. Refer to the ternary Fe–Cr–Ni composition isotherms in Figure 6.50. In each of these isotherms locate the position of the alloy of composition 70% Fe–20% Cr–10% Ni and list the phases present at each temperature.

6.27. Oxygen-free high-conductivity copper (called OFHC) contains substantial quantities of oxygen in the form of oxides. Why is it labeled as OFHC?

## SUGGESTED READING

American Society for Metals, *Metals Handbook*, 10th ed., Vols. 1 and 2. ASM International, Metals Park, Ohio, 1990.

American Society for Metals, *Metals Handbook*, Desk ed., ASM International, Metals Park, Ohio, 1985.

American Society for Metals, *Metals Handbook*, 9th ed., Vol. 6, *Welding, Brazing and Soldering*, ASM International, Metals Park, Ohio, 1983.

Brandes, Eric A., Ed., *Smithells' Metals Reference Book*, 6th ed., Butterworth, Stoneham, Mass., 1983.

Dieter, G. E., *Mechanical Metallurgy*, 3rd ed., McGraw-Hill, New York, 1983.

Guy, A. G., *Elements of Physical Metallurgy*, 2nd ed., Addison-Wesley, Reading, Mass., 1959.

Massalski, T. B., Ed., *Binary Alloy Phase Diagrams,* ASM International, Metals Park, Ohio, 1986.
Smith, W. F., *Principles of Materials Science and Engineering,* McGraw-Hill, New York, 1990.

# 7

# Ceramics: Processing and Properties

## 7.1 INTRODUCTION

### 7.1.1 Traditional Ceramics

The word *ceramics* originated from the Greek word *keramos,* which translates to "potter's clay." Ceramics have a long history. Many centuries ago, dating back to about 6000 B.C., clay products, made by burning naturally occurring clays, were probably the first manufactured ceramics. The silicate glasses date back to a few thousand years B.C. Many centuries later, porcelain ceramic products were introduced that were made by mixing 50 to 60% of the mullite-type clays $(3Al_2O_3 \cdot 2SiO_2)$ with flint (ground $SiO_2$) and feldspar, an anhydrous alumina silicate containing $K^+$, $Na^+$, or $Ca^{2+}$ ions. When fired, this mixture resulted in crystalline phases bonded by a complex glassy phase. Figure 7.1 shows the traditional whiteware products resulting from various combinations of these three ingredients. The third group of traditional ceramics to emerge were the brick-type products, which initially used the sedimentary clay minerals that came from deposits in oceans, lakes, and swamps. These clays were of high silica and alkali content since they were mixed with sand and shale. Present-day bricks are vastly superior and contain a glassy or vitrified bond in the product. The ASTM has strength specifications for three grades of brick that vary from 10.3 to 20.7 MPa (1.5 to 3.0 ksi) in compressive strength. But we still place bricks in the traditional category of ceramics.

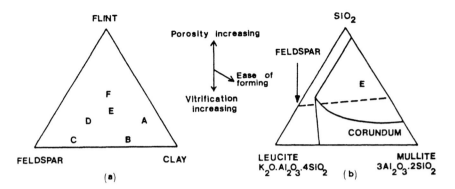

**Figure 7.1.** Composition of raw materials to produce certain ceramic products: A, wall tiles; B, vitreous whiteware; C, floor tiles; D, electrical porcelain; E, hard porcelain; F, semivitreous porcelain. (From I. J. McColm, *Ceramic Science for Materials Technologies,* Blackie & Son Ltd., 1983, p. 1. Distributed in the United States by Chapman & Hall.)

The latest emerging category of the traditional ceramics are the concretes, which are really composites consisting of rock, gravel, and sand, bonded together by some type of cement. Portland cements, which have been around for a little more than 100 years, contain compounds of calcium, silicon, aluminum, and oxygen [e.g., tricalcium aluminate $(Ca_3Al_2O_6)$]. The chemical reactions and resulting properties of concretes are well understood and form a technology rather than an art, as compared to some of the older traditional ceramics.

Other traditional ceramics include the refractories, both clay and the nonclay types used for heat insulation, and the abrasive cutting, grinding, and polishing group of materials. Most of the traditional ceramics contain silicates (i.e., the silicon–oxygen–metal combinations).

## 7.1.2 Nontraditional Ceramics

During the past 30 years or so, newer, more scientifically developed ceramics have emerged that have been labeled with a number of different titles. In the classic ceramic science textbook by Kingery et al. (see the Selected Reading), they are called *physical ceramics.* Others have applied terms such as *technical ceramics; fine ceramics,* because of the very small particle size of the powders used in processing the ceramic product, and modern; or *advanced ceramics,* to separate the new from the old. Such a distinction, according to raw materials, processing methods, properties, and applications, is presented in Table 7.1. The applications part of this classification probably serves best to distinguish

**Table 7.1** Distinction Between Conventional and Advanced Ceramics

|  | Traditional | Advanced |
| --- | --- | --- |
| Raw materials | Natural clays, $SiO_2$ (sand, quartzite), feldspar, anhydrous alumina silicate, gravel, sand, kalolinite | Prepared powders[a]: $Al_2O_3$, $Si_3N_4$, $ZrO_2$, SiC, Mgo, $B_4C$, $BaTiO_3$ |
| Processing Methods | Slip casting, baking, cementing, sun-dried | Slip casting, HIP, pressing-sintering |
| Products | Whiteware, bricks, concrete, glasses, abrasives, insulators | Insulators, lasers, dielectrics, superconductors, substrates, piezoelectrics, magnetics, electro-optics, structural |

[a] Powders are prepared by mechanical attrition, precipitation, sol gel techniques, plasma techniques, calcining, and spray techniques.

these two groups. The newer ceramics that have been developed to date have been designed either for electronic and electro-optic applications or for elevated-temperature use in jet engines, although significant use in the latter case has not yet materialized.

In previous chapters the chief advantages of structural ceramics—their high-temperature strength and relative inertness to attack by gaseous elements—have been emphasized. From the physical and chemical property viewpoint, we have noted their good electrical and thermal insulation characteristics and corrosion resistance. Their negative feature, particularly for structural applications, is their brittleness. Some success has been made in developing tougher and less brittle ceramics, but to steal a familiar Wall Street phrase, these developments must still be treated with "cautious optimism." General processing difficulties also arise because of their brittle nature.

### 7.1.3 Review of the Mechanical Behavior of Ceramics

Both crystalline and noncrystalline glassy ceramics are brittle. They have a low fracture toughness. They are *notch sensitive* (i.e., their strengths are extremely sensitive to scratches and notches). These stress concentrations can magnify a small applied stress to the point where the stress at the notch tip is sufficient to initiate a crack. Furthermore, the strain energy stored during the stress application is more than that required to create two new surfaces, so the crack propagates instantaneously and catastrophically to failure when the stress at the crack tip attains a critical value. This is what is meant by a low fracture toughness. The fracture toughness of plate glass is on the order of 0.7 Mpa $\sqrt{m}$, compared to common steels of around 50 to 200 MPa $\sqrt{m}$, and even higher values for the more ductile metals, such as aluminum and copper. Typical values for crystalline ceramics are not much higher than that for glass. Commercial $Al_2O_3$ has about a 1 MPa $\sqrt{m}$ fracture toughness. All the fracture toughness data for ceramics show a lot of scatter since the values depend very strongly on processing methods and surface conditions. But these values are still low. One can obtain much higher fracture toughness numbers on glass above its transition temperature $T_g$ (Figure 4.28). But polycrystalline ceramics do not really have a transition temperature. They are brittle even at temperatures near their melting points, due to the grain boundary sliding and concomitant cracking that occurs at high temperatures.

The mechanisms of plastic flow in crystalline and amorphous ceramics are different. Plasticity in crystalline ceramics occurs by dislocation motion, the same as in metals. But the amount of dislocation flow prior to fracture in ceramics is very limited. At room temperature the stress to move dislocations in ceramics is much higher than that in metals, because of their higher bond strength, complex crystal structures, and limited-slip systems. In 1958 many materials scientists were excited by papers published by Parker and co-workers (see the Suggested Reading, paper by R. J. Stokes) which suggested that ductile ceramics were a distinct possibility, particularly at high temperatures, higher than that of the melting point of metals. Such an achievement would mean higher operating temperatures and marked improvement in the efficiency and performance of jet aircraft. During World War II, the Germans built some ceramic jet engine components, but apparently they were never tested in jet aircraft.

During the period 1958–1972, it was established that carefully prepared defect-free surfaces of single-crystalline ceramic specimens could be deformed extensively at slow stain rates in a plastic manner (Table 7.2 and Figure 5.24).

**Table 7.2**   Relative Brittleness of Ceramic Materials

|  | Dislocation behavior | | Typical materials | |
|---|---|---|---|---|
| Type | Mobile | Maneuverable | Room temperature | High temperature |
| Completely brittle | No | No | TiC, $SiO_2$, $Al_2O_3$, Si, $TiO_2$ | |
| Semibrittle | Yes | No | $CaF_2$, MgO, LiF, NaCl, CsCl | $Al_2O_3$, $TiO_2$, MgO |
| Ductile | Yes | Yes | AgCl, AgBr | $CaF_2$, LiF, NaCl, CsCL |

*Source*: R. J. Stokes, Chapter 4 in *Fracture*, Vol. 7, H. Liebowitz, Ed., Academic Press, New York, 1972, p. 159.

But they were not tough (i.e., cracks, once formed, readily propagated to fracture). In surface-polished single crystals, cracks formed internally at intersecting slip bands, and these also readily propagated to failure under relatively low stresses. These cracks at the intersecting slip bands formed via dislocation motion, but the amount of motion was very small before the cracks formed. This behavior is not found in metals since dislocation motion is easier, allowing plastic flow to relieve the stresses at intersecting slip bands. But more important, in polycrystalline ceramic materials cracks formed at grain boundaries at low temperatures as a result of dislocation pileups, and at higher temperatures (e.g., $0.6T_m$), grain boundary sliding caused cracks to form in polycrystalline ceramics. Figure 7.2 shows the appearance of MgO bicrystals which were subjected to resolved shear stresses of 1060 g/mm$^2$ (Specimen No. 3) and 428 g/mm$^2$ (Specimen No. 5). The load was applied in axial compression at a temperature of 1373°C ($0.54T_m$). At temperatures in the vicinity of 1400°C ($0.64T_m$) this sliding occurred over a period of time in a discontinuous fashion (Figure 7.3), indicating that a diffusion-controlled incubation period was required to initiate each continuous sliding segment. Each cataclysmic sliding event is initiated by the migration of grain boundary jogs and terminated when another grown-in jog is reached. Fracture occurs when a criti-

Specimen        Specimen
No. 3            No. 5

**Figure 7.2.**  Grain boundary sliding displacement in MgO bicrystals at temperatures near $0.5T_m$. (From M. A. Adams and G. T. Murray, "Direct Observations of Grain-Boundary Sliding in Bi-crystals of Sodium Chloride and Magnesia," *J. Appl. Phys.,* Vol. 33, No. 6, 1962, p. 2126.)

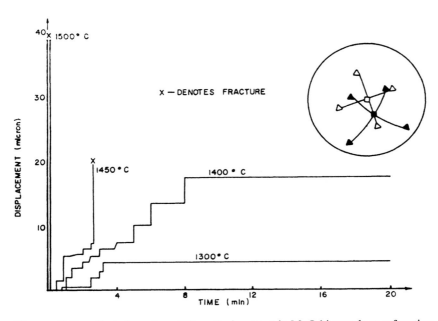

**Figure 7.3.** Grain boundary sliding displacement in MgO bicrystals as a function of time and temperature. The bicrystals were loaded in compression such that a resolved shear stress of 1450 g/mm² was applied along the boundary. (From G. T. Murray, J. Silgailis, and A. J. Mountvala, "Creep-Rupture Behavior of MgO Bicrystals," *J. Am. Ceram. Soc.*, Vol. 47, No. 10, 1964.)

cal crack is formed, owing to the stress created around an existing jog. In tricrystals fracture occurred at the meeting point of the three boundaries. These cracks propagated to failure because of the low fracture toughness of ceramics. In metals, grain boundary sliding also occurs at high temperatures, but metals can plastically deform and relieve the stresses created by the grain boundary sliding. In summary, ceramics are brittle because of their strong covalent and ionic bonds, thereby restricting dislocation motion (plastic deformation), because of grain boundary effects and also because of their complex structures, many being of a hexagonal type that provide only a limited number of slip systems. In Section 5.6 it was pointed out that five independent slip systems were required to deform a body to any arbitrary shape. In conclusion, ceramics are inherently brittle.

During the past 10 to 15 years, efforts have been made, with some degree of success, to achieve a higher fracture toughness in ceramic materials. These efforts have centered around processes known as *transformation toughening,* in which crystallographic transformations reduce the stress at the crack tip and slow its rate of propagation. Similar methods deflect the crack path via second-phase particles. It is interesting that fracture toughness values on the order of 20 MPa $\sqrt{m}$ have been reported in transformation-toughened zirconia. This is quite an achievement. But for comparison purposes, a high-strength steel with a bainite or tempered martensite microstructure has a fracture toughness on the order of 80 MPa $\sqrt{m}$. Also, this improved fracture toughness in ceramics does not necessarily lead to higher ductilities.

## 7.2 SILICATE STRUCTURES

### 7.2.1 SiO$_4$ Tetrahedra and Chain Structures

In Figure 4.18 the more common cubic crystal structures were displayed and discussed to some extent. The $BaTiO_3$ and the $Al_2O_3$ corundum hexagonal structures shown previously (Figure 4.20) are more typical of the complex ceramic crystalline structures that we encounter, and to some extent account for their brittleness. Many of the more common structures and their compounds are summarized in Table 7.3. However, the silicates comprise the largest segment of the traditional ceramic industry. Many of the silicate ceramics are in the form of glasses. Because of their importance and their numerous combinations, they warrant a separate section.

The silicates have rather unique structures in that the basic building block, the $SiO_4^{4-}$ tetrahedron (Figure 4.30), forms all sorts of structures that include many widely used ceramic materials. Silicon and oxygen have a cation/anion ratio of 0.29, thus per Table 4.2 a coordination number of 4 is required, yielding a tetragonal structure of four oxygen ions bonded to a single central ion (i.e., the $SiO_4^{4-}$ tetrahedron). The $-4$ valency in large part affords the possibility of structural variety by the numerous ways in which this valency can be neutralized. First, the $-4$ tetrahedron can be connected to positive metallic ions, such as $K^+$, $Na^+$, $Ca^{2+}$, $Mg^{2+}$, $Al^{3+}$, $Si^{4+}$, and $Fe^{2+}$ to form a large number of compounds. Second, the $SiO_4^{4-}$ tetrahedra can connect to themselves when two oxygen atoms share a pair of electrons with a second silicon ion. This oxygen is the bridging oxygen. In this fashion ring, chain and sheet structures, which may or may not be bonded to metallic ions, can be formed. When they are not connected to metallic ions, they connect to each other in such a manner as to form one of the basic $SiO_2$ (silica) structures, quartz, tridymite, or cristobalite.

The $SiO_4^-$ tetrahedra can be linked such that corners are shared in a number of ways, resulting in several different valencies for these linked structures, each valency depending on how they are linked. Figure 7.4 shows a number of chainlike structures and their resulting valencies as a result of the different tetrahedra linking methods. These structures then combine with metallic ions to

**Table 7.3**  Some Common Ceramic Structures

| Name of structure | Packing of anions | Coordination of anions | Coordination of cations | Examples |
|---|---|---|---|---|
| Rock salt | Cubic close-packed | 6 | 6 | NaCl, MgO, CaO, LiF, CoO, NiO |
| Zinc blende | Cubic close-packed | 4 | 4 | ZnS, BeO, SiC |
| Perovskite | Cubic close-packed | 6 | 12, 6 | $BaTiO_3$, $CoTiO_3$, $SrZrO_3$ |
| Spinel | Cubic close-packed | 4 | 4, 6 | $FeAl_2O_4$, $MgAl_2O_4$, $ZnAl_2O_4$ |
| Inverse spinel | Cubic close-packed | 4 | 4 (6, 4)[a] | $FeMgFeO_4$, $MgTiMgO_4$ |
| CsCl | Simple cubic | 8 | 8 | CsCl, CsBr, CsI |
| Fluorite | Simple cubic | 4 | 8 | $CaF_2$, $ThO_2$, $CeO_2$, $UO_2$, $ZrO_2$, $HfO_2$ |
| Antifluorite | Cubic close-packed | 8 | 4 | $Li_2O$, $Na_2O$, $K_2O$, $Rb_2O$ |
| Rutile | Distorted cubic close-packed | 3 | 6 | $TiO_2$, $GeO_2$, $SnO_2$, $PbO_2$, $VO_2$ |
| Wurtzite | Hexagonal close-packed | 4 | 4 | ZnS, ZnO, SiC |
| Nickel arsenide | Hexagonal close-packed | 6 | 6 | NiAs, FeS, CoSe |
| Corundum | Hexagonal close-packed | 4 | 6 | $Al_2O_3$, $Fe_2O_3$, $Cr_2O_3$, $V_2O_3$ |
| Ilmenite | Hexagonal close-packed | 4 | 6, 6 | $FeTiO_3$, $CoTiO_3$, $NiTiO_3$ |
| Olivine | Hexagonal close-packed | 4 | 6, 4 | $Mg_2SiO_4$, $Fe_2SiO_4$ |

*Source:* D. W. Richerson, *Modern Ceramic Engineering*, 2nd ed., Marcel Dekker, New York, 1992, p. 11; as adapted from Kingery et al.

[a] First Fe in tetrahedral coordination, second Fe in octahedral coordination.

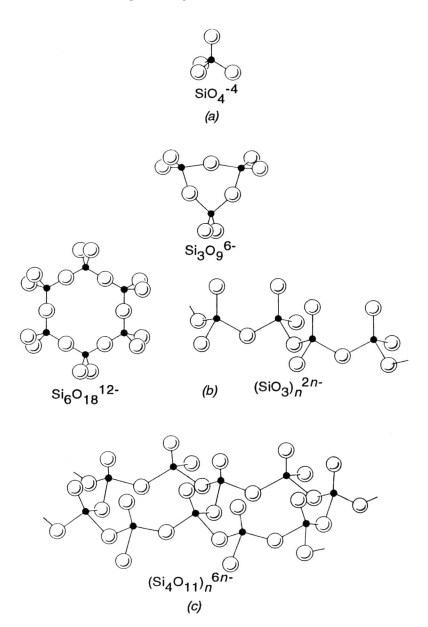

**Figure 7.4.** Silicate ions and chain structures. (From K. M. Ralls, T. H. Courtney, and J. Wulff, *Introduction to Materials Science*, Wiley, New York, 1976, p. 137.)

**Table 7.4** Bonding of Silicate Structures

| Bonding of tetrahedra | Structure classification | Schematic | Examples |
|---|---|---|---|
| Independent tetrahedra | Orthosilicates | $SiO_4^{4-}$ | Zircon ($ZrSiO_4$), mullite ($Al_6Si_2O_{13}$), forsterite ($Mg_2SiO_4$), kyanite ($Al_2SiO_5$) |
| Two tetrahedra with one corner shared | Pyrosilicates | $Si_2O_7^{6-}$ | Ackermanite ($Ca_2MgSi_2O_7$) |
| Two corners shared to form ring or chain structures | Metasilicates | $Si_6O_{18}^{12-}$  $(SiO_3)_n^{2n-}$ | Spodumene [$LiAl(SiO_3)_2$], wollastonite ($CaSiO_3$), beryl ($Be_3Al_2Si_6O_{18}$), asbestos [$Mg_3Si_2O_5(OH)_4$] |
| Three corners shared | Layer silicates | See Kingery, et al., pp. 75–79 | Kaolinite clay [$Al_2(Si_2O_5)(OH)_4$], mica (muscovite var.) [$KAl_2(OH)_2(AlSi_3O_{10})$], talc [$Mg_3(Si_2O_5)_2(OH)_2$] |
| Four corners shared | Framework silicates | See Kingery, et al., pp. 71–74, for $SiO_2$ and derivatives | Quartz, cristobalite, tridymite ($SiO_2$), feldspars (albite var.) ($NaAlSiO_3$), zeolites (mordenite var.) [$(Ca, Na_2)Al_2Si_9O_{22} \cdot 6H_2O$] |

*Source:* D. W. Richerson, *Modern Ceramic Engineering*, 2nd ed., Marcel Dekker, New York, 1992, p. 17. *Note:* The last two items of this table were taken from Kingery et al.

form a number of different ceramic compounds, some of which are listed in Table 7.4.

## 7.2.2 Sheet Silicate Structures

The layer silicate structures are made up of the $SiO_4^{4-}$ tetrahedra with three shared corners, resulting in a $(SiO_5)_n^{2n-}$ valency (Figure 7.5). These sheet or layer silicates include talc, mica, and the clay minerals, such as kaolinite (Table 7.4). Al substitutes for some of the Si in the sheet and also $(OH)^-$ ions are often incorporated. In many of the sheet silicate structures, weak van der Waals bonds exist between the layers. These bonds are easily broken. Clays become slippery when water is introduced, which acts to break the weak bonds. The layers in talc can easily slide past each other, as these weak bonds are broken under slight stress. The layers of mica are bonded together with $K^+$ ions and are somewhat more difficult to separate. Other layered structures can be found in the book by Kingery et al.

## 7.3 GLASSES

### 7.3.1 Fused Quartz

In the various forms of silica, $SiO_2$, four corners of the tetrahedra are shared. Crystalline $SiO_2$ exists in several crystalline forms. Three basic structures,

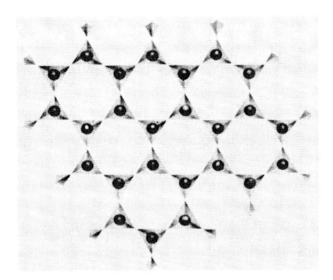

**Figure 7.5.** Arrangements of $SiO_4$ tetrahedra in a sheet silicate structure. (From W. G. Moffatt, G. W. Pearsall, and J. Wulff, *The Structure and Properties of Materials*, Vol. 1, Wiley, New York, 1964.)

quartz, tridymite, and crystobalite, each exist in two or three different modifications. The room-temperature stable structure of quartz was shown in Figure 4.31. It is formed by bridging every oxygen atom between two silicon atoms. This is the structure of quartzite rock, quartz sand, and flint. When this form of quartz is melted, somewhere between 1701 and 1734°C (the reported melting point varies), and cooled extremely slowly, a single crystal can be formed. These crystals are used in the electronic industry for frequency control. Synthetic quartz crystals are now often manufactured by a hydrothermal process. But if the fused silica, which has a viscosity of about $10^7$P at its melting point, is solidified with a crystallization velocity (i.e., the rate of growth of the liquid–solid interface, greater than $2.2 \times 10^{-7}$ cm/s), no appreciable amount of crystalline cristobalite will be formed. The resulting product is an amorphous glass that is commonly referred to as fused or vitreous quartz.

## 7.3.2  Network Modified Glasses: Plate Glass

Most silicate glasses contain network modifiers such as CaO and $Na_2O$, which supply cations to the molten silica to further discourage crystallization. The addition of $Na_2O$ produces two $Na^+$ ions, which serve to break up two bridging oxygen atoms. One ion of $Ca^{2+}$ would do the same job. The network structure of these glasses were depicted in Figure 4.32. The lack of crystallinity is evident by the presence of numerous nonbridging oxygen atoms. This bridging is necessary for crystallinity. These glasses are often referred to as soda-lime glasses (the $Ca^{2+}$ coming from lime) and are ingredients in common plate glass. In commercial plate glasses the modifiers constitute about 20 wt % of the total glass. Because there is not a great difference between weight and atom percent here, it means that less than three-fourths of the oxygen atoms are bridged.

The glass transition temperature characteristics were presented in Figure 4.28 in our discussion of polymers and will not be repeated here. The schematic drawings of the glass transition temperature for polymers and the common glasses are identical. On cooling the liquid there is a discontinuous change in volume (density) at the melting point if crystallization occurs. If no crystallization occurs, the volume decreases at the same rate in the liquid state as it does when a rigid solid glass is formed. This "solid" glass has a viscosity of $10^{12}$ to $10^3$ P, sufficient to be called a solid, although the term *supercooled liquid* is also appropriate. Eventually, it attains its maximum density at the glass transition temperature $T_g$, any further decrease being due to a decrease in the expansion (contraction) coefficient. Below $T_g$ the glass is in a true solid glassy state rather than being a supercooled liquid. The glass transition temperature $T_g$ is somewhat dependent on the cooling rate, increasing with increasing cooling rate.

### 7.3.3 Other Glasses

Lead–Alkali Glasses

Lead-alkali glasses are produced by substituting lead oxide for some of the lime in the soda-lime glasses. This addition causes the dispersion of light, resulting in a more brilliant ornamental-type glass. Higher lead contents are used for glass shielding of gamma and x-rays.

Borosilicate Glasses

When boric acid ($B_2O_3$) is substituted for some of the alkali and $SiO_2$ content, the coefficient of thermal expansion is reduced by about one-third, the softening temperature is increased somewhat, and the resulting glass is more resistant to chemical attack. This glass is known by the trade name Pyrex (Corning).

96% $SiO_2$ Borosilicates

The 96% $SiO_2$ borosilicate glass contains about 4% $B_2O_3$ and is even more heat resistant than Pyrex-type glasses. It is known by the trade name Vycor (Corning).

Glass Ceramics

Glass ceramics are polycrystalline materials, not true glasses, although they have been defined in that manner. Typical compositions are $2MgO \cdot 2Al2O_3 \cdot 5SiO_2$ and $Li_2O \cdot Al2O_3 \cdot 4SiO_2$. Their strengths are four to five times that of plate glass and they have low coefficients of thermal expansion. We know them as Corningware. The chemical compositions and applications of these glasses are listed in Table 7.5.

### 7.3.4 Processing and Properties of Glasses

The viscosity of a glass determines its melting conditions, working and annealing temperatures, upper use temperature, and devitrification rate. Viscosity is a measure of the resistance of Newtonian behavior materials to deformation. The viscosity $\eta$ is the ratio of the applied shear stress to the change in velocity of the flow of the material. If two parallel planes of the material of area $A$, a distance $d$ apart, are subjected to a tangential force difference $F$, the viscosity can be expressed as

$$\eta = \frac{Fd}{Av} \qquad (7.1)$$

The viscosity varies exponentially with temperature as

$$\eta = \eta_0 \exp\left[\frac{Q}{RT}\right] \qquad (7.2)$$

**Table 7.5**  Properties, Compositions, and Uses of Important Commercial Silicate Glasses

| Type | Major components (%) | | | | | | | | Comments |
|---|---|---|---|---|---|---|---|---|---|
| | $SiO_2$ | $Al_2O_3$ | CaO | $Na_2O$ | $B_2O_3$ | MgO | PbO | Other | |
| Fused silica | 99 | | | | | | | | Very low thermal expansion, very high viscosity |
| 96% silica (Vycor) | 96 | | | | 4 | | | | Very low thermal expansion, high viscosity |
| Borosilicate (Pyrex) | 81 | 2 | | 4 | 12 | | | | Low thermal expansion, low ion exchange |
| Containers | 74 | 1 | 5 | 15 | | 4 | | | Easy workability, high durability |
| Flat (float) | 73 | 1 | 9 | 14 | | 4 | | | High durability |
| Lamp bulbs | 74 | 1 | 5 | 16 | | 4 | | | Easy workability |
| Lamp stems | 55 | 1 | | 12 | | | 32 | | High resistivity |
| Fiber (E-glass) | 54 | 14 | 16 | | 10 | 4 | | | Low alkali |
| Thermometer | 73 | 6 | | 10 | 10 | | | | Dimensional stability |
| Lead glass tableware | 67 | | | 6 | | | 17 | $K_2O$ 10 | High index of refraction |
| Optical flint | 50 | | | 1 | | | 19 | $\left\{\begin{array}{l}BaO\ 13\\ K_2O\ 8\\ ZnO\ 8\end{array}\right.$ | Specific index and dispersion values |
| Optical crown | 70 | | | 8 | | | 10 | $\left\{\begin{array}{l}BaO\ 2\\ K_2O\ 8\end{array}\right.$ | Specific index and dispersion values |

*Source:* L. H. Van Vlack, *Materials for Engineering*, Addison-Wesley, Reading, Mass., 1982, p. 352.

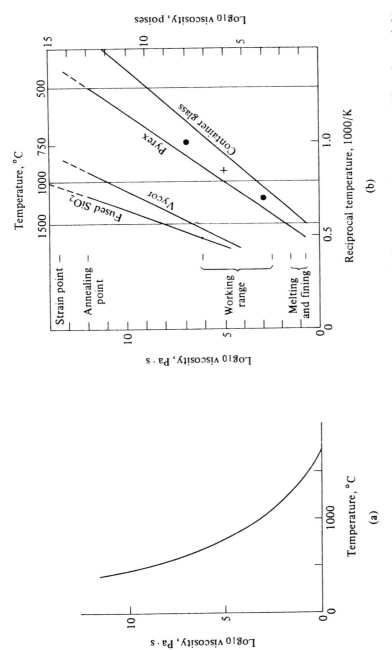

**Figure 7.6.** Viscosities and temperatures of interest for some commercial silicate glasses (a) common soda-lime glass and (b) from Equation 7.2. (From L. H. Van Vlack, *Materials for Engineering*, Addison-Wesley, Reading, Mass., 1982, p. 358.)

and will also vary considerably with composition. $Q$ is the activation energy for the flow process. The unit of viscosity is dyne-s/cm$^2$, which reduces to grams per centimeter per second and is called a poise (P) (10 P = 1 Pa-s). The working point of a glass (i.e., where it can readily be formed) is that where the viscosity is about $10^4$ P. The annealing point, where a glass is essentially strain free, is $10^{14}$ to $10^{15}$ P. These conditions for the commercial silicate glasses are illustrated in Figure 7.6.

Forming Glass Products

Plate or flat glass used for windows and architectural purposes is manufactured by a "float process," as depicted in Figure 7.7. A sheet of glass up to 4 m wide flows from the forehearth of the melting furnace onto a bath of molten tin. The glass is initially soft enough to form a uniformly thick layer. As the glass moves at rates up to 20 cm/s, the temperature drops along the furnace's length, with the result that the glass is completely rigid and nearly stress free at the exit.

Container glass, such as bottles, are produced by glass blowing, but in a somewhat more controlled fashion than the craftsman glass blower, who is also somewhat of an artist. The process has been mechanized as shown schematically in Figure 7.8.

Continuous glass fibers are manufactured by drawing molten glass at velocities of the order of 400 m/s through multiple-orifice spinnerettes onto a high-speed takeup drum. These types of fibers are used for glass cloth, electrical

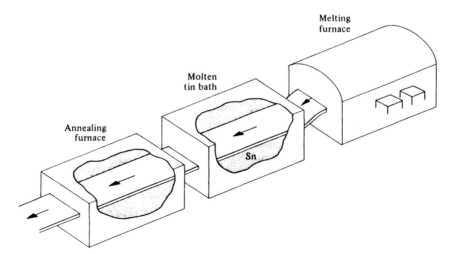

**Figure 7.7.**  Plate glass production made by floating soft glass on a molten tin bath. (From L. H. Van Vlack, *Materials for Engineering,* Addison-Wesley, Reading, Mass., 1982, p. 348.)

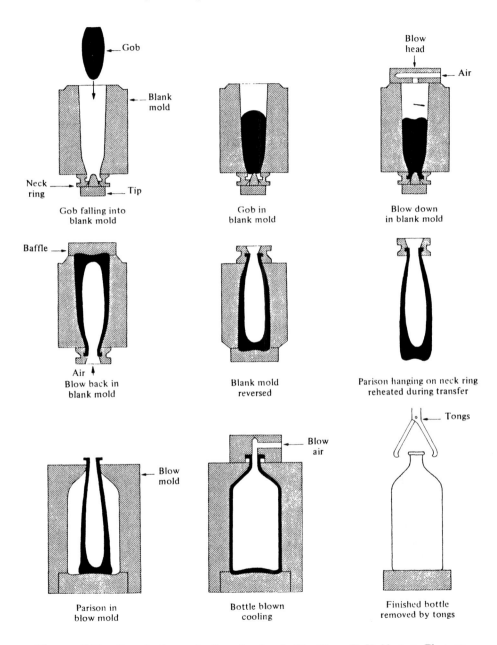

**Figure 7.8.** Steps in the production of a glass bottle. (From F. H. Norton, *Elements of Ceramics*, Addison-Wesley, Reading, Mass, 1952, p. 186.)

insulators and communication applications, and for glass–polymer composites. Continuous fibers can also be made by viscous drawing of soft glass through a single die. Glass fibers are frequently coated with polymers to protect them from surface abrasion. Discontinuous fibers, the randomly oriented glass wool type, are made by centrifugal spraying of molten glass. These fibers are on the order of 10 $\mu$m in diameter and are usually in mat form, also protected by polymers.

## 7.4 PROCESSING OF CERAMICS

The processing of crystalline ceramics is quite different from that employed for the amorphous glasses. For many years ceramics were processed by the firing of clays and clay mixtures, as mentioned in the section on traditional ceramics. As the common ceramics such as nitrides, borides, oxides, carbides, and so on, began to be used, the pressing and sintering route was the accepted method and is still used for many ceramic products. In Chapter 6 the processing of metal powders was discussed in some detail, and in general all of these processes have now been applied to ceramic powders. Sintering involves the heating of pressed powders to temperatures at which diffusion, sufficient to bond the particles together, occurs. The sintering temperatures for ceramics are somewhat higher than those used in powder metallurgy, necessitating some modifications in the pressing equipment. These differences will be pointed out as we go along. Densities in the range 90 to 95% are common for pressed and sintered ceramics, somewhat less than those obtained in powder metal compacts. Around 1960, hot pressing, although requiring more elaborate and expensive equipment, began to be used, particularly for small parts. Pressing coupled with simultaneous sintering increased the densities significantly, some approaching the 100% theoretical density values. The hot-pressing development was followed by introduction of the cold (room-temperature) isostatic pressing process, where pressure is applied equally in all directions. This type of pressure application resulted in better product uniformity. It was obvious, then, that *hot isostatic pressing* (HIP) would achieve even better-quality products, so its development followed closely that of the cold isostatic process. These processes, along with the more conventional slip casting of ceramics, are discussed in the following. A large segment of *Modern Ceramic Engineering* by D. W. Richerson is devoted to ceramic processing. Some of the following was extracted and condensed from that book.

### 7.4.1 Powder Processing (Raw Materials)

There exist a number of ways to obtain raw material powders for further processing into final shapes. Often, the method used is dictated by the material of interest. Most of these methods have been covered in detail in the book by Richerson.

Alumina ($Al_2O_3$) occurs in nature in the corundum crystal structure. Corundum powder is produced in large quantities from the mineral bauxite by the Bayer process, which involves selective leaching of the alumina by caustic soda and precipitation of aluminum hydroxide. The hydroxide is thermally converted to alumina powder. Most of the alumina produced by the Bayer process is used for aluminum metal production (the aluminum is electrolytically plated out from molten cryolite, $Na_3AlF_6$). But substantial quantities of the powders are also used to manufacture polycrystalline alumina-based ceramics, which include metal casting molds, high-temperature cements, wear-resistant parts, abrasives, and refractories.

Magnesia (MgO) powders are produced from $MgCO_3$ or from seawater. Magnesium hydroxide is extracted from seawater and then converted to MgO by heating. Although much of the hydroxide is converted to magnesium metal, the MgO powders are used extensively for high-temperature insulation and in refractory brick.

Silicon carbide (SiC) powders are produced by the Acheson process, which involves the passing of an electric current through a mixture of $SiO_2$ (sand) and coke. The mixture attains temperatures of about 2200°C, at which the coke reacts with the $SiO_2$ to produce SiC plus CO gas. SiC, one of the more widely used ceramics, is found in the form of abrasives, heating elements, wear-resistance applications and is being evaluated as a leading contender for use as a structural ceramic component in heat engines. SiC can also be prepared from other sources of silicon and carbon, and it has been prepared from rice hulls.

Silicon nitride powders have been made by the reaction of silicon metal powders with nitrogen in the temperature range 1250 to 1400°C, resulting in $\alpha$ and/or $\beta$ polymorphs of $Si_3N_4$. The reacted product is crushed and sized. Higher-purity $Si_3N_4$ powders have been made by the reduction of $SiO_2$ with carbon and by the reaction of $SiCl_4$ with ammonia.

Zirconia powders have been prepared by the hydrolysis of zirconium tetrabutoxide in an alcohol solution. Zirconia, partially stabilized with $Y_2O_3$ (yttrium), is being studied intensively for improvement in the fracture toughness of ceramics via transformation from the tetragonal to the monoclinic phase. It has been used as extrusion dies for brass rod and as a plasma-sprayed coating for gas turbine components.

Titanium dioxide ($TiO_2$) has been widely used in the paint, plastics, and paper industries as well as for ceramic components. It occurs in nature in the minerals ilmenite and rutile ($\approx 97\%$ $TiO_2$), from which it is processed to powder form by a variety of complex chemical processes.

There are a number of other powders that are used in smaller quantities, such as the borides, BeO, and the titanates. Information on the preparation of these powders can be found in the references listed in the Suggested Reading. More recently, ceramic powders have been made by pyrolysis of polymeric

metallorganic compounds and by vapor-phase synthesis, which include car-
bothermal reduction and plasma synthesis.

## 7.4.2  Pressing

As discussed earlier, there are several pressing methods that are used to convert ceramic powders to their final shape. These processes are also applicable
for metal powders. But whereas metal powders are usually pressed as such,
ceramic powders are mixed with suitable binders and lubricants prior to pressing. Most binders are organic materials that can be removed thermally and are
selected to be compatible with the chemistry of the ceramic and purity requirements of the application.

Uniaxial Pressing

The oldest method of powder compaction is by applying pressure along a single axial direction into a rigid die. Pressing may be done with 0 to 4% moisture plus binder, called *dry pressing,* or with large additions of moisture (e.g.,
10 to 15%). The latter is used frequently with clay-containing products and is
labeled *wet pressing.* Presses are of either the mechanical or hydraulic type.
Some are single-stroke-sequence presses, while others are of the rotary type,
where several dies are placed on a rotary table. Uniaxial pressing is widely
used for small parts, especially for electronic ceramics such as magnets and
dielectric and insulating components.

Isostatic Pressing

This hydrostatic process involves application of pressure equally to all sides of
the powder, resulting in much better uniformity of density. A schematic of an
isostatic pressing system is depicted in Figure 7.9 and is essentially the same as
those used in the powder metallurgy industry. There are two types of isostatic
presses, the wet-bag and dry-bag types. In the *wet-bag process,* shown in Figure 7.9a, the powder is sealed in a watertight flexible die. As the pressure of
the liquid is increased by hydraulic pumping, the walls of the die deform and
transmit the pressure uniformly to the powder, resulting in compaction. In the
*dry-bag process* (Figure 7.9b), the tooling is built with channels into which the
high-pressure fluid is pumped. The advantages of the latter technique are
higher production rates and closer dimensional tolerances. Parts such as spark
plug insulators have been pressed at rates in excess of 1000 parts per hour.
Production presses have pressure capabilities on the order of 407 MPa (30 ksi).
Isostatic pressing in conjunction with green (unfired) machining is often used
for shapes that are difficult to form by uniaxial pressing. These parts include
large radomes, cone classifiers, and cathode-ray-tube envelopes.

  The compacts resulting from both these types of pressing operations must be
densified by heating to temperatures where sufficient diffusion takes place to

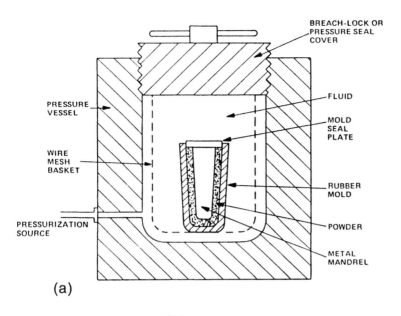

(a)

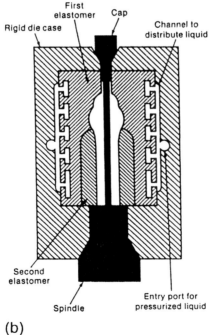

(b)

**Figure 7.9.** Schematic of isostatic pressing systems: (a) wet-bag process; (b) dry-bag process. (From D. W. Richerson, *Modern Ceramic Engineering,* 2nd ed., Marcel Dekker, New York, 1992, pp. 440 and 442, copyright ASM International.)

cause the particles to bond together. This "sintering" will usually produce densities on the order of 90 to 95%. $Si_3N_4$ must be sintered in a closed system with an overpressure of nitrogen in order to prevent decomposition of the compound. Some sintering operations take advantage of low-melting-point phases where the compact is heated above the melting temperature of this phase. This *liquid-phase sintering* occurs much more rapidly, allowing the liquid to flow into the narrow channels between particles, where it essentially cements them together. High densities, approaching 100%, are sometimes obtained. This is the major firing process for the silicate systems, and in these systems it is referred to as a vitrification process because the liquid results in a glassy phase. The *cemented carbides* used for the high-speed cutting tools are made by liquid-phase sintering, the most popular one being that of tungsten carbide particles that are cemented together by cobalt (mp 1498°C). The microstructure of a typical cemented carbide is shown in Figure 7.10. Powder mixtures of the carbide and metal are cold pressed at pressures up to 30 tons/in.$^2$ and then sintered for about an hour around 1480°C (2000°F). These sintered carbides are

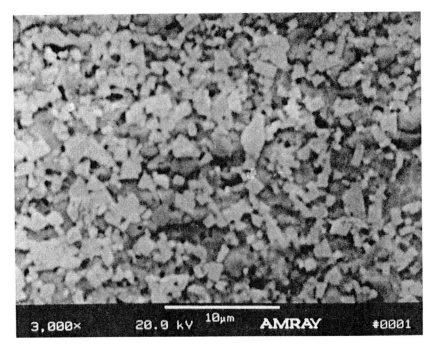

**Figure 7.10.**   Microstructure of 94% WC–6% Co prepared by the liquid-phase sintering technique. × 1500. (Courtesy of J. Lasiter, Cal Poly.)

also used for wear-resistant surfaces, where the cemented carbide plate is bonded to another surface. Surgical clamps, lathe cutting tools, and extrusion die inserts are examples of the use of braze-bonded carbide sections.

Hot Pressing

Hot pressing is merely a combination of pressing and sintering in one operation. The application of pressure while simultaneously sintering increases the contact area of the particles, thus providing more area over which diffusion and bonding can occur. The densities achieved, >98%, are higher than those obtained from consecutive pressing and sintering operations. Hot-pressing temperatures are on the order of $0.5T_m$. A schematic of a uniaxial hot-pressing unit is depicted in Figure 7.11. The same presses are used for both metals and ceramics. The temperature desired dictates the furnace type. For the higher temperatures, 2000°C (3632°F), induction heating of a graphite susceptor is frequently employed. Graphite, silicon carbide, and other resistance-heating elements have also been used. Most hot-pressing operations use pressures in the range 6.9 to 34.5 MPa (1 to 5 ksi). Graphite dies and pistons are suitable

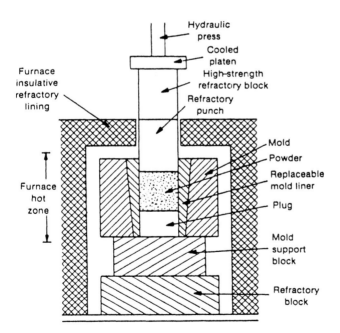

**Figure 7.11.** Schematic showing the essential elements of a hot press. (From D. W. Richerson, *Modern Ceramic Engineering,* 2nd ed., Marcel Dekker, New York, 1992, p. 553.)

for these high temperatures. Ceramic sputtering targets for use in thin-film deposition are almost always hot pressed. Densities near 100% are needed here to avoid gas evolution and concomitant contamination of the microcircuits. A comparison of the strengths and densities achieved by hot pressing versus the pressing followed by the sintering route is shown for $Si_3N_4$ in Table 7.6.

Hot Isostatic Pressing (HIP)

Apparatus that could provide uniform pressure at high temperatures was almost an inevitable approach after the separate advantages of these two parameters were realized. But such apparatus must have some medium to provide the uniform pressures at high temperatures. A hot isostatic press was shown in Figure 6.59. Whereas glycerine or hydraulic oil was used for room-temperature isostatic pressing, inert gases proved to be the best medium for transmitting pressures at high temperatures. Resistance-heating elements of molybdenum or graphite are frequently employed. Induction heating would ionize the inert gases and cause possible plasma formation that could result in reactions with press parts. The ceramic preform must first be sealed in a gas-impermeable envelope in order to prevent any of the pressing medium from entering the ceramic. At present, glass appears to be the preferred sealant material for ceramics. In contrast, stainless steel cans are frequently used to seal metal powders. The current HIP equipment consists of a high-temperature furnace enclosed in a water-cooled autoclave capable of withstanding pressure to 242

**Table 7.6**  Comparison of Densities and Strengths Achieved by Hot Pressing Versus Sintering

| Material | Sintering aid | Density (% theoretical) | RT MOR[a] | | 1350°C MOR | |
|---|---|---|---|---|---|---|
| | | | MPa | kpsi | MPa | kpsi |
| Hot pressed $Si_3N_4$[b] | 5% MgO | >98 | 587 | 85 | 173 | 25 |
| Sintered $Si_3N_4$[b] | 5% MgO | ~90 | 483 | 70 | 138 | 20 |
| Hot pressed $Si_3N_4$[c] | 1% MgO | >99 | 952 | 138 | 414 | 60 |
| Sintered $Si_3N_4$[d] | $BeSiN_2 + SiO_2$ | >99 | 560 | 81 | — | — |
| Sintered $Si_3N_4$[e] | 6% $Y_2O_3$ | ~98 | 587 | 85 | 414 | 60 |
| Hot-pressed $Si_3N_4$[e] | 13% $Y_2O_3$ | >99 | 897 | 130 | 669 | 97 |

*Source*: D. W. Richerson, *Modern Ceramic Engineering*, 2nd ed., Marcel Dekker, New York, 1992, p. 558.

[a] Room-temperature modulus of rupture.

[b] G. R. Terwilliger, *J. Am. Ceram. Soc.*, Vol. 57, No. 1, 1974, pp. 48–49.

[c] D. W. Richerson, *Am. Ceram. Soc. Bull*, Vol. 52, 1973, pp. 560–562, 569.

[d] C. D. Greskovich and J. A. Palm, *U.S. DOE Conf. 791082*, 1979, pp. 254–262.

[e] Data from C. L. Quackenbush, GTE Laboratories, Waltham, Mass.

MPa (45 ksi). HIPing makes possible the net-shape forming of products of high uniform density. This process has been used for some electronic parts (e.g., the ferrites and $BaTiO_3$), in applications such as magnetic recording heads. It is also being used on an experimental basis for high-temperature ceramic gas turbine components. Incidentally, the HIP process is probably used in the metal industry as much or more than for ceramic processing. Superalloy parts were made by this process in the early 1980s. Some of these parts, formerly pressed from superalloy powders, are now being manufactured from ceramic materials by the HIP process (Figure 7.12).

**Figure 7.12.** $Si_3N_4$ gas turbine dense rotor blades of 13 + cm diameter made by the HIP process. (Courtesy of Garrett Ceramic Components Division of Allied Signal Aerospace Co.)

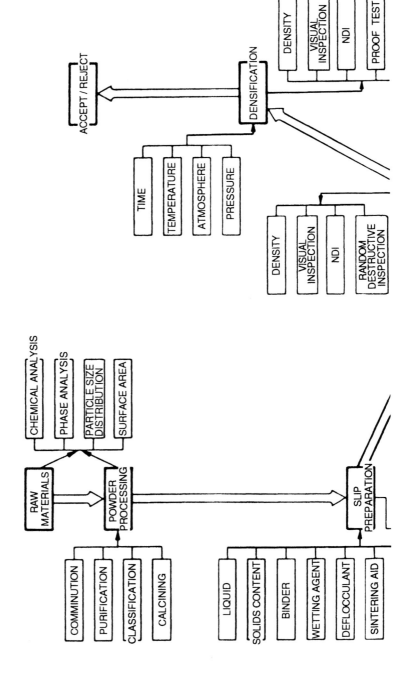

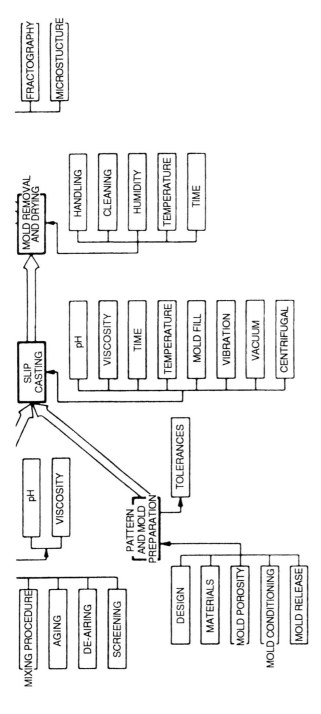

**Figure 7.13.** Critical process steps in slip casting that must be carefully controlled. (From D. W. Richerson, *Modern Ceramic Engineering*, 2nd ed., Marcel Dekker, New York, 1992, p. 445.)

### 7.4.3 Casting Methods

Slip casting is a very old technique that uses no binders to hold the compact together. A slurry (the slip) is made by suspending the ceramic particles in water prior to pouring into porous plaster molds. Primarily, the asymmetry of the powder in the slip holds the entire structure together. Van der Waals forces of attraction are also involved. There are a lot of variables in the process that must be carefully controlled to optimize strength and other desired properties. Figure 7.13 illustrates the steps and the parameters that must be considered.

Although clays are the most used raw materials in the slip-casting process, many oxides, particularly $Al_2O_3$, have also been slip cast. The slip preparation can be done by a variety of techniques, but the most common is wet ball milling, where the mixture is tumbled in a container containing steel balls.

The mold must have controlled porosity so that it can remove the fluid from the slip by capillary action. Simple casting by pouring is a common method, sometimes combined with suction to extract the water. Centrifugal casting, much like that used for the investment casting of metals, where the mold is spun rapidly to facilitate filling the mold completely, particularly for intricate shapes, is also employed for slip casting.

Once the casting is complete, the part dries and shrinks from the mold wall. The product is then removed from the mold and fired at a high temperature, sufficient for some sintering action to occur. Slip casting is used for small runs of a few items. It is the preferred method for artistic products and some more common items, much as thin-walled tubes, dishes, heavyware in the plumbing industry, and pots, but it has also been used for high-tech gas turbine stators.

Tape Casting

Tape, in the context used here, means sheet- and strip-shaped products on the order of a fraction of an inch in thickness and several inches in width. It consists of casting a slurry, much as that made for mold slip castings, onto a moving carrier surface, most often made of some type of polymer (e.g., Teflon, Mylar). The slurry is spread to a controlled thickness with a knife edge called a doctor blade. The slurry is carefully dried, resulting in a green tape that is subsequently cut or stamped to size and then fired to sinter the particles together. A schematic of a tape caster is shown in Figure 7.14. One of the most frustrating problems in this process is the control of warpage. This process is well known for its use in the production of fine-grained alumina substrates for use in manufacturing both thick- and thin-film hybrid microcircuitry.

### 7.4.4 Other Forming Processes

Molding

Injection molding is somewhat of a ubiquitous process that has been used on many materials: polymers, ceramics, metals, and composites. Die casting and

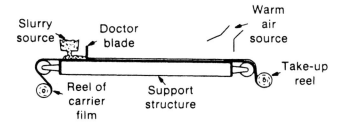

**Figure 7.14.** Schematic illustrating the doctor blade tape casting process. (From D. W. Richerson, *Modern Ceramic Engineering,* 2nd ed., Marcel Dekker, New York, 1992, p. 472.)

injection molding of metals are discussed in Chapter 6. Just as in metals, it is also a low-cost, high-volume production technique for making near net shape ceramic products. Ceramic injection molding consists of forcing a mixture of ceramic powders and a polymer binder into a die cavity, from which it is then extracted and the binder removed by thermal treatments. The ceramic powder is fed into the polymer, which is in a near-liquid state and contained in the "barrel" of the injection molder. For ceramic parts, the dies are made of harder, more wear-resistant metals than are dies for polymer or metal injection molding. The ceramic will often make up about 75% of the mixture. Pressures used are on the order of 13.8 MPa (2 ksi).

Compression molding is similar to injection molding except that a slug of the material is used as the feed stock rather than a free-flowing powder. The shape is built into the platens of the press or the faces of the tool, and the mix plastically deforms under pressure to fill the cavity. (This is analogous to the metal forging process.) After molding, injection, or compression, the shaped product, after binder removal, is fired to bind the ceramic particles together.

Extrusion

Extrusion is another process widely used for metals and polymers and has also been applied to ceramic plastic mixtures. The extrusion process for both ceramics and polymers differs slightly from that used in metals. In metals, an ingot is fitted into the extrusion chamber, where it is pressed forward by a plunger and forced through a die (Figure 6.56). In both plastic ceramic and polymer mixtures, the material is carried forward by an auger and forced through the die, as depicted in Figure 7.15. Extrusion of clay mixtures have been used extensively for fabrication of bricks, tile, tubes, and rods. Alumina, silicon carbide, and magnesia have also been extruded on a more limited basis. Like all the other ceramic processing methods, the extruded shape is sintered at high temperatures after the binder has been removed by heating at a relatively low temperature.

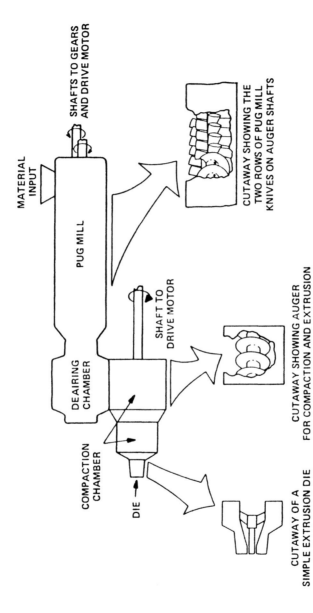

**Figure 7.15.** Schematic of a ceramic paste extruder. (From D. W. Richerson, *Modern Ceramic Engineering*, 2nd ed., Marcel Dekker, New York, 1992, p. 479.)

## 7.5 MECHANICAL PROPERTIES OF CERAMICS

### 7.5.1 Test Methods

Ceramics and other brittle materials are not normally tested in tension for the determination of their mechanical properties. Any slight misalignment of the tensile specimen in the grips introduces a bending moment. Brittle materials are very sensitive to surface flaws, which are magnified by bending moments and their related stress concentrations. This sensitivity accounts for the large scatter in tensile property data. On a laboratory scale, this has been circumvented by surface polishing and by attempts to improve alignment by special jigs and fixtures. More recently, a self-aligning specimen gripping system has been developed at Oak Ridge National Laboratory (*Advanced Materials and Processes,* ASM, September 1990, p. 59) that has been used to evaluate the behavior of ceramics to temperatures of 1600°C. We can expect to see more tensile data on ceramics in future years, but for the present, data reported in compression and flexure will be more common.

Despite the effects of surface conditions, many bend tests have been performed on ceramics, probably because of its simplicity. More often we find strength data reported in compression. This can lead to some confusion, since compressive strengths of ceramics are much larger than those measured in tension. Since cracks propagate by tensile rather than compressive stresses, small cracks forming in the compression test may not readily propagate to failure. R. W. Rice (see the Selected References) has reviewed the literature and suggested that the microplastic yield stress can be estimated by dividing the measured microhardness (Vickers or Knoop) by 3, and that the compressive strength of well-fabricated ceramic materials ranges from about one-half to three-fourths of this yield stress. Why does the fracture stress not exceed or equal the microhardness-derived yield stress? There are two reasons for this apparent discrepancy. One which we have just mentioned is that the compressive fracture stress depends on how one defines it (i.e., at the appearance of a crack or at the stress where the part fails to sustain this stress). Second, the microhardness values are determined under very small loads, on the order of 100 g. Such localized hardness impressions probably do not represent the material behavior as a whole. Porosity and other defects will lower the overall compressive strength. Certainly, a good factor of safety must be used when applying published strengths to engineering designs.

We have already mentioned that hardness test methods for brittle materials must be done at small loads to prevent cracking. Conventional microhardness measurements, where the indentation is viewed and measured under the microscope, is suitable.

Although ceramics have little ductility, for those that do, the ductility can be measured in tensile tests in the same manner as that for metals (see Chapter 2).

In bend tests the outer fiber strain at fracture is an indication of ductility. It is doubtful if the ductility measured in compression tests have much validity.

## 7.5.2 Typical Mechanical Properties

The values of the moduli of elasticity, ultimate tensile strength, and compressive strength for some typical ceramic compounds are listed in Table 7.7. A number of sources for these data were examined, and as one might expect, a wide range of values were found, particularly for the strength numbers. Unfortunately, for design purposes it is not possible to use published strength values. There are too many variables involved. Strength varies with density (rather markedly), purity, surface condition, and test methods. Also, some strength values were reported as *modulus of rupture,* which is derived from strength-of-materials equations dealing with the bending of beams. It has been defined as the nominal stress at fracture in a bend or torsion test. The design engineer must have strength values that were determined on the material that will be used in the final structure. Published moduli of elasticity values are somewhat more reliable than the strength values.

Since ceramics are being seriously considered for structural members at elevated temperatures, the strength–temperature relationship is very important. Strengths versus temperature relationships for oxide and silicate ceramics are shown in Figure 7.16 and for SiC and $Si_3N_4$ (compared to superalloys) in Figure 7.17. Creep and stress–rupture data are even more important than the decrease in strength with increasing temperature. The torsional creep rate of a number of ceramic materials at 1300°C and 12.4 MPa (1.8 ksi) are listed in Table 7.8. The creep rate of $Si_3N_4$, SiC, mullite, and $Al_2O_3$ at various temperatures is plotted versus log stress in Figure 7.18. SiC appears a clear-cut winner based on these data. NC-132 (manufactured by the Norton Company) is hot-pressed $Si_3N_4$ with MgO additives. It contains a glassy phase at the grain boun-

**Table 7.7** Strength and Elastic Moduli of Selected Ceramics

| Material | Modulus of elasticity (psi × 10$^6$) | UTS (ksi) | Compressive tensile strength (ksi) |
|---|---|---|---|
| $Al_2O_3$ | 50–57 | 39–70 | 400–650 |
| MgO | 30–40 | 15–20 | 120–200 |
| SiC | 50–68 | 25 | 150 |
| $ZrO_2$ | 20–25 | 10–12 | 70–290 |
| WC | 65–94 | 130 | 650 |
| BeO | 45–50 | 20–40 | 185–360 |
| $B_4C$ | 42 | 22 | 414–420 |

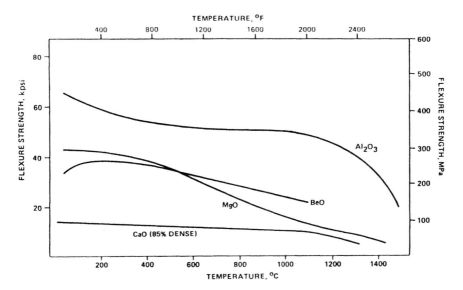

**Figure 7.16.** Strength versus temperature for oxide and silicate ceramics. (From D. W. Richerson, *Modern Ceramic Engineering,* 2nd ed., Marcel Dekker, New York, 1992, p. 187.)

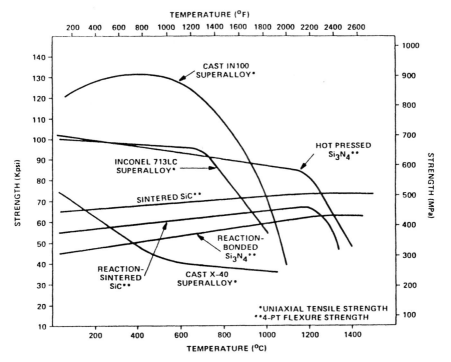

**Figure 7.17.** Strength versus temperature for carbide and nitride ceramics, and some superalloys. (From D. W. Richerson, *Modern Ceramic Engineering,* 2nd ed., Marcel Dekker, New York, 1992, p. 188.)

**Table 7.8** Torsional Creep of Several Ceramics

| Material | Creep rate at 1300°C, 1800 psi (in./in./hr) |
|---|---|
| Polycrystalline $Al_2O_3$ | $0.13 \times 10^{-5}$ |
| Polycrystalline BeO | $(30 \times 10^{-5})^a$ |
| Polycrystalline MgO (slip cast) | $33 \times 10^{-5}$ |
| Polycrystalline MgO (hydrostatic pressed) | $3.3 \times 10^{-5}$ |
| Polycrystalline $MgAl_2O_4$ (2–5 $\mu$m) | $26.3 \times 10^{-5}$ |
| Polycrystalline $MgAl_2O_4$ (1–3 mm) | $0.1 \times 10^{-5}$ |
| Polycrystalline $ThO_2$ | $(100 \times 10^{-5})^a$ |
| Polycrystalline $ZrO_2$ (stabilized) | $3 \times 10^{-5}$ |
| Quartz glass | $20{,}000 \times 10^{-5}$ |
| Soft glass | $1.9 \times 10^9 \times 10^{-5}$ |
| Insulating firebrick | $100{,}000 \times 10^{-5}$ |
| | Creep rate at 1300°C, 10 psi (in./in./hr) |
| Quartz glass | 0.001 |
| Soft glass | 5 |
| Insulating firebrick | 0.005 |
| Chromium magnetite brick | 0.0005 |
| Magnesite brick | 0.00002 |

*Source*: W. D. Kingery, H. K. Bowen, and D. R. Uhlmann, *Introduction to Ceramics,* Wiley, New York, 1976, p. 755.
[a] Extrapolated.

daries that enhances grain boundary sliding above 1100°C. This is a prominent creep mechanism at high temperatures, $\approx 0.7T_m$. NCX-34 is hot-pressed $Si_3N_4$ with $Y_2O_3$ additives that necessitates much higher temperatures for the onset of grain boundary sliding. NC-350 (manufactured by the Norton Company) was prepared by the reaction bonding process, involving the reaction of silicon with nitrogen to form $Si_3N_4$, resulting in a considerable improvement in creep resistance. NC-435 (manufactured by the Norton Company) is SiC prepared in such a way that no grain boundary phases are present. The SiC (prepared by the Carborundum Co.) also is lacking a grain boundary glassy phase. The absence of glassy phases at grain boundaries promote creep resistance. However, even pure ceramics (and metals) without grain boundary phases will exhibit grain boundary sliding at sufficiently high temperatures. One must always be concerned about the fracture strength and fracture toughness of ceramics when grain boundary sliding occurs, particularly in view of their notch sensitivity. Stress–rupture data for hot-pressed $Si_3N_4$ showing the typical scatter in results

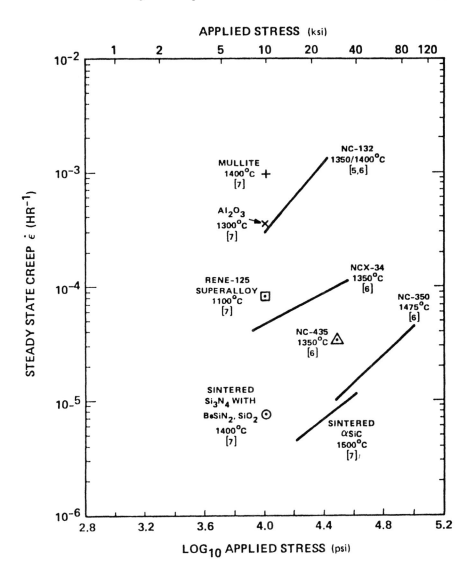

**Figure 7.18.** Creep of $Si_3N_4$, SiC, Mullite, and Rene 125 superalloy. (From D. W. Richerson, *Modern Ceramic Engineering*, 2nd ed., Marcel Dekker, New York, 1992, p. 332, as adapted from M. S. Seltzer, *Bull. Am. Ceram. Soc.*, Vol. 56, No. 4, 1977, p. 418, and D. C. Larsen and G. C. Walther, *Interim Report 6 FAFML*, Contract F33615-75-C-5196, 1978.)

is listed in Table 7.9. The creep rate of several oxides at 50 psi as a function of temperatures is shown in Figure 7.19.

### 7.5.3  Fracture Toughness and New Approaches

Fracture toughness was discussed in Chapter 2, and perhaps a quick review of that section would be beneficial for the reader in understanding the following discussion. Basically, when the stress intensity at a crack tip reaches a certain critical value, the crack will propagate to fracture. The stress intensity is called $K_{1c}$ and is the fracture toughness usually expressed in the units of MPa $\sqrt{m}$. In brittle materials, such as ceramics, the crack will absorb little energy as it propagates. It is the energy that is absorbed during crack propagation that determines the fracture toughness.

Glass has a fracture toughness of about 0.7 MPa $\sqrt{m}$, conventional ceramics approximately 1 MPa $\sqrt{m}$, cast iron on the order of 20 MPa $\sqrt{m}$, most other useful metals greater than 40 MPa $\sqrt{m}$ and many metals three or four times that figure. Since about 1970 it has generally been recognized that the lack of

**Table 7.9**  Stress–Rupture Data for Hot-Pressed $Si_3N_4$ Illustrating Typical Scatter

| Temperature | | Stress | | Time of test (hr) | | Source of data |
|---|---|---|---|---|---|---|
| °C | °F | MPa | kpsi | No failure | Failure | |
| 1066 | 1950 | 310 | 45 | 50 | | |
| 1066 | 1950 | 310 | 45 | | 0.46 | |
| 1066 | 1950 | 310 | 45 | | 0.18 | a |
| 1066 | 1950 | 310 | 45 | 50 | | |
| 1066 | 1950 | 310 | 45 | 50 | | |
| 1200 | 2192 | 262 | 38 | | 380 | |
| 1200 | 2192 | 276 | 40 | 480 | | |
| 1200 | 2192 | 262 | 38 | | 180 | |
| 1200 | 2192 | 262 | 38 | | 105 | b |
| 1200 | 2192 | 262 | 38 | | 52 | |
| 1200 | 2192 | 276 | 40 | | 32 | |

*Source*: D. W. Richerson, *Modern Ceramic Engineering*, 2nd ed., Marcel Dekker, New York, 1992, p. 327.

[a] AiResearch Report 76-212188(10), *Ceramic Gas Turbine Engine Demonstration Program*, Interim Report 10, Aug. 1978, prepared under contract N00024–76–C–5352.

[b] G. D. Quinn, *Characterization of Turbine Ceramics After Long-Term Environmental Exposure*, AMMRC TR 80–15, Apr. 1980, p. 18.

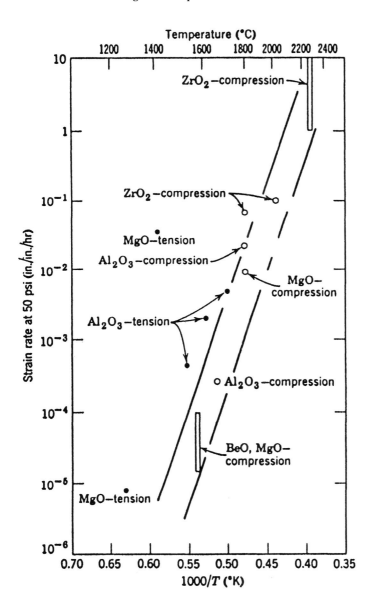

**Figure 7.19.** Creep rate as a function of temperature for several polycrystalline oxides at 50 psi. (From W. D. Kingery, H. K. Bowen, and D. R. Uhlmann, *Introduction to Ceramics,* Wiley, New York, 1976, p. 756.)

ductility and fracture toughness in ceramics will prohibit their use for most structural applications.

In the past 15 years or so, a different approach has been taken toward improving fracture toughness, which does not necessarily ensure higher ductility, but improved fracture toughness will tend to reduce the rate of crack propagation, and that factor, coupled with other improvements, such as surface condition, porosity, and uniformity, could possibly permit ceramics to be used in more applications, particularly those at elevated temperatures. These current efforts center around processes known as transformation toughening, microcrack toughening, and crack deflection toughening.

Transformation toughening, which appears to be the more potent of the three, is apparently a crack shielding, or equivalently, a residual stress-related toughening, in which the crystallographic transformations of certain constituents reduces the stress at the crack tip and slows its rate of propagation. $ZrO_2$ goes through a martensitic phase transformation from the tetragonal to the monoclinic form while cooling through a temperature of approximately 1150°C. With small additions of certain compounds (e.g., $Y_2O_3$), some of the tetragonal phase can be retained at room temperature. When a stress is applied, the tetragonal zirconia grains near the crack tip transform to the stable monoclinic form. This martensitic transformation is accompanied by a 3% volume increase, which places the crack in compression and halts, or at least slows, the propagation process. Fracture toughness levels in the range 20 to 25 MPa $\sqrt{m}$ have been reported in $ZrO_2$-$Y_2O_3$, known as partially stabilized zirconia (PZT), the yttrium stabilizing the cubic form at room temperature in some of the material. Mixtures of the stabilized cubic phase and unstable monoclinic phase bring about the increase in fracture toughness.

Ceramics usually contain either surface or internal flaws, and in most commercial ceramics, both are likely to be present in the finished product. Surface flaws can result from a surface glaze crack, a machining or handling defect, a pit, or some other type of surface discontinuity that may arise during processing. An examination of the fracture surface will often reveal if a surface flaw initiated the fracture. Internal flaws can often be traced to porosity or inclusions. In a very sound ceramic material the crack may simply be created by dislocation intersections (intersecting slip bands) or by dislocation pileups at a grain boundary. Again, fracture surface examination can assist in determining the source or cause of the fracture. The fracture features are most easily discernible when viewed with a scanning electron microscope (SEM). Figure 7.20 shows SEM fractographs of silicon nitride, where both surface and internal flaws initiated the crack.

One recent excitement in the ceramic field has been the discovery of superplasticity in some ceramics, which occurs by the same grain boundary sliding mechanism as in metals. These ceramics are ductile under certain conditions,

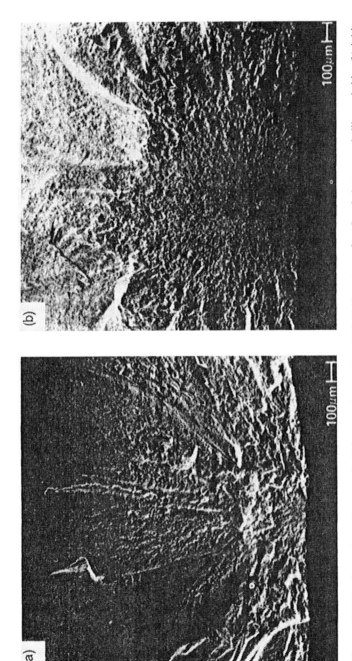

**Figure 7.20.** Examples of fractures in high-strength ceramics: (a) initiation at an internal flaw in reaction-sintered silicon nitride; (b) initiation at a surface flaw in hot-pressed silicon nitride. (From D. W. Richerson, *Modern Ceramic Engineering*, 2nd ed., Marcel Dekker, New York, 1992, p. 685.)

namely very small grain size ($< 10$ $\mu$m), and deformation at very slow strain rates ( $\approx 10^{-4}$ per second) and at temperatures above $0.5T_m$. Ductilities of 78% in compression and 160% in tension in yttrium-stabilized tetragonal zirconia (Y-TZP) have been achieved. Other ceramics have shown this same phenomenon. Although this may help in forming ceramics, it remains to be seen if it will result in significantly better fracture toughness under ordinary strain rates and at lower temperatures.

Despite the lack of ductility in ceramics per se, we must not write them off as structural members. Considerable improvements have been made in ceramic-based composites, particularly these containing SiC fibers. This subject is covered in Chapter 9.

## 7.6  PHYSICAL PROPERTIES

### 7.6.1  Thermal Properties

The good thermal insulating properties of ceramics is one of their attractive features. Heat is conducted in metals by the free-valence electrons. But due to the ceramic covalent–ionic bond, free electrons are not available for heat conduction. In such covalently bonded materials the heat is conducted by the atom vibrations, which act in unison to produce an atomic lattice wave called a *phonon*. These elastic wave phonons collide with other phonons and in so doing transmit heat. The thermal conductivity is expressed by the equation

$$K = \tfrac{1}{3}cv\lambda \tag{7.1}$$

where $c$ = heat capacity
   $v$ = average phonon velocity
   $\lambda$ = mean free path between phonon collisions

Reasonably reliable values of the thermal conductivities for ceramics can be found in a number of books. Porosity decreases the conductivity, but for high-density materials, it is not a significant factor. For temperatures well above the Debye temperature, only $\lambda$ varies with temperature. The mean free path varies inversely as temperature, so

$$k_t \propto \frac{1}{T}$$

Kingery et al. list the thermal conductivities for a number of ceramic materials and have summarized their variation with temperature in Figure 7.21. Note that these conductivities are plotted on a log scale, so even though ceramics in general are poor conductors of heat, at 1200°C, BeO, for example, has about 10 times the conductivity of $ZrO_2$. The best insulators are the firebricks.

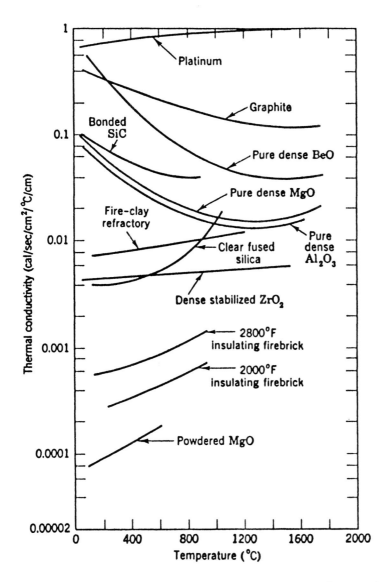

**Figure 7.21.** Thermal conductivity versus temperature for ceramic materials, a metal, and a polymer. (From W. D. Kingery, H. K. Bowen, and D. R. Uhlmann, *Introduction to Ceramics,* Wiley, New York, 1976, p. 643.)

Thermal Expansion, Thermal Stresses, and Thermal Shock

Ceramics have low-thermal-expansion coefficients (Figure 7.22) because again the strong bond restricts the amplitude of atom vibrations as the temperature is increased. This high bond strength is also responsible for their high elastic moduli and high strength. These factors, accompanied with their low thermal conductivity, make ceramics highly susceptible to thermal shock. What is thermal shock? It is the cracking of glass when it is nonuniformly heated, or heated and cooled rapidly. It is the "spalling" or flaking off of surface layers of brick when one side is heated while the other remains cool. It is caused by the high stresses set up by high thermal gradients. Thermal gradients can easily occur in materials with low thermal conductivities.

When thermal expansion is restrained, serious thermal stresses can develop within the material. Restraint may take two forms: external bodies connected to one of a vastly different coefficient of expansion (e.g., a metal–ceramic sealed joint) or nonuniform expansion within the body itself. It is the latter situation with which we are now concerned.

When heat is applied locally (e.g., to one surface of a solid), two things happen: The solid absorbs heat locally at a rate depending on its heat capacity per unit volume $\rho c$ and the thermal conductivity $K$. Here $\rho$ is the density and $c$ the specific heat capacity (or specific heat):

$$c = \frac{1}{m} \frac{dE}{dT} \tag{7.2}$$

where $m$ = mass
$\phantom{where }E$ = total energy content
$\phantom{where }T$ = temperature

The specific heat capacity is measured in metric units as cal/g/K or in engineering units as Btu/lb/°F. It is the quantity of heat that must be added to unit mass to raise the temperature 1 degree. The *specific heat* of a material is the ratio of the specific heat capacity of that material to that of water, measured in the same units. The specific heat of water is unity when measured in either units. At the same time that the solid is absorbing heat, the heat is conducted away at a rate depending on its thermal conductivity. Thus a high conductivity reduces thermal gradients. These opposing effects are combined in the term *thermal diffusivity*, that is,

$$\text{thermal diffusivity} = \frac{K}{\rho c} \tag{7.3}$$

A high diffusivity is desirable. It reduces local stresses.

Consider a slab of a material where heat is applied to one surface only as depicted in Figure 7.23. Because the surface is prevented from expanding by

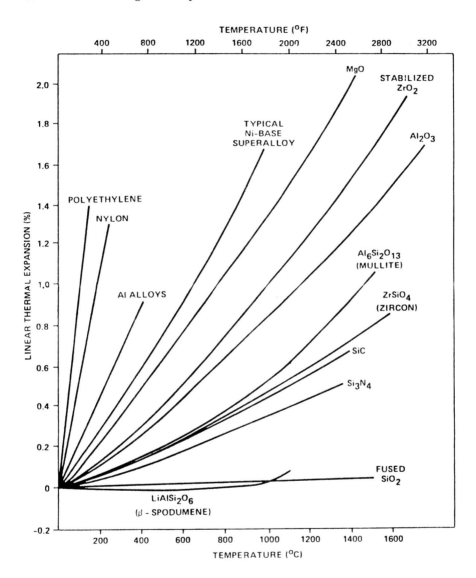

**Figure 7.22.** Thermal expansion coefficients of typical ceramics compared to metals and polymers. (From D. W. Richerson, *Modern Ceramic Engineering,* 2nd ed., Marcel Dekker, New York, 1992, p. 147.)

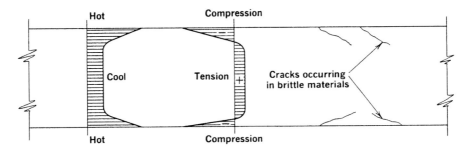

**Figure 7.23.** Results of surface heating of a hot slab. (From C. W. Richards, *Engineering Materials Science,* Brooks/Cole, Monterey, Calif., 1961, p. 477.)

the restraint of the interior, the surface is placed in a state of compressive stress. If the material is ductile, it will yield and relieve the stress. But brittle materials, if the stress exceeds the fracture stress, will crack. The reverse effect can be obtained by cooling a uniformly heated body on its surface. Thermal cycling can therefore set up alternating stresses. These stresses have been known to cause thermal fatigue failure, crack formation, and propagation in superalloy turbine blades. Brittle ceramic materials are even more susceptible to fracture under thermal cycling.

*Thermal shock* is the effect of a sudden change of temperature on a material. Thermal diffusivity is an important factor that if large causes the effects of shock to be greatly reduced. Other factors are the coefficient of thermal expansion $\alpha$, the modulus of elasticity $E$, the mechanical strength $\sigma_u$, and Poisson's ratio $\mu$. A number of thermal shock parameters have been devised, including all these variables in one parameter that expresses a material's resistance to thermal shock. The quantity $\sigma u/E\alpha$ appears to be an important factor in thermal shock resistance and is incorporated in all the parameters proposed.

The peak thermal stress at the surface can be expressed by

$$\sigma_{th} = \frac{E\alpha\Delta T}{1 - \mu} \tag{7.4}$$

Various thermal shock parameters have been devised according to whether crack initiation or crack propagation is the controlling factor. One frequently used thermal shock resistance parameter $R$, based on resistance to crack initiation, is expressed as

$$R = \frac{\sigma_u(1 - \mu)}{\alpha E} \tag{7.5}$$

The values for $R$ for several common structural ceramics are presented in Table 7.10. A high value of $R$ is desirable.

**Table 7.10**  Calculated Values for the Thermal Shock Parameter $R$ for Various Ceramic Materials Using Typical Property Data

| Material | Strength,[a] $\sigma$ (psi) | Poisson's ratio, $\nu$ | Thermal expansion, $\alpha$ (in./in. $\cdot$ °C) | Elastic modulus E (psi) | $R = \dfrac{\sigma(1-r)}{\alpha E}$ (°C) |
|---|---|---|---|---|---|
| $Al_2O_3$ | 50,000 | 0.22 | $7.4 \times 10^{-6}$ | $55 \times 10^6$ | 96 |
| SiC | 60,000 | 0.17 | $3.8 \times 10^{-6}$ | $58 \times 10^6$ | 230 |
| RSSN[b] | 45,000 | 0.24 | $2.4 \times 10^{-6}$ | $25 \times 10^6$ | 570 |
| HPSN[b] | 100,000 | 0.27 | $2.5 \times 10^{-6}$ | $45 \times 10^6$ | 650 |
| LAS[b] | 20,000 | 0.27 | $-0.3 \times 10^{-6}$ | $10 \times 10^6$ | 4860 |

*Source*: D. W. Richerson, *Modern Ceramic Engineering*, 2nd ed., Marcel Dekker, New York, 1992, p. 363.

[a] Flexure strength used rather than tensile strength.

[b] RSSN, reaction-sintered silicon nitride; HPSN, hot-pressed silicon nitride; LAS, lithium aluminum silicate ($\beta$-spodumene).

## 7.6.2  Electrical and Magnetic Properties

The electrical and electronic industries have become such large users of ceramic materials that the term *electronic ceramics* has been coined to label this group of materials. Semiconductor devices such as rectifiers, photocells, transistors, thermistors, detectors, and modulators have become an important part of the electronics industry. Similarly, dielectrics, capacitors, thick- and thin-film electronic devices, substrates, insulators, ferroelectrics, and magnetic ceramics play a major role in the electronic ceramics group. It is beyond the scope of this book to go into these materials in any significant detail, but many are mentioned in Chapter 10. What we attempt to do in the following is group these ceramics according to their applications and to some extent, their electronic properties.

Electrical Conductivity

Because of their covalent–ionic bond and lack of free electrons, ceramics in general have been considered as insulators. In Figure 2.22, they fall in the low range of conductivity (high range of resistivity). These are the electrical insulating ceramics such as $Al_2O_3$, with conductivities on the order of 10 to 15 $(\Omega \cdot m)^{-1}$. Other insulating ceramics include $SiO_2$, brick, $Si_3N_4$, porcelains, and MgO, all having this similar ballpark value with respect to electrical conductivity. The electrical carriers are the ions. But many ceramics are semiconductors where a small number of electrons have sufficient energy to conduct electricity. They include SiC, $B_4C$, ZnO, and $Fe_3O_4$, with conductivities in the midrange of Figure 2.22 [i.e., around $10^{-2}$ to $10^2$ $(\Omega \cdot m)^{-1}$]. Yet some transition metal oxides, such as $CrO_2$, $TiO_2$, and $ReO_2$, conduct by free electrons, which result from the overlap of wide unfilled $d$ or $f$ bands. Their conductivities fall between semiconductors and metals, with conductivities around $10^{-7}$ $\Omega^{-1} m^{-1}$.

Most ceramics show an increase in conductivity with increasing temperature (Figures 7.24 and 7.25). $ZrO_2$ is a good high-temperature oxide conductor, being about $10^{-3}$ $(\Omega \cdot m)^{-1}$ at 2000°C compared to around $10^{-6}$ at 2000°C

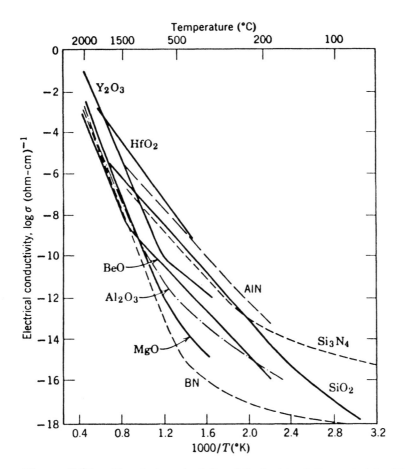

**Figure 7.24.** Electrical conductivity of the best insulating materials. (From A. A. Bauer and J. L. Bates, Battelle Mem. Inst., *Report 1930,* July 1974.)

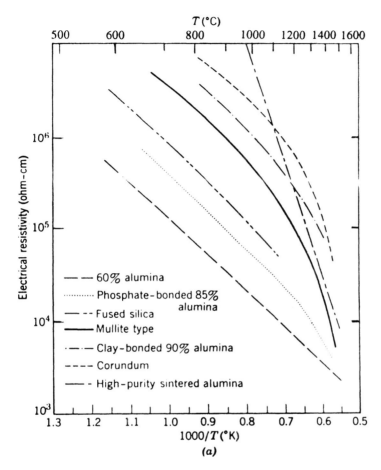

**Figure 7.25.** Electrical resistivity of several refractories. (From R. W. Wallace and E. Ruh, *J. Am. Ceram. Soc.*, Vol. 50, 1967, p. 338.)

for the good insulating oxides in Figure 7.24. At lower temperatures the difference between $ZrO_2$ and the other oxides becomes even larger. This ionic conductivity of $ZrO_2$ is due to diffusing oxygen ions and is very sensitive to oxygen pressure. This characteristic sensitivity has been used for oxygen-detection devices.

Semiconducting ceramics show an even greater dependency of conductivity with temperature. Some of these (e.g., $Fe_3O_4$) are mixed with nonconducting ceramics to achieve a material of controlled temperature coefficient of resistivity. They are called *thermistors*.

Dielectric Properties

When a voltage is applied between two spaced electrodes in vacuum (Figure 7.26), a charge accumulates on each electrode. The electric field $\epsilon$ between them becomes

$$\epsilon = \frac{E}{d} \tag{7.6}$$

If a material is inserted in the place of the vacuum the charge increases because of the polarization of the ions in the material. The energy stored in a capacitor is thus increased by the dielectric constant of the material. The total polarization resulting from electronic, ionic, and dipole orientations is referred to as the dielectric constant $K'$ (to distinguish it from the thermal conductivity $K$). It is the ratio of the charge density with the material in place to that in vacuum. For a parallel-plate capacitor the capacitance $C$ can be expressed as

$$C = \frac{K'\epsilon_0 A}{d}$$

where $\epsilon_0$ is the permittivity of vacuum ($8.854 \times 10^{-12}$ F/m) and $A$ is the area of plate. Because of the presence of the dielectric material, a positive ion will attract more negative charges to the plate, and vice versa. The dielectric constant for a number of ceramics used for this purpose is given in Table 7.11. The dielectric strength is also an important value of dielectric materials since it is the capability of a dielectric to withstand an electric field without breakdown. Generally, the dielectric strength for the titanates are in the range 100 to 300 V/mil and phenolics approximately 2000 V/mil.

Most ceramic dielectrics can be classified as insulators when $K' < 12$ and capacitors when $K' > 12$. Ferroelectrics have a very high $K'$, of about 2000 to 10,000. These are the titanates (e.g., $BaTiO_3$). Ferroelectricity is defined as the spontaneous alignment of electric dipoles in the absence of an electric field. These are the materials that are capable of converting mechanical to electrical energy, and vice versa. Their permanent dipoles can be altered by mechanical pressure, creating an alteration of the charge (dipole) distribution, hence producing an electric signal.

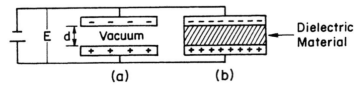

**Figure 7.26.** Change in charge density on a capacitor by the insertion of a polarizable dielectric material.

**Table 7.11**  Dielectric Constants for Ceramic and Organic Materials

| Material | $K'$ |
|---|---|
| NaCl | 5.9 |
| Mica | 2.5–7.3 |
| MgO | 9.6 |
| BeO | 6.5 |
| $Al_2O_3$ | 8.6–10.6 |
| $TiO_2$ | 15–170 |
| Porcelain | 5.0 |
| Fused $SiO_2$ | 3.8 |
| High-lead glass | 19.0 |
| $BaTiO_3$ | 1600 |
| $BaTiO_3$ + 10% $CaZrO_3$ + 1% $MgZrO_3$ | 5000 |
| $BaTiO_3$ + 10% $CaZrO_3$ + 10% $SrTiO_3$ | 9500 |
| Rubber | 2.0–3.5 |
| Phenolic | 7.5 |

*Source*: D. W. Richerson, *Modern Ceramic Engineering*, 2nd ed., Marcel Dekker, New York, 1992, p. 256.

## Magnetic Behavior

Although magnetism will be covered in a later chapter, suffice it to mention here that many ceramics exhibit ferromagnetism (i.e., the ability to possess a magnetic field in the absence of an applied field). These ceramics are commonly referred to as *ferrites*. They include the iron oxides, the rare earth oxides, and mixtures thereof. They are used as permanent magnets, memory units in computers, and other electronic devices.

Some more recently reported properties, both physical and mechanical, of some common ceramics are given in Table 7.12.

## 7.7 CONCRETE

Cements have been used for over 100 years and are one of the largest group of materials consumed (Figure 1.1). The most common commercial cement is portland* cement, which contains $Ca_2SiO_4$, $Ca_3SiO_5$, and $Ca_3Al_2O_6$. In a somewhat simplified form, the hydration reaction that occurs when wet cement hardens can be written as

$$Ca_2SiO_4 + xH_2O \rightarrow Ca_2SiO_4 \cdot x(H_2O)$$

*Portland cement was given that name because it resembled a limestone found on the Isle of Portland on the coast of England.

**Table 7.12**  Typical Property Values of Some Common Ceramics

| Property | Units | ALN | SiC | BN[a] | BeO | Al$_2$O$_3$ (99%) |
|---|---|---|---|---|---|---|
| Thermal conductivity | W/m · K | 170 | 70 | 25 | 260 | 25 |
| Electrical resistivity | Ω · cm | ≥10$^{14}$ | ≥10$^{11}$ | ≥10$^{11}$ | ≥10$^{14}$ | ≥10$^{14}$ |
| Dielectric constant | RT-1 MHz | 8.6 | 40.0 | 4.1 | 6.7 | 9.9 |
| Dielectric loss | RT-1 MHz | 0.0005 | 0.0500 | 0.0045 | 0.0004 | 0.0004 |
| Thermal expansion coefficient | 10$^{-6}$ °C$^{-1}$ (RT-400°C) | 4.7 | 3.8 | 0.0 | 7.2 | 7.1 |
| Density | g/cm$^3$ | 3.30 | 3.21 | 1.90 | 2.85 | 3.89 |
| Bending strength | MPa | 276 | 441 | 53 | 221 | 441 |
| Hardness (Knoop) | GPa | 11.8 | 27.5 | 2.7 | 9.8 | 19.6 |
| Young's modulus | GPa | 331 | 407 | 43 | 262 | 345 |

*Source*: *Ceram. Ind.*, July 1991, p. 29.

[a] HP grade.

where $x$ represents the molecules of water. Another hydration reaction occurs when tricalcium silicate hydrates.

$$Ca_3SiO_5 + (x + 1)H_2O \rightarrow Ca_2SiO_4 \cdot x(H_2O) + Ca(OH)_2$$

This occurs rapidly, giving a fast initial set.

Thus cement does not harden by drying but by hydration. The *heat of hydration* permits cement to harden below 0°C, the heat warming the water above its freezing point. The setting time is determined in large part by the calcium silicate content; the higher the content, the faster the setting time.

The ASTM has classified cements into four categories, as listed in Table 7.13.

## DEFINITIONS

*Annealing point*: with respect to glasses, temperature at which a rapidly cooled glass is essentially strain-free.

*Ball mill*: rotating container that contains hard steel balls (sometimes ceramic balls), used to crush and grind ceramic materials to small particles.

**Table 7.13** ASTM Cement Classification

| ASTM type | Characteristics | Composition$^a$ (wt %) | | | | |
|---|---|---|---|---|---|---|
| | | $C_3S$ | $C_2S$ | $C_3A$ | $C_4AF$ | Others |
| I | Standard | 45 | 27 | 11 | 8 | 9 |
| II | Reduced heat of hydration and increased sulfate resistance | 44 | 31 | 5 | 13 | 7 |
| III | High early strength (coupled with high heat of hydration) | 53 | 19 | 11 | 9 | 8 |
| IV | Low heat of hydration (lower than II and especially good for massive structures) | 28 | 49 | 4 | 12 | 7 |
| V | Sulfate resistance (better than II and especially good for marine structures) | 38 | 43 | 4 | 9 | 6 |

*Source*: R. Nicholls, *Composite Construction Materials Handbook*, Prentice-Hall, Englewood Cliffs, N.J., 1976, and the Portland Cement Association.

$^a$ Shorthand notation is used for cement technology: $C_3S$, $3CaO$; $C_2S$, $2CaO \cdot SiO_2$; $C_3A$, $3CaO \cdot Al_2O_3$; $C_4AF$, $4CaO \cdot Al_2O_3 \cdot Fe_3O_3$.

*Bridging oxygen*: location at which two oxygens share electrons to form a bridge between the $SiO_4^{4-}$ tetrahedra [e.g., in $SiO_2$ (quartz)].

*Borosilicate glass*: glass in which some $B_2O_3$ has been substituted for some $SiO_2$ to make the glass more heat resistant.

*Corundum*: normal room-temperature form of alumina.

*Feldspar*: anhydrous alumina silicate containing alkali ions.

*Fused quartz*: quartz that has been melted and solidified to the glassy (vitreous) noncrystalline state.

*Glass ceramics*: polycrystalline ceramics that are frequently classified in the glass family (e.g., Corningware).

*Glass temperature* ($T_g$): temperature below which further contraction of a glass or polymer occurs only by the contraction of the interatomic distance of individual atoms.

*Kaolinite*: alumina silicate mineral of the $Al_2(Si_2O_5)(OH_4)$ formula.

*Lead-alkali glasses*: glasses in which some alkali content has been replaced by lead oxide to give a more brilliant appearance to the glass.

*Mullite*: type of clay $(3Al_2)_3 \cdot 2SiO_2$.

*Perovskite*: type of crystal structure possessed by many complex oxides (e.g., $BaTiO_3$).

*Phonons*: elastic wave produced by the atom motion (vibration) in a crystalline solid. Phonon–phonon collisions account for the thermal conductivity in most crystalline ceramics.

*Portland cements*: mortar type of binder for gravel and sand. They are mostly calcium aluminates.

*Silicates*: minerals and compounds formed by connecting $SiO_4^{4-}$ tetrahedra to themselves and to metallic ions.

*Soda-lime glass*: fused $SiO_2$ to which calcium and sodium oxides have been added to interrupt the bridging oxygen of the $SiO_4^{4-}$ tetrahedra.

*Titanates*: $BaTiO_3$ type of compounds that often exhibit piezoelectric behavior.

## QUESTIONS AND PROBLEMS

7.1. Estimate the temperatures that should be used for hot pressing the following ceramic materials and explain the reasons for your selection; $M_gO$ and $SiO_2$.

7.2. What are the advantages of hot pressing over the cold pressing/sintering combination?

7.3. What would be an acceptable sintering temperature for the materials in Problem 7.1?

7.4. List four factors that affect the density of pressed products.

7.5. Would you expect a hot-pressed SiC gas turbine component to be superior to a cast and sintered SiC? Why?

7.6. Compute the thermal shock resistance based on fracture initiation for MgO and fused quartz at a temperature of 1000°C.

7.7. Why does SiC have a lower thermal-expansion coefficient than MgO at all temperatures above room temperature?

7.8. Why does porosity decrease the thermal conductivity of ceramics?

7.9. Why does porosity decrease the strength of ceramics?

7.10. Assuming similar particle size, density, surface condition, and processing parameters, why does polycrystalline $Al_2O_3$ have a compressive strength more than three times that of MgO? Would you expect this same big difference in strength for the two materials in single-crystal form?

7.11. List three reasons why ceramic materials are more brittle than metals.

7.12. From the data of Figure 7.18, compute the extension of a 1-ft length of a hot-pressed bar of $Si_3N_4$ with $Y_2O_3$ additives (NCX-34) at a stress of 10 ksi for a period of 10 h at 1350°C.

7.13. What is the difference between slip casting and compression molding? What processing method is analogous to these processes in metals?

7.14. What are the purposes of sintering aids?

7.15. What ceramic material and process would you select for a gas turbine component to be used at 1000°C and a stress of 45 ksi? What factors did you consider in your selection?

7.16. What are the primary mechanisms of creep in (a) polycrystalline ceramics; (b) ceramic single crystals?

7.17. What is the difference between creep and stress–rupture data? In what situation or under what conditions would you base your materials selection on the latter data compared to the former?

7.18. What is meant by the term "reaction sintering" and how does it differ from ordinary sintering?

## SUGGESTED READING

American Society for Metals, *Engineered Materials Handbook,* Vol. 4, ASM International, Metals Park, Ohio, 1992.

Doremus, R. H., *Glass Science,* Wiley, New York, 1973.

Kingery, W. D., Bowen, H. K., and Uhlmann, D. R., *Introduction to Ceramics,* 2nd ed., Wiley, New York, 1976.

McColm, I. J., *Ceramic Science for Materials Technologies,* Chapman & Hall, London, 1983.

Rice, R. W., "The Compressive Strength of Ceramics," in *Materials Science Research,* Vol. 5, W. W. Kriegel and H. Palmour, Eds., Plenum Press, New York, 1971, p. 195.

Richerson, D. W., *Modern Ceramic Engineering,* Marcel Dekker, New York, 1982.

Stokes, R. J., in Chapter 4 of *Fracture,* Vol. 7, H. Liebowitz, Ed., Academic Press, New York, 1972, p. 159.

# 8

# Processing and Properties of Polymers

## 8.1 INTRODUCTION

The term *polymer* is derived from the Greek roots *meras,* meaning parts, and *poly,* meaning many. Also, the term *macromolecule* is often used to define or describe polymers. In Chapter 1 we classified polymers as thermoplastic polymers (thermoplasts), thermosetting polymers (thermosets), and elastomers. The term *thermoplast* is not often used, but it is a convenient way to combine two words, so we will use this term in that light. Thermoplasts solidify on cooling from the melt and can be repeatedly reheated, re-formed, and remelted. Thermosets are chemically active in the molten state and harden by the further reaction, called *cross-linking,* between groups of nearby chains forming a three-dimensional network structure. They may soften somewhat on reheating but will decompose without attaining a formable or a molten state. Although thermosets and thermoplasts can be clearly defined and separated, it is not so with the elastomers. Elastomers have been described as a general term for natural and synthetic rubbers. The ASTM has defined an elastic material as one that can be stretched repeatedly to twice its original length and, upon release of the stress, return immediately to the approximate original length. Many have considered this to be an appropriate definition of an elastomer. By this definition an elastomer does not necessarily have to be a rubber, that is, of course, unless we define a rubber as a material that exhibits high elastic deformation, as does the ASM International *Handbook on Engineering Plastics.* The trouble arises when the term *hard rubber* is used because such materials are not very elastic. But for now we describe elastomers as a family of rubberlike materials, many of which are formed by cross-linking long-chain thermoplastic molecules. But when they are heavily cross-linked they appear and behave more like the three-dimensional thermosets, and these hard rubbers are no longer elastomers. Elastomers can be either thermoset polymers, the older and more common type, or they can be of the more recently developed thermoplastic kind. The thermoplastic elastomers do not require cross-linking. In commercial practice, thermoset elastomers, including natural and synthetic rubber, are considered part of the rubber industry, while thermoplastic elastomers are considered part of the plastics industry. A thermoplastic elastomer is a material that combines the processibility of a thermoplast with the functional performance and properties of a conventional thermoset rubber. They have been used extensively only since the 1970s. Their hardness ranges from 54 Shore A to 50 Shore D and as the hardness increases, their rubberlike characteristics decrease and the plastic-like properties increase.

In Chapter 4 we examined briefly the structures of the thermoplasts and introduced the concepts of *mer, monomer, addition,* and *condensation* (or *stepwise*) polymerization methods. The addition method forms only thermoplasts, while the stepwise method can form either thermoplasts or thermosets.

The history of polymer development is a very interesting subject and merits a short review. Some authors place the time of the discovery of polymerization at about 1838, when the conversion of styrene into a gelatinous mass was first reported. But the plastics industry actually started early in the nineteenth century with the use of natural rubber and was further enhanced by the vulcanization process, developed in 1839 by Charles Goodyear, who also invented ebonite, a high-sulfur-content rubber in 1851. It could be considered as the first thermoset polymer. But the materials first deserving the name plastic occurred about 1868 when celluloid was produced by mixing cellulose nitrate with camphor, producing a plastic molded material that became very hard when dried. The first synthetic polymers appeared in the early twentieth century when Leo Bakeland invented Bakelite, a condensation three-dimensional network thermoset polymer formed by combining the two monomers phenol and formaldehyde. In 1911, Lebedev polymerized butadiene and shortly thereafter proposed that both natural rubber and polybutadiene were chain structures. Others proposed that the elastic properties of vulcanized rubber arose from the covalent cross-links made by sulfur between the naturally occurring long-chain rubber molecules, something that was not fully understood by Goodyear. In 1920, Staudinger published a key paper that proposed chain formulas for polystyrene and polyoxmethylene, and he was awarded the Nobel prize in 1953 for his work in establishing polymer science. In 1934, W. H. Carothers demonstrated that chain polymers could be formed by condensation reactions, and invented nylon in 1935 by polymerization of pyrehexamethydenediamine and adipic acid by such a reaction. This new fiber was hailed as one of the greatest discoveries in chemistry. Carothers died in 1937, at the early age of 41, without realizing the full impact of his discovery. Commercial nylon was placed on the market in 1938 by the DuPont Company.

Carothers also invented neoprene, which some call the first successful synthetic rubber. He established several important principles: (1) that the monomers must be at least bifunctional, (2) that monomers must be relatively pure, and (3) that reactions must be such as to achieve nearly 100% yield. By the late 1930s, DuPont scientists also had developed polymethyl methacrylate (plexiglass), and both polystyrene and polyvinyl chlorides (PVC) were in commercial production.

The key to the further development of linear condensation polymers lay in recognizing that natural fibers such as rubber, sugars, and cellulose were giant molecules of high molecular weight. Cellulose is the main component of all plants (including cotton fibers). These are natural condensation polymers, and understanding their structure paved the way for the development of the synthetic condensation polymers, such as the polyesters, polyamides, polyimides, and polycarbonates.

The development of long-chain molecules by the addition method somewhat paralleled that of the condensation process. In the latter process the macromolecules being formed grow steadily throughout the reaction period. Both ends of the growing molecule remain functional, allowing the growth to proceed in more than one direction. A condensation molecule can always join with another molecule for continued growth. In contrast, most addition polymerization follows a method of chain growth, where each chain, once initiated, grows at an extremely rapid rate until terminated, and then, once terminated, cannot grow any more except by side reactions. By 1935, laboratory production of the low-density branched polyethylene had been developed. But it was not until 1954 that polymerization of polyethylene by catalysts had been achieved. This polyethylene was more crystalline, and possessed a higher density, strength, and melting point (127°C) than those of the branched polyethylene. The chronological order in which the major polymers emerged is listed in Table 8.1.

The term *engineering polymers*, which emerged rather recently, is subject to a variety of interpretations. It has been used interchangeably with the terms *high-performance polymers* and *engineering plastics*. The ASM *Handbook* has defined *engineering plastics* as synthetic polymers of a resin-based material that have load-bearing characteristics and high-performance properties, which permit them to be used in the same manner as metals and ceramics. Others have limited the engineering term to thermoplasts only. Elastomers are cross-linked linear thermoplastic polymers and many fall into the *engineering*

**Table 8.1**   Dates of the Commercialization of Some Major Polymers

| Date | Material |
|------|----------|
| 1907 | PF resins (Bakelite; Baekeland) |
| 1926 | Alkyd polyester (Kienle) |
| 1931 | PMMA plastics |
| 1938 | Nylon, 6,6 fibers (Carothers) |
| 1942 | Unsaturated polyester (Ellis and Rust) |
| 1943 | Fluorocarbon resins (Teflon; Plunkett) |
| 1943 | Silicones |
| 1943 | Polyurethanes (Baeyer) |
| 1947 | Epoxy resins |
| 1948 | Copolymers of AN, butadiene, and styrene (ABS) |
| 1950 | Polyester fibers (Whinfield and Dickson) |
| 1956 | POM (acetals) |
| 1957 | PC |
| 1962 | PI resins |
| 1965 | PPO |
| 1965 | Polysulfone |
| 1970 | PBT |
| 1971 | Polyphenylene sulfide |
| 1978 | Polyarylate (Ardel) |
| 1979 | PET-PC blends (Xenoy) |
| 1981 | Polyether block amides (Pebax) |
| 1982 | PEEK |
| 1983 | Polyetheramide Ultem |
| 1984 | Liquid-crystal polymers (Xydar) |
| 1985 | Liquid-crystal polymers (Vectra) |
| 1986 | Polyacetal-elastomer blends (Delrin) |
| 1987 | Engineering polymer |
| 1988 | PVC-SMA blend |

*Source*: R. B. Seymour, *Engineering Polymer Sourcebook*, McGraw-Hill, New York, 1990, p. 19.

category. On the other hand, the major products of the polymer industry, which include polyethylene, polypropylene, polyvinyl chloride, and polystyrene, are not considered as engineering polymers because of their low strength. Many engineering polymers are reinforced and/or *alloy polymers,* a blend of different polymers.

## 8.2 THERMOPLASTS

### 8.2.1 Chemistry and Products of the Addition Process

The catalysts first used for the addition process consisted of a complex formed from aluminum triakyl and $TiCl_4$. Later, transition metal oxides were used. Today, the catalysts, often called the *initiators,* are organic peroxides such as $H_2O_2$ (hydrogen peroxide) or benzoyl peroxide.

The formation of polyethylene is the classic example used for the addition process. Ethylene gas has a double-bonded carbon structure in the $C_2H_4$ monomer molecule, that is,

$$\begin{array}{ccc} H & & H \\ | & & | \\ C & = & C \\ | & & | \\ H & & H \end{array}$$

which is said to be *unsaturated,* because each carbon is not bonded to the maximum number of four atoms (as in diamond). Ethylene is also *bifunctional* since the transfer in position of one of these double bonds provides a site for another atom to become attached. This is the role of the initiator. Thus the peroxide supplies free radicals and in the straight-line notation for covalent bonds, the reaction can be written as

$$\underset{H_2O_2}{H - O - O - H} \rightarrow \underset{\text{free radicals}}{2H - O \cdot}$$

To initiate polymerization, the ·OH free radical combines with one of the double-bonded electrons (a pi electron) of the ethylene molecule (the monomer) to form a single bond with the monomer but leaving an unshared electron at the other end, as follows:

$$\begin{array}{ccccccc} & H & H & & & H & H \\ & | & | & & & | & | \\ R + & C & : : & C & \longrightarrow & R : C : C \\ & | & | & & & | & | \\ & H & H & & & H & H \end{array}$$

where R is the free ·OH radical. The dot at the end indicates a free electron, which is ready to react with another monomer that also has experienced a broken double bond by the initiator. This active site is transferred to each successive end monomer to form a linked chain. The radical is capable of reacting with an olefinic monomer (olefins are unsaturated hydrocarbons such as $C_2H_4$, $C_3H_6, \ldots, C_nH_{2n}$) to generate a chain carrier that can retain its activity long enough to propagate a macromolecular chain. This is a fast zipperlike reaction of mer bonding. The average lifetime of a growing chain is short. A chain of over 1000 units can be produced in $10^{-2}$ to $10^{-3}$ s. In the case of polyethylene, this leads to a chain of mers that form the macromolecule as depicted in the following:

Mer Unit

Note that the mer unit is not the smallest repeating unit, but is the smallest unit with the same chemical formula as the monomer, in this case $C_2H_4$.

The complete polymerization proceeds in three distinct stages: (1) *initiation,* when the active center that acts as the chain carrier is created; (2) *propagation,* involving growth of the macromolecular chain by repeatedly adding monomers; and (3) *termination,* where the kinetic chain is brought to a halt by the neutralization or transfer of the active center. In theory the chain could propagate until all the monomer in the system is consumed, but the free radicals are a particularly reactive species and interact as quickly as possible to form inactive covalent bonds. The more initiator radicals that are present competing for the available monomers, the shorter the chains will be. On the other hand, the more monomer molecules in the vicinity of the chain radical, the higher the probability of another monomer addition—hence a longer chain is produced. Termination can take place in several ways: (1) the interaction of two active chains; (2) the reaction of an active chain end with an initiator radical; (3) termination by transfer of the active center to another molecule, which may be solvent, initiator, or monomer; or (4) interaction with impurities (e.g., oxygen) or inhibitors. The most important termination reaction is the first, a bimolecular reaction between two chain ends. The free radical is not destroyed in this process but is merely transferred. Organic peroxide initiators are particularly susceptible to chain transfer.

In the linear polymers, where the molecular structure is somewhat like a string of beads, a greater degree of complexity is introduced when branches attached to and projecting from this string are produced. These branches contain several mer units, and some long branched chains may have only a few branches, like rivers with tributaries, while others may have elaborate branching comparable to that of mature trees. When branched chains are not attached to other chains, the polymer is considered to be a thermoplast.

The key chemical difference between linear and branched molecules is that the former contain only bifunctionally linked chemical blocks, while the latter contain trifunctionally linked building blocks at the point of branching. Remember that the number of bonding sites is referred to as the *functionality*. It is the number of bonds that a mer can form with other mers in a reaction. Reactants that are monofunctional cannot form polymers. If the reactant is difunctional, allowing it to react at each end, it can form a linear chain polymer, by the additive or condensation method. Molecules with higher degrees of functionality will also lead to polymers. Since branching requires a few points of tri (or higher)-functionality, as the number of these trifunctional points increase along linear chains, the probability that the branches will finally reach from chain to chain increases. The entire polymer mass then becomes one tremendous molecule, a *cross-linked* network polymer, and would be classified as a thermoset polymer. The vulcanization of rubber by cross-linking linear chains with a substantial quantity of sulfur is an example of a network polymer. The higher the degree of cross-linking, the higher the modulus of elasticity. The range of moduli and stiffness varies from a low value in soft rubber bands through the intermediate values for tires to hard rubber items such as combs and cases.

   The several architectural forms that linear molecules can take are demonstrated schematically in Figure 8.1. The two most useful synthetic branched polymers are those of low-density polyethylene and the polyvinyl acetates. Branched polymers are less crystalline than true linear chains. The higher the crystallinity, the greater the stiffness and yield stress. There are other ways of creating network polymers, the traditional one being that of the condensation method, where we start with reaction masses containing sufficient amounts of trifunctional monomers to form a three-dimensional network.

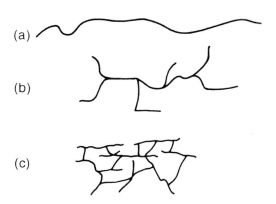

(a)

(b)

(c)

**Figure 8.1.**   Architectural forms of linear, branched, and cross-linked network polymers.

Molecular Weight of Chain Polymers
Since the chains all have different lengths, they will also all possess different molecular weights. The mass of the large molecules is simply the product of the mer mass times the number of mers in the molecule. We define the degree of polymerization (DP) as the number of mers per molecule.

$$DP = \frac{\text{molecular weight}}{\text{mer weight}}$$

However, since all molecules are not the same size, we must speak of a molecular distribution. The *mass average* molecule, $\bar{M}$, can be expressed as

$$\bar{M} = \frac{\Sigma f_i M_i}{\Sigma f_i} \tag{8.1}$$

where $\bar{M}$ = average molecular mass
    $M_i$ = mean molecular mass for each particular mass range selected
    $f_i$ = mass fraction of the material having molecular masses of a selected molecular mass range

The DP for commercial polyethylene usually falls in the range 3500 to 25,000. Since the mer mass is

2 carbons × 12   = 24 g
2 hydrogens × 1   =   4 g
                  28 g/mer

the molecular weights must vary from 28 × 3500 to 28 × 25,000 = 98,000 to 784,000 g/molecule.

*Example 8.1.* Polypropylene has the chemical formula $C_3H_6$. If its average chain molecular weight is 500,000 amu, what is the average degree of polymerization?

Solution. The mer weight = $(3 \times 12) + (6 \times 1)$ = 42 g/mer.

$$DP = \frac{\text{molecular mass}}{\text{mer mass}} = \frac{500,000}{42} = 11,905$$

Vinyl Polymers
The vinyl group of useful polymers is formed by replacing one hydrogen in the polyethylene chain by an atom or group of atoms. These polymers are frequently written in formula form as follows:

$$n \left[ \begin{array}{c} \text{H} \quad \text{H} \\ | \qquad | \\ \text{C}=\text{C} \\ | \qquad | \\ \text{H} \quad \text{H} \end{array} \right] \longrightarrow \left[ \begin{array}{c} \text{H} \quad \text{R}_1 \\ | \qquad | \\ -\text{C}-\text{C}- \\ | \qquad | \\ \text{H} \quad \text{H} \end{array} \right]_n$$

where $R_1$ represents the replacement group of atoms as shown in Table 8.2.

The values of the glass transition temperature (and other properties) will often vary with density and whether the material is atactic or isotactic. These linear polymers are not as linear nor as ordered as one might think based on the diagram above for the polyethylene chain. In addition to being kinked, the side groups may be irregular. This phenomenon of irregularity is covered by the term *stereoisomers*. As illustrated in Figure 8.2, *isotactic* ordering occurs when the side groups are always at the same position within the mer. In the *atactic* arrangements the side groups are randomly positioned on the polymer chain. There also exists a *syndiotactic* configuration where the R groups alternate on each side of the chain.

Vinylidene Polymers

In the vinylidene chain polymers, two hydrogen atoms are replaced by two other groups of atoms $R_2$ and $R_3$:

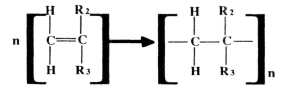

Three of these types of polymers are listed in Table 8.3.

Copolymers

*Copolymers* are large-chain molecules formed by the addition reaction of two or more monomers. They are formed by the addition mechanism described earlier. In essence they are chains in which one mer has been substituted with another mer, somewhat analogous to the substitutional solid solutions in metallic alloys. The polymers discussed in the preceding section were *homopolymers,* where the chains are made up of single repeating units. The mers in copolymers are different mers. They can be arranged at random along the chain, as regularly alternating mers, or as block polymers in which blocks of similar mers alternate with blocks of the other mers; finally, a block of like mers can be grafted onto a chain consisting of another set of different mers (Figure 8.3).

**Table 8.2**  Vinyl Polymer Group

| $R_1$ | Mer | Polymer | $T_m$ (°C) | $T_g$ (°C) |
|---|---|---|---|---|
| H | H H<br>–C–C–<br>H H | Polyethylene | 115 to 130 | –110 |
| $CH_3$ | H H<br>–C–C–<br>H H–C–H<br>H | Polypropylene | 176 | –10 to –18 |
| Cl | H Cl<br>–C–C–<br>H H | Polyvinyl chloride | 212 | 87 |
| OH | H OH<br>–C–C–<br>H H | Polyvinyl alcohol | 212 | 87 |
| $C_2H_3O_2$ | H H<br>O=C–C–H<br>H<br>H O<br>–C–C–<br>H H | Polyvinyl acetate | — | 29 |
| $C\equiv N$ | H H<br>–C–C–<br>H $C\equiv N$ | Polyacrylonitrile | 317 | 104 to 130 |
| Benzene ring | H H<br>–C–C–<br>H (benzene ring) | Polystyrene (PS) | 245 | 100 to 105 |

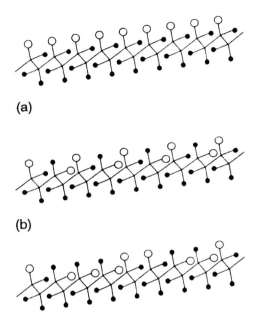

(a)

(b)

(c)

**Figure 8.2.** Stereoisomers of polyproylene: (a) isotactic; (b) syndiotactic; (c) atactic. Solid dots, H; open dots, $CH_3$. (From *Engineered Materials Handbook,* Vol. 2, ASM International, Metals Park, Ohio, 1988, p. 58, article by L. L. Clements.)

There are many useful copolymer materials. Probably the best known are the commercial polyvinyl chloride–acetates. The weight percent of the vinyl chloride varies from about 85 to 95%, corresponding to about 12 to 4% of acetate mers. The products include lacquers, injection-molded polymers, synthetic fibers, and substitute rubbers as the percent acetate mers decrease in the order in which these products are listed. The ABS copolymers, formed from acrylo-

**Table 8.3** Some Vinylidene Polymers

| Polymer | $R_2$ | $R_3$ | $T_m$ (°C) | $T_g$ (°C) |
|---|---|---|---|---|
| Polyvinylidene chloride (PVDC) | Cl | Cl | 198 | −17 |
| Polymethacrylate (PMMA) | $C_2H_3O_2$ | $CH_3$ | 105 to 120 | 3 |
| Polyisobutylene | $CH_3$ | $CH_3$ | 120 | −60 to −70 |

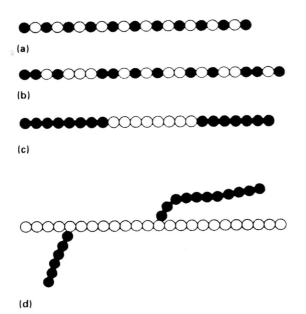

(a)

(b)

(c)

(d)

**Figure 8.3.** Types of copolymers: (a) alternating mers; (b) random; (c) block; (d) graft. (From *Engineered Materials Handbook*, Vol. 2, ASM International, Metals Park, Ohio, 1988, p. 58, article by L. L. Clements.)

nitrile, butadiene, and styrene, are used extensively for pipes, tanks, and artificial rubbers. Many other copolymers are products of the addition of the vinyl and the diene monomers. Some of the best known names are nytril, vinyon, and saran. Many elastomers are copolymers: for example, butyl rubber, which consists of the isobutylene–isoprene mers and the styrene–butadiene copolymer elastomers, which are used for tires, shoe soles, and electric insulation.

## 8.2.2 Long-Chain Thermoplastic Molecules and Products Made by the Stepwise (Condensation) Polymerization

As we stated earlier, the condensation process of polymerization can produce either the long thermoplast chain molecules or the three-dimensional network arrangement of atoms characteristic of the thermoset polymers. The chemistry of synthesizing linear condensation polymers is the chemistry of simple organic reactions. However, certain special conditions are required of the monomers: (1) the monomer must have two reactive groups (i.e., be bifunctional), (2) the material must be free of impurities that might cause termination, and (3) there must be a stoichiometric balance. With respect to the kinetics of the reaction, there are also two conditions: the monomers must be functionality capable of proceeding indefinitely to very high yields and at the same time must be free of side reactions. These stringent conditions limit the condensation thermoplasts to a relatively few organic reactions in the industrial synthesis of linear condensation polymers. It is now preferable to replace the term *condensation* with *stepgrowth* or *step reaction*. Such reclassification now permits the inclusion of polymers such as polyurethanes, which grow by a step mechanism without the elimination of a molecule such as water, hence no condensation.

Other linear thermoplast polymers made by this process include the heterochain polymers, where the chain backbone contains other atoms in addition to carbon, or carbon with multiple bonds. These include the nylons (the generic name for polyamides), polycarbonates, polyethylene terephthalate (PET), polyimides, and some polyesters (some polyesters are also thermosets). These structures, which are rather complex, are listed in Table 8.4 together with their melting and glass transition temperatures. [There are few polymers (e.g., polyethylene oxide) that can be formed by either the catalyzed chain method or by the stepwise reaction.] The chemistry of these reactions is fairly complex and beyond the scope of this book. A good reference for step-growth polymerization is Chapter 2 of the book by J. M. G. Cowie listed in the Suggested Reading.

## 8.2.3 Thermoplastic Polymers for High-Temperature Service

There is an important class of polymers that contain what is commonly referred to as *aromatic rings,* so called because many of these organic compounds have a distinctive aroma. These compounds are also known as the benzene ring or phenyl group. The structure consists of a ring of six carbon atoms with alternating single and double bonds between them with the electrons resonating such that their characteristics are somewhere between single and double bonds, as depicted in Figure 8.4. Such rings, which can occur either in the backbone of the polymer chains or attached as a side group, are characteristic of high-temperature thermoplastic polymers.

The *high-temperature* polymers are becoming very important in applications such as under the hood of automobiles. One of the chief drawbacks to extending the degree of polymer usage in many engineering designs has been a result of their degradation and loss of strength at slightly elevated temperatures. Since most polymers undergo a dimensional change at their glass transition temperatures, a loss of stiffness and dimensional stability occurs near $T_g$. The high-$T_g$ thermoplastic polymers generally possess bulky side groups, which improve their stiffness, and also have a rather strong hydrogen intermolecular bond between chains. Crystallinity in general improves the structural strength

**Table 8.4** Some Typical Heterochain Thermoplastic Polymers

| Chemical name | $T_g$ | | $T_m$ | | Mer chemical structure |
|---|---|---|---|---|---|
| | °C | °F | °C | °F | |
| Polytheylene oxide | −67 to −27 | −90 to −15 | 62 to 72 | 145 to 160 | $\left[\begin{array}{c} \text{H} \quad \text{H} \\ -\text{C}-\text{C}-\text{O}- \\ \text{H} \quad \text{H} \end{array}\right]$ |
| Polyoxymethylene | −85 | −120 | 175 | 345 | $\left[\begin{array}{c} \text{H} \\ -\text{C}-\text{O}- \\ \text{H} \end{array}\right]$ |
| Polyamide Nylon 6 | 50 | 120 | 215 | 420 | $\left[\begin{array}{c} \text{H} \quad \text{O} \\ -\text{N}-(\text{CH}_2)_5-\text{C}- \end{array}\right]$ |
| Nylon 6,10 | 40 | 105 | 227 | 440 | $\left[\begin{array}{c} \text{H} \quad \text{H} \quad \text{O} \quad \text{O} \\ -\text{N}-(\text{CH}_2)_6-\text{N}-\text{C}-(\text{CH}_2)_8-\text{C}- \end{array}\right]$ |

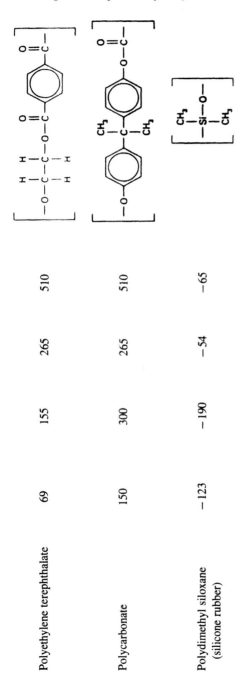

| | | | | |
|---|---|---|---|---|
| Polyethylene terephthalate | 69 | 155 | 265 | 510 |
| Polycarbonate | 150 | 300 | 265 | 510 |
| Polydimethyl siloxane (silicone rubber) | −123 | −190 | −54 | −65 |

*Source: Engineered Materials Handbook,* Vol. 2, ASM International, Metals Park, Ohio, 1988, p. 53, article by L. L. Clements.

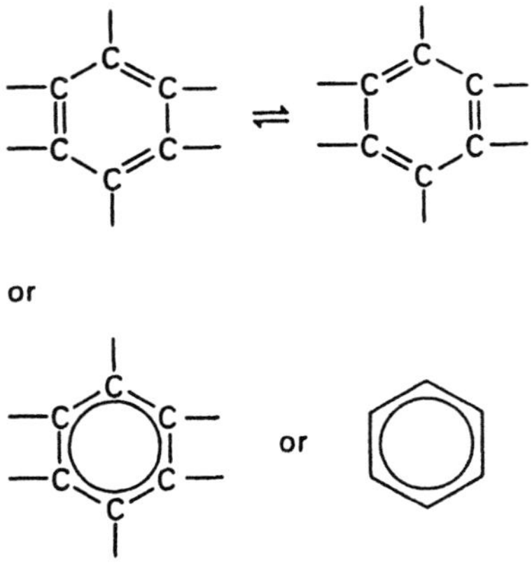

**Figure 8.4.** Aromatic rings with single and double resonating bonds. (From *Engineered Materials Handbook,* Vol. 2, ASM International, Metals Park, Ohio, 1988, p. 52, article by L. L. Clements.)

of many linear polymers. The thermoplastic polymers with the greatest regularity have the highest degree of crystallinity. Less branching and less bulky side groups promote crystallinity. For elevated temperature usage the bulky side groups provide less crystallinity but also higher stiffness and better elevated temperature properties. In this case the advantage of the stiffness of the side groups outweighs the benefits of crystallinity. However, the best high-temperature thermoplasts are those that have both inflexible bulky rings and are heterochain polymers having sites for the stronger hydrogen bond, yet have enough flexibility to permit a high degree of crystallinity. These include polyether ether ketone (PEEK) and polyphenylene sulfide (PPS). They obtain the flexibility needed for crystallinity from the flexible ether and sulfide linkages. Another group of high-temperature thermoplasts are the thermotropic or melt-processed liquid-crystal polymers (LCPs). These contain densely packed, rodlike molecular chains that provide self-reinforcement. The main thermotropic LCPs are the aromatic polyesters. The best polymers of the high-temperature thermoplasts have the high temperature and chemical resistance of thermosets but the processing ease of thermoplasts. The structures and glass temperatures are listed in Table 8.5.

Most of the thermoplastic polymers have been improved in room temperature and in high-temperature strength by the addition of about 30% glass fibers. These are composite materials (covered in Chapter 9) that are being used for under-hood automotive components as well as nonautomotive applications.

## 8.2.4 Properties and Applications of Thermoplastic Polymers

Mechanical Properties

The mechanisms of deformation of thermoplastic polymers will assist in understanding their mechanical properties. Of equal importance are the mechanical test methods required to truly portray their mechanical behavior. Thermoplasts deform elastically by the bending, unkinking, and stretching of the molecular bonds under stress. This elastic extension and recovery are in step with the stress, much as in elastic deformation in metals. All of these materials respond very rapidly to stress in the elastic region, although the crystalline metals and ceramics deform by a small displacement of the atoms, less than 10% of their interatomic distances. Thermoplasts, on the other hand, have much larger elastic extensions, yet the molecules recoil almost immediately from the load, although some polymers may exhibit a slight delay in the elastic recoil that can be easily detected on most test equipment.

The plastic deformation behavior of crystalline metals and ceramics is not only different in mechanism from the thermoplasts, but also in the time dependency between stress and strain. The former deform plastically by dislocation motion, which is in step timewise with the applied stress. The thermoplasts are made to deform plastically under stress by the sliding of long-chain molecules past one another via the breaking of the weak secondary intermolecular bonds. This type of plastic deformation increases with time without an increase in stress. Some portion of the early "plastic" strain somewhat reduces the bending and stretching of the molecules and can be considered as a retarded elastic strain. Further deformation is all plastic but is time dependent, like the creep strain in metals. Below $T_g$ all polymers are brittle. We will return to this point later. But between $T_g$ and $T_m$ the thermoplasts most often show a stress–strain curve as depicted in Figure 8.5. Point *a* represents zero time on both plots. When stress is applied, there is a rapid elastic response of strain from *a* to *b*, at which point time-dependent creep begins. Creep in both metals and thermoplasts is similar to the flow of high-viscosity liquids. This is true viscous behavior. Viscous materials obey the equation

$$\frac{d\epsilon}{dt} = \frac{1}{\phi}\sigma$$

where $d\epsilon/dt$ = strain rate
$\phi$ = coefficient of viscosity
$\sigma$ = stress

Viscosity is usually expressed in poise, where 1 poise = 1 $dyn/cm^2 \cdot s$. Viscous behavior is also said to be *newtonian*.

When the stress is released, there is an instantaneous strain recovery from point *c* to point *d*, followed by a time-dependent recovery. Materials that ex-

**Table 8.5**  Thermoplastic Polymers for High-Temperature Service

| Chemical name | $T_g$ | | $T_m$ | | Mer chemical structure |
|---|---|---|---|---|---|
| | °C | °F | °C | °F | |
| Poly(p-phenylene terephthalamide) (aromatic polyamide or aramid) | 375 | 705 | ~640[a] | ~1185 | |
| Polyaromatic ester | — | — | 421 | 790 | |
| Polyether ether ketone | 143 | 290 | 223 | 635 | |
| Polyphenylene sulfide | 85 | 185 | 285 | 545 | |
| Polyamide-imide | 277–289 | 530–550 | | [b] | |

| Polyether sulfone | 225 | 435 | b |
| Polyether-imide | 215 | 420 | b |
| Polysulfone | 193 | 380 | b |
| Polyimide (thermoplastic) | 280–330 | 535–625 | b |

*Source: Engineered Materials Handbook*, Vol. 2, ASM International, Metals Park, Ohio, 1988, p. 54, article by L. L. Clements.

[a] $T_c = 500°C$ (930°F). R contains at least one aromatic ring.

[b] Polymer is generally 95% or more noncrystalline. $T_m$ is a function of % crystallinity.

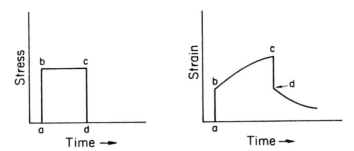

**Figure 8.5.** Typical stress–strain–time behavior for thermoplasts between $T_g$ and $T_m$.

hibit elastic and viscous behavior and time-dependent relaxation are said to be *viscoelastic*. It should be added that all of these effects are also strongly temperature dependent. Superimposed on this is the degree of crystallinity. Polymers in the range 20 to 60% of crystallinity have been referred to as being *leathery* and tough. In general, the strength modulus and density all increase with the degree of crystallinity while the toughness decreases. The steric factors play a large role in crystallinity and $T_g$. Large side groups impede crystallization except when the arrangements are regular, as in isostatic chains.

The stress–strain behavior of thermoplasts are controlled to a large extent by the temperature and strain rate (Figure 8.6). Low strain rates and high temperatures promote ductility, just as in metals and ceramics. In the case of ductile behavior the stress–strain curves may take one of several forms, as depicted in Figure 8.6. In Figure 8.6c, marked by curves $d_1$, $d_2$, and $d_3$, curve $d_1$ shows yielding at point *y*, but not as pronounced a yield point as one obtains in metals. In $d_2$ an unstable neck forms at point *y*, followed by inhomogeneous yield behavior. In $d_3$, a neck forms at point *y* followed by some strain hardening, which stabilizes the neck, giving rise to a drawing out of the necked region. The necked region grows to very long extensions as demonstrated by the schematic in Figure 8.7.

As stated previously, at low temperatures, below $T_g$, brittle fracture with little ductility is characteristic behavior. But the fracture process is somewhat different from the transgranular brittle cleavage type of failure of metals. In a glassy polymer the initial crack propagates by the formation of a *craze* that runs ahead of the crack tip. The craze material comprises *fibrils* (molecules in threadlike strands or bundles) interspersed with small interconnecting microvoids. Craze lines are easily observed by the scattering of light from the discontinuities produced by the microvoids (Figure 8.8). The craze thickness is on the order of 5 $\mu$m. The craze provides a path for crack propagation, but in

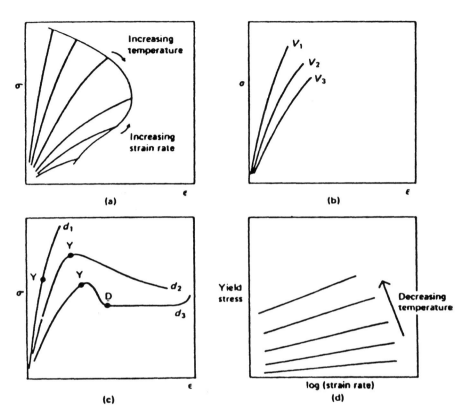

**Figure 8.6.** Effect of temperature and strain on the stress–strain curves of poly-mers: (a) failure envelope showing effect of strain rate and temperature; (b) influence of straining rate on a fairly brittle polymer ($V_1 > V_2 > V_3$); (c) different types of ductile behavior; (d) influence of testing speed on the tensile yield stress of an amorphous poly-mer at several temperatures. (From C. Hall, *Polymer Materials: An Introduction for Technologists and Scientists,* 2nd ed., Wiley, New York, 1989, p. 79.)

so doing it absorbs energy and thereby contributes to the fracture toughness. Brittle thermoplasts are somewhat tougher than the silicate glasses. Fracture eventually occurs by the breaking of bonds along the main chain. Brittle frac-ture in crystalline thermoplasts appears to occur more by separation along the amorphous boundary between crystalline regions. In some cases the crystals rotate toward the stress axis and break into smaller blocks but retain their crys-talline structure. Microfibrils form in the tensile direction, but crazing is nor-mally absent, although some rare cases have been reported.

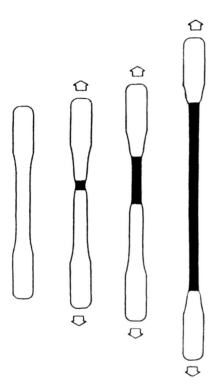

**Figure 8.7.** Ductile behavior in a cold-drawing polymer. (From C. Hall, *Polymer Materials: An Introduction for Technologists and Scientists*, 2nd ed., Wiley, New York, 1989, p. 80.)

Ductile thermoplasts show a lot of extension via the molecular sliding of the molecules prior to fracture. But eventually, the stress will exceed the covalent bond strength of the chain and fracture occurs. It is the strong bond that determines the strength rather than the weaker intermolecular bond. Both brittle and ductile modes are possible in the thermoplasts. Below $T_g$ they will fracture in a brittle manner, while above this temperature both ductile and brittle fractures are possible. Factors that favor ductile fracture are increasing temperatures, low strain rates, and absence of notches. Almost all polymers have low fracture toughness, whether measured in $K_{1c}$ values by the fracture mechanics method or in impact energy (see Tables 8.6 and 8.8). Schematics of both brittle and ductile fractures are shown in Figures 8.9 and 8.10.

In addition to all the other variables mentioned above, additives can also affect polymer properties. An *additive* is a substance usually introduced to

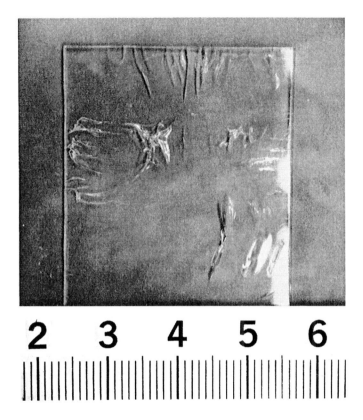

**Figure 8.8.** Crazing in polymethyl methacrylate.

affect the color, flame resistance, oxidation resistance, flexibility, and cost. The cost factor is affected by fillers (often called inert fillers), which reduce the volume of the total polymer content but at the same time will have an effect on the mechanical properties almost in proportion to the filler content. The common fillers are wood flour (a very fine sawdust), silica flour (a finely ground quartz that also improves abrasion resistance), and numerous fibers, such as glass, organic textile, and mineral fibers. The fibers increase the overall strength. At some point, the fiber content becomes sufficient to classify the product as a composite. We treat these materials as *reinforced polymers* in Chapter 9.

Plasticizers are probably the additive that will most often be present in the thermoplastic polymers (excluding glass fibers, which we treat under compos-

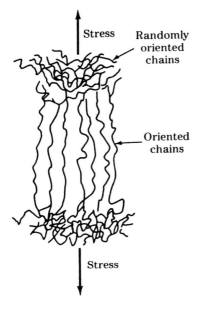

Stress   Randomly oriented chains

Oriented chains

Stress

**Figure 8.9.**   Plastic yielding and fracture of a thermoplast. After substantial plastic deformation via the molecules sliding past one another, the molecules break by fracture of the chain covalent bonds. (From W. F. Smith, *Principles of Materials Science and Engineering,* 2nd ed., McGraw-Hill, New York, 1990, p. 405.)

ites). Plasticizers are usually small liquid polymer molecules of low molar mass and high boiling points added to improve flexibility and reduce the glass temperature. PVC is a good example, where 35% by weight of a plasticizer reduces the $T_g$ from about 80°C to 0°C.

In view of all of the variables involved, it is difficult to list the mechanical properties of polymers without specifying in detail all pertinent factors. Interpreting handbook- or computer-based property surveys requires all the facts, including test methods. In addition to the usual room-temperature strength, moduli, hardness, and so on, one will frequently see listed the *heat deflection temperature.* This temperature is defined in the ASTM D648 test methods as the temperature at which a 125-mm (5-in.) bar deflects 0.25 mm (0.10 in.) under a specified stress. Two stresses are used, either 0.4 MPa (0.064 ksi) or 1.82 MPa (0.264 ksi), the latter being the more often reported. This temperature should not be used as the thermal stability of the material.

The mechanical properties of typical nearly additive-free thermoplastic polymers are listed in Table 8.6. The ASTM mechanical test method specification numbers used in most tests are listed as follows:

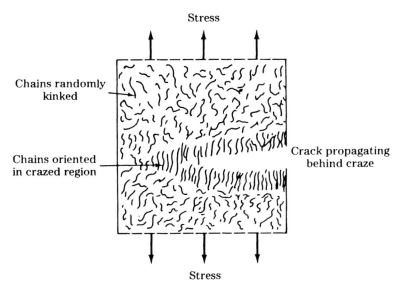

Stress

Chains randomly
kinked

Chains oriented
in crazed region

Crack propagating
behind craze

Stress

**Figure 8.10.**  Schematic illustration of a crack propagating along a craze. (From W. F. Smith, *Principles of Materials Science and Engineering,* 2nd ed., McGraw-Hill, New York, 1990, p. 404.)

D256   Izod impact strength
D638   Tensile properties
D695   Compressive properties
D785   Rockwell hardness
D790   Flexural properties
D953   Bearing strengths
D648   Heat deflection temperature

Figure 8.11 summarizes some strength and modulus data for a number of common polymers. These property values were taken from a number of sources (see the Suggested Reading) and *should be used as a guide only.* Precise properties should be obtained from the manufacturer (or supplier) and verified by the user.

Due to the viscoelastic behavior of thermoplasts, creep properties are of prime importance. The creep strain after 1000 h versus stress for several thermoplasts is shown in Figure 8.12 at room temperature. Just as in metals, the stress–rupture data are also important. Figure 8.13 shows the time for rupture at 20°C for several polymers.

**Table 8.6** Typical Mechanical Properties of Selected Thermoplastic Polymers at Room Temperature

| Material | Tensile strength [MPa (ksi)] | Elongation (%) | Tensile modulus [MPa (ksi)] | Hardness | Izod impact energy [J (ft-lb)] | HDT (°C)[a] |
|---|---|---|---|---|---|---|
| **Polyethylene** | | | | | | |
| High density | 40–44 (3–5) | 21–35 | 15–100 (100–200) | 700–1400 $R_r$ | 65 (1–5) | 1.4–6.8 |
| Low density | 7–21 (1–3) | 50–800 | 100–250 (14.5–36.2) | $R_r$ 10 | 21.6–33.7 (16–25) | — |
| Polypropylene | 30–40 (4.5–5.8) | 150–600 | 1150–1550 (166–225) | $R_r$ 90 | 1.35–14.8 (1–11) | 49–60 |
| Polystyrene | 33–55 (5.1–8.0) | 1–4 | 2400–3350 (348–486) | $R_r$ 75 | 0.15–0.54 (0.2–0.4) | 76–94 |
| Polyvinyl chloride | 35–63 (5–9) | 2–30 | 2000–4200 (286–600) | $R_r$ 115 | 1.0–2.7 (0.7–2.0) | 60–77 |
| ABS | 35–48 (5–7) | 15–80 | 1750–2500 (250–357) | $R_r$ 90–110 | 5.4–10.8 (4–8) | 88–107 |
| **Polyamides** | | | | | | |
| Nylon 6,6 | 84 (12) | 60–100 | 2070–3245 (300–470) | $R_r$ 118 | 1.35–2.7 (1–2) | 79–93 |
| Nylon 6,12 | 62 (8.8) | 150–340 | 2100 (3000) | $R_r$ 114 | 1.35–2.7 (1–2) | 58 |
| Polycarbonates | 63 (9) | 110 | 2400 (348) | $R_m$ 70–118 | 16–22 (11.7–16.2) | 132 |
| PMMA, acrylic | 55–75 (8–10.9) | 5 | 2400–3100 (348–450) | $R_r$ 130 | 0.68 (0.5) | — |
| Polyesters | 56 (8) | 300 | 2400 (348) | $R_m$ 117 | 1.63 (1.2) | 50–85 |
| **Acetals** | | | | | | |
| Homopolymer | 69 (10) | 50 | 3588 (520) | $R_m$ 92 | 1.89 (1.4) | 154 |
| Copolymer (celcon) | 62 (9) | 60 | 3105 (450) | $R_m$ 85 | 1.75 (1.3) | 128 |
| Polyimides | 97 (14) | 8 | 2070 (300) | $R_m$ 97 | 2.0 (1.5) | 154 |
| Polysulfones | 70.3 (10.2) | 5–6 | 2482 (360) | $R_m$ 69 | 1.75 (1.3) | 192 |
| Poly ether ketone (PEEK) | 100 (14.5) | >40 | 3900 (565) | — | 2.15 (1.6) | 1.67 |
| Polyethylene terephthalate | 45–145 (6.5–21.0) | | 2300–10300 (330–1500) | $R_r$ 120 | 0.21–2.0 (0.4–1.5) | 50–210 |
| Polyvinylidene chloride (PVDC) | 19 (2.8) | 350 | 345–552 (50–80) | $R_m$ 60–65 | 0.17–1.35 (0.13–1.0) | 54–65 |

[a] At 264 psi.

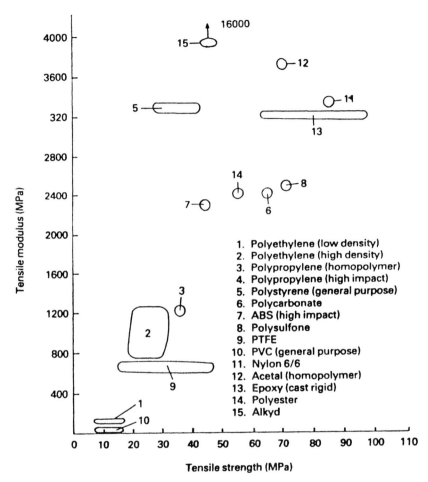

**Figure 8.11.** Comparison of some commonly used polymers on the basis of tensile strength and tensile modulus. Polymers numbered 13, 14, and 15 are thermosets. All others are thermoplasts. (From M. M. Farag, *Selection of Materials and Manufacturing Processes for Engineering Design*, Prentice Hall, Englewood Cliffs, N.J., 1989, p. 81.)

Fatigue of polymers is one mode of failure that has occurred frequently and has been studied rather extensively. Such failure in polymers may be induced either by large-scale hysteretic heating, which causes the polymer to soften and even melt, or by fatigue crack initiation and propagation to final failure. Evidence has been presented in the literature (see Hertzberg and Manson book listed in the Selected Reading) that both "clamshell" markings and "striations"

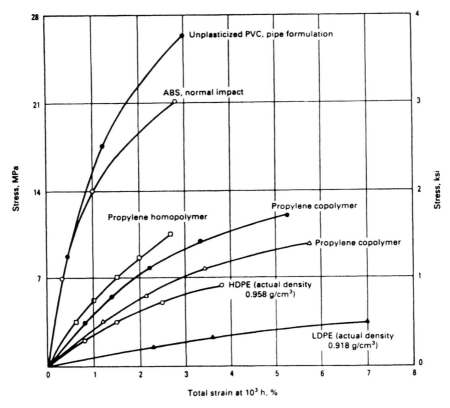

**Figure 8.12.** Tensile creep strain versus stress of various thermoplasts after 1000 h at room temperature. (From R. Horsely, in "Mechanical Performance and Design in Polymers," O. Delatychi, Ed., Vol. 17, *Proceedings of the Applied Polymer Symposium.*)

such as those observed in metal fatigue can also occur in polymers. Striations represent crack advancement per stress cycle and require very high magnifications for detection, while clamshells can be seen without the need for any magnification. While the mechanisms of polymer fatigue are still controversial, there are some cases where it appears to be associated with crazing. The stress–life data (*S–N* curves) for several polymers are shown in Figure 8.14. The life cycle decreases with stress amplitude, and below a certain stress level some polymers appear to have an infinite life.

Physical Properties
Like ceramics, polymers do not have free electrons roaming around to provide electron and heat conductivity. They are basically insulators and are used in

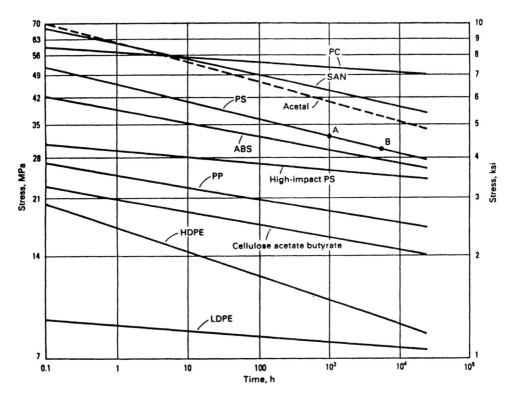

**Figure 8.13.** Tensile stress for rupture versus time for several polymers. (From R. Kahl, paper presented at the *Principles of Plastic Materials Seminar*, Center for Professional Advancement, 1979.)

large quantities for such applications at temperatures up to their maximum use temperature, which for the best performance thermoplastic polymers (listed in Table 8.5) is about 200°C. There are a number of ways in which the long-term temperature resistance may be expressed. One is the Underwriters Laboratories' (UL) thermal index, defined as that temperature when the properties drop to 50% of their initial values. Initial values presumably are specified at a temperature near room temperature. The maximum continuous temperature (MCUT) is the temperature at which the strength decreases to 50% of its original value in 11.4 years. The continuous-use temperature (CUT) has not been clearly defined, and it is pretty much up to the manufacturer to specify what is meant by this temperature. The heat deflection temperature (HDT) defined previously is used frequently as an indicator for mechanical strength versus temperature. The UL index temperature is almost always lower than the HDT. It is

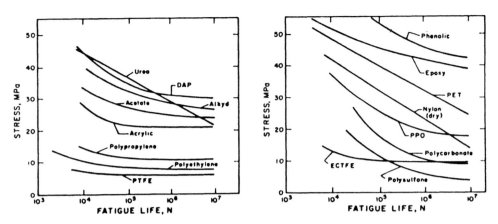

**Figure 8.14.** Representative *S-N* curves for several polymers. (From M. N. Ridell, *Plast. Eng.*, Vol. 30, No. 4, 1974, p. 71.)

difficult to locate tabulated CUT values. One should resort to the manufacturers' literature here since definitions may vary.

The other physical properties—thermal conductivity, $T_g$ and $T_m$, electrical resistivity, and dielectric strength—are more readily available and are listed in Table 8.7 for some common thermoplasts.

Forming of Thermoplasts

In our discussion of the processing of metals and ceramics we have used such terms as pressing (molding), injection molding, blow molding (glasses), extrusion, casting, and drawing. All of these processes apply to thermoplastic polymers as well, the chief difference between the materials being that the processing temperatures for ceramics fall at the high-temperature end of the forming temperature scale and thermoplasts at the low end. Some thermoplasts can be formed at room temperature, the forming temperature being dependent on their glass temperatures. Forming of thermoplasts must be done somewhere above $T_g$, and often it is done above $T_m$.

The most important and widely used forming techniques for thermoplasts are extrusion and injection molding. As was discussed for ceramics, the difference in extrusion details between ceramics and metals are the methods into which the materials are fed into the compression chamber and forced through the die. Whereas metals usually are in the form of cylindrical ingots, ceramics are fed in powder form. Polymers are more akin to ceramics here, being fed into the chamber in pellet form. The pellets are carried through the extrusion chamber by an auger type of screw mechanism. The chamber is normally heated externally, and coupled with the material being further heated by

**Table 8.7** Typical Physical Properties of Selected Thermoplasts

| Material | Specific gravity | Thermal conductivity (W/m·K) | Thermal expansion coefficient [$10^{-5}K^{-1}$ ($10^{-5}°F^{-1}$)] | Volume resistivity short time[a] (Ω·m) | Dielectric strength[b] [V/cm (V/mil)] |
|---|---|---|---|---|---|
| Polyethylene | | | | | |
| High density | 0.96 | 0.44 | 11–13 (6.1–7.2) | $10^{15}$–$10^{16}$ | 200–190 (510–480) |
| Low density | 0.92 | 0.35 | 13–20 (7.2–11.1) | $10^{15}$–$10^{16}$ | 390–180 (990–460) |
| Polypropylene | 0.9 | 0.12–0.14 | 6–10 (1.0–5.5) | $10^{15}$ | 260–200 (660–510) |
| Polyvinyl chloride | 1.4 | 0.16 | 9–18 (5–10) | $10^{13}$ | 240 (610) |
| Nylon 6,6 | 1.14 | 0.24 | 8–9 (4.4–5) | $10^{13}$ | 160 (410) |
| Polycarbonates | 1.2 | 0.20 | 6.8 (3.8) | $2 \times 10^{14}$ | 160 (410) |
| PMMA (acrylic) | 1.18 | 0.19 | 4.5 (2.5) | $10^{13}$–$10^{14}$ | 200–160 (510–410) |
| Acetal copolymer | 1.4 | 0.23 | 8–10 (4.4–5.5) | — | — |
| Polyimides | 1.43 | — | 5.4–8 (3–4.5) | $10^{14}$–$10^{15}$ | 220 (560) |

[a] Volume resistivity increases with time.
[b] From *Engineered Materials Handbook*, Vol. 2, ASM International, Metals Park, Ohio, 1988, p. 469.

the mechanical action of the screw, it may melt or be in a very high viscosity state. In the extrusion process the thermoplasts usually reach the die orifice in a molten state. The temperatures of the barrel (chamber) are precisely controlled along the barrel zones because different polymers soften at different temperatures. Low-density polyethylene, which is amorphous, softens as the temperature increases, whereas highly crystalline nylon softens abruptly at a temperature in the vicinity of $T_m$.

In the injection molding process, the open die of the extrusion process is replaced with a closed die. This is more akin to the die casting of metal. Here a polymer melt is forced into a mold, where it solidifies. The mold is cooled, thus solidifying the polymer; opens into two halves, whence the product is ejected; and then the mold closes for the next cycle. Low melt viscosity promotes faster solidification, but the viscosity must be such to obtain good flow into the mold cavities. Thermoplastic resins are more commonly formed by compression molding, very analogous to the forging of metals. A metered quantity of pellets is fed directly into the bottom half of the mold die and the mating half is pressed with hydrostatic pressures on the order of 15 MPa (2.2 ksi). The product can be removed hot and a new cycle quickly repeated.

Blow molding was borrowed from the glass industry. For restricted neck bottles, a soft plastic tube (the *parison*) is surrounded by a mold. Air pressure is used to expand it to the configuration of the mold. In a modification of this process called *rotational molding,* a charge of polymer powder is introduced into a hot mold. When the charge becomes molten, the mold is rotated to coat the mold walls to a uniform thickness.

Polymeric materials in sheet form are often produced and reduced in thickness by passing them through heated rolls, an operation known as *calendering.* This is not very different in principle from the sheet rolling of metals. Some thermoplasts can be shaped without heating in a number of cold-forming operations, such as the stamping and forging operations applied to metals. Typical forming methods for thermoplasts are depicted in Figure 8.15.

### Applications

Next we discuss, in the order of prominence, the most commonly used thermoplasts. The polyethylenes are the most widely used of all polymers, comprising about 30% of the total polymer market. They are relatively cheap, tough, chemically resistant, have good dielectric properties, and are easy to process. They are widely used in sheet form. They are classified as low, medium, and high density, with their strength and moduli increasing in that order. Their physical properties and cost are less affected by density.

Polystyrenes account for about 17% of the total market. Various modifications create a wide range of properties. They are generally rigid and in clear form are often used for plastic containers. The very well known foamed insulating polystyrene may soon be replaced by other insulating materials, due to environmental concerns.

The vinyls occupy about the same market segment as that of the polystyrenes. They are versatile, low-cost materials that can vary in flexibility from rigid to rubbery, depending on the quantity of plasticizer. Their maximum use temperature is about 77°C (175°F). The most widely used are the PVC and PVC–acetate copolymer types. They are used for flexible tubing, footwear, upholstery, and chemical tanks, as well as for a host of smaller applications.

Polypropylene is next in line of use ($\sim 5\%$) and competes to some extent with the high-density polyethylenes. There also exist a number of copolymers of polyproylene with other olefin monomers. Typical applications include housewares, tubes and pipes, housings, and many small parts.

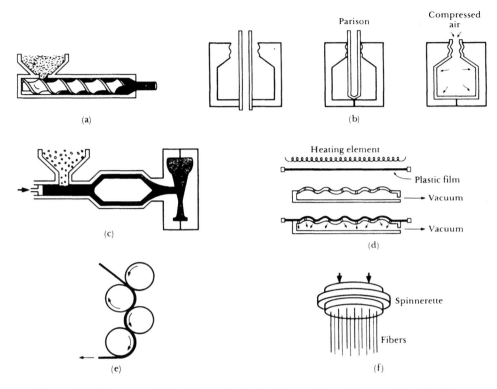

**Figure 8.15.** Typical forming methods for thermoplasts: (a) extrusion; (b) blow molding; (c) injection molding; (d) thermoforming; (e) calendaring; (f) spinning. (Adapted from D. R. Askeland, *The Science and Engineering of Materials,* 2nd ed., PWS-Kent, Boston, 1989, p. 269.)

The ABS copolymers have a good balance of toughness, hardness, and rigidity up to temperatures of about 107°C (225°F). They compete with PVC for pipes, with both PVC and vinyls for chemical tanks, and are used as fume hoods, ducts, and appliance housings. They are somewhat more expensive than the more popular polymers.

The acrylic polymers, based on the PMMA polymers, modified by copolymerization, are strong, hard, and available in a variety of colors. High-impact grades are produced by blending with rubber stock. They can be obtained in just about all forms and can be processed by machining, molding, and extrusion. Typical applications include lighting fixtures, goggles, optical lenses, control handles, control pump parts, and tool handles.

Polyamides (nylons) are linear molecules with a high degree of crystallinity and are among the high-strength thermoplasts. They have excellent electrical insulating properties and chemical resistance but are among the more expensive polymers. Typical applications include those requiring high wear resistance and low friction: as, for example, bushings, rollers, fasteners, zippers, housings, gears, and electrical parts.

Acetals are also highly crystalline and among the strongest thermoplasts. They have good creep resistance, dimensional stability, fatigue resistance, and are useful to temperatures of about 105°C (220°F). This means that they can be used in boiling-water environments. They also have good wear resistance and find application in the same arenas as the nylons (i.e., gears, bushings, cams, rollers, latches, and bearings). They are also competitive pricewise with the nylons.

Polycarbonates are high-temperature noncrystalline transparent plastics. Their maximum service temperature is 140°C (285°F) and they have good creep resistance. Their only drawbacks are their relatively high cost and susceptibility to attack by organic solvents. These hard polymers find use as helmets, windshields, load-bearing electrical insulators, battery cases, toolhouses, and parts requiring dimensional stability. It is somewhat surprising that they account for less than 1% of the polymer market.

The fluoroplastics, which we have ignored up to now because they are not hydrocarbons, are linear molecules in which fluorine has replaced hydrogen. Teflon, the best known of this group, has a high service temperature (260°C) and is known for its low-friction surface. It is ideal for bearings, bushings, and linings for chemical process equipment. It had its popular beginning as coatings for home cookware, but has since found some strong competition from other products. High-temperature insulation for wire and cable is another important application.

Polyesters of the thermoplastic type have good mechanical and electrical properties and good dimensional stability, but a relatively low service temperature, in the neighborhood of 50°C.

Polyurethanes are tough, with excellent abrasion and impact resistance. They have found use in lacquer form as wood coatings. Some polyurethanes are also thermosets.

Cellulose acetate, ethyl cellulose, and cellulose acetate butyrate are tough and hard with good optical properties. They can be processed by most thermoplastic processing methods. They are used as knobs, tool handles, steering wheels, toys, and packaging.

Finally, we have the more recently developed high-temperature thermoplasts, including polyimide, polysulfone, polyether ether ketone, and polyphenylene sulfide. Many of these have service temperatures in excess of 200°C. Although they are difficult to process and high in price, they find use in heat-resistant under-hood automotive components.

Most thermoplasts are furnished in completely polymerized pellet form ready to mold or extrude. Some are available as liquid resins for casting or injection molding.

## 8.3 THERMOSETS

### 8.3.1 Chemistry and Products of the Network Polymers

Metals and crystalline ceramics are basically three-dimensional structures in which the pattern of atom arrangement repeats itself precisely throughout the three-dimensional lattice. There are several materials with network structures that sort of fit in between these precise arrangements and that of the network polymers. For example, regular two-dimensional sheet structures occur in graphite, boron nitride, and many silicates. The latter groups are crystalline because the sheets are all connected in the same fashion to form a three-dimensional network. The common network polymers are not that regular in structure. Such polymer materials are thermosets.

The thermoset polymers are formed by a large amount of cross-linking of linear prepolymers (a small amount of cross-linking produces elastomers) or by the direct formation of networks by the reaction of two monomers. The latter is the more prominent of the two methods. It is the stepwise or condensation method, which has been defined as *the reaction of two monomers to produce a third plus a by-product, usually water or alcohol.* This definition is probably no longer exactly correct, since in some cases a by-product is not produced. The reaction is now called a *stepwise* polymerization, which in some reactions can result in a by-product, such reactions being called *condensation reactions.* In other words, the latter is a subset of the former. Also, in the stepwise process chain thermoplasts as well as networks may be the final polymer structure. In the network case one of the chemical reactants must have more than two reaction sites. The classic reaction of this type, that of phenol–formaldehyde (phenolic), is illustrated in Figure 8.16, and the resulting three-dimensional structure is shown in Figure 8.17. The chemistry is generally more complex than that of thermoplastic polymerization. As the reactants combine, random three-dimensional networks are formed via branched structures until the entire mass constitutes a single giant molecule. These structures are incapable of melting or being dissolved. They cannot return to a fluid condition by heating but decompose at an elevated temperature. Often, these thermosets are partially polymerized and shipped as a powder in this state to the manufacturer, who completes the polymerization (curing) in the mold. In other cases, the polymerization and shaping occur simultaneously.

The principal thermosets are the phenolics, the amino resins, the unsaturated polyesters, the epoxies, and the cross-linked polyurethanes. In both the epoxy and unsaturated polyester cases, the first state of the action is the construction

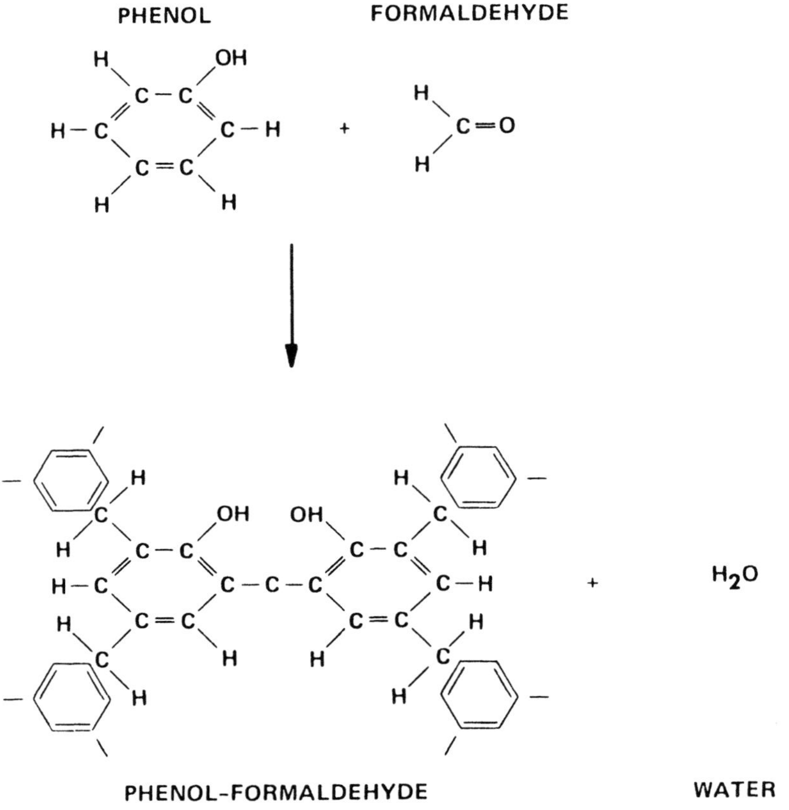

PHENOL                          FORMALDEHYDE

PHENOL-FORMALDEHYDE                                   WATER

**Figure 8.16.** Condensation polymerization reaction that produces phenol–formaldehyde (Bakelite) plus the water by-product. (From D. W. Richerson, *Modern Ceramic Engineering*, Marcel Dekker, New York, 1982, p. 27.)

of short linear chains by condensation. These uncured resins are stable and are made available to the user in liquid or paste form. The final curing is achieved by the addition of a hardener or curing agent immediately prior to molding.

## 8.3.2  Properties and Applications of the Thermosets

Although thermosetting polymers are fewer in number than thermoplastic polymers, they do make up about 14% of the total polymer market. They are more brittle, stronger, harder, and generally more temperature resistant than the thermoplasts. Other advantages are better dimensional stability, creep resistance, chemical resistance, and good electrical properties. However, thermosets are more difficult to process and generally more expensive.

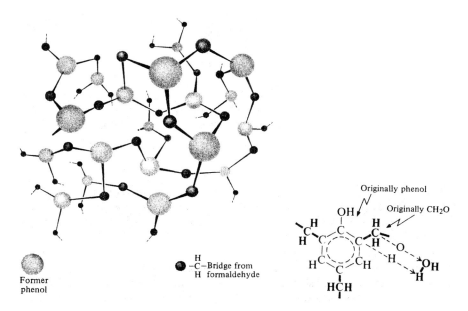

**Figure 8.17.** Three-dimensional structure of phenol-formaldehyde. (From L. H. Van Vlack, *Materials for Engineering,* Addison-Wesley, Reading, Mass., 1982, p. 186.)

The phenols are the most widely used, occupying about 6% of the total polymer market and thus about 43% of the thermoset use. They are relatively inexpensive and are readily molded with good stiffness. Most contain wood- or glass-flour fillers and often glass fibers. Typical applications include pulleys, electrical insulators, connector plugs, housings, handles, and ignition parts.

Amino (NH or $NH_2$ groups)-, urea-, and melamine-containing thermosets are the second largest thermoset group. Their properties vary widely but they are generally hard and abrasion resistant. Applications of the ureas are in electrical and electronic components, while the melamines are used in dinnerwares and handles.

The epoxies are highly versatile but also somewhat more expensive than the other thermosets. They have excellent bonding properties and good chemical resistance and electrical properties. In addition to bonding applications, they find use in encapsulation of electronic components and for some tools and dies.

Polyester thermosetting resins are usually copolymers of a polyester and a styrene. Most are used in the reinforced condition as boat hulls, swimming pool fixtures, and lawn chairs. They have low moisture absorption and relatively high strength.

Polyurethane thermosets can be rigid or flexible, depending on the formulation. The flexible ones have better toughness. They are often used as foamed parts.

Not all of the processing methods listed for the thermoplasts are applicable to the thermosets. Since thermosets polymerize in the mold, the common processing methods are pretty much limited to compression and injection molding, although the chemically reactive epoxies and polyesters can be mixed and poured to cast shapes without the addition of heat. Transfer molding, where the material is heated in one mold and then forced under pressure into a closed mold, is sometimes used for thermosets.

Mechanical Properties

Some of the most common thermoset polymers are listed as follows: epoxies, unsaturated polyester, phenolic (phenol–formaldehyde), melamine–formaldehyde, urea–formaldehyde, polyurethane, and bismaleimides (polyimides). Accordingly, these polymers were selected for the mechanical property listings provided in Table 8.8. These property values were obtained from many compilations (see the Selected Reading). Many of these compiled values in turn were obtained from such handbooks as the *Modern Plastics Encyclopedia* (McGraw-Hill), *Polymer Handbook* (Wiley, 1975), *Polymers: An Encyclopedia Sourcebook of Engineering Properties* (Wiley, 1987). The properties listed in Table 8.8 will show a considerable range of values because of different processing methods as well as different sources. Hardness values are given in many different hardness scales, making comparisons difficult. All values are at room temperature (although the HDT is given). In all cases the pure or *neat resins*, presumably absent of additives and reinforcements, are the ones listed. One should be aware that many, if not most, polymers are used with reinforcements, usually glass fibers. In some polymers it was difficult to locate the *neat* properties. The Izod impact values of all thermosets are very low (0.5 to 2.0 ft-lb).

Physical Properties

Most of the same physical properties that were of interest for the thermoplasts also apply for the thermoset polymers. Again they are poor conductors of heat and electricity. Moisture can affect the electrical breakdown, as well as time and temperature. Any thermoplast or thermoset polymer can be made conductive by incorporating conductive additives (e.g., copper powder). The physical properties of several thermosets are listed in Table 8.9.

## 8.4  ELASTOMERS

### 8.4.1  Introduction to Elastomer Chemistry

Despite the fact that polymer chemists and engineers themselves have a clear understanding of elastomers and their properties, elastomers are described

**Table 8.8** Typical Mechanical Properties of Selected Thermoset Polymers at Room Temperature

| Material | Tensile strength [MPa (ksi)] | Elongation (%) | Tensile modulus [MPa (ksi)] | HDT (°C) | Hardness | Comments |
|---|---|---|---|---|---|---|
| Epoxies | 50–140 (7.2–20.3) | 1.5–1.8 | 3100–3800 (450–550) | 150–240 | $R_m$ 106 | Cast values are lower than molded |
| Unsaturated polyesters | 40–70 (4–8) | 1.3–3.3 | 2400–3450 (348–500) | 80–130 | Barcol 40, $R_m$ 117 | The saturated polyesters are thermoplasts |
| Phenol–formaldehyde | 35–63 (5–9) | 0 | 800 (114) | 174 | $H_r$ 125, $R_e$ 95 | |
| Melamine–formaldehyde | 52 (7.5) | — | 9650 (1400) | — | $R_m$ 120 | Used as adhesives in pure form |
| Urea–formaldehyde | 41–69 (6–10) | 0.5 | 6900 (1000) | — | $R_m$ 115 | |
| Polyurethane high density | 20–70 (2.9–10.1) | 180–300 | 3000–6000 (438–870) | 120 | Shore D 39–64 | Higher values are RIM material |
| Bismaleimides (polyimides) | 97 (14) | 1–8 | 4100 (594) | 360 | $R_f$ 45–58 | Izod impact 80 J/m |

**Table 8.9**  Typical Physical Properties of Selected Thermosets

| Material | Specific gravity | Thermal conductivity (W/m · K) | Thermal expansion coefficient $(10^{-5}\,\mathrm{K}^{-1})$ | Volume resistivity, short time $(\Omega \cdot m)$ | Dielectric strength (V/mil) |
|---|---|---|---|---|---|
| Epoxies | 1.1–1.4 | 0.19 | 4.5–6.5 | $10^{14}$–$10^{16}$ | 300–500 |
| Unsaturated polyesters | 1.04–1.20 | 0.17 | 11.3 | — | 380–500 |
| Phenol–formaldehyde | 1.24–1.32 | 0.09 | 6.8 | $10^{12}$–$10^{14}$ | 250–400 |
| Polyurethane | 1.2 | 0.17 | 5.7 (3.2) | — | — |
| Polyimides | 1.2 | 0.09 | 3.1 | — | — |

and/or defined in various ways in textbooks and related literature. From a strictly mechanical point of view they are polymers that exhibit large-scale reversible extensions of 100% or more at room temperature. By this definition, the hard rubbers have much smaller elongations and are not true elastomers but are more akin to network polymers. A typical stress–strain curve of an elastomer is depicted in Figure 8.18. It does not really have a significant straight-line elastic section, as a true Hookean solid would show. Moduli are a few hundred psi and difficult to measure. For those elastomers that show a small straight-line segment near the origin, the modulus is obtained from the slope of the extension of this segment. For those elastomers that do not conform to Hooke's law even near the origin, the slope of a tangent near the origin is taken to obtain what is called a *secant modulus* (consult ASTM test specification D638). The elastomers usually deviate from a straight line near the origin as the

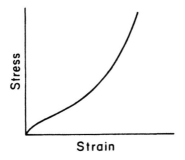

**Figure 8.18.**   Typical stress–strain curve of an elastomer.

molecules uncoil but approach a straight line again after a very large extension. The entire extension is recoverable. The *modulus* thus changes with extension and is not a Hooke's law modulus. Fracture occurs after large extensions, most often with little or no plastic deformation. They exhibit this elasticity above their glass temperature. Necessary but not sufficient conditions for a polymer to be an elastomer are that it be noncrystalline at room temperature and that its $T_g$ be well below room temperature. The unstretched molecules of an elastomer are amorphous and coiled like cooked spaghetti.

Probably the reason for the confusing definitions is that some refer to elastomers as rubbers, and vice versa. The rubber industry was well established by 1900 using natural rubber as its raw material, well before the modern polymer industry appeared. Today, some definitions place natural and synthetic rubbers into two distinct categories and refer to the latter as *elastomers*. But let's not dwell on semantics. Elastomers should really include natural rubber, polybutadiene, styrene–butadiene, butyl rubber, polychlorprene, some polyurethanes, and synthetic polyisoprene [isoprene, $(C_5H_8)_n$, is natural rubber], to name a few.

The butadiene-type molecules form many of the rubbers. These are rubber-type molecules where one hydrogen is replaced by another group of atoms. Butadiene is a gas widely used in styrene, vinyl chloride, and acrylonitrile copolymers, ABS being the most common. Butadiene has the mer

$$
\begin{array}{ccccc}
\text{H} & \text{R} & & \text{H} & \text{H} \\
| & | & & | & | \\
\text{C} & = & \text{C} - \text{C} & = & \text{C} \\
| & & & & | \\
\text{H} & & & & \text{H}
\end{array}
$$

When R = H, we have polybutadiene rubber, $T_g = -90°C$, $T_m = 154°C$; when R = Cl, polychloroprene (neoprene), $T_g = -50°C$, $T_m = 80°C$; and when R = CH$_3$, polyisoprene (natural rubber), $T_g = -100°C$, $T_m = 28°C$.

All of the synthetic rubbers were modeled on natural rubber, being linear-chain polydienes and random copolymers with C=C double bonds to permit irreversible cross-linking. Natural rubber latex has a very high molar mass with average chain links on the order of 100,000 carbon atoms. By mechanically working the rubber (mastication) the chain links can be reduced by a factor of 10. Sulfur is then added, resulting in structural changes that involve the sulfur atom cross-links between the chains, as depicted in Figure 8.19. Stiffness can be controlled by the number of cross-links, which is dependent on the sulfur content. Cross-linking at about 10% of the possible sites gives the rubber mechanical stability (which natural rubber does not have), yet it retains its flexibility. The sulfur atoms act as pinning points, which jerks the rubber chain back into its unstressed position when the load is released. Large

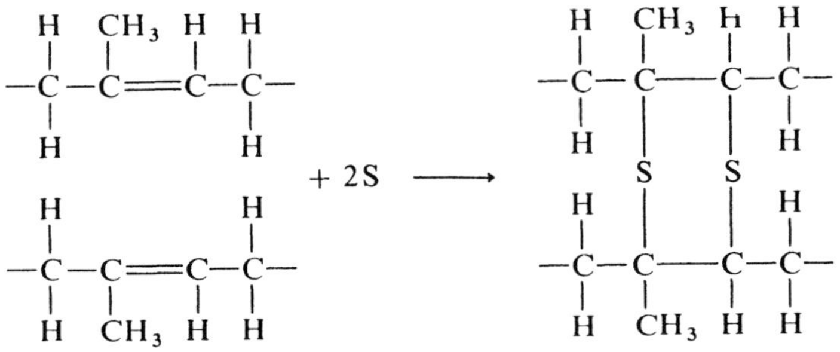

**Figure 8.19.** Vulcanization of rubber by cross-linking with sulfur.

amounts of sulfur produce the hard rubbers. About 40% sulfur content yields the *ebonites,* which are highly cross-linked, rigid nonelastomers. These are more like the three-dimensional network polymers. The effect of sulfur has been enhanced by the addition of accelerators, which increase the rate of cross-linking. These rubbers also contain stabilizers, such as carbon black, which inhibit the oxidative aging.

The technology that has been applied to natural rubbers has been extended to the polydiene and polyolefin synthetic rubbers. Vulcanization by sulfur is applied to many of the synthetic rubbers. Zinc and magnesium oxide are the normal vulcanizing agents for polychloroprene, while the ethylene propylene rubbers and the fluoroelastomers can be cross-linked by various organic substances.

The polyurethane elastomers form a group of heterochain polymer elastomer materials that have a variety of structures. The solid polyeurethane elastomers are strong and have exceptional abrasion and tear resistance, and good oxidation resistance. The silicone rubbers are also heterochain elastomers with service temperatures as high as 200°C.

During the 1970s the thermoplastic elastomers (TPE) began to be used in significant quantities and have been growing at a very rapid rate, approximately 9% per year. In 1989, 600,000 tons were used worldwide. These elastomers contain sequences of hard and soft repeating units in the polymer chain. Elastic recovery occurs by the hard segments acting to pull back the more soft and rubbery segments. Cross-linking is not required. The copolymer styrene–butadiene–styrene has become the classic example. The butadiene is the soft segment, while the styrene prevents the sliding plastic deformation common to thermoplasts. This is in direct contrast to styrene–butadiene which is cross-linked. These six generic classes of TPEs are, in order of increasing cost and performance, the styrene block copolymers; polyolefin blends, elastomeric alloys, thermoplastic polyurethanes, thermoplastic copolyesters, and thermoplastic polyamides. Some of the commercially important elastomers and their applications are listed in Table 8.10.

## 8.4.2  Properties of Elastomers

Natural rubbers, with all of their additives, account for about 20% of all present elastomer usage. Less than 1% of natural rubber today comes from wild rubber trees. The rest comes from the few thousand seedlings smuggled out of Brazil in the late nineteenth century and cultivated in Ceylon. The plantations of Malaysia and Indonesia now account for over 70% of the world's production of natural rubber. Major uses of natural rubber include conveyor belts, gaskets, and automobile tire products.

The styrene–butadiene rubbers are copolymers of butadiene and styrene and are the most widely used group of rubbers because of their low cost and their

**Table 8.10**   Commercial Elastomers and Their Applications

| Material | ASTM Class | Applications |
|---|---|---|
| Natural rubber | R | Rubber |
| Styrene–butadiene rubber | R | Synthetic rubber: used in place of natural rubber tires, belts, and mechanical goods |
| Polybutadiene | R | Synthetic rubber |
| Polychloroprene | R | Synthetic rubber: gaskets, V-belts, cable coatings |
| Ethylene–propylene rubbers | M | Saturated carbon chain elastomer: copolymer-resistant to ozone and ultraviolet radiation, good chemical resistance to acids but not to hydrocarbons; cable coating, hoses, roofing, automotive parts |
| Poly(propylene oxide) rubbers | O | Polyester elastomer |
| Silicone elastomers | Q | Useful from −101 to 316°C: tubing, gaskets, molded products |
| Polyurethane polyester | Y | Thermoplastic elastomers: combination of a rubbery and a hard phase |
| Polyester polyether | Y | Thermoplastic elastomers: footwear, injected-molded parts |
| Elastomeric alloys | Y | Extruded, molded, blow-molded and calendered goods (thermoplastic) |

wide use in automobile tires. The butadiene–acrylonitrile rubbers have better resistance to oil and gasoline and a higher service temperature than those of the butadiene-styrene rubbers.

Neoprene (polychloroprene) is a synthetic rubber that is chemically and structurally similar to natural rubber but has better resistance to oils, chemicals, and sunlight. Major uses include V-belts, heavy-duty conveyor belts, footwear, and motor mounts.

Butyl rubbers consist of copolymers of isobutylene and a few percent of isoprene and are one of the lowest-priced synthetic rubbers. They have excellent dielectric strength and are widely used for cable insulation and encapsulation of electronic components. They are also used as steam hoses, machine mounts, and inner-tube linings.

Silicone rubbers are the most stable of all rubbers and have been used to temperatures as high as 270°C. They are high-performance, high-priced materials used for seals, gaskets, O-rings, encapsulation, and wire and cable insulation.

The elastomer properties listed in Table 8.11 were taken from a number of sources, and in some cases average or approximate values were used.

## 8.5  ADHESIVES

Adhesives are not polymers that are consumed in large quantities; nevertheless, they occupy an important role in our select group of engineering materials. ASM International has just published its third volume of the *Engineered Materials Handbook,* and it is totally devoted to adhesives. Over 100 types and several thousand products are listed in another *Adhesives Handbook* (Butterworth, 1984). Adhesives with strengths higher than some metals have been developed. Adhesives have been separated into five types (from C. Hall, *Polymer Materials,* p. 179);

1.  Solutions of thermoplastics (including unvulcanized rubber) which bond by loss of solvent
2.  Dispersions of thermoplastics in water or organic liquids which bond by loss of the liquid phase
3.  Thermoplastics without solvents (hot-melt adhesives)
4.  Polymeric compositions which react chemically after joint assembly to form cross-linked thermoset polymers (e.g., epoxies)
5.  Monomers which polymerize in situ

As one can observe from the above, adhesives do not fall into clearly defined categories and could be considered as subclasses of the other three (i.e., thermoplasts, thermosets, and elastomers). The use of adhesives covers the range from aircraft and missile skins to food and beverage cans. Chemical bonding between the adhesive and adhered surface does not usually take place, although some believe that there exists the possibility of some type of van der

**Table 8.11**    Room-Temperature Mechanical Properties of Some Typical Elastomers

| | Pure gum vulcanizates | | Carbon-black reinforced vulcanizates | |
|---|---|---|---|---|
| | Tensile strength $(kg/cm^2)$ | Elongation (%) | Tensile strength $(kg/cm^2)$ | Elongation (%) |
| Natural rubber (NR) | 210 | 700 | 315 | 600 |
| Styrene–butadiene rubber (SBR) | 28 | 800 | 265 | 550 |
| Acrylonitrile–butadiene rubber (NBR) | 42 | 600 | 210 | 550 |
| Polyacrylates (ABR) | — | — | 175 | 400 |
| Thiokol (ET) | 21 | 300 | 85 | 400 |
| Neoprene (CR) | 245 | 800 | 245 | 700 |
| Butyl rubber (IIR) | 210 | 1000 | 210 | 400 |
| Polyisoprene (IR) | 210 | 700 | 315 | 600 |
| Ethylene–propylene rubber (EPM) | — | 300 | — | — |
| Polyfluorinated hydrocarbons (FPM) | 50 | 600 | — | — |
| Silicone elastomers (SI) | 70 | 600 | — | — |
| Polyurethane elastomers (AU) | 350 | 600 | 420 | 500 |

*Source*: Adapted from R. B. Seymour and C. E. Carraher, Jr., *Polymer Chemistry,* Marcel Dekker, New York, 1988, p. 519.

Waals bonding taking place in addition to some mechanical interlocking on rough surfaces. It is essential for the adhesive to wet the adhered surface during joint assembly. Surface cleanliness is important but not surface smoothness. Intentional roughening by abrasion, such as grit blasting, is recommended.

There are many polymers and polymer systems that may be used for adhesives. Table 8.12 lists the properties and preparations of some common adhesives.

## 8.6    COMPARATIVE MECHANICAL PROPERTIES

The strength and moduli of thermoplasts, thermosets, and elastomers are compared in Figures 8.20 and 8.21. Actually, the moduli of elastomers were excluded from Figure 8.21 because it is not a meaningful value since it changes

**Table 8.12**   Typical Adhesives and Their Properties

| Material | Curing temp. [°C (°F)] | Service temp. [°C (°F)] | Lap-shear strength [MPa (ksi)] | Room temp. peel strength [N/cm (lb/in.)] |
|---|---|---|---|---|
| Acrylics | RT (RT) | To 120 (to 250) | 17.2–37.9 (2.5–5.5) | 17–105 (10–60) |
| Anaerobics | RT (RT) | To 166 (to 330) | 15.2–27.6 (2.2–4.0) | 17.5 (10) |
| Cyanoacrylates (thermosetting) | RT (RT) | To 166 (to 330) | 15.2–27.6 (2.2–4.0) | 17.5 (10) |
| Epoxy RT cure | RT (RT) | −51 to 82 (−60 to 180) | 17.2 (2.5) | 7 (4) |
| Epoxy HT cure | 90–175 (195–350) | −51 to 75 (−60 to 350) | 17.2 (2.5) | 8.8 (5.5) |
| Epoxy–nylon alloy | 120–175 (250–350) | −250 to 82 (−420 to 180) | 41 (60) | 123 (70) |
| Polyurethanes | 149 (300) | To 66 (to 150) | 24 (3.5) | 123 (70) |
| Silicones | 149 (300) | To 260 (to 500) | 0.5 (0.04) | 43.8 (25) |

with the amount of elastic strain, being very small at low strains and quite large at high strains. In general, the thermosets are stronger and harder than the thermoplasts. But if we take the top materials in each group we find that PEEK and the polyimide bismaleimides have about the same strength and moduli. If the operating temperatures are of concern, certainly PEEK has the advantage, and in this case the polysulfones and PET must also be considered. The chemical reactivity may also be a factor for consideration, a topic that we will address when we take up the subject of environmental effects. What is noticeable is the low values of strength and moduli of the polymers compared to that of metals.

## 8.7 SOME MORE RECENTLY DEVELOPED TECHNOLOGIES FOR POLYMERS

Blending of polymers, although not a real new technology, is on the upslope of the polymer growth curve. All sorts of mixtures are being attempted, some by more or less trial-and-error procedures. But some worthwhile blends have emerged, particularly those containing mixtures of crystalline and amorphous polymers. Nylon–phenolic blends have been used as ablative heat shields on

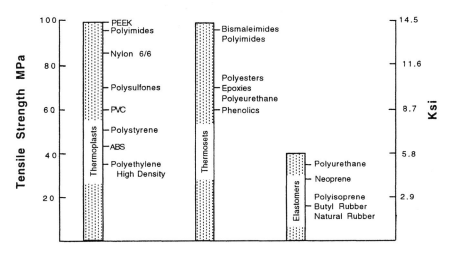

**Figure 8.20.** Comparison of room-temperature tensile strengths for thermoplast, thermosets, and elastomers.

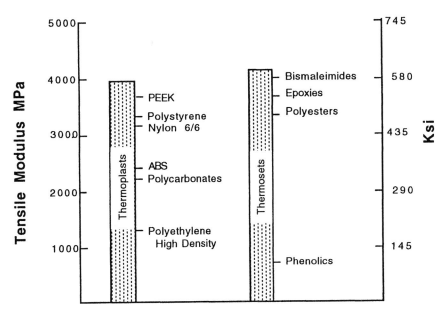

**Figure 8.21.** Comparison of room-temperature tensile moduli for thermoplasts and thermosets.

space vehicles during reentry when temperatures in the thousands of degrees are attained. In the 1970s the term *polymer alloys* emerged to describe rubber-toughened materials in which the matrix was a mixture of polymers types. The term *hybrids* has also been used for some polymer blends. Many of the new developments in elastomers are occurring via the alloying route.

Interpenetrating polymer networks (IPNs) consist of a combination of two polymers where at least one is synthesized and/or cross-linked in the immediate presence of the other. Silicone thermosets are being added to a variety of thermoplasts including nylon, PBT, and polypropylene to form IPNs, which results in improved elevated temperature strength.

High-performance polymers such as polysulfone, polyarylate, and high-strength nylons are being molded into porous forms for filters and automotive vents. Porous polymers, somewhat different from foams, have a network of open-celled, omnidirectional pores running through the polymers. Fluids travel through the interconnecting pores and experience a filtering action.

Liquid-crystal polymers, described earlier, are finding new uses. Among these include grades for injection molding of electrical and electronic parts.

Around 1977 it was found that the conductivity of polyacetylene could be increased by more than a factor of 10 by doping it with various elements to give it semiconducting properties. The crystalline polymers have energy band structures similar to semiconductors with energy gaps, $E_g$, on the order of 1.5 eV. Low concentrations of dopants such as $ClO_4^-$ or $I_3^-$ in polyacetylene and poly($p$-phlene) achieve conductivities similar to that found in semiconductors. With a large concentration of dopants, on the order of several percent, conductivities approaching that of metals have been attained. This phenomenon has been observed in a number of polymers. This new class of materials may also have some desirable optical, thermal, and chemical properties. The conducting polymers are being considered for use in applications such as solar cells, fuel cell catalysts, and inexpensive lightweight batteries.

## DEFINITIONS

*Ablative polymer*: material that absorbs heat through a decomposition process (material is expendable).
*ABS*: acrylonitrile–butadiene–styrene copolymer. They are thermoplastic resins that are rigid and hard but not brittle.
*Acetal copolymers*: family of highly crystalline thermoplastic polymers.
*Acetal homopolymer*: highly crystalline chain polymer formed by the polymerization of formaldehyde with acetate end groups.
*Acetal resins*: copolymer thermoplastics produced by addition polymerization of aldehydes by means of the carbonyl function (polyformaldehyde and polyoxymethylene resins). These are strong thermoplastics.

*Acrylic polymer*: thermoplastic polymer made by polymerization of esters of acrylic acid or its derivatives.

*Acrylic resins*: polymers of acrylic methacrylic esters, sometimes modified with nonacrylic monomers such as the ABS group.

*Addition polymerization*: chemical reaction, usually requiring an initiator, where unsaturated (bifunctional) double-carbon-bond monomer molecules are linked together.

*Additive*: substance such as plasticizers, fillers, colorants, antioxidants, and flame retardants added to polymers for specific objectives.

*Adhesive*: bonding agent for bonding two surfaces of like or unlike materials together, or to seal a surface from its environment. Most are polymers. (Solders and brazing alloys are not normally defined as adhesives.)

*Alkyd polymer*: thermoset polymer based on resins composed principally of polymeric esters in which the ester groups are an integral part of the main chain and serve as the cross-links.

*Alloy*: in polymers, a blend of polymers or copolymers with other polymers.

*Amino resins*: resins made by polycondensation of amino groups such as urea or melamine.

*Aramid*: manufactured fiber of a long-chain synthetic aromatic polyamide.

*Aromatic*: unsaturated hydrocarbon with one or more benzene rings in the molecule.

*Atactic stereoisomerism*: chain molecule in which the side groups are positioned in a random fashion.

*Bakelite*: proprietary name for phenolics often phenol–formaldehyde. Named after L. H. Bakeland (1863–1944).

*Bismaleimide (BMI)*: polyimide that is formed by addition rather than condensation.

*Block copolymer*: linear copolymer consisting of repeated sequences of polymeric segments of different chemical structures.

*Branched polymer*: main chain with attached side segments.

*Butadiene*: gas ($CH_2:CH \cdot CH:CH_2$) widely used in forming copolymers.

*Butene (1-butene)*: member of the olefin family used in the production of HDPE.

*Butylene polymers*: polymers based on resins made by polymerization or copolymerization of butene.

*Butyl rubber*: synthetic elastomer made by copolymerizing isobutylene with small amounts of isoprene.

*Calendering*: passing of sheet material between rollers.

*Catalyst*: substance that changes the rate of a reaction.

*Cellulose* (and its derivatives): natural polymer occurring in plant tissues. Modified cellulose includes *cellulose nitrate (celluloid), cellulose acetate, cellulose ester,* and *cellulose acetate butyrate.*

*Cis stereoisomer*:   stereoisomer in which the side groups are on the same side of a double bond present in a chain of atoms.

*Condensation polymerization*:   reaction of two monomers to produce a third, plus a by-product, usually water or alcohol.

*Crazing*:   regions of ultrafine cracks that form in some polymers under sufficient stress.

*Degree of polymerization (DP)*:   number of mers per molecule.

*Dielectric constant*:   ratio of the capacitance between two electrodes containing the subject material to that in air.

*Dielectric strength*:   potentials per unit thickness (volts/mil) at which failure occurs by electric breakdown.

*Engineering plastics*:   general term covering all plastics with or without fillers or reinforcements that have mechanical, chemical, and thermal properties suitable for use as construction materials.

*EPM/EDPM rubbers*:   synthetic ethylene–propylene copolymer rubbers.

*Epoxy resins*:   large chemically complicated class of thermosetting resins of great value as adhesives, coatings, and in composites.

*Ethylene polymers*:   polymers or copolymers based on ethylene ($C_2H_4$).

*Fluorocarbon plastics*:   polymers made with monomers composed of fluorine and carbon only (Teflon).

*Fluorocarbons (general)*:   include polytetrafluoroethylene, polychlorotrifluoroethylene, polyvinylidine, and fluorinated ethylene propylene.

*Foamed plastics*:   resins in sponge form.

*Glass transition*:   reversible change in an amorphous polymer at which the molecular mobility becomes sufficiently small to promote brittleness. Below $T_g$ polymers are brittle.

*Hardener*:   additive to enhance the curing action (mostly for adhesives).

*Hardness scales*:   Barcol, Knoop, Mohs, Rockwell, and Shore (see Chapter 2).

*HDPE*:   high-density polyethylene.

*HDT*:   heat deflection temperature (see ASTM D648).

*Hydrocarbon polymers*:   those based on polymers composed of hydrogen and carbon only.

*Impact test*:   measure of toughness (see Chapter 2).

*Interpenetrating polymer networks*:   polymer $X$ (or its monomer) interpenetrating into an already cross-linked polymer network $Y$ and then itself becoming cross-linked.

*Injection molding* and *injection blow molding*:   methods of forming heat-softened polymers to desired shapes.

*Inomer resins*:   group of linear chain thermoplasts that contains up to 20% of an acid monomer which is neutralized by a metal or quarternary ammonium ion. Contains both covalent and ionic bonds.

*Kevlar*: organic polymer composed of aromatic polyamides (aramids) often used as fibers in composites (DuPont trademark).

*Liquid-crystal polymers*: linear polymers with a stiff primary chain structure that align their molecular axes in solution in the melt to form liquid crystals. First commercial one was an aromatic polyamide.

*Olefine*: group of unsaturated hydrocarbons of the general formula $C_nH_{2n}$ and named after the corresponding paraffins by the addition *-ene* or sometimes *-ylene* to the root.

*Polyamides (nylons)*: tough crystalline polymers with repeated nitrogen and hydrogen groupings which have wide acceptance as fibers and engineering thermoplastics.

*Polybutylene terephthalate (PBT)*: member of the polyalkylene terephthalate family, similar to polyethylene terephthalate.

*Polycarbonate (PC)*: thermoplastic polymer derived from direct reaction between aromatic and aliphatic dihydroxy compounds with phosgene or by the ester reaction.

*Polyether ether ketone (PEEK)*: linear aromatic crystalline with a continuous-use temperature as high as 250°C.

*Polyether-imide (PEI)*: amorphous polymer with good thermal properties and continuous-use temperatures of about 170°C.

*Polyether sulfone (PESV)*: high-temperature engineering thermoplastic consisting of repeating phenyl groups linked by thermally stable ether and sulfone groups.

*Polyethylene terephthalate (PET)*: saturated thermoplastic resin made by condensing ethylene glycol and terephthalic acid and used for fibers, films, and injection-molded parts.

*Polyurethane (PUR)*: large family of polymers based on the reaction product of an organic isocyanate with compounds containing a hydroxyl group.

*Polymethyl methacrylate (PMMA)*: thermoplastic polymer synthesized from methyl methacrylate.

*Polyimide (PI)*: polymer produced by reacting an aromatic dianhydride with an aromatic diamine. It is a highly heat resistant resin (about 315°C).

*Polyolefins*: polymers made from monomers of olefins only.

*Polyoxymethylene (POM)*: acetal plastics based on polymers in which oxymethylene is the sole repeating unit.

*Polypropylene (PP)*: tough, lightweight polymer made by polymerization of propylene gas.

*Polystyrene (PS)*: homopolymer thermoplast produced by the polymerization of styrene (vinyl)–benzene.

*Polysulfide*: polymer containing sulfur and carbon linkages produced from organic dihalides and sodium polysulfides.

*Polyvinyl acetate (PVAC)*: thermoplastic polymer material composed of polymers of vinyl acetate.

*Polyvinyl chloride (PVC)*: thermoplastic material composed of polymers of vinyl chloride.

*Polyvinyl chloride acetate*: thermoplastic material composed of copolymers of vinyl chloride and vinyl acetate.

*Polyvinylidene chloride (PVDC)*: thermoplastic material composed of polymers of vinylidene chloride.

*Reaction injection molding*: molding process whereby liquids are mixed and reacted under pressure and then forced into a mold.

*Resin*: solid or pseudosolid organic material, usually of high molecular weight that tends to flow under stress. Most resins are polymers.

*Styrene–butadiene rubber (SBR)*: copolymer elastomer that is the most important of the synthetic rubbers (world production about 5 million tons in 1986). Competes with vulcanized natural rubbers.

## QUESTIONS AND PROBLEMS

8.1. If the average molecular mass of a PVC molecule is 25,000 g/mol, what is the degree of polymerization? What is the mechanism of polymerization?

8.2. What is the molecular weight of phenol? Of formaldehyde? Write the polymerization equation for the combination of these two to produce a giant molecule of phenol–formaldehyde. What is the mechanism of polymerization?

8.3. If a completely reacted mass of phenol and formaldehyde was found to be 1 kg, what would be the weight of water produced?

8.4. If a unit cell of polyethylene consists of two mers, compute its theoretical density. Compare with the values listed in Table 8.7. Assuming that the difference is due to lack of crystallinity, estimate the percent crystallinity of LDPE. What is the percent free space in LDPE?

8.5. Figure 4.2 shows a sketch of the molecular arrangement in one form of crystallinity. Sketch a different arrangement that is also found in crystalline polymers.

8.6. Synthetic fibers of the polyvinyl acetate copolymers have molecular weights ranging from 16,000 to 23,000 g. What is the range of the wt % of chloride mers to the total mers?

8.7. Which of the polyvinyls listed in Table 8.2 has the largest mer weight? What is this weight?

8.8. Sketch the density (not free-space) variation with temperature for amorphous and crystalline polymers. Label the points $T_g$ and $T_m$. Describe the molecular rearrangements that are occurring between $T_g$ and $T_m$.

8.9. Using the references given (or any other source), list the typical values of high-, medium-, and low-strength polymers and list one polymer for each of these strength ranges. Also indicate the type (i.e., thermoplast, thermoset, elastomer) of polymer you have selected.

8.10. Using an elastomer stress–strain curve, show how the modulus is determined.

8.11. Assuming that a particular application requires a high strength-to-weight ratio and that other factors are not important, which polymers would you consider? Now assume that this part must also withstand temperatures as high as 150°C for long periods of time; which polymers would you consider?

8.12. Assume that two applications require high strength/weight polymers that can withstand temperatures of 150°C but that, in addition, application A requires a nonbrittle tough polymer, while for application B toughness is not important. Select a polymer for each application.

8.13. Sketch stress–strain curves for a (a) typical thermoplast ($T > T_g < T_m$); (b) typical thermoset; (c) typical elastomer.

8.14. Why would you expect the degree of crystallinity to change in a thermoplast or an elastomer when stress is applied? Would the crystallinity increase or decrease?

8.15. What actually happens to the molecules when plastic deformation occurs in thermoplast polymers? What would you expect to occur within thermoset and elastomer molecules at or just after their respective yield stresses have been attained?

8.16. What occurs on a molecular scale during creep of a thermoplastic polymer?

8.17. Does creep occur in elastomers and thermosets? Explain your answer.

8.18. Does crystallinity increase the strength of thermoplasts?

8.19. What is the mechanism on a molecular level of the elastic behavior observed in elastomers?

8.20. Explain the fracture mechanism in each of the three types of polymers.

8.21. Explain what is meant by the term *secant modulus* and in what type of polymer mechanical behavior it might be used.

8.22. Describe how the heat deflection temperature is measured. List the ASTM methods for the other pertinent mechanical properties of polymers.

8.23. Why are both the strength and modulus of polymers much lower than those found in metals? Give typical values for each.

8.24. What is the basic difference between a rubber band type of elastomer and a thermoplastic elastomer?

8.25. What ratio of acetate to vinyl chloride mers in this copolymer would be most suitable for an electrical insulation material?

8.26. What is the difference between a saturated and an unsaturated hydrocarbon molecule?

8.27. How much sulfur must be added to 1 kg of butadiene rubber to cross-link 10% of the possible sites?

8.28. What is the difference between stepwise and condensation polymerization?

8.29. In general, how do the strengths and ductilities of the thermosets compare to that of the thermoplasts?

8.30. We treat elastomers as a third type of polymer material, yet they can also have thermoplastic or thermoset characteristics. In what way are they distinct?

## SUGGESTED READING

American Society for Metals, *Engineered Materials Handbook,* Vol. 2, *Engineering Plastics,* ASM International, Metals Park, Ohio, 1988.

Bolker, H. I., *Natural and Synthetic Polymers,* Marcel Dekker, New York, 1974.

Cowie, J. M. G., *Polymers: Chemistry and Physics of Modern Materials,* Intext Educational Publishers, New York, 1973.

Farag, M. M., *Selection of Materials and Manufacturing Processes for Engineering Design,* Prentice Hall, Englewood Cliffs, N.J., 1989.

Hall, C., *Polymer Materials, An Introduction for Technologists and Scientists,* 2nd ed., Wiley, New York, 1989.

Hertzberg, R. W., and Manson, J. A., *Fatigue of Engineering Plastics,* Academic Press, New York, 1980.

MacDermott, C. P., *Selecting Thermoplastics for Engineering Applications,* Marcel Dekker, New York, 1984.

*Modern Plastics Encyclopedia,* McGraw-Hill, New York, published annually.

Moore, G. R., and Kline, D. E., *Properties and Processing of Polymers,* Society of Plastic Engineers, and Prentice Hall, Englewood Cliffs, N.J., 1984.

Seymour, R. B., *Engineering Polymer Sourcebook,* McGraw-Hill, New York, 1990.

Seymour, R. B., and Carraher, C. E., Jr., *Polymer Chemistry,* 2nd ed., Marcel Dekker, New York, 1988.

# 9

# Composites

## 9.1 INTRODUCTION

In Chapter 1 a composite was defined as "a mixture of two or more materials that are distinct in composition and form, each being present in significant quantities (e.g., >5 vol %)." Another source defines a composite as "the union of two or more diverse materials to attain synergistic or superior qualities to those exhibited by individual members." There are about as many definitions of composites as there are textbook authors on the subject of materials. Volume 1 of the ASM *Engineered Materials Handbook* has elected the following definition: *"a combination of two or more materials differing in form or composition on a macroscale. The constituents retain their identities ... and can be physically identified."* The words *macroscale* and *identities* or *distinct* are the key words in any composite definition. This automatically rules out solid solutions as well as most metallic alloys that have their properties enhanced via the presence of microconstituent phases. The dual-phase steel and the duplex stainless steel are exceptions in the metallic family. They are composed of two distinct and identifiable (although requiring a $50 \times$ or so optical microscope) phases both being present in excess of 20% by volume, and the properties of those metallic composites are directly related to the fractional amount of each phase present. These alloys are considered to be composites since their mechanical properties obey the rule of mixtures. The tungsten carbide particles in a cobalt matrix, shown in Figure 7.10, is an example of a particulate ceramic–metal matrix composite. In composites consisting of identifiable constituents, each possessing its own individual properties, the strength and moduli can be expressed by the well-known *rule of mixtures,* that is,

composite strength = (strength constituent 1 × fraction consitituent 1)

+ (strength constituent 2 × fraction constituent 2)

or in symbol form

$$\sigma_c = \sigma_1 V_1 + \sigma_2 V_2 \tag{9.1}$$

where $V_1$ and $V_2$ are the volume fractions of the two constituents and $\sigma$ is the strength. Although we have not yet said anything about the orientation and distribution of these phases, constituents, components, or whatever we like to call them, whenever they are present on a macroscale and retain their identifiable properties, they will obey the rule of mixtures or some modification thereof, such modification being to account for their respective shapes, orientations, and bond-interface characteristics. In most structural or load-bearing composites, the strong phase (e.g., a covalently bonded ceramic material) is designed to carry most of the load, while the weaker phases, such as a polymer or ductile metal, provide for toughness and ductility of the composite.

There exist many natural composites. Wood and many other plant tissues contain the cellulose chain polymer in a matrix of lignin, a phenol–propane polymer, and other organic compounds in a type of cellular structure (Figure 9.1). The load-bearing parts of the human body are in essence composites composed of the fibrous tissue, ligaments, which are cords as strong as rope, and which are attached to the bones, thereby binding the bone segments together.

We do not attempt here to trace the origin of composites historically because the variety of composites and the varied definitions of them prohibit such a monumental task. The first seriously manufactured composites were developed because no single homogeneous material could be found that had all of the properties sought for a given application. Composites are the first *engineered materials.* Fiberglass, which was developed during and immediately after World War II, is considered by many to be our first engineered composite. These glass-fiber-reinforced polymers were immediately put to use in filament-wound rocket motor cases as well as small boat hulls and a myriad of smaller applications.

There are a number of ways in which one might classify structural composites. One is according to the geometry of the load-bearing component, which can be in fiber, particulate, or laminar form, the latter often being made up of filaments or tows of fibers in matrixes that are shaped in a laminar arrangement. Particulate composites have no distinct arrangement of the phases or components and the properties tend to be more isotropic than the fiber or lam-

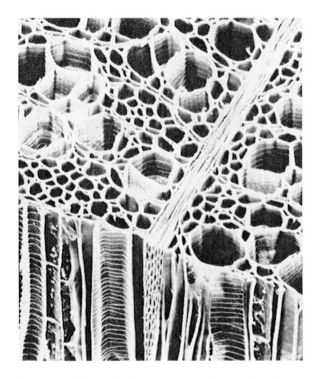

**Figure 9.1.** Cellular structure of basswood. (Courtesy of Prof. Tim O'Keefe, Cal Poly.)

inar composites. Examples of particulate composites include the above-mentioned two-phase steels (dual-phase and duplex stainless) as well as the cemented carbides and concrete. A more recently developed and commercially used particulate composite is that of SiC particles in an aluminum alloy matrix. Laminar composites are layered materials. They include the clad metals, often where a more corrosion-resistant metal is bonded to a stronger one, fiber-reinforced laminates with different fiber orientations, and plywood and sandwich panels, where two outer layers are separated by a less dense and lower-strength material. The latter include the honeycomb core, which is placed in between panels to form a sandwich-like structure. Figure 9.2 shows schematically a number of composite configurations, including a laminate honeycomb sandwich and a unidirectional fiber composite component contained in a single rotor blade.

Fibrous composites are probably the largest composite category, and accordingly we concentrate on this class of composites, particularly with respect to the mathematical stress and modulus analysis. Another classification

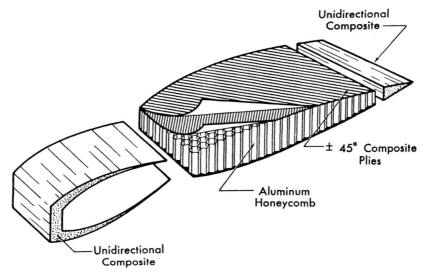

**Figure 9.2.** Composite rotor blade construction containing honeycomb, laminates, and unidirectional composites. (From R. L. McCullough, *Concepts of Fiber-Resin Composites,* Marcel Dekker, New York, 1971, p. 15.)

could be developed based on the materials used, and here we could have a subdivision, depending on whether the material is the reinforcing member or the matrix material. The matrix material is generally the softer and continuous phase and often present in the largest quantity. It usually surrounds the stronger load-bearing components, as, for example, the polymer matrix materials that are reinforced by glass fibers. We will use the latter classification—based on the matrix material. The mathematical analysis will be confined to fiber composites based on the rule of mixtures. The rule of mixtures can also be applied to particulate composites in a simple fashion that does not involve the complexities of fiber length.

## 9.2  THE RULE OF MIXTURES

The rule of mixtures will be used throughout our discussions, so let's examine it in more detail. We use the fiber model for this purpose (i.e., where fibers are used to reinforce a weaker matrix material). There are several conditions that must be met for our model to be of any analytical value. These are listed as follows:

  1.  The fiber must be securely bonded to the matrix. We generally refer to this bond as a chemical bond in the sense that atoms of each constituent react

and bond together. However, the analysis will still hold for a mechanically bonded interface as long as the interface is not the region of failure when the composite is subjected to a stress that exceeds the maximum value that the composite will sustain. Unfortunately, the interface is often the weakest region of the composite, and much research has been devoted toward the objective of obtaining a strong chemical bond between the fiber and matrix. In general, a chemical bond is stronger than a mechanical bond, the latter being where the two components are interlocked by the penetration of the matrix material into the surface irregularities of the fibers.

2.   The fibers must be either continuous or overlap extensively along their respective lengths. Let's examine the extreme case of discontinuous fibers that do not overlap in one region of the composite, as depicted in Figure 9.3a by the region between the dashed lines. Since a chain is no stronger than its weakest length, the region containing no fibers will fail under a critical axial stress that equals the matrix fracture stress and the fibers will not contribute one iota to the strength of the composite. If, on the other hand, the fibers overlap, the load can be transferred to an adjacent fiber. This is the major role of the matrix: to transfer the load from the end of one fiber to an adjacent overlapping fiber (Figure 9.3b). Secondary roles of the matrix are to protect the fibers from surface damage and to blunt cracks that arise from fiber fractures.

3.   There must be a critical fiber volume, $V_{f\text{crit}}$, for fiber strengthening of the composite to occur.

4..   There must be a critical fiber length for strengthening to occur. This critical length is also dependent on the fiber diameter $d_f$. The ratio of the critical length $l_c$ to the critical diameter for that length, $d_{fc}$, is called the *aspect ratio*.

 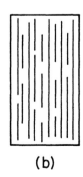 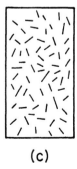

(a)          (b)          (c)

**Figure 9.3.**   Schematics of a fiber composite (a) in which the parallel aligned fibers do not overlap in all regions of the composite; (b) in which the longitudinal aligned fibers overlap extensively (continuous fibers); and (c) of randomly oriented short fibers.

## 9.2.1 Mathematical Analysis of Fiber Composites

Continuous Longitudinal Fibers: Equal Strain and Elastic Conditions
(Figure 9.4a)

When the fibers are continuous, and in this case we mean that the fibers are of sufficient length for extensive overlap, and when the composite is uniaxially stressed in the fiber alignment direction, the stress in the composite, $\sigma_c$, is simply that stated by the rule of mixtures:

$$\sigma_c = V_f\sigma_f + v_m\sigma_m \tag{9.2}$$

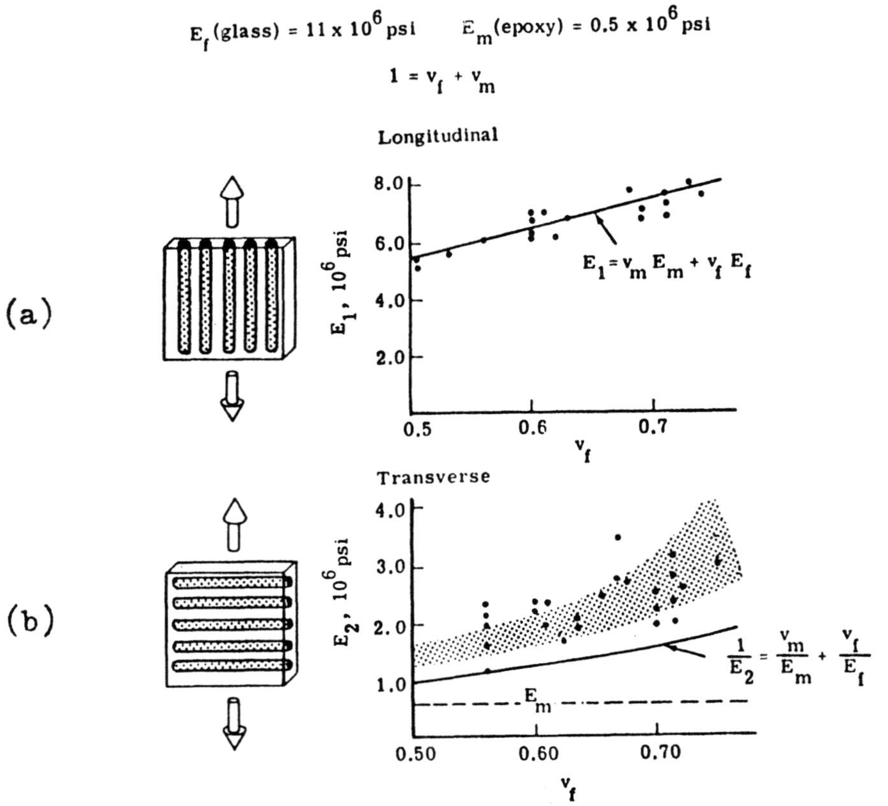

**Figure 9.4.** Comparison of experimental results with theory for the moduli of longitudinal and transverse loading of fibers. $E_1 = E_c$ for (a) and $E_2 = E_c$ for (b). (From R. L. McCullough, *Concept of Fiber-Resin Composites*, Marcel Dekker, New York, 1971, p. 31. Experimental data from L. C. Chamis and G. P. Sendeckyi, *J. Composite Mater.*, Vol. 2, 1968, p. 332.)

where $V_f$ = volume fraction of fiber
$\sigma_f$ = stress in fiber
$V_m$ = volume fraction of matrix
$\sigma_m$ = stress in matrix

Since $V_m = 1 - V_f$, equation (9.2) becomes

$$\sigma_c = V_f\sigma_f + (1 - V_f)\sigma_m \tag{9.3}$$

In most composites the $V_f\sigma_f$ term is the predominate factor.

The strain in the case above is equal in both components by the conditions imposed. If both components are stressed only elastically, $\sigma_f = E_f\epsilon_f$, $\sigma_m = E_m\epsilon_m$, and $\sigma_c = E_c\epsilon_c$. Thus $E_c\epsilon_c = V_fE_f\epsilon_f + V_mE_m\epsilon_m$. But since all strains are equal, the composite modulus becomes

$$E_c = \sigma_f V_f + \epsilon_m V_m = E_f V_f\epsilon_f + (1 - V_f)E_m \tag{9.4}$$

The longitudinal arrangement of fibers in Figure 9.4a has been called the parallel reaction and applies just as well to thermal or electrical loads. Each fiber and the matrix will experience the same strain due to the uniform load applied to the entire composite cross section. Again the $E_f V_f$ is the major contributing term to the composite modulus $E_c$. If the fiber properties are two orders of magnitude greater than the matrix, the matrix contribution is negligible. The equations above were derived for continuous fibers, but when the fibers are long and overlapping, as in Figure 9.3b, the equations are good approximations of the composite elastic strength and modulus. This is because the load (or stress) at the end of the fiber is transferred by the matrix to adjacent fibers. One can see this more clearly in Figure 9.5a, where the tensile stress $\sigma$ and the shear stress $\tau$ at the fiber–matrix interface along the fiber length are depicted. At the fiber end points the normal stress $\sigma$ is zero and the high shear stresses cause the tensile stresses to be transferred to the adjacent overlapping fibers. Initial fiber fractures (Figure 9.5b) do not result in complete composite failure because of a coupling action between the fibers and the matrix. As the two broken fiber ends attempt to pull away from each other, the plastic flow of the matrix parallel to the stress (Figure 9.5c) resists this pulling action, thereby building up shear stresses in the matrix. However, the stress $\sigma$ in the nonbroken fibers increase because they must carry a larger part of the load.

Continuous Fibers in a Transverse Arrangement (Figure 9.4b): Elastic Condition, Equal Stress

The equal stress state occurs when the stress is applied in a direction perpendicular to the fiber axes. Actually, in the case of fibers the cross-sectional area presented to the load in such a geometry is sufficiently small that the fibers may not make a significant contribution to the composite strength, except for composites of large volume percent fiber content (e.g., >30%). The stress is the

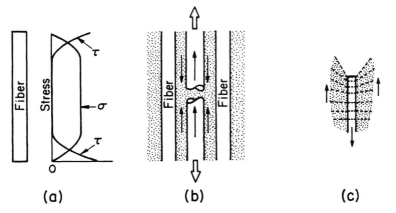

**Figure 9.5.** (a) Variation of the normal and shear stresses along the fiber when it is stressed along the fiber axis; (b) and (c) mechanism of load transfer from a broken fiber to adjacent fibers.

same in both components, but the strain will differ. The composite strain, $\epsilon_c$, can be expressed as

$$\epsilon_c = \epsilon_f V_f + \epsilon_m V_m \qquad (9.5)$$

where $\epsilon_f$ = strain in the fiber
$V_f$ = volume fraction of fiber
$\epsilon_m$ = strain in matrix
$V_m$ = volume fraction of matrix

Again we can apply Hooke's law and express the composite modulus as

$$\frac{\sigma_c}{E_c} = \frac{\sigma_f}{E_f} V_f + \frac{\sigma_m}{E_m} V_m \qquad (9.6)$$

But since $\sigma_c = \sigma_f = \sigma_m$ we can write

$$E_c = \frac{E_f E_m}{V_f E_m + V_m E_f} \qquad (9.7)$$

In the transverse arrangement the experimental results deviate considerably from the simple rule of mixtures. The numerical analysis shows that the behavior is dependent on the shapes of the fibers, their packing geometry, and the distribution of spacings between fibers. There have been different mathematical analyses and modifications of the simple rule of mixtures, but the comparison with experimental data shows good agreement with the simple rule of mixtures for the longitudinal situation, but not for the transverse fiber geometry. However, the simple rule of mixtures gives a conservative estimate of the transverse moduli.

Continuous Fibers: Fiber Elastic and Matrix Plastic, Equal Strain Condition

In most cases of fiber composites the strong fibers will show little plastic deformation when the yield strength of the matrix has been exceeded. This is particularly true in the case of ceramic fibers in a metal matrix. Then the ultimate strength of the composite, $\sigma_{cu}$, is given by

$$\sigma_{cu} = \sigma_{fu} V_f + \sigma_m'(1 - V_f) \tag{9.8}$$

where $\sigma_{fu}$ is the ultimate tensile strength of the fiber and $\sigma_m'$ is the flow stress of strain-hardened matrix. To obtain any strengthening via the fiber content, the ultimate strength of the composite, $\sigma_{cu}$, must be greater than the ultimate strength of the matrix, $\sigma_{cu}$ (i.e., $\sigma_{cu} \geq \sigma_{mu}$). Thus

$$\sigma_{fu} V_f + \sigma_m'(1 - V_f) \geq \sigma_{mu} \tag{9.9}$$

This leads to a *critical fiber volume*, $V_{f\text{crit}}$, for strengthening to occur. $V_{f\text{crit}}$ can be obtained by setting $V_f = V_{f\text{crit}}$ in equation (9.9) and solving as follows:

$$V_{f\text{crit}} = \frac{\sigma_{mu} - \sigma_m'}{\sigma_{fu} - \sigma_m'} \tag{9.10}$$

A situation may arise from small fiber volumes where all fibers are stressed to fracture, and the strain-hardened matrix carries the load. Assuming such a condition, we can compute a minimum volume of fibers as follows:

$$\sigma_{cu} \geq \sigma_{mu}(1 - V_f) \tag{9.11}$$

which is arrived at by setting $\sigma_{fu} = 0$ in equation (9.8). This expression serves to define the minimum volume fraction that must be exceeded for reinforcement to occur; thus

$$V_{f\min} = \frac{\sigma_{mu} - \sigma_m'}{\sigma_{fu} + \sigma_{mu} - \sigma_m'} \tag{9.12}$$

However, $V_{f\min} < V_{f\text{crit}}$; therefore, $V_{f\text{crit}}$ is the fiber content that really must be exceeded for fiber strengthening to occur. This analysis was intended for metal matrix composites; however, although polymers matrices do not strain harden much and the strength of polymer fibers may be much less than that of ceramic fibers, the equations above are still applicable to such composites.

## 9.2.2 Load Transfer

For the load to be transferred to the adjacent fibers as illustrated in Figure 9.5, the matrix must be plastically deformed at the fiber endpoints. The transfer mechanism is the same whether the fiber breaks or whether it is merely the endpoint of the original fiber. The high shear stresses provide for this deformation. The fiber strength can be utilized properly only if the plastic zone in the matrix does not extend to the fiber midlength before the stress in the fiber reaches the ultimate strength of the fiber. Equilibrating the forces that cause the shear stresses along the fiber over a length of ½ (the fiber is being loaded from both ends) to the force causing the tensile stress in the fiber, $\sigma_f$, we obtain

$$\sigma_f \frac{\pi d^2}{4} = \tau \pi d \frac{l}{2}$$

and solving for $l$ yields

$$l_c = \frac{\sigma_f d}{2\tau_c} \tag{9.13}$$

This $l$ becomes the critical fiber length $l_c$ and $\tau$ becomes $\tau_c$, which is the critical fiber–matrix bond strength, or the matrix yield strength, whichever is the smaller. When $l > l_c$ the fiber can be loaded to its fracture stress, which is the objective of strength gain through fiber reinforcement by means of the load transfer via the plastic deformation of the matrix around the fiber endpoint. The fiber diameter, $d$, in equation (9.13) is also the critical diameter $d_{fc}$ for this critical length, and the ratio of $l_c/d_{fc}$ is the *critical aspect ratio*, which can now be expressed in equation form as

$$\frac{l_c}{d_{fc}} = \frac{\sigma_{fu}}{2\tau_c} \tag{9.14}$$

When $l \gg l_c$, somewhere in the vicinity of $l = 15l_c$, and if $l_c$ is also on the order of 100 fiber diameters (i.e., $l_c > 100d_f$), the fibers behave essentially as continuous fibers. At the point in the composite where a fiber ends, the load is transferred to an adjacent fiber. Clearly, not all fibers break at the same stress or length. The average fiber length must be used in this analysis. But as long as $l > l_c$ and the bond holds, the load will be transferred as illustrated in Figure 9.5. Fiber breakage will continue, however, and probably at a faster rate as the load increases, since for every fiber fracture additional load is transferred to adjacent fibers. Eventually, a sufficient number of fibers will, by virtue of their fracture, have lengths of $l < l_c$, which leads to incomplete load transfer and thus less contribution of the fibers to the composite strength. Continuation of fiber breakage will eventually lead to failure of the composite.

Discontinuous fibers are much shorter than the continuous ones and are used as both aligned and randomly oriented geometries, more often in the latter arrangement (Figure 9.3c). These randomly oriented short fibers include the chopped glass fibers, aramid fibers, whiskers, and most carbon fibers. A correction factor based on the $l/l_c$ ratio is needed for the case of discontinuous aligned fibers. When $l < l_c$ the average stress is simply $\sigma_{max}/2$. When $l > l_c$ a fraction of the fibers carry the same load as a continuous fiber, and the remaining fibers carry, on the average, half of this stress. The various average stresses that the fibers carry, based on their lengths compared to $l_c$ are depicted in Figure 9.6. For discontinuous aligned fibers, when $l > l_c$ and $\sigma_f = \sigma_{fu}$, the strength of the composite can be expressed as

$$\sigma_{cu} = \sigma_{fu}V_f \left( 1 - \frac{1}{2l} \right) + \sigma'_m V_m \qquad (9.15)$$

Composites of this type are mostly those of the short-chopped glass fibers; however, carbon and aramid aligned discontinuous fibers are also used to some extent. In the metal matrix composites $\sigma_m$ should be replaced by $\sigma'_m$, the stress in the strain-hardened matrix, and $\sigma_f$ by $\sigma_{fu}$, the tensile strength of the whisker.

For most aligned fiber composites the lengths are often hundreds of times $l_c$. For example, a fiber having an $l/l_c$ ratio of 10 would in a composite exhibit an average fracture strength of only 5% less than if the fiber were continuous. Thus appreciable reinforcement can be provided by discontinuous fibers when their lengths are much greater than the usual small critical lengths.

For the case of discontinuous randomly oriented fibers a different correction factor must be applied and equation (9.4) for the composite modulus becomes

$$E_c = \alpha E_f V_f + E_m V_m \qquad (9.16)$$

where $\alpha$ is a fiber efficiency parameter that is dependent on $V_f$, the $E_c/E_m$ ratio and fiber orientation. The efficiency of fiber reinforcement is often taken to be unity for an oriented fiber composite in the alignment direction, and zero per-

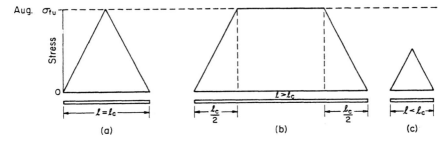

**Figure 9.6.** Relationship between the stress in the fiber and the fiber length. In (a) and (b) the fibers are being fully utilized in that they carry at some point along their length stresses equal to their ultimate tensile strengths (using average fiber strengths), while in (c) the load-carrying capacity of the fibers is not fully utilized because of their short length.

pendicular to it. Experimentally, it is found to fall between about 0.2 and 0.6. The use of discontinuous randomly oriented fibers are quite common in glass-fiber-reinforced polymers. They achieve isotropic properties in the composite and often ease of processing. Intricate shapes can be formed by a variety of processes. In metal matrix composites of this type the fibers can be added to the molten metal prior to casting, or a fibrous mass can be infiltrated by the molten metal.

## 9.3 FIBERS AND THEIR PROPERTIES

There are many different types of fibers used in commercially available composites. The common ones include glass, boron, carbon, silicon carbide, and the aramid polymer fibers. In the more advanced state of composite development we can also include alumina ($Al_2O_3$), alumina silicates, mullite, and tungsten. There are also small fine fibers called whiskers which have been used in metal matrix whisker reinforced composites. Whiskers have been produced from a very large number of materials, including metal whiskers of iron, nickel, and copper.

The stress–strain curves for some common fibers are shown in Figure 9.7. The carbon (HM = high modulus and HT = high tensile strength) and boron fibers have the higher moduli and in addition are lightweight and high-strength materials. The aramid fiber Kevlar 49 has a good combination of strength modulus and low density and accounts in part for its popularity. The aramid fibers are also less expensive than the carbon and boron fibers. The specific moduli and specific strength of several fiber types, which takes into account the fiber densities, are depicted in Figure 9.8. Some property data on the more common fibers are summarized in Table 9.1. The most detailed property data can be found in the *Engineered Materials Handbook,* Vol. 1, Section 2 (ASM International, Metals Park, Ohio, 1987).

### 9.3.1 Fiber Processing

Carbon and graphite fibers were originally developed for the aerospace industry but are now being found in other applications, such as sports equipment. All carbon fibers produced to date start with organic precursors, rayon, polyacrylonitrile (PAN), and isotropic and liquid crystalline pitches. The high-modulus (HM) fibers are made from liquid crystalline pitch precursors or from PAN, with the latter being the most common source. The PAN fibers are first oxidized at 200 to 250°C to thermoset the fibers and then carbonized around 800°C to drive off the hydrogen and nitrogen. The resulting carbon fiber is passed through another higher-temperature (to 3000°C) furnace to graphitize the fiber. The modulus increases with the graphitization temperature, but the strength tends to maximize around 1500°C. Carbon fibers are available in tow

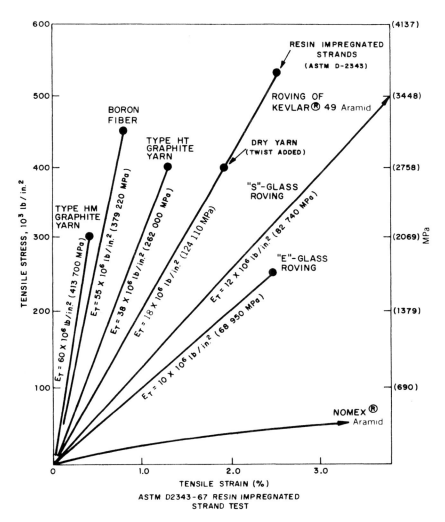

**Figure 9.7.** Stress–strain behaviors of various types of fibers. (From *DuPont Bulletin K-5,* Sept. 1981.)

(untwisted) bundles or rovings (yarns, strands, or tows collected in a parallel bundle with no twist) containing many thousands of fibers. The PAN fibers vary in modulus from about 203 to 320 GPa and the strength values from 2400 to 5500 MPa.

Kevlar aramid fibers are made from the condensation reaction of para-phenylene diamine and terepthaloyl chloride (Figure 9.9). They contain the

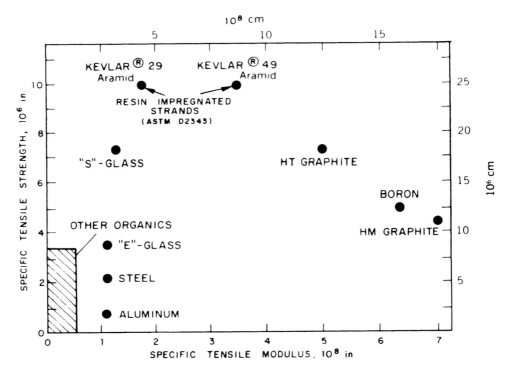

**Figure 9.8.** Specific tensile strength and specific tensile moduli of typical fibers. (From *DuPont Bulletin K-5,* Sept. 1981.)

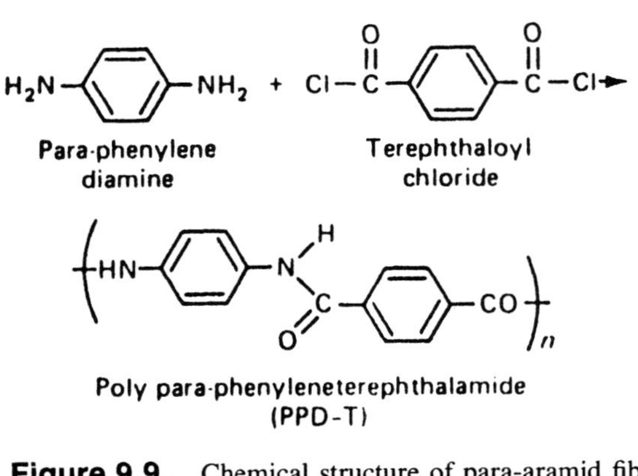

**Figure 9.9.** Chemical structure of para-aramid fibers. (From *Engineered Materials Handbook,* Vol. 1, ASM International, Metals Park, Ohio, 1987, p. 54, article by J. J. Pigliacampi.)

**Table 9.1**  Properties of Selected Fibers from a Variety of Sources

| Material | UTS [MPa (ksi)] | Modulus [MPa (ksi)] | Density (g/cm³) |
|---|---|---|---|
| $\alpha$-Al$_2$O$_3$ | 1380 (200) | 2 (0.3) | 3.95 |
| FP Al$_2$O$_3$ silica coated, continuous | 1900 (275) | 2 (0.3) | 3.95 |
| SiC, continuous (Textron) | 3450 (500) | 400 (58) | 3.0 |
| Boron continuous (Textron) | 3600 (520) | 400 (58) | 2.57 |
| Kevlar 49 (DuPont) | 2930 (420) | 117 (17) | 1.87 |
| Nylon 728 | 985 (143) | 5.5 (0.8) | 1.1 |
| Stainless steel | 1720 (250) | 200 (29) | 7.9 |
| Tungsten | 3900 (565) | 350 (50.7) | 19.3 |
| E-glass | 2410 (350) | 69 (10) | 2.55 |
| Fused silica | 3450 (500) | 69 (10) | 2.2 |
| SiC whiskers | 6900 (1000) | 690 (100) | 3.2 |
| Discontinuous oxides (average of several types) | 1500 (217) | 175 (25) | 2.75 |
| Graphite | 3200 (464) | 490 (71) | 1.9 |

aromatic rings, which lead to stiffness and high thermal stability. These fibers also belong to the rigid rodlike liquid-crystalline polymer class, because in solution they can form ordered domains in parallel arrays. When the solutions are extruded through a spinneret and drawn, the liquid crystalline domains can align, giving an anisotropic high strength and high modulus in the fiber direction. There are different grades of Kevlar, with Kelvar 49 being the most popular. They are available in a range of yarn and tow counts as well as short forms. Tensile strengths are in the range 3400 to 4100 MPa and the moduli in the range 83 to 186 GPa.

Boron fibers are made by a chemical vapor deposition (CVD) process whereby a tungsten wire is continuously drawn through a glass tube containing a mixture of boron trichloride and hydrogen. The BCl$_3$ vapors deposit boron on the hot wire, the fiber diameter increasing as it passes through the reactor tube and usually attaining 100 to 140 μm in diameter. The tungsten core is small (~2 μm), thus yielding predominantly the lightweight boron of around 2.3 g/cm³ density. Tensile strengths of 3600 MPa and moduli of 400 GPa are typical mechanical property values obtained on this wire. Most boron fibers are available in the form of a boron–epoxy preimpregnated tape (called a prepreg) that the composite manufacturer uses to build ply layups and part fabrication. Boron–epoxy composites have been used in several military aircraft, the space shuttle and helicopters. However, these fibers have also been used to reinforce metal matrices such as the titanium–boron and aluminum–boron composites.

Silicon carbide fibers are also produced by the CVD process, where a heated carbon monofilament of about 33 $\mu$m diameter reacts with a mixture of hydrogen and chlorinated alkyl silicones. The resulting 140 $\mu$m-diameter fiber has a density of 3 g/cm$^3$, a tensile strength in the range 2800 to 4600 MPa and a modulus around 400 GPa. Several variations of SiC fibers have been produced, with carbon-containing silicon coatings to prevent reaction with aluminum matrices and others to be compatible with titanium matrices. It is projected that in production volumes the cost will be on the order of $100 per pound, less than that for boron and carbon fibers. These fibers have been used on a development basis for reinforcing aluminum and titanium and to some extent for organic and ceramic matrices.

Many ceramic fibers have been developed in the past 30 years, driven primarily by the aerospace industry needs. Commercially available continuous oxide fibers are listed in Table 9.2. Discontinuous oxide fibers are also commercially available from a number of sources listed in Table 9.3. The high alumina content fibers have strengths in the range 1700 to 2200 MPa and moduli in the range 155 to 380 GPa. The high-silica fibers, except for the quartz fibers, are considerably weaker than the alumina fibers. The oxide fibers of interest today have been prepared by a variety of patented methods. A 85% $Al_2O_3$–15% $SiO_2$ high-performance fiber has been prepared by the Sumitomo Chemical Co. of Japan in continuous multifilament form by spinning a polymerized organoaluminum compound that was blended with a compound containing silicon. It is reported to have a tensile strength of 1450 MPa and a modulus of 193 GPa. Discontinuous oxide fibers were made in the 1950s from clay and slag (from the steel industry) and were generally made in low-cost fibrous wool or batts. Kaowool, an alumina–silica insulating material made by Babcox and Wilcox, has been used for reinforcing aluminum. Sources for commercially available carbide and nitride fibers are given in Table 9.4.

## 9.4  POLYMER MATRIX COMPOSITES

Not many years ago, 10 to 15 perhaps, polymers were considered as inexpensive, corrosion-resistant, non-load-bearing replacements for metals. Today, polymers reinforced by glass, carbon, and aramid fibers are the materials of choice for many auto body panels, aircraft structural components, appliances, and a host of other industrial and consumer products. Almost all of the neat polymers that were discussed in Chapter 8 are also available with glass rein-

**Table 9.2** Commercially Available Continuous Oxide Fibers

| Composition (wt %) | Identification | Company | Forms[a] |
|---|---|---|---|
| $Al_2O_3$, > 99 | Fiber FP | E. I. DuPont de Nemours & Co., Inc. | C, Y, F |
| $Al_2O_3$, 85<br>$SiO_2$, 15 | High-performance alumina fiber | Sumitomo Chemical Co., Ltd., Japan; distributed by Avco Specialty Materials, Textron, Inc. | C, Y, F |
| $Al_2O_3$, 80<br>$SiO_2$, 20 | Long alumina fiber | Denki Kagaku Kogyo K. K., with Nichibi Co. Ltd. | C, Y |
| $Al_2O_3$, 70<br>$B_2O_3$, 2<br>$SiO_2$, 28 | Nextel 440 and Nextel 480 | Minnesota Mining & Manufacturing Co. | C, R, Y, F |
| $Al_2O_3$, 62<br>$B_2O_3$, 14<br>$SiO_2$, 24 | Nextel 312 | Minnesota Mining & Manufacturing Co. | C, R, Y, F, |
| $SiO_2$, 99, 95 | Astroquartz II | Distributed by J. P. Stevens & Co. Inc. | C, R, Y, F |
| $SiO_2$, 98 | Refrasil | Hitco Materials Div., Armco Inc. | Y, F |
| $SiO_2$, 98 | Siltemp | Ametek, Inc. | F |
| $ZrO2$, 68<br>$SiO_2$, 32 | Nextel Z-11 | Minnesota Mining & Manufacturing Co. | C, R, Y, F |

*Source*: *Engineered Materials Handbook*, Vol. 1, ASM International, Metals Park, Ohio, 1987, p. 60, article by D. D. Johnson and H. G. Sowman.

[a] C, continuous; Y, yarn; F, fabric, R, roving

forcement and many with carbon and/or aramid fiber support. Figures 9.10 and 9.11 show the increase in strength and moduli, respectively, of several polymers as a function of glass content. The data were adapted from the 1990 issue of *Modern Plastics Encyclopedia.* The data listed showed a range of values, but for simplicity, average values were used for Figures 9.10 and 9.11. These graphs were constructed to show the general trend of how the strength and moduli increase with glass fiber content. These data should not be used for design purposes. Some data were listed for both long and short fibers. The long fiber composites tend to have the higher strengths for the same fiber volume content, but not by a large amount.

In the case of fibers in polymers one should be aware that the equations for the critical fiber length must be modified to account for the fact that in polymer matrix composites the matrix is not capable of extensive plastic deformation.

**Table 9.3**  Commercially Available Discontinuous Oxide Fibers

| Composition (wt %) | Identification | Company | Forms[a] |
|---|---|---|---|
| $Al_2O_3$, 95<br>$SiO_2$, 5 | Saffil | Imperial Chemical Industries, PLC Ltd., England; distributed by Babcock & Wilcox | D, B, M, Ch |
| $Al_2O_3$, 72<br>$SiO_2$, 28 | Fibermax | Sohio Engineered Materials (formerly Carborundum) | D, B |
| $Al_2O_3$, 70<br>$B_2O_3$, 2<br>$SiO_2$, 28 | Nextel 440 and Nextel 480 Ultrafiber | Minnesota Mining & Manufacturing Co. | D, M, Ch |
| $Al_2O_3$, 60–68<br>$B_2O_3$, 4–9<br>$SiO_2$, 23–32 | Staple fiber | Nichias Corp. | D, B |
| $Al_2O_3$, 62<br>$B_2O_3$, 14<br>$SiO_2$, 24 | Nextel 312 Ultrafiber | Minnesota Mining & Manufacturing Co. | D, M, Ch |
| $Al_2O_3$, 52<br>$SiO_2$, 48 | Fiberfrax | Sohio Engineered Materials | D, B, M, F |
| $Al_2O_3$, 49–50<br>$SiO_2$, 50–51 | Innswool | A. P. Green Refractories | D, B, M |
| $Al_2O_3$, 52–55<br>$SiO_2$, 41–44 | Cer-wool | Combustion Engineering | D, B, M |
| $Al_2O_3$, 47<br>$SiO_2$, 53 | Cerafiber | Manville Corp. | D, B, M |
| $Al_2O_3$, 42.5<br>$Cr_2O_3$, 2.5<br>$SiO_2$, 55 | Cerachrome | Manville Corp. | D, B, M |
| $Al_2O_3$, 40<br>$SiO_2$, 50<br>CaO, 5<br>MgO, 3.5<br>$TiO_2$, 1.5 | Cerawool | Manville Corp. | D, B, M |
| $ZrO_2$, 92<br>$Y_2O_3$, 8 | Zircar | Zircar Products, Inc. | D, B, M, F |

*Source*: *Engineered Materials Handbook,* Vol. 1, ASM International, Metals Park, Ohio, 1987, p. 61, article by D. D. Johnson and H. G. Sowman.

[a] D, discontinuous; B, bulk; M, mat or blanket; Ch, chopped; F, fabric

**Table 9.4**  Commercially Available Carbide and Nitride Fibers

| Composition | Identification | Company | Forms[a] |
|---|---|---|---|
| SiC | Nicalon | Nippon Carbon Co. Ltd. | C, Y, Ch, F, M, R |
| Si-Ti-C | Tyranno | Ube Industries Ltd. | C |
| Si-Zr-C | Tyranno | Ube Industries Ltd. | C |
| SiC on C core | CVD SiC | Tokai Carbon Co., Ltd.; distributed by Avco Specialty Materials, Textron, Inc. | C |
| SiC | Tokawhisker | Tokai Carbon Co., Ltd.; distributed by Avco Specialty Materials, Textron, Inc. | D |
| SiC | Silar | Arco Metals Co. | D |
| SiC | Tateho | Tateho Chemical Industry Co., Ltd.; distributed by ICD Group Inc. | D |
| $Si_3N_4$ | Tateho | Tateho Chemical Industry Co., Ltd.; distributed by ICD. | D |

*Source*: *Engineered Materials Handbook,* Vol. 1, ASM International, Metals Park, Ohio, 1987, p. 61, article by D. D. Johnson and H. G. Sowman.
[a] C, continuous; Y, yarn; Ch, chopped; F, fabric; M, mat or blanket; R, roving; D, discontinuous

Usually, the interfacial stress will cause delamination at the fiber end. Subsequent to delamination the matrix flows by the fiber and the tensile stress is transferred by a frictional force associated with this displacement. Thus $\tau_c$ in equations (9.13) and (9.14) should be replaced by $\mu P$, where $\mu$ is the frictional coefficient between the fiber and matrix and $P$ is an internal pressure resulting from the shrinkage of the polymer matrix around the fiber during curing or incompatible lateral deformation. The product $\mu P$ in polymer composites is on the order of one-tenth that of $\tau_c$ for metal matrix composites.

In addition to increasing the strength and moduli of polymers, the glass fibers also improve their high-temperature capabilities, significantly increasing the heat-deflection temperature as illustrated in Figure 9.12, and the continuous-use temperature as depicted in Figure 9.13.

There has been a tendency in the literature to classify polymer matrix composites into low-performance and high-performance categories based primarily

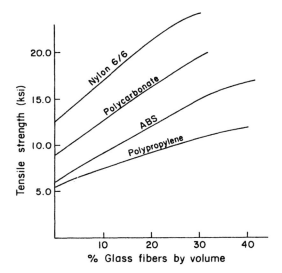

**Figure 9.10.** Tensile strength versus volume percent of glass fibers for some common polymer–glass composites. (Adapted from *Modern Plastics Encyclopedia*, McGraw-Hill, New York, 1990.)

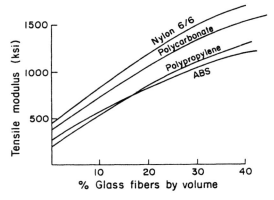

**Figure 9.11.** Tensile moduli versus volume percent glass fiber for some typical polymer–glass composites. (Adapted from *Modern Plastics Encyclopedia*, McGraw-Hill, New York, 1990.)

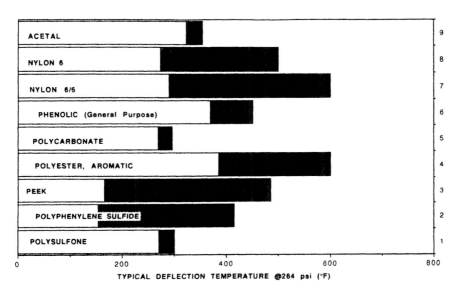

TYPICAL DEFLECTION TEMPERATURE @264 psi (°F)

**Figure 9.12.** Increase in the heat deflection temperature of polymers by glass fiber aeditions. Shaded areas indicate percent performance increase. (From "Guide to Materials Selection," *Adv. Mater. Process.*, June 1990, p. 23.)

on their strength and use temperatures. Polyester and vinyl ester resins are the most widely used resin matrices. They are usually glass filled since they do not adhere well to carbon and aramid fibers (which are also more expensive). These composites fall into the low-performance category and are used in a multitude of applications, including automotive parts, bathtubs, showers, and piping. They do not require high-temperature strength. Most matrices are thermoset polymers, which have historically been the major matrix material for composites, although thermoplastics are making inroads.

In the high-performance category, there are three major matrices, all thermosets: the epoxies, phenolics, and polyimides. The epoxies are the most common resin matrix for the high-performance composites and have been used with just about all fibers, but the development of boron, carbon, and aramid fibers made the fabrication of advanced polymer matrix composites possible. Their strengths, reinforced with boron, carbon, and glass, are compared with the alloys of steel, titanium, and aluminum in the stress–strain curves of Figure 9.14. The actual alloys were not given but the strengths are approximately those of a 2024-T6 aluminum alloy, a Ti–6% Al–4% V alloy and sort of a midrange value for steels. These curves show that boron composites are about equally stiff, but the boron composites are stronger and the weights of the com-

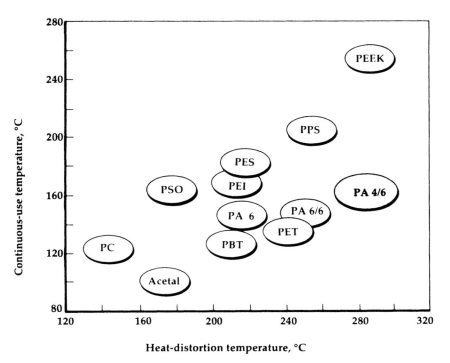

**Figure 9.13.** Continuous-use temperatures (10,000 h) and heat-distortion temperatures of glass-reinforced (30 vol %) thermoplastic polymers. (From D. F. Baxter, Jr., "Green Light to Plastic Engine Parts," *Adv. Mater. Process.*, May 1991, p. 28.)

posite materials are much less than those of the metals, even for the lightweight aluminum alloys. The epoxy matrix composites have been used in military aircraft structural components since 1972. They are also utilized in sporting goods and printed circuit boards. They are currently limited to temperatures of about 120°C (250°F).

The reinforced phenolic (phenol–formaldehyde resins) are not used as much as the epoxy composites, one reason being their processing difficulties, involving the water by-product that is released as a result of the condensation reaction, resulting in voids and loss of strength. But phenolics have high heat resistance (200 at 500°F) and good char strengths. They also produce less smoke and toxic by-products on combustion and are often used as aircraft interior panels. Their good char strengths make them desirable materials for ablative shields for reentry vehicles, rocket nozzles, nose cones, and rocket motor chambers. They have been used with glass, high silica, quartz, nylon, and graphite fibers.

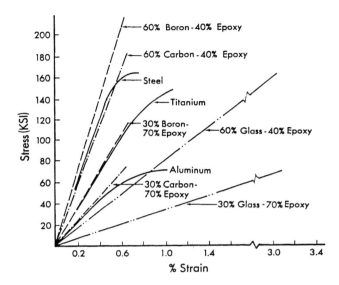

**Figure 9.14.** Strengths of epoxy composites compared to steel, titanium, and aluminum alloys. (From R. M. McCollough, *Concepts of Fiber-Resin Composites*, Marcel Dekker, New York, 1971, p. 9.)

The polyimide thermoset matrices with boron, glass, and carbon fibers have been used to 150°C (302°F). They are candidate materials with carbon fibers for use to 316°C (600°F). They cost more than epoxies, primarily because of processing problems.

Although the thermosetting polymer matrices have been the favorite for high-performance polymer composites, inroads are being made by certain thermoplasts. These consist primarily of the high-temperature thermoplasts such as PEEK, PPS, PEK, and polyamideimide (PAI). The PAI composites are molded as thermoplasts but then postcured in the final composite to produce partial thermosetting characteristics and high-temperature resistance 190°C (375°F). They have better temperature resistance than the polyvinyl and polyvinyldene polymers and reportedly can be used continuously to temperatures as high as those of most epoxy composites. PEEK thermoplastic composites have shown continuous-use temperatures to 250°C (480°F). Another drawback to the thermoset composites is their brittleness—just as in nonreinforced thermosets. Some gains have been made to improve their toughness by the addition of about 20 wt % thermoplasts to the composite.

Recently, quite a number of crystalline and amorphous carbon and glass-reinforced thermoplastic composites have been tested to 260°C (500°F). In addition, the chemical resistance to some organic solvents and selected automotive fluids was also determined. The tensile strengths and the flexural moduli are summarized in Figures 9.15 and 9.16. Here CF and Gl denote carbon and graphite fibers, respectively. The tensile strengths decrease markedly with temperature. Reinforced amorphous resin composites are stronger near their $T_g$ temperature, but the crystalline resin composites retain measurable strengths above $T_g$. Thus $T_g$ becomes a critical design factor for elevated temperature use. Of the composites evaluated, 30 wt % glass/PEK had the best overall chemical resistance.

## 9.5 METAL MATRIX COMPOSITES

The development of metal matrix composite (MMC) materials lags that of PMC materials by somewhere between 15 and 20 years. While the foundations

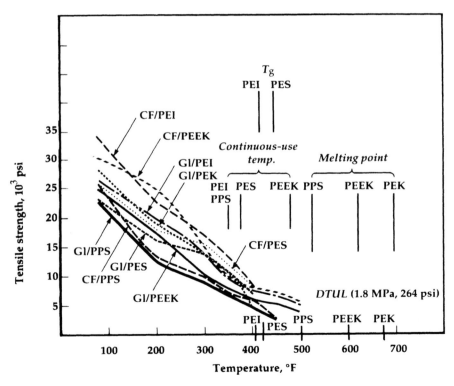

**Figure 9.15.** Tensile strengths of some selected thermoplastic composites as a function of temperature. (From K. R. Quinn and C. A. Carreno, *Adv. Mater. Process.*, Aug. 1991, p. 27.)

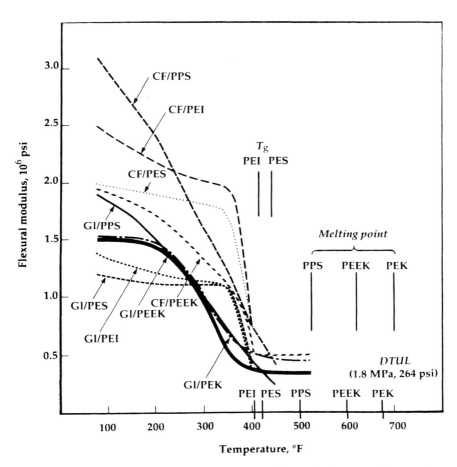

**Figure 9.16.** Flexural moduli of some selected thermoplastic composites as a function of temperature. (From K. R. Quinn and C. A. Carreno, *Adv. Mater. Process.*, Aug. 1991, p. 27.)

for MMCs date back to the early 1960s, the first commercial trials for such composites were the ceramic fiber–aluminum alloy pistons used by Toyota in their diesel engines in the early 1980s. Even as late as 1991, the only commercial source of castable MMCs was the fine ceramic particles in molten aluminum composites made by Duralcan USA, a subsidiary of Alcan Aluminum. The risks associated with using most MMCs is still very high, due to an immature technology.

MMCs can be divided into two subgroups: those reinforced by discontinuous whiskers and particulates, such as the one mentioned above, and those reinforced by continuous fibers such as carbon, alumina, silicon carbide, and boron. The discontinuous and particulate composites have a decided advantage in cost and versatility. The Duralcan aluminum MMC composites have been made in ingots as large as 600 kg (1320 lb) and have subsequently been fabricated into a wide variety of shapes. These composites are composed of commercial aluminum alloys reinforced with either SiC or alumina particulates. They can be cast, extruded, die cast, rolled, forged, and machined. This composite material has also been used as metallizing wire in the thermal spray process for wear-resistance applications. The yield strength and modulus increases for 10 to 20 vol % of alumina in 6061-T6 aluminum alloy over that of 6061-T6 alloy alone are depicted in Figure 9.17 and for 10 to 20 vol % of silicon carbide in a A356-T6 casting alloy in Figure 9.18. But as in all MMC composites, the ductilities decrease to a rather low value with increasing percent of reinforcing material. The fracture toughness of the alumina and SiC composites is the order of 20 and 15 ksi√in. respectively. This material has been used for bicycle frames and is being tested in a number of automotive applications (Figure 9.19).

The cost-effective processing techniques used for the discontinuous particles should provide MMCs for consumer goods, automotive parts, and electromechanical machinery. Other Al alloy–10 vol % SiC fiber composites formed by *squeeze casting* have reported strengths and moduli of 200 MPa (20 ksi) and 82 GPa (11.9 ksi), respectively. Squeeze casting involves the use of pressure ($\sim 13$ GPa; 200 ksi) via forging presses during the infiltration of the porous fiber compact. Considerable development work is still needed for this process because careful control of the many variables, such as preheat temperature, alloying elements, tooling temperature, and pressure levels and duration, are required to obtain reproducible properties. However, this process holds considerable promise for the future manufacturing of MMCs.

The combination of cold and hot isostatic pressing (CHIP) process discussed in the powder metallurgy section is also an attractive process for fabrication of particulate MMCs. Prototype titanium alloy–10% TiC particulate-reinforced MMC parts have been CHIP processed to near net shapes of 99% densities and yield strengths of 40 ksi at 649°C (1200°F). Aluminum reinforced by silicon carbide particulates up to approximately 70% by volume of SiC have been formed to near net shape by a pressureless metal infiltration process (Figure 9.20). The molten metal matrix alloy wets the filler (reinforcement) so well that the matrix infiltrates the filler spontaneously to form the composite. Compared to conventional aluminum alloys these composites have low thermal expansion coefficients and high moduli (270 GPa) with little or no loss in thermal conduction and density. As a result they are attractive candidates for packages, substrates, and support structures for electronic components.

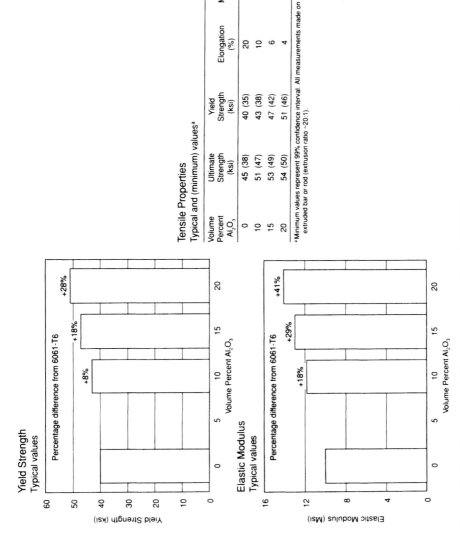

**Figure 9.17.** Increase in yield strength and modulus of 6061-T6 aluminum by the addition of alumina particulates. (Courtesy of Duralcan, USA.)

Tensile Properties

Typical and (minimum) values[a]

| Volume Percent Al$_2$O$_3$ | Ultimate Strength (ksi) | Yield Strength (ksi) | Elongation (%) | Elastic Modulus (Msi) |
|---|---|---|---|---|
| 0 | 45 (38) | 40 (35) | 20 | 10.0 |
| 10 | 51 (47) | 43 (38) | 10 | 11.8 |
| 15 | 53 (49) | 47 (42) | 6 | 12.9 |
| 20 | 54 (50) | 51 (46) | 4 | 14.1 |

[a] Minimum values represent 99% confidence interval. All measurements made on extruded bar or rod (extrusion ratio ~20:1).

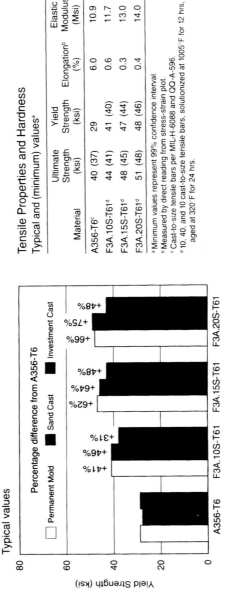

### Tensile Properties and Hardness
Typical and (minimum) values[a]

| Material | Ultimate Strength (ksi) | Yield Strength (ksi) | Elongation[b] (%) | Elastic Modulus (Msi) | Rockwell Hardness, HRB |
|---|---|---|---|---|---|
| A356-T6[c] | 40 (37) | 29 | 6.0 | 10.9 | 55 |
| F3A.10S-T61[d] | 44 (41) | 41 (40) | 0.6 | 11.7 | 66 |
| F3A.15S-T61[d] | 48 (45) | 47 (44) | 0.3 | 13.0 | — |
| F3A.20S-T61[d] | 51 (48) | 48 (46) | 0.4 | 14.0 | 74 |

[a] Minimum values represent 99% confidence interval.
[b] Measured by direct reading from stress-strain plot.
[c] Cast-to-size tensile bars per MIL-H-6088 and QQ-A-596
[d] 10, 40, and 10 cast-to-size tensile bars, solutionized at 1005°F for 12 hrs, aged at 320°F for 24 hrs.

**Figure 9.18.** Increase in yield strength and modulus of an aluminum casting alloy (A356-T6) by the addition of silicon carbide particulates. (Courtesy of Duralcan, USA.) F3A is an Al-7% Si alloy and the 10S, 15S, and 20S represent 10, 15, and 20 vol. % SiC.)

**Figure 9.19.**   Brake disk rotor cast from A356 aluminum containing 20 vol % silicon carbide particulates. (Courtesy of Texas Metal Casting Co. and Duralcan, USA.)

The continuous-fiber MMCs are expensive and have been, until recently, limited primarily to aerospace applications. Carbon, boron, and silicon carbide fibers have now been used in MMC sporting goods, including golf club shafts, tennis rackets, fishing rods, skis, and bicycles. These parts tend to have anisotropic properties compared to the isotropic nature of particulate and chopped fiber forms. Aluminum, titanium, and magnesium matrices have been studied rather extensively because of their light weight and potential aerospace applications. The strength–temperature relationships for experimental Al/SiC and Ti/SiC composites are compared to unreinforced titanium and aluminum in Figure 9.21. SiC, SiC-coated boron, and boron have been the reinforcement continuous fibers of greatest importance. A 40% boron fiber content in a 2024-T6 aluminum matrix composite has strength values as high as 1459 MPa (211.5 ksi) and a modulus of 220 GPa. These composites have been used for structural tubular struts as the forms and truss members in the shuttle midsection fuselage. This composite is also being used as a heat dissipator for multilayer-

**Figure 9.20.**  Parts made of SiC particulates in an aluminum alloy matrix by a pressureless metal infiltration process. (Courtesy of Lanxide Corporation.)

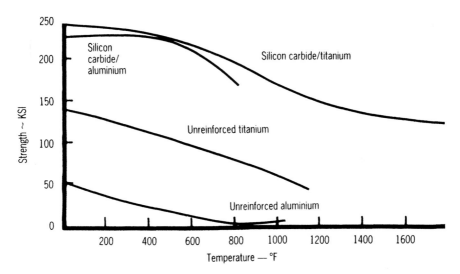

**Figure 9.21.**  Strength versus temperature for experimental aluminum and titanium MMCs compared to the unreinforced metals. (From F. H. Froes, "Structural Aerospace Materials," *Mater. Edge,* May/June 1989, p. 39.)

board microchip carriers using its high thermal conductivity in conjunction with its lower thermal expansion coefficient. SiC fibers manufactured by the CVD process and used at a 47 vol % level in 6061-T6 aluminum have achieved strengths of 1378 MPa (200 ksi) and a modulus of 207 GPa ($30 \times 10^6$ psi). The boron and SiC composites properties were both measured in the fiber direction. They are very anisotropic materials. A micrograph of a boron–aluminum matrix and a schematic of boron–metal matrix composites are shown in Figure 9.22.

MMCs based on a matrix of titanium aluminide ($Ti_3Al$) reinforced by continuous silicon carbide fibers have good high-temperature oxidation and strength, thus offering a use temperature in excess of 800°C (1472°F). Titanium aluminide matrix composites are fabricated by consolidating a foil/fiber/foil layup using hot isostatic pressing. This material is being considered for use in advanced aircraft turbine engines and as a candidate for the National Aero-Space Plane (NASP).

The room-temperature mechanical properties of some MMCs are summarized in Table 9.5. Not only are the room-temperature values higher than that of the nonreinforced materials, but the composites tend to hold this strength at elevated temperatures (Figure 9.21). The primary advantage of boron–aluminum over boron–epoxy composites, for example, is that the MMC offers useful mechanical properties to 510°C (950°F) compared to 190°C (375°F) for the PMC. The ductilities of all MMCs are low. Data on fracture toughness for all MMCs fall approximately in the range 1 to 20 MPa$\sqrt{m}$.

## 9.6 CERAMIC MATRIX COMPOSITES

Although many ceramics themselves could be considered as composites in that they contain significant quantities of two or more ceramic phases, we will not consider these ceramics here as composites. Most of the current ceramic matrix composite (CMC) work is directed toward improving the ductility and toughness of ceramics by the addition of modest amounts of short fibers. One other direction that is being taken is to utilize the high moduli and strengths of carbon and ceramic continuous fibers in high-volume fractions in ceramic matrixes, the same as for polymers and metals, in an attempt to provide higher-temperature capability. Some success has been achieved in producing composites with improved properties for use in the low to intermediate temperature range, but a good high-temperature ceramic matrix composite that can withstand oxidizing conditions and have some significant degree of toughness remains to be developed.

Whisker- and particulate-reinforced ceramics are relatively easy to process but can fail catastrophically. By comparison, continuous fiber-reinforced ceramics fail noncatastrophically, if fiber-layup architecture and fiber–matrix

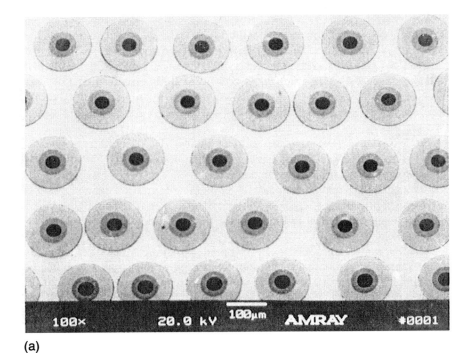

(a)

(b)

**Figure 9.22.** (a) SiC fibers in an aluminum matrix; (b) schematic of a boron fiber–metal matrix composite. (Courtesy of Textron Corp.)

**Table 9.5** Properties of Selected MMCs

| Composite | Ultimate tensile strength [MPa (ksi)] | Modulus [GPa ($10^6$ psi)] | Ductile elongation (%) | Density (g/cm$^3$) | Source |
|---|---|---|---|---|---|
| | | Particulates | | | |
| 6061Al–20 vol % SiC Al$_2$O$_3$ | 345 (50) | 97 (14.1) | 4 | — | Duralcan |
| A356Al–20 vol % SiC | 352 (51) | 97 (14.1) | 0.4 | 2.76 | Duralcan |
| Al–SiC % not given | 210 (30.5) | 220 (31.9) | — | — | Lanxide |
| Al–4.5% Cu–10 vol % | 109 (15.8) | 72 (10.4) | 6 | — | P. Rohatgi, *Adv. Mater. Process.*, Feb. 1990 |
| | | Continuous or nearly continuous filaments | | | |
| Al–60 vol % SiC | 250 (36) | 190 (27.6) | 0.3 | 2.9 | Lanxide |
| Al–50 vol % Al$_2$O$_3$ | 450 (ksi) | 150 (27.6) | 0.7 | 3.2 | Lanxide |
| 6061 Al–48 vol % SiC, laminate 0° | 1550 (225) | 193 (28) | — | 2.84 | Textron |
| 6061 Al–48 vol % SiC, laminate 90° | 12 (0.2) | — | — | 2.84 | Textron |
| 6Al–4V Ti–35 vol % SiC 0° | 1725 (250) | 193 (28) | — | 3.86 | Textron |
| 6Al–4V Ti–35 vol % SiC 90° | 415 (60) | — | — | 3.86 | Textron |
| 6061 Al–50 vol % graphite tows, wire | 1385 (201) | 420 (61) | — | — | ASM HB |
| AZ91 Mg–54 vol % graphite filament wound | 1240 (180) | 390 (57) | — | — | ASM HB |
| Mg–40 vol % graphite filament wound 0° | 720 (105) | 172 (25) | — | — | ASM HB |

interfacial bonding is optimized so that mechanical failure is fiber dominated. These two categories are represented schematically in terms of increasing cost and toughness versus decreasing likelihood of noncatastrophic failure in Figure 9.23.

The toughening mechanism in CMCs depend on the fiber type. For brittle fibers it comes about by the fiber–matrix debonding energy when the composite failure occurs via fiber pullout. When the reinforcement fibers are ductile metal wires, the plastic deformation of the wires during fracture absorbs energy and thereby increases the toughness. In laminate composites delamination between plies can absorb energy during fracture and can add to the toughness. In fact, in CMCs, opposite to that in metal matrix composites, it is more advantageous to have a weak fiber–matrix bond; otherwise, a matrix crack once initiated will propagate right through the fibers. The weak interfacial bond allows the fiber to slip so that the crack is deflected along the length of the fiber. Crack deflection increases toughness since more energy is absorbed. It is also similar to the crack deflection mechanism for improved toughness of zirconia.

Silicon carbide in massive forms, often prepared by powder metallurgy techniques, is a hard but brittle refractory material that resists both oxidation and corrosion. But SiC fibers in a SiC matrix allow the fibers to slip and deflect a crack. At the same time both forms have the same coefficient of thermal expansion, where that of the fiber matches that of the matrix, and eliminates cracking due to thermal cycling, a common source of cracking in composites of widely varying thermal expansion coefficients. SiC/SiC composites, formed by

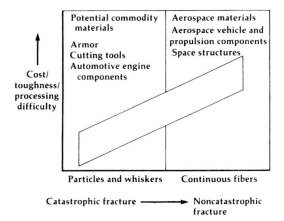

**Figure 9.23.** CMCs divided into two categories based on failure mode and processibiliy. (From S. V. Grisofe, "Ceramic Matrix Composites," *Adv. Mater. Process.*, Jan. 1990, p. 43.)

chemical vapor infiltration, have room-temperature strengths of 520 MPa (75 ksi), but they maintain a high strength at elevated temperatures. At 1000°C (1830°F) they have a strength of 280 MPa (40 ksi), comparable to that of most steels. Most recently, composites made by hot isostatic pressing have demonstrated room-temperature strengths of 750 MPa (110 ksi). SiC whiskers have also been introduced into a silicon nitride matrix by injection molding to produce gas turbine rotors that operate at temperatures of 1250°C. The microstructure is shown in Figure 9.24. SiC monofilament-reinforced silicon nitride (Textron) has a room-temperature strength of 414 MPa (60 ksi) and a UTS of 310 MPa (45 ksi) at 1200°C.

Coating fibers with certain materials weakens the fiber–matrix interfacial bond strength. SiC has been vapor deposited onto carbon and alumina fibers and has improved fracture toughness of CMCs by a factor of 2 to 4 times that of the uncoated fibers. Boron nitride has also been used as a coating material to improve toughness. SiC/SiC composites have demonstrated fracture toughness values of 25 MPa$\sqrt{m}$ to temperatures of 1400°C (2550°F). Single-crystal SiC whiskers added to a $Si_3N_4$ matrix and hot pressed to a high density have pro-

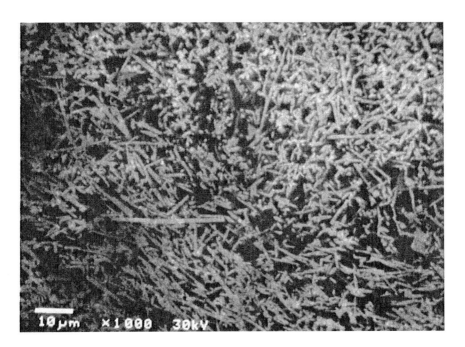

**Figure 9.24.** Microstructure of gas turbine rotors made by injection molding silicon carbide whiskers into a silicon nitride matrix. (Courtesy of GTE.)

duced composites with a fracture toughness up to 56 MPa√m̄, ten times that of $Si_3N_4$ alone. $Si_3N_4$ components have been formed into parts for electrical components, wear-resistant materials, cutting tools, and engine components.

Crystalline ceramic fibers have also been used to reinforce amorphous ceramic glasses. Borosilicate glass, reinforced with SiC fibers, has a strength of 830 MPa (120 ksi), four times that of the glass alone. Whisker-reinforced glass ceramics serve in low- to moderate-stress, high-temperature applications.

Other CMCs include $SiC/TiB_2$ composites, which are hard and abrasion resistant and used for nozzles and bearings for oil field components, and transformation-toughened zirconia reinforced with SiC fibers. The zirconia fibers are coated with a layer of boron nitride.

CMCs have also been manufactured by a patented process (Lanxide Corp.) that involves the accelerated oxidation of molten metals to grow ceramic matrixes around preplaced filler replacement materials. Typically, the reinforcement is shaped into a preform of the same size and shape of the final part. Cloth layup, braiding, weaving, and filament winding can be used to fabricate preforms. Oxidation occurs outward from the metal surface and into the preform such that the reaction product becomes the matrix surrounding the reinforcement in the preform. The matrix usually contains some residual metal, which provides added toughness. $SiC/Al_2O_3$ composite systems have been made that show strength on the order of 480 MPa (70 ksi) at temperatures of 1200°C (2190°F). Fracture toughness values at room temperature and at 1200°C are both around 25 MPa√m̄. Typical products, which have been made in prototype quantities, are shown in Figure 9.25.

All of the discussion above of successes in the CMC arena would lead one to believe that we are just a step away from having high-temperature materials so badly needed, and sought after for many years, for utilization in the hot regions of gas turbine engines. But such is not the case. Ceramic products that can be substituted for superalloys remain elusive, despite widespread claims of new ceramic components that will solve all of our needs. Although reinforcement appears to be the most desirable and effective means for toughening ceramics, even considerably better than partially stabilized zirconia, problems still exist. Cost is one factor, but for aerospace applications, the lack of reliability and consistency is far more important. As pointed out in a recent article (C. T. Sims, *Advanced Materials and Processes,* June 1991) after 20 years of extensive research no aircraft or industrial gas turbines have incorporated nonmetallic materials in a structural capacity. Superalloys are still being pushed to new levels via alloying and cooling methods. Turbine inlet temperatures have been increased to 1700°C (3000°F) by cooling superalloy parts with redirected compressor air. The nonmetallic parts developed to date do not have the ductility nor toughness that come close to that of the superalloys. Other problems include creep resistance, high-temperature stability, and chemical

**Figure 9.25.** Prototypes of SiC-reinforced $Al_2O_3$. (Courtesy of Lanxide Corp.)

reactivity. There also appears to be a lack of consistency in test methods and data reporting procedures that prohibit a meaningful comparison of ceramic data to that for metallic structural materials.

## 9.7 GRAPHITE MATRIX COMPOSITES

### 9.7.1 Carbon–Carbon Composites

The carbon–carbon composites consist of an amorphous carbon matrix reinforced with carbon fibers. The fibers are based on rayon, pitch, or polyacrylonitrile (PAN) and increase in cost in the order listed. The chief reason for using carbon–carbon composites is that their strength is still maintained at temperatures to 2500°C (4532°F). They are biocompatible and can be made with directional properties. They do oxidize readily and must have some type of protective coating. But their high-temperature properties, coupled with the fact that they do not degrade or outgas the way that organic resins do in the vacuum of outer space, outweigh the necessity of protective coatings for reentry vehicles.

One of the most notable applications of carbon–carbon composites is in the nose cone and leading edge components of the space shuttle. Reentry temperatures are around 1650°C (3000°F). Not only can carbon–carbon composite materials withstand the heat but remain unscathed mission after mission in space and actually increase in strength. The nose cone is made by a two-dimensional layup. Graphite cloth is preimpregnated with phenolic resin, then pyrolyzed, driving off the gases and moisture as the phenolic resin converts to a graphite matrix. This relatively soft composite is then impregnated with furfural alcohol and pyrolyzed three more times, with the density, strength, and modulus increasing each time. A ceramic coating of silica and alumina is applied, which after firing results in a SiC surface layer. The surface is further protected by impregnation with tetraethylothosilicate, which when cured leaves a silicon dioxide residue throughout the coating.

Carbon–carbon composites are also used as aircraft brakes, where their high thermal conductivity and light weight are advantageous; as prosthetic devices, where its compatibility with body tissue and directional properties are desirable traits; and in other spacecraft components. Advanced carbon–carbon (ACC) is used for turbine wheels on hypersonic aircraft. Continuous PAN fibers are used for this product. The manufacturing method is often tailored to fit the application. Carbon–carbon composites have also been used as lightweight housings for satellite solar cells and are currently being flight-weight tested by NASA for the X-30 National Aero-Space Plane.

Carbon–carbon composites have a density of about 1.3 $g/cm^3$, compared to 2.7 and 4.5 for aluminum and titanium, respectively. But the densities and strength of carbon–carbon composites vary considerably with manufacturing methods. Room-temperature strengths of around 450 MPa (65 ksi) are typical [although 2100 MPa (304 ksi) has been reported for ACC material], with 600 MPa (87 ksi) for continuous fiber composites. Moduli in the range 125 to 175 GPa (18 to 25 × $10^{16}$ psi) are common. The carbon–carbon bond is stronger than that in ceramic matrix composites. Fracture occurs via debonding as well as fiber fracture. But carbon–carbon composites are the strongest materials known at temperatures of 2204°C (4000°F). Figure 9.26 shows the specific strength versus temperature characteristics for carbon–carbon composites compared to those of superalloys.

In the physical properties category the coefficient of expansion is of little concern since both the fiber and matrix form have essentially the same values. Thermal conductivities are on the order of 240 W/m · K (0.142 Btu/ft-hr-°F).

## 9.7.2 Graphite–Boron Composite

Graphite matrices have been used with other fibers, notably boron. This composite is currently a very popular material for sports equipment, including golf

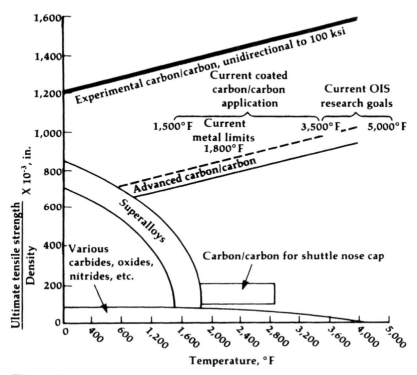

**Figure 9.26.** Specific strength versus temperature for carbon–carbon composites compared to superalloys. (From A. J. Klein, "Carbon/Carbon Composites," *Adv. Mater. Process.*, Nov. 1986, p. 66; data supplied by LTV Aerospace and Defense.)

club shafts, skis, and bicycles (Figure 9.27). Textron claims to have proof that golf clubs with boron–graphite shafts hit balls up to 20 m farther than clubs having titanium, graphite, or steel shafts.

## 9.8 FABRICATION OF COMPOSITES

With the large variety of composites available today, together with the almost unique way in which each composite is manufactured, a complete coverage of all the fabrication methods would be a volume in itself. Furthermore, as new composites are developed they must be accompanied by new fabrication methods. Generally, we can simplify the fabrication methods by grouping them not according to materials but to the geometry of the reinforcing material.

The small randomly oriented fibers, whiskers, and particulate composites can be grouped together because the fabrication methods, exclusive of fiber

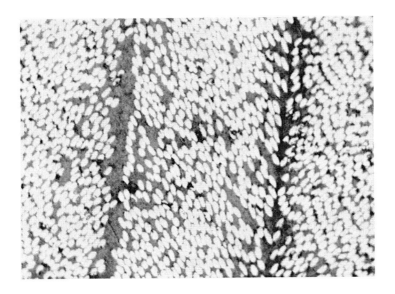

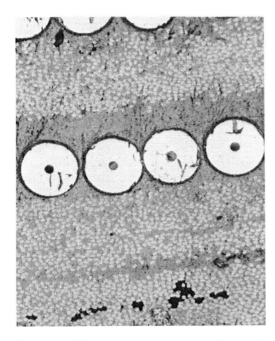

**Figure 9.27.**   Longitudinal section of a graphite–boron fiber golf club shaft. Note that the layers are stacked such that the boron fibers run in different directions in adjacent layers. (Courtesy of Aldila Corp.)

preparation, which was discussed earlier, are very similar if not identical to the very common metal- and polymer-forming processes. The polymer matrix composites of these type can be formed in the same manner as for the nonreinforced polymers. These methods include injection and pressed molding, extrusion, and sheet molding. Also, the metals and some ceramics matrix composites are formed by similar methods whether or not small reinforcement particulates are present. Thus we will not repeat all of the various pressing, casting, forging, extrusion, rolling, die casting, and molding processes. The flexibility provided by the large variety of forming methods is one of the more attractive features of these small-particle composites. Most of these forming methods, with a possible exception in some cases of extrusion, yield a composite material with isotropic properties. Some preforms and prepregs of these composites are also used in laminates, permitting additional design versatility.

The continuous and nearly continuous long fiber composites as well as the laminar type are fabricated in a large variety of methods to give shapes with intended directional properties. These methods include filament winding, braiding, ply layup methods, pultrusion, impregnation and coating of fibers, and honeycomb structures.

Fiber Weaving

To obtain more isotropic properties in many fiber composites, multidimensional fiber weaving methods were developed. For simplicity, the industry concentrated on three-dimensional configurations, and today, automated computer-controlled three-dimensional weavers are extensively used. The objective of three-dimensional weaving is to produce orthogonal preforms as pictured in Figure 9.28. This weaving machinery is described in the ASM *Handbook*. Fibers that have been woven include carbon, glass, silica, alumina, silicon carbide, and aramid.

Fiber Braiding

Braiding is a textile process similar to weaving in which two or more yarns are interwoven in a variety of patterns. Braiding has many similarities to filament winding in that dry or prepreg yarns, tapes, or tow can be braided over a rotating removable mandrel. Near net forms of structural shapes made by three-dimensional braiding are shown in Figure 9.29.

Fiber Preforms

Fiber preforms for resin impregnation consist of fabric made by weaving, braiding, rovings, laid-up parts, and almost any configuration that lends itself to being impregnated by polymer resins. Fiber forms that are preimpregnated with matrix resin in the uncured state are known as *prepregs*. Prepregs consist of both continuous and chopped fibers that the manufacturer uses to fabricate parts by molding, laminating, or rearranging in ways to form a finished product. The manufacturer subjects these parts to heat and pressure so as to cure

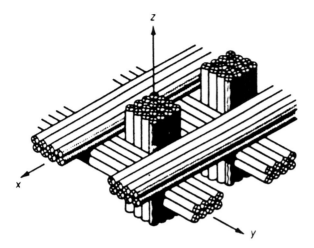

**Figure 9.28.** Geometry of a three-dimensional orthogonal weave preform. (From *Handbook of Engineered Materials,* Vol. 1, ASM International, Metals Park, Ohio, 1987, p. 130, article by F. P. Magin.)

them to their final desired properties. The manufacturers of parts usually have nothing to do with the pregs, which are supplied by the larger composite fabricators. Often, the prepregs are divided into high-performance composites for the aerospace industry that have less stringent demands, such as those for such consumer products as appliance housings and automotive parts. The former often include the carbon composites, of strengths around 1600 MPa (230 ksi) and moduli of 140 GPa (20 × 10$^6$ psi). The latter are most often the sheet molding compounds, which contain polymer or glass fibers in polyester or vinyl resins, whose properties are more in the neighborhood of 50% of the high-performance prepreg composites. In all of these products it must be realized that properties will vary markedly depending on whether the resulting composite is formed in more of a unidirectional manner or as a combination of multidirectional layers, the latter having lower but isotropic properties compared to the higher unidirectional properties of a continuous-tape prepreg.

Filament Winding

Filament winding is one of the older methods of fabricating PMCs and was used to fabricate large rocket motor cases in the 1960s. Resin-impregnated rovings or tows are fed to a rotating mandrel, around which they are continuously wound (Figure 9.30). The mandrels are usually made of water-soluble sand, or the mandrels are collapsible for easy removal. In some cases, such as in pressure vessels, a metal mandrel remains as part of the load-bearing structure. The

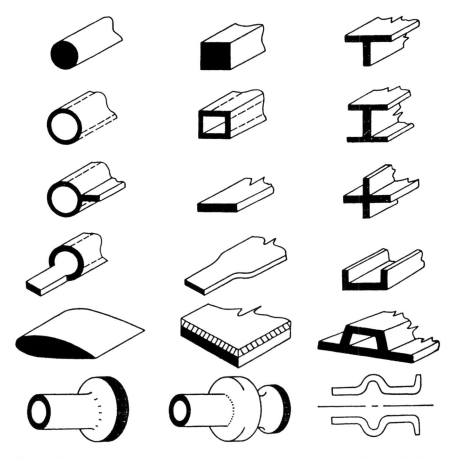

**Figure 9.29.** Net shapes produced by three-dimensional braiding. (From *Handbook of Engineered Materials,* Vol. 1, ASM International, Metals Park, Ohio, 1987, p. 524, article by F. K. Ko.)

fibers can be laid down in a variety of patterns, each layer often being oriented in different directions. This is a rather economical process because the fibers and resin are essentially combined at the same time without the need for a prepreg.

### Boron and SiC Fiber Composites

As stated earlier, both boron and silicon carbide fibers are made by depositing boron and SiC onto a hot wire by a chemical vapor deposition process. Once the fibers are formed, they are available to be turned into whatever composite

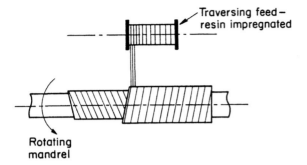

**Figure 9.30.**   Filament winding fabrication process for continuous filaments.

is of interest. Figure 2.22 showed these fibers in metal matrices. Basically, the metal foil is hot pressed around a fiber layer until it completely surrounds the fiber, forming a composite sandwich layer. These layers or preforms can then be stacked as layup plies to form a more extensive composite (Figure 9.31). Most often the fibers run in the same direction, but if desired they can be placed at right angles for high strength in two directions. These types of preforms have also been fabricated by spraying the fibers with plasma.

Pultrusion

Pultrusion is a rather recently developed process that consists of pulling continuous fibers through a resin bath and then through a die into a curing chamber. Generally, the resin is a thermosetting polymer that reacts during the

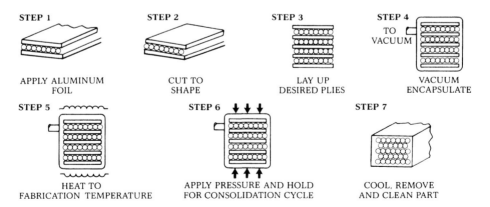

**Figure 9.31.**   Typical fabrication process for boron and SiC fibers in metal matrices. (Courtesy of Textron Corp.)

fabrication process to produce a shape that cannot be further altered or reshaped. This is a fairly complex process. The machinery must be capable of pulling, heating, infiltrating and cutting the profiles to the desired length. In general, pultrusions can be produced in nearly any constant cross-sectional shape that can be extruded. It is a very versatile process that is expected to penetrate an increasingly large number of markets as it becomes better developed and recognized for its versatility and economy. For example, aluminum extrusions account for about 15% (and is growing) of all aluminum products; pultrusion now accounts for only about 5% of the reinforced polymer market. Pultrusions have been fabricated into I-beams, channels, and tubular shapes, much as in the metal industry. The number of materials that are amenable to this process are numerous. A typical pultrusion setup is shown schematically in Figure 9.32.

Layup Methods

The layup method was probably the first used to make a modern composite structure. It has a number of similar names, such as layup molding, wet layup, or contact laminating. Basically, a mat of woven or braided fabric is saturated with resin, then mats are laid on top of one another to obtain the desired thickness. The impregnation is done *in process* as the layers are placed in the mold. The wet layup process involves laying the dry fiber mat into the mold and then applying the resin. Another modification is the spray gun method, where a mixture of chopped resin and fibers is sprayed into a mold. The prepreg method is a modification of the wet layup process. Usually, the prepreg is purchased in an impregnated tape or woven fabric that is cut and fitted into a mold. The prepreg method usually involves *vacuum bagging* to assist in compressing the plies while simultaneously withdrawing excess volatiles. This method is presently used only for PMCs. *Laminates* are the result of multilayers of chopped or continuous fibers that have been woven, braided, or by some other method been formed into prepregs such that the layers can be

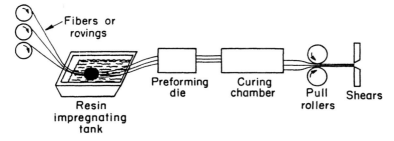

**Figure 9.32.** Pultrusion process for fabrication of continuous-filament PMCs.

bonded together by resins. Usually, several layers are placed in different orientations in order to resist stress in all directions. Excess resin is removed by vacuum bagging, which simultaneously presses the layers together and removes the volatiles and entrapped air.

## 9.9  FAILURE OF COMPOSITES

Again in discussing composite failure it is convenient to divided them into two types: (1) the continuous filament or laminate type, and (2) the discontinuous particulate, whisker, or small randomly oriented fiber-reinforcement composites.

Fracture of continuous fiber composites can occur in a number of ways. When the fibers are oriented in the direction of the load, the fibers are placed in tension and thus the best strength characteristics are realized. In such loading the fracture can occur by fiber fracture, matrix failure, or by delamination at the fiber–matrix interface, commonly called the *fiber pullout* (Figure 9.33). This fracture occurs after a rather severe treatment that consisted of thermal cycling the composite between 300 and 800°C for 75 cycles at a stress of 82.3 ksi. Even then the composite did not fracture until it was elongated an additional 0.5 in. at the high temperature and stress. The matrix nonreinforced alloy (Ti–6% Al–4% V) lasted only 6 cycles at a stress of 52.8 ksi.

Many continuous fiber composites are made by alternating layers in which the fibers run in different directions, as illustrated in Figure 9.28. In such situations the failure then takes place by what has been called intralaminar (Figure 9.34a), interlaminar (Figure 9.34b), and translaminar (Figure 9.34c). The first, intralaminar failures are those occurring internally within a ply and are very close to, and may often be, failure along the fiber–matrix interface. Interlaminar failures, on the other hand, are fractures that occur between the lamela used in the layups. This is more of a construction problem in actually obtaining a good bond between the matrices of the respective laminates. Translaminar fractures involve fiber breakage, and although the matrix is fracturing at the same time, this type of fracture suggests that either the fiber was weak, contained too many defects, or that the lamela contained something less than the critical volume of fibers. Although the matrix also fractures in this case, in general one is relying on the fibers to carry the load, and if this were the situation, the matrix cannot be blamed.

So far we have been concerned with the failure of the composite in terms of its weaker components. Actually, in many cases it may be desirable, for example, to have a certain amount of fiber pullout, even though it may reduce the overall strength of the component. This is because fracture energy is absorbed during fracture via the pullout mechanism, and any increase in energy absorption increases the fracture toughness, or in other words, it leaves less energy

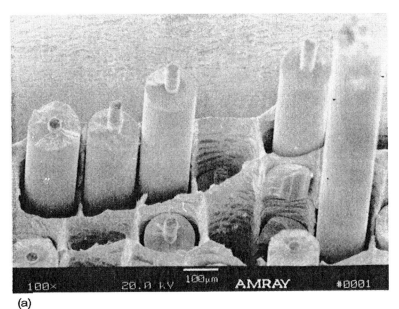

(a)

(b)

**Figure 9.33.** Composite fractures: (a) fiber pullout (Glenn Wensloff, M.S. thesis, Cal Poly, 1991); (b) compression delamination failure (courtesy of R. Leonesio, Cal Poly.)

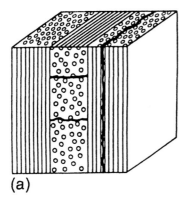

(a)

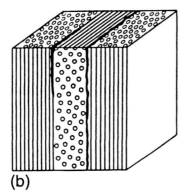

(b)

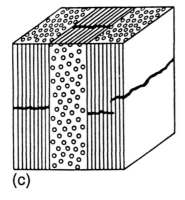

(c)

**Figure 9.34.**   Different pr.anes of separation in continuous-fiber-reinforced composites: (a) intralaminar, (b) interlaminar; (c) translaminar. (From *Handbook of Engineered Materials*, Vol. 1, ASM International, Metals Park, Ohio, 1987, p. 787, article by B. W. Smith.)

for crack propagation. Cottrell (*Proc. Roy. Soc.* A282, 1964) suggested that the fiber length $l$ should be less than $l_c$, so that the fibers will pull out and prevent a fracture from propagating. Of course, this must be done with a sacrifice of strength since maximum strengthening occurs when $l > l_c$. But in metals we often sacrifice strength for toughness, the tempering of martensite being the best example.

We must also consider fatigue in composites—that deteriorating effect of cyclic stress with time (or number of cycles) that occurs in metals and polymers and is now known to occur in composites. Metal matrix composites develop significant matrix cracking when cycled above the fatigue limit. Perhaps this is just metal fatigue multiplied or enhanced by an increase in the number of crack sites that could be present at matrix–fiber interfaces. Although the classical *S-N* curves can be established, extensive testing is required and at present not many data exist (Figure 9.35). From the data we have it appears that the fatigue mechanism is very much associated with the composite type, which implies that the design engineer should have fatigue data for the specific composite to be used in the design. Duralcan has generated fatigue data on their composites that show the composite to have fatigue strength superior to that of the nonreinforced aluminum alloy. Some fatigue data are presented in the ASM composite handbook. Other references are listed in the Suggested Reading and also in the ASM *Handbook.*

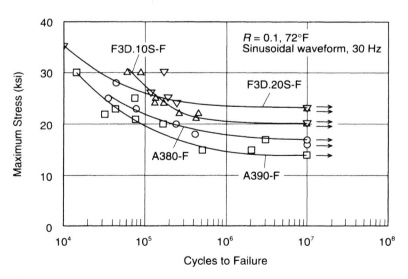

**Figure 9.35.** Fatigue *S-N* curves for the aluminum casting alloy 380 and 390 compared to similar alloys containing 10 and 20 vol % SiC particulates. (Courtesy of Duralcon Corp.)

Fracture in discontinuous composites is also very complex and dependent on the type of composite. Fracture in particulate composites will be different from that in whisker and short-fiber composites. Fracture failure in Al/SiC composites, one of the more advanced particulate materials, appears to be dominated by matrix failure. Fracture mechanics, which as yet has not really dealt with composite structure, appears to be a suitable method for obtaining $K_{Ic}$ values that can be used for design purposes. What are needed at the moment are more data. If the fracture mode extends through both matrix and fiber in a more-or-less continuous manner, fracture mechanics should be a useful approach.

## DEFINITIONS

*Anisotropic laminate*:  one in which the properties vary with direction in the laminate.

*Aramid*:  highly oriented organic material derived from polyimide but containing aromatic rings. Kevlar and Nomex are examples of aramid fibers.

*Aromatic*:  unsaturated hydrocarbon with one or more benzene rings.

*Borsic*:  boron fiber that has an outer layer of SiC to prevent reaction with certain matrix materials.

*Braiding*:  process where two or more systems of yarn are intertwined in the bias direction to form an integrated structure.

*Carbonization*:  process of pyrolyzing certain polymers in the range 800 to 1600°C to obtain carbon fibers.

*Cermet*:  composite containing inorganic compounds (ceramics) with a metallic binder.

*Char*:  to heat a composite in air until it is reduced to ash.

*Chemical vapor deposition (CVD)*:  process of reacting metal compounds, usually halides, with a hot surface whereby the compound decomposes and the metal deposits on the hot substrate.

*Critical fiber length*:  length necessary to cause fiber overlap and transfer of load to adjacent fibers. It is also a function of fiber diameter.

*Cross-ply laminate*:  laminate with plies usually oriented at 0° and 90° only.

*Cure*:  to change the properties of a thermosetting resin irreversibly by chemical reaction.

*Delamination*:  separation of the materials in a laminate. Also applied to separation of fiber from the matrix.

*Disbond*:  area of a bonded interface between two adherents where adhesion has been lost.

*E-glass*:  family of glasses of calcium aluminoborosilicate with about 2% alkali content.

*Felt*:  fibrous material made of interlocked fibers.

*Fiberglass*:  composite of glass fibers in a polymer matrix.

*Filament*: smallest unit of a fibrous material, usually less than 25 $\mu$m (1 mil) in diameter.

*Filament winding*: process for fabricating a composite in which continuous filaments gathered into strands (or yarn or tape) are placed over a rotating removable mandrel to form a cylindrical composite when the filaments are impregnated with a resin.

*Graphite fiber*: fiber made from a polymer precursor by a carbonization or graphitization process.

*Graphitization*: process of pyrolyzation of certain carbon-containing compounds in an inert atmosphere at temperatures in excess of 2500°C, which thereby converts carbon to its crystalline allotropic graphite form.

*Hand layup*: process of placing successive plies of reinforcing material in resin-impregnated reinforcement in a mold by hand.

*Helical winding*: winding in which a filament advances along a helical path, not necessarily at a constant angle except for some cylinders. A shape frequently used in filament wound structures.

*Honeycomb*: hexagonal cells made from numerous lightweight materials (paper, metal foil, glass fabric).

*Hybrid laminate*: composite laminate consisting of laminae of two or more different composite materials.

*Impregnate*: to force resin, metal, or other matrix material into a fibrous body. Similar to *infiltration*.

*Interface*: region between two different adhered materials.

*Interlaminar*: event or process (e.g., fracture) occurring between two or more adjacent laminae.

*Kaowool*: fiber derived from clay for insulation, but also used in composites. Most often discontinuous. A Babcock and Wilcox trademark.

*Kevlar*: organic polymer composed of aromatic polyimides having a parallel chain extending bonds from each aromatic nucleus (a DuPont trademark).

*Knitted fabrics*: fabrics produced by interlooping chains of yarn.

*Lamina*: single ply or layer in a *laminate*.

*Layup*: process of placing reinforcing material into a mold, usually resin-impregnated.

*Macro*: scale of materials that pertain to their individual nature and of a size that can normally be seen without the aid of a microscope. Applies to macroscale.

*Mat*: discontinuous fiber reinforced material available in sheet, blanket, or ply form.

*Matrix*: continuous phase of a composite that acts as the binder to hold together the reinforcing materials.

*Particulate*: small metallic or inorganic reinforcing materials of a *composite*, often as chopped fibers or small particles.

*Pitch*:   high-molecular-weight material left as a residue from the distillation of coal and petroleum products.

*Ply*:   fabrics or felts consisting of one or more layers. The layers that make a stack. A single layer of prepreg.

*Precursor*:   material preceding or used to make something else. For carbon or graphite fibers it is the rayon, PAN, or pitch that is *pyrolyzed* to form the fibers.

*Preform*:   preshaped fibrous reinforcement formed by distribution of chopped fibers or cloth over the surface of a perforated screen to the approximate thickness desired in the finished product. Also, a preshaped fibrous reinforcement of mat or cloth formed to the desired shape on a mandrel or mockups before being placed in a mold.

*Prepreg*:   either ready-to-mold materials in sheet form or ready-to-wind materials in roving form (e.g., mat or cloth) subjected to unidirectional fiber impregnation with resin and stored for use.

*Pultrusion*:   continuous process for manufacturing composites that consists of pulling a fiber-reinforced material through a resin impregnation bath and subsequently through a die and curing chamber.

*Pyrolysis*:   thermal process by which organic precursor fiber materials are chemically changed by heating into a fiber.

*Reinforcement material*:   strong material bonded into a matrix usually to add strength, but could also be used for other purposes, such as improving thermal or electrical conductivity. Usually in the form of long or chopped fibers, whiskers, or particulates.

*Resin*:   solid or pseudoidal organic material, usually of high molecular weight, that tends to flow when subjected to stress. Most resins are polymers.

*Roving*:   number of yarns, strands, or tows or ends collected into a parallel bundle with little or no twist.

*Tack*:   stickiness of an adhesive- or filament-reinforced resin prepreg.

*Tow*:   untwisted bundle of continuous filaments.

*Translaminar*:   going through the laminae, such as in crack propagation.

*Vacuum bag molding*:   process in which a sheet of flexible materials is placed over a mold and sealed at the edges. A vacuum is applied to suck out entrapped air and volatiles and, simultaneously, to apply pressure.

*Weaving*:   manner in which a fabric is formed by interlacing yarns.

## QUESTIONS AND PROBLEMS

9.1.   For an epoxy resin with reinforced continuous fibers, what fiber would you select if:
   (a) The tensile strength were the most important factor?
   (b) The tensile modulus were the most important factor?

(c) The modulus and strength were of equal significance?

9.2. For a 2024 aluminum alloy reinforced with discontinuous fibers of SiC:
(a) Estimate the composite strength for 10, 20, and 30 vol % fibers.
(b) Plot a graph of strength versus fiber content and extrapolate it to 65 vol % fiber content. Would you expect to obtain the strength that the 65% point shows on the graph? What problems would you encounter in fabricating such a composite?

9.3. Suppose that you used discontinuous fibers for Problem 9.2(a). Estimate the strength change compared to continuous fibers.

9.4. Describe how graphite (carbon) fibers are produced, including starting materials and process temperatures. How does the cost vary with the starting material?

9.5. Boron and SiC fibers are both made by the chemical vapor deposition method. Compare the advantages and disadvantages of the two fibers.

9.6. Compute the critical volume of E-glass for reinforcing pure aluminum. Assume that the composite will be strained until the aluminum has been plastically strained 10%. (You may need Chapter 6 for this one.)

9.7. What could you do to increase the strength of the aluminum composite of Problem 9.6 by a factor of 1.5 while holding the matrix strain to 10%? First list the variables that could be involved, including other fiber shapes and materials. Only the 10% strained aluminum matrix remains constant.

9.8. For a 10% strained aluminum matrix reinforced with "continuous" E-glass fibers, what is the critical fiber length for a 1-mil fiber diameter?

9.9. In Equation (9.15), what factors affect the efficiency coefficient $\alpha$? Why is the range of $\alpha$ so large? Could it be the controlling factor for the moduli of discontinuous fiber composites?

9.10. Name two each of low- and high-performance PMCs.

9.11. Why do most commercial PMCs made to date have thermoset rather than thermoplastic matrices? What is one advantage of the latter?

9.12. Most composites have very low ductilities. Could you suggest a composite where the matrix is considerably strengthened by fibers, yet the composite will have ductilities in excess of 20% elongation?

9.13. Ceramics are brittle. Why would CMCs be considered for structural purposes?

9.14. What is the one primarily advantage or characteristic of carbon–carbon composites over other material combinations? What is a current application of carbon–carbon composites?

9.15. What is the difference between a preform and a prepreg?

9.16. Equation (9.10) expresses the fact that a critical volume of fibers is required for reinforcement to occur. Would a critical volume be required even if the matrix did not strain harden? Write a similar equation for a polymer matrix that does not strain harden.

9.17.   Many CMCs use fibers in a thermosetting epoxy that does not readily
        deform plastically. How is the load transferred for overlapping fibers in
        this situation?

## SUGGESTED READING

American Society for Metals, *Engineered Materials Handbook: Composites*, ASM
International, Metals Park, Ohio, 1987.
*DuPont Bulletin K-5,* Sept. 1981.
McCullough, R. L., *Concepts of Fiber–Resin Composites,* Marcel Dekker, New York,
1971.

# 10

# Electronic Materials

## 10.1 INTRODUCTION

The electron conductivity of solid materials gives us almost an unambiguous way to classify them. Simply put, on the basis of electrical conductivity, they are either conductors, semiconductors, superconductors, or insulators. In Figure 2.22 the range of conductivities shown for all materials was quite large, being about $10^7$ to $10^8$ $(\Omega \cdot m)^{-1}$ for conductors (metals), $10^3$ to $10^{-6}$ for semiconductors, and less than $10^{-6}$ for insulators. Where do we draw the line between these groups? On the scale of Figure 2.22 it is somewhat arbitrary. But for crystalline solids we can define these three categories fairly precisely in terms of the number of electrons available for conduction. This number can be computed using the energy band structure for the valence electrons, a subject covered in the following section. For polymers the situation is not quite so well defined. In Figure 2.22 the polymers listed are insulators, and most polymers do have a low conductivity, due to their strong covalent bonds and the absence of free-valence electrons. But in some polymers a conducting powder is mixed with the polymer to form a conducting composite. In a few others, the so-called conducting polymers, there exist some free electrons within the polymer structure creating conductivity on the order of that found in crystalline semiconductors, and in some conducting polymers the conductivity approaches that of metals. There exist a tremendously large variation in the conductivities of solids, being about a factor of $10^{25}$ from conductors to insulators and about $10^{40}$ from superconductors to insulators.

In Section 2.3 we used Ohm's law to express conductivity and its reciprocal, resistivity, which are not a function of specimen dimensions, and the conductance and resistance, which are a function of specimen dimensions. Equation (2.21) is restated here for convenience.

$$R = \rho \frac{l}{A} \tag{2.21}$$

where $R$ = resistance
  $\rho$ = resistivity (usually expressed in $\Omega \cdot m$)
  $l$ = specimen length
  $A$ = specimen area

and in terms of conductivity $\sigma$,

$$\sigma = \frac{1}{\rho} = \frac{1}{RA} \qquad [\text{units are } (\Omega \cdot m)^{-1}]$$

For semiconductors it is more desirable and convenient to express the conductivity in terms of the number of charge carriers and their mobility. To accomplish this feat, we state Ohm's law in a different fashion:

$$J = \sigma \xi \tag{10.1}$$

where $J$ = current density or flux $= \dfrac{\text{Amp}}{m^2}$

  $\sigma$ = conductivity
  $\xi$ = electric field strength $= E/l$ = V/m

Now the current density can also be written as

$$J = nqV_d \tag{10.2}$$

where $n$ = number of charge carriers
  $q$ = charge on the carrier, usually that of the electron ($1.6 \times 10^{-19}$ C)
  $V_d$ = average drift velocity of carriers

$V_d$ is the only expression above that may be foreign to the reader and require an explanation. In any material there exists some resistance to the electron flow. In metals this resistance is primarily that due to electron collisions with the lattice ions. It is related to the acceleration of the electron as a result of the collision and the time between collisions. The collisions and scattering are depicted in Figure 10.1. $V_d$ is proportional to the electric field strength $\xi$:

$$V_d = \mu \xi \tag{10.3}$$

where $\mu$, the proportionality constant, is called the mobility of the carrier. It has the units

$$\frac{\text{velocity}}{\text{field strength}} = \frac{\text{m/s}}{\text{V/m}} = \frac{m^2}{V \cdot s}$$

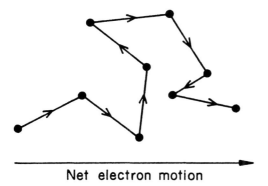

Net electron motion

**Figure 10.1.** Schematic diagram of the scattering path for an electron as it passes through a crystal.

Combining equations (10.1), (10.2), and (10.3), we have

$$\sigma = nq\mu \qquad (10.4)$$

This is the equation that we will be using in computations involving semiconductors. The mobility, $\mu$, is somewhat different in metals than in semiconductors. In metals it is related to the mean free path and the average velocity between electron collisions. It can be shown that in metals

$$\sigma = \frac{ne^2\lambda}{2mV_d} \qquad (10.5)$$

where $n$ = number of carriers
     $e$ = charge on electron
     $\lambda$ = mean free path between collisions
     $m$ = mass of electron
     $V_d$ = average velocity between collitions = $\Delta V/2$

Since $\sigma = nq\mu$ ($q = e$), $\mu$ becomes

$$\mu = \frac{e\lambda}{2mV_d} \qquad (10.6)$$

In Cu at room temperature ($\mu = 4.4 \times 10^{-3}$ m$^2$/V · s. In semiconductors the mobility is much larger (0.14 m$^2$/V · s for silicon) and more difficult to explain. The electron mass behaves as if it has a smaller mass, called the effective mass $m^*$, than the rest mass. In metals $m \approx m^*$, but in semiconductors $m^* < m$. In physical terms it is related to the density of energy states into which the electron can go. The mobility is determined experimentally by measuring the conductivity.

There is only one convenient way of explaining semiconducting behavior and that is via the *band theory of solids*. A course in quantum mechanics might assist in the understanding of this theory, but for our purposes it is not necessary. All that is required is the acceptance of the fact that there are discrete and well-defined energy levels allowed for electrons in a solid and that all other energy levels are forbidden. A somewhat oversimplified picture is that electrons behave like waves and have an energy $h\nu$, where $h$ is Planck's constant and $\nu$ is the frequency of the wave of the electron, the frequency being the reciprocal of the wavelength $\lambda$. As the electrons pass through a solid, under the effect of some potential gradient, if at any time they possess a certain restricted energy they will be diffracted. The spacing and achitectural arrangement of the atoms (the crystal structure) determine the values of the allowable energies or energy levels. This situation creates *bands* or *zones* of energy levels that the electron is forbidden to occupy. Perhaps the term *zone* is a better word for expressing this theory. The more advanced texts refer to it as the *Brillouin zone theory*, named after Leon Brillouin, who pioneered this concept and dealt with it in three-dimensional reciprocal lattice space. But in two-dimensional space we know it as the *band theory of solids*. We adhere to this more simplified and more easily understood approach.

## 10.2  BAND THEORY OF SOLIDS

Since the band theory really deals only with electron energy levels, let's begin with an examination of the energy levels of a single atom, sodium for example. Sodium has an atomic number of 11 and an electron configuration expressed in the shorthand notation as $1s^2 2s^2 2p^6 3s^1$. These energy levels are separated as depicted in Figure 10.2a and are those for a single atom. In Figure 10.2b we show what happens to the energy levels as the atoms are brought together to form a crystal of sodium. Sodium has a melting point of 97.8°C and is a crystalline solid at room temperature. Let's start with 1 mol of sodium atoms in the vapor phase that are suddenly brought together at ambient conditions such that they want to form a solid crystal. We will select one atom to be at the zero position on an $x$-$y$ coordinate system, as depicted in Figure 10.2b. It is not really necessary to deal with three dimensions at this point. As the vapor starts to condense, an atom from some far distance moves toward the one at the origin. As they come to their equilibrium interatomic spacing of $4.28 \times 10^{-8}$ cm, not much happens to the inner electron energy levels (i.e., the $1s$, $2s$, and $2p$ levels). For practical purposes these bands of the $1s$, $2s$, and $2p$ electrons at close interatomic distances do not play any significant role in the chemical behavior of sodium. But what about the $3s$-level valence electrons? Here, due to the influence of the electrons and nuclei of adjacent atoms, the allowable lev-

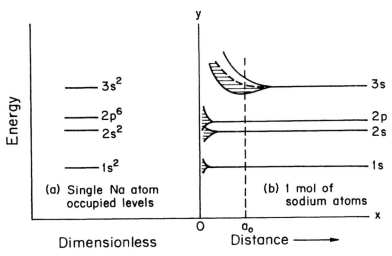

**Figure 10.2.** (a) Electron energy levels in a single sodium atom; (b) spread of the energy levels for the valence electrons when many atoms are brought together to form a solid crystal.

els spread into a tremendously large number of allowed levels, and they also do this at the normal interatomic distance denoted in Figure 10.2b by $a_0$. Usually, $a_0$ denotes the interatomic distance at 0 K, but it does not change much between 0 and 300 K (near room temperature). Problem 10.1 asks you to compute this small change. Thus the valence electrons exist in a band of allowable energy levels. The energy levels are discrete but are so close together that their spacing cannot be measured. We refer to this spacing as a "continuum of levels." The valence electrons of all the other crystalline elements that are solids in the vicinity of room temperature have similar energy-level bands. But concentrating on sodium for the moment, remember that each atom has one valence electron. *It can be shown that when a total of N atoms are brought together to form a crystal, the number of allowable energy states in any valence band is equal to the total number of states for the corresponding level in the N atoms.* This means that if we are considering the $s$ valence band, which is the present case for sodium, each of the $N$ atoms in the crystal contributes one energy level to the crystal. If we are dealing with an element whose valence band was a $p$ level, each of the $N$ atoms in the crystal would contribute three allowable energy levels to the valence band. If we bring 1 mol of sodium atoms together to form a crystal, there will exist $6 \times 10^{23}$ allowable $3s$ energy levels within this crystal. These allowable levels are not associated with any one atom but with the crystal as a whole. Since sodium is univalent, there are $6 \times 10^{23}$ electrons available to occupy these $6 \times 10^{23}$ allowable levels. According to the Pauli exclusion principle, each state can hold two electrons but of opposite spin. Accordingly, the 3s band will only be half full of electrons. Also, the lowest energy levels will be filled first. Thus only one half of the levels are filled. This situation is depicted in Figure 10.2b by the shaded band. This band represents the filled levels and we have elected for simplicity and also to comply with convention to represent the empty allowable levels by the absence of shading, the latter not being shown in the empty section of the 3s band within the crystal, which is called the conduction band. If an electric potential is applied to the conducting crystal, an electric field is set up within it. As a result, the valence electrons of the crystal, which are not associated with any one atom but are free to roam throughout the crystal, will be accelerated. By virtue of this acceleration, they gain energy. Fortunately, empty energy levels are available in the conduction band into which these electrons can move and occupy. This gives us an unambiguous way to define a metal. *A material in crystalline form that contains empty electron energy levels immediately adjacent to the filled levels will conduct electricity when an electric potential is applied. Such conductors of electricity will be called metals.*

It is of little or no significance to the present explanation of the valence bands, but for sake of completeness the reader should recognize that even the innermost energy levels expand into a band or zone. But the atom spacing at which this happens is very small, being only a fraction of the equilibrium interatomic distance at room temperature. Extremely high pressures would be required to force atoms this close together. The repulsive energy of the inner electron shells plus that of their positive nuclei prohibit such spacing. To my knowledge no one has pressed atoms this close together. These inner-level bands, which we show in Figure 10.2b for the 1s level electrons, is more of a hypothesized band, or possibly a calculated band, and then with some very liberal assumptions involved in the calculations.

*Example 10.1.* Copper has an atomic number of 29. According to the definition above, are we justified in calling copper a metal?

Solution. The electron configuration of copper is $1s^2 2s^2 2p^6 3s^2 3p^6 3d^{10} 4s^1$. All levels are filled with electrons except the 4s level. The number of allowable energy levels for the 4s valence band is equal to the number of atoms brought together. Since each atom has one valence electron, $N/2$ levels will be occupied. This analysis applies to polycrystalline solids as well as single crystals. Although the atoms in the grain boundaries are not at the same equilibrium state as are the other properly positioned atoms, their number is sufficiently small that any alteration in the conductivity by virtue of this disorder is negligible.

The 4$s$ band can then be constructed as follows

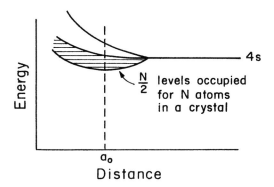

$\frac{N}{2}$ levels occupied for N atoms in a crystal

Thus copper fits our description of a metal.

Now that we see that univalent elements are metals (i.e., by our definition according to the band theory), let's examine a divalent metal. Magnesium, atomic number 12, is an appropriate divalent metal for this example. Its electron configuration is $1s^22s^22p^63s^2$. When we bring 1 mol of magnesium atoms together to form a crystal, again the inner levels play no particular role. The 3$s$ level forms a continuum of levels at the equilibrium interatomic distance as depicted in Figure 10.3a. Just as in the case for copper and sodium, there are $6 \times 10^{23}$ allowable energy levels in the valence band of this crystal. But each magnesium atom within this crystal contributes two valence electrons. Now we have $N$ levels and $2N$ electrons available. These electrons will completely fill the valence band. Thus magnesium should not be able to conduct electricity and therefore should act as an insulator. But our experience has taught us otherwise. The answer, as shown in Figure 10.3a, is that the empty 3$p$ level, which also forms a band of allowable but not necessarily filled energy levels at the equilibrium interatomic distance, $a_0$, overlaps the 3$s$ band. In such a situation there exists allowable empty levels in the 3$p$ band into which the 3$s$ electrons can move when they gain energy by virtue of the application of an electric field to the material. Without this overlap of the 3$s$ and 3$p$ levels, magnesium would not be a metal. The crystal structure and the atom spacings dictate whether these bands will overlap.

Exploring metals further, let's examine the band structure of aluminum (Figure 10.3b), which we know from experience to be a good conductor of electricity. Aluminum has an atomic number of 13 and an electron configuration of $1s^22s^22p^63s^23p^1$. Since the 3$s$ level is filled, one might expect aluminum to be univalent, and just as sodium and copper, have a half-filled 3$p$

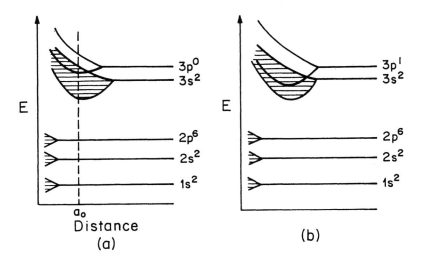

**Figure 10.3.** (a) Valence electron energy levels for a large number (e.g., 1 mol) of
(a) crystalline magnesium atoms; (b) aluminum atoms.

band. Such is not the case however. Aluminum behaves as a 3-valency metal.
Both of the 3s electrons and the one 3p electron become involved in bonding
and also in electron conduction. This is analogous to the $sp$ hybrid bonding
orbitals that we encountered in carbon and carbon compounds in Chapter 3. It
could be called an $sp^2$ hybrid orbital, but it may be simpler to view this
phenomenon as that of overlapping bands, just as we did for the metal magnesium.

There are many other elements that are metals because of the overlapping of
an empty band with a filled band. This overlapping situation fulfills our
definition of a metal in that there exist allowable empty or unoccupied energy
levels immediately adjacent to the filled levels of the outermost valence band.
All of the transition metals possess this type of band overlap as well as that of
other divalent metals. So we really have four different groupings of metals, the
univalent with partially filled bands, the divalent with overlapping bands, the
transition metals in the middle of the periodic chart also with overlapping
bands, and finally, the trivalent metals, such as Al, Ga, and In. But not all
trivalent elements are conductors of electrons. Boron is of a $1s^2 2s^2 2p^1$ electron
configuration and is classified as a semiconductor on our scale of Figure 2.22,
having a conductivity of about $10^{-4}$ $(\Omega \cdot m)^{-1}$ at room temperature. It does
not possess empty levels adjacent to filled levels characteristic of some of the
other trivalent atoms.

Actually, one does not need the band theory to explain the conductivity in metals. The older free electron theory that described the metallic bond as being one of positive ions immersed in a sea or cloud of free valence electrons could account for the high metallic electron conductivity in a qualitative fashion (see Chapter 3). These valence electrons roamed freely throughout the metal crystals and were not attached to any one atom. The application of a potential would cause these free electrons to flow toward the positive potential. The band theory is needed to explain semiconducting and insulating behavior, however. It also allows for a more quantitative explanation of the difference between the relatively low conductivity of some metals compared to the high conductivity of the univalent metals. The univalent metals have the highest conductivity of the metallic group because they have the highest number of carriers. The conductivity is a function of the number of carriers, in this case the electrons, and their mobility. We will return to the mobility later. For now we can explain the high conductivity of the univalent metals (Au, Cu, Ag) based on the number of carriers. All of the electrons in the valence band are available for conduction since there exists an equal number of empty levels into which they can move. In the case of the divalent metals such as Mg, Ca, and Zn, only a fraction of the electrons spill from the filled band into the overlapping empty band; hence a smaller number of carriers are available for conduction.

After the univalent metals, aluminum is the next-best conductor. It is trivalent and has an approximately half-filled $3p$ band, due to the number of electrons available and to the extent of the overlap of bands. In general, those metals that conduct electrons via overlapping bands have a smaller number of carriers available for conductivity than do the metals that conduct electrons due to partially filled bands. The relative conductivity of some selected metals corrected for temperature is given in Table 10.1. We will return to the other factors affecting the conductivity of metals after we have examined the band structure of semiconductors and insulators.

The energy of the electron in the highest filled state at 0 K is called the *Fermi energy*, $E_f$. This is not of any great significance in metals, but we will find that we will need an expression for the Fermi energy for semiconductors. The Fermi energy is the energy at which half of the possible energy levels within a band are occupied. When the temperature is increased, some of the electrons in a metal have energies slightly higher than $E_f$. The Fermi distribution gives the probability that a certain level is occupied by an electron and is given by

$$f(E) = \frac{1}{1 + \exp[(E - E_f)/kT]} \tag{10.7}$$

where $k$ is the Boltzmann constant ($8.63 \times 10^{-5}$ eV/K) and $T$ is the temperature in Kelvin. The value of $kT$ at room temperature is about 1/40 of an electron volt.

**Table 10.1**  Relative Conductivity $[\Omega \cdot cm)^{-1} \times 10^{-4}]$ of Monovalent, Divalent, and Transition Metals, Corrected for Temperature

| Monovalent | | Divalent | | Transition | |
|---|---|---|---|---|---|
| Li | 12.9 | Be | 2.0 | Ti | 0.21 |
| Na | 24.0 | Mg | 8.1 | Cr | 0.51 |
| K | 15.3 | Ca | 11.1 | Fe | 1.14 |
| Cu | 9.1 | Ba | 1.0 | Co | 1.7 |
| Ag | 12.4 | Zn | 6.1 | Ni | 1.9 |
| Au | 8.1 | Cd | 4.5 | | |
| Avg. | 13.6 | Avg. | 5.46 | Avg. | 1.09 |

Most semiconductors, both elemental and compound, have a valency of 4, crystallize in the diamond structure, and have an energy band structure like that depicted in Figure 10.4. Silicon has an atomic number of 14 and the electron configuration of $1s^2 2s^2 2p^6 3s^2 3p^2$. When silicon atoms are brought together to form the diamond cubic crystal, the $s$ and $p$ states spread into a band such that at the equilibrium interatomic spacing $a_0$ there exists an energy gap $E_g$ lying between the filled valence band and the empty conduction band. These crystals are covalently tetrahedrally bonded by $sp^3$ hybrid orbitals (the same as that for carbon in diamond form described in Section 3.2.1 and Figure 3.9). When a total of $N$ atoms are brought together, the number of allowable

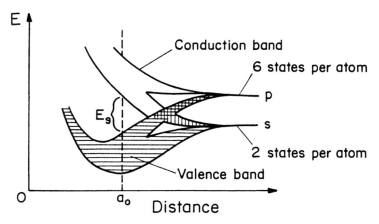

**Figure 10.4.**   Energy-level band structure for semiconductors and insulators.

energy states in any band is equal to the total number of sites for the corresponding level in $N$ atoms. Thus in the $3p$ atomic level of silicon there are three states per atom, or a total of $3N$ states in the solid for the $p$ level. Similarly, for the $3s$ band there are $N$ states in the solid. Silicon has a valency of 4, one electron from the $3s$ level being excited to the $3p$ state to form the $sp^3$ hybrid orbitals. Now we have $4N$ electrons available. Each of the hybrid bands can contain $4N$ electrons, and since each level can contain two electrons of opposite spin, all $4N$ electrons go into the lower valence band, leaving the upper conduction band empty but not adjacent to the filled band at the equilibrium atomic spacing $a_0$.

Generally, when we show the band structure for a crystalline material we just take a small slice near $a_0$ from Figure 10.4 and sketch it in a magnified fashion, as shown in Figure 10.5. In this graph the ordinate represents energy. The abscissa essentially has no meaning since we are using only an incremental slice of distance near $a_0$, although we expand this distance in the sketch for purposes of clarity. The filled valence band is represented by the shaded area, while the conduction band is shown as being empty, similar to the procedure for metals. But there is one major difference between semiconductors and metals. In the semiconductors there exists a region between the filled and conduction bands which we call the forbidden region or energy gap $E_g$. No electron within the crystal is permitted to possess energy that falls within this region. An electron wave of this energy destructively interferes with itself. Carbon, silicon, germanium, tin, and certain compounds have such forbidden regions. These materials where the energy gap $E_g$ value is on the order of 1 eV will have a slight conductivity at room temperature. The thermal energy is sufficient at room temperature to cause a few electrons to be excited to the conduction band (i.e., they jump across the forbidden region). Silicon has an $E_g$ value of 1.1 eV and germanium of 0.67 eV at 20°C. Both of these elements are

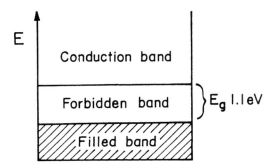

**Figure 10.5.** Energy band structure for silicon at the equilibrium interatomic spacing $a_0$.

semiconductors. On the other hand, carbon, in diamond form, has a large forbidden region with an $E_g$ value of about 5.4 eV at 25°C. The thermal excitement of the electrons at room temperature is not sufficient to cause a measurable quantity of electrons to jump this gap. Therefore, diamond is an insulator. The only difference, then, between a semiconductor and an insulator is the size of the energy gap. The $E_g$ division between semiconductors and insulators is about 4 eV.

## 10.3 FACTORS THAT AFFECT THE CONDUCTIVITY OF METALS

Following equation (10.4), the conductivity of metals is dependent on the number of carriers and their mobility. We cannot alter the charge $q$ of the electron. The band theory explained the number of carriers and their variation among the different metals. The factors that affect mobility are those that cause scattering of the electrons. We mentioned the scattering caused by collisions of the electrons with the atoms. Theoretically, at the absolute zero of temperature (0 K), the atoms are motionless, and in the absence of any defects the metal would have zero electrical resistivity. Copper of commercial purity has a resistivity of $10^{-12}$ $\Omega \cdot$ m at 4.2 K, and this resistivity is due primarily to impurity atoms. According to quantum mechanics, there is no scattering of electrons in a perfect crystal at 0 K. As the temperature increases the atoms vibrate about their equilibrium positions (these equilibrium distances increase with temperature via thermal expansion), causing an interaction of the electrons with the atoms, thereby giving rise to an increase in electrical resistivity as depicted in Figure 10.6. Also shown in this figure is the curve of $\rho$ versus $T$ for a real metal containing defects.

The electrical resistivity of metals can be divided into three parts as follows:

$$\rho_t = \rho_{\text{th}} + \rho_d + \rho_i \tag{10.8}$$

where $\rho_t$ = total resistivity

$\rho_{\text{th}}$ = thermal atom vibration contribution

$\rho_d$ = resistivity due to lattice defects, primarily vacancies, dislocations, and grain boundaries

$\rho_i$ = resistivity due to impurity atoms

Basically, anything that alters the perfect periodicity of the atomic lattice structure will cause some electron–lattice interaction to occur and thus a contribution to the resistance. At temperatures near absolute zero the $\rho_{\text{th}}$ varies approximately as $T^5$, but at higher temperatures it varies linearly with temperature. The $\rho_d + \rho_i$ contribution is independent of temperature. A measurement of the resistance or resistivity near 0 K is therefore a good indication of the $(\rho_d + \rho_i)$ contribution. Usually, this measurement is done in liquid helium at 4.2 K.

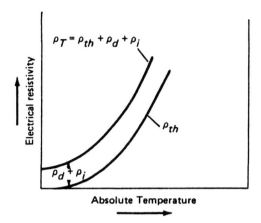

**Figure 10.6.** Resistivity versus temperature for an idealized perfect metal compared to one containing crystalline defects such as vacancies, dislocations, and impurity atoms. (From ASM *Metals Handbook,* Desk Edition, Article by G. T. Murray, 1985, p. 14.2.)

Since $\rho_i \gg \rho_d$ this measurement can be used as a qualitative or relative value of metal purity. Such measurements are frequently done for metals used in microcircuits since purity is very important. Most often one simply measures the resistance at 4.2 K and again at room temperature, the ratio being reported as an indicator of the degree of purity. For ultrahigh-purity aluminum prepared by zone refinement, this resistance ratio is about 40,000.

The linear portion of the $\rho$ versus $T$ curve is used to obtain the temperature coefficient of resistivity or resistance (TCR). The resistivity $\rho_T$ at temperature $T$ can be expressed as

$$\rho_T = \rho_{\text{ref}} + \frac{\Delta\rho}{\Delta T(\rho_{\text{ref}})}(T - T_{\text{ref}}) \tag{10.9}$$

where $\rho_{\text{ref}}$ = resistivity at some reference temperature, usually 0°C (sometimes 20°C is used, which is near room temperature). The $\Delta\rho/\Delta T(\rho_{\text{ref}})$ number is called the temperature coefficient of resistivity. Note that it is not simply the slope of the straight-line segment in Figure 10.6, but the slope divided by $\rho_{\text{ref}}$. Also note that the same reference temperature must be used throughout equation (10.9). The *Handbook of Chemistry and Physics* defines the temperature coefficient as the ratio of the change in resistance or resistivity per °C (or K) to its resistance at 0°C:

$$\text{TCR} = \frac{\Delta R}{\Delta T R_{0\,\text{K}}}$$

Since $\rho = RA/l$, we can substitute this term throughout equation (10.9) and arrive at

$$R_T = \rho_{ref} + \frac{\Delta R}{\Delta T R_{ref}}(T - T_{ref})  \tag{10.10}$$

It should become apparent now that the temperature coefficient of resistivity is the same numerical value as the temperature coefficient of resistance, because the specimen dimensions cancel out in equation (10.10).

*Example 10.2.* If the resistance of a metal bar of 0.1 cm diameter by 46.5 m length is 10 $\Omega$ at 0°C and 20 $\Omega$ at 250°C, what is its temperature coefficient of resistance? What metal might this be?

Solution

$$\text{TCR} = \frac{\Delta R}{\Delta T R_{ref}} = \frac{10}{240(10)} = 0.004 \text{ per } 0 \text{ K}$$

To find out what this metal is, let's compute its resistivity and compare it to handbook values.

$$\rho = \frac{RA}{l} = \frac{10 \ \Omega \times 7.85 \times 10^{-3} \ cm^2}{4650 \ cm} = 1.694 \times 10^{-6} \ \Omega \cdot cm$$

$$= 1.694 \ \mu\Omega \cdot cm$$

If we check *Smithells' Metals Reference Book,* page 14-2, we find that this is the resistivity of copper. Many handbooks list resistivity as $\mu\Omega \cdot cm$ because it gives approximately one- or two-digit numbers in these units. Note that copper is listed as having a TCR value of 0.0043 $K^{-1}$. Many of the univalent metals have TCR values around 0.004, which means that they double in resistance for a 250°C increase in temperature. The effect of temperature on the resistivity of a select group of metals is shown in Figure 10.7 and of impurities on the resistance of copper in Figure 10.8.

## 10.4   SEMICONDUCTORS

### 10.4.1   Intrinsic Semiconductivity

The electron conductivity in a pure nondoped semiconducting element or compound is referred to as *intrinsic* conductivity; that is, it is a function of this substance and temperature only. From the Brillouin zone theory, which we did not develop herein, it can be shown that the Fermi energy $E_F$ for semiconductors lies in the middle of the forbidden region. Now the energy level at the bottom of the conduction band becomes

$$E = E_F + \frac{E_g}{2}  \tag{10.11}$$

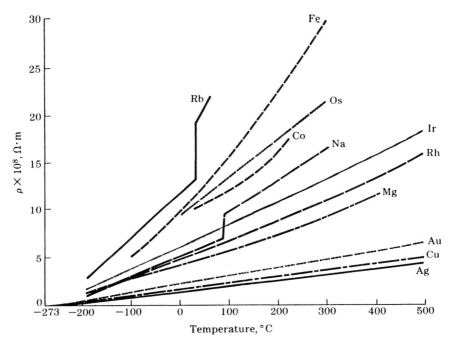

**Figure 10.7.** Effect of temperature on the electrical resistivity of selected metals. (Adapted from C. Zwikker, *Physical Properties of Solid Materials*, Pergamon Press, London, 1954, pp. 247, 249.)

Accordingly, equation (10.7), which was developed for metals, can be revised for semiconductors to read

$$f(E) = \frac{1}{1 + \exp\left[(E_F + E_g/2 - E_F)/kT\right]} = \frac{1}{1 + \exp E_g/2kT} \qquad (10.12)$$

Using this function, it can be shown that the number of electrons $n$ that can jump the forbidden region gap is

$$n = n_0 \exp\left[-\frac{E_g}{2kt}\right] \qquad (10.13)$$

where $n_0$ is a constant that is a function of the effective mass of the electron and of temperature. (In essence it is related to the density of states in the conduction band.) When an electron jumps to the conduction band, a deficiency of negative charge is left behind in the valence band, as illustrated in Figure 10.9. This acts as a positive charge and is often referred to as a hole. It moves in an electric field in the direction opposite to that of the electron. (This is somewhat

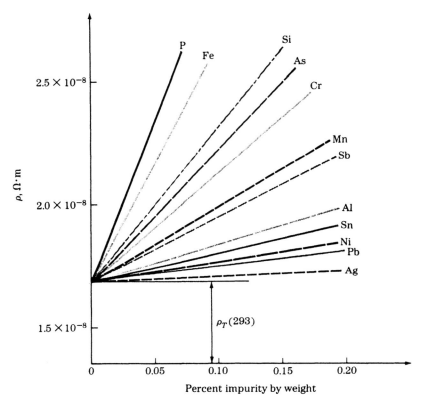

**Figure 10.8.**  Effect of small amounts of impurity atoms on the electrical resistivity of metals. (From F. Pawlek and K. Reichel, *Z. Metallkd.*, Vol. 47, 1956, p. 347.)

analogous to the substitutional diffusion mechanism, where a vacancy moves in a direction opposite to that of the diffusing atoms.) Now the conductivity becomes

$$\sigma = n_e q_e m_e + n_h q_h m_h \qquad (10.14)$$

where the subscripts, $e$ and $h$ represent electrons and holes, respectively. Now since $n_e = n_h$ and $q_e = q_h$, we can write

$$\sigma = nq(\mu_e + \mu_h) = n_0 q(\mu_e + \mu_h) \exp\left[-\frac{E_q}{2kt}\right] \qquad (10.15)$$

Table 10.2 lists the values of $E_g$ and the mobilities of electrons and holes for a number of materials. Note that the mobility of the electrons is larger than that

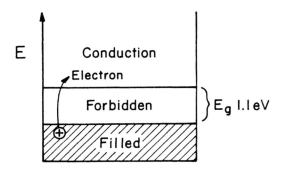

**Figure 10.9.** Electron energy band structure for an intrinsic semiconductor such as silicon. At room temperature there is sufficient thermal energy to excite a small fraction of the electrons from the filled valence band to the conduction band, leaving behind a positive charge (hole) in the valence band.

of the holes in each substance. Also note that Sn is listed. Theoretically, Sn is a semiconductor, but the energy gap is so small that it essentially becomes a conductor at room temperature (Problem 10.3). The room-temperature resistivity of Sn is 6.4 times that of Cu.

*Example 10.3.* Estimate the probability that an electron in Si at 27°C can have sufficient energy to enter the conduction band.

**Table 10.2** Values of the Band-Gap Energy and Carrier Mobility for Selected Semiconductors [a]

| Material | $E_g$ (eV) | $\mu_e(m^2/V \cdot s)$ | $\mu_h(m^2/V \cdot s)$ |
|---|---|---|---|
| C | 5.3 | 0.18 | 0.16 |
| Si | 1.1 | 0.14 | 0.48 |
| Ge | 0.7 | 0.39 | 0.19 |
| GaAs | 1.4 | 0.85 | 0.40 |
| GaP | 2.3 | 0.11 | 0.007 |
| InSb | 0.2 | 8.0 | 0.075 |
| CdTe | 1.5 | 0.03 | 0.006 |
| ZnS | 3.54 | 0.02 | 0.005 |
| CdS | 2.4 | 0.03 | — |
| ZnTe | 2.26 | 0.03 | 0.01 |

[a] Mobility values vary considerably from source to source and must obviously be very structure sensitive and thus a function of processing history.

Solution

$$f(E) = \frac{1}{1 + \exp\left[(E - E_F)/kT\right]}$$

$$E = E_F + \tfrac{1}{2}E_g = E_F + \tfrac{1}{2}(1.1) = E_F + 0.55$$

$$E - E_F = 0.55$$

$$f(E) = \frac{1}{1 + \exp\left(0.55\ \text{eV}/0.025\ \text{eV}\right)} = \frac{1}{1 + \exp\left(22\right)} = 2.5 \times 10^{-10}$$

Even though the probability in Example 10.3 is a small number, there are a sufficient number of electrons that enter the conductive band to make the conductivity measurable. In solving problems in semiconductor technology one must consider the number of carriers per unit volume. In that sense the units of $(\Omega \cdot m)^{-1}$ for conductivity should be well understood. In dimensional terms it can be expressed as follows:

$$\sigma = \frac{\text{carriers}}{m^3} \times \frac{\text{coulombs}}{\text{carrier}} \times \frac{A \cdot s}{C} \times \frac{m^2}{V \cdot s} = (\Omega \cdot m)^{-1}$$

The two middle terms represent the $q$ in the expression

$$\sigma = nq\mu$$

## 10.4.2 Extrinsic Semiconductors

The intrinsic semiconductors have a very small conductivity, which is measurable but has found little practical application other than as temperature sensors. The conductivity is very sensitive to temperature changes, and with a sensitive meter (e.g., an electrometer or picoammeter) it is a very good method for making temperature measurements. But to fully exploit the semiconducting properties of these materials, a higher conductivity is needed. This conductivity is brought about by the introduction of minute quantities of foreign elements called dopants. Depending on our choice of dopant, we can make the charge carrier become either the negative electron or the positive hole. These extrinsic semiconductors are referred to as $N$ and $P$ type, respectively.

N-Type Semiconductors

If we add elements of a valency of 5 (e.g., P, As, Sb, and Bi) to silicon or germanium, we introduce an extra electron for each atom added. This electron can be viewed as lying outside the covalent bond (Figure 10.10) and is somewhat free to move under the influence of an electric field. Each 5-valency dopant atom added introduces an extra electron, so that the total number of charge carriers is almost temperature independent but instead is dependent on the concen-

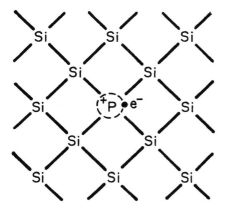

**Figure 10.10.** Extra electron bound to a 5-valence donor impurity which is added to the silicon crystal.

tration of the impurity element. Such a semiconductor is an extrinsic or impurity semiconductor. Note that the dopant is a substitutional impurity atom.

Instead of the physical picture depicted in Figure 10.10, we usually view extrinsic semiconductors in terms of the energy band structure sketched in Figure 10.11 for an *N*-type extrinsic semiconductor. The dopant charge is *screened* out by the silicon ions, which redistribute their electron states slightly in response to the local positive charge. As a result of this screening action, the extra electron is bound rather loosely to its ion. This binding energy is referred to as the ionization energy of the dopant (i.e., the energy required to remove the extra electron from the impurity atom). For phosphorus in germanium, this

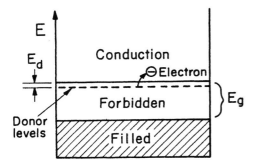

**Figure 10.11.** Energy band structure for an *N*-type extrinsic semiconductor showing the donor impurity energy levels.

ionization energy is 0.012 eV. As a result, this electron energy state resides in the forbidden region of the semiconductor energy band structure very close to the conduction band. This ionization energy is represented in Figure 10.11 as $E_d$, the donor energy-level gap, since it can, by thermal energy at room temperature, donate an electron to the conduction band. For arsenic in silicon $E_d$ is only 0.05 eV. Again, this electron energy state lies in the forbidden region of the semiconductor, very close to the conduction band. For all donor dopants $E_d$ is very small, 10% or less of the $E_g$ value in an intrinsic semiconductor. Now many electrons can jump the gap from the donor levels to the conduction band and we can write

$$\sigma \approx n_d q \mu_e \approx n_{0d} \exp \left( -\frac{E_d}{kT} \right) q \mu_e \tag{10.16}$$

Note that we do not have an equivalent hole or positive charge for each electron that enters the conduction band as was the case for the intrinsic semiconductors. Furthermore, the entire contribution of the intrinsic electrons and holes is insignificant at room temperature compared to the extrinsic effect, when $N_d \gg N$-intrinsic. Also, in the exponential term we have $kT$ instead of $2kT$ since for the intrinsic case the factor 2 arose because the Fermi level $E_F$ occurred in the middle of the forbidden gap. Furthermore, the ionization energy in the exponential term is the small $E_d$ value rather than the larger $E_g$ value of the intrinsic semiconductor.

Now let's examine the plot of conductivity for the $N$-type extrinsic semiconductor as a function of temperature, as shown in Figure 10.12 for germanium. At the lower temperature just a slight increase in temperature will result in many donor electrons moving into the conduction band. But at a certain temperature and a fixed concentration of dopant atoms, all of the impurity electrons will have entered the conduction band. This is portrayed in Figure 10.12 as the *exhaustion* region. There may be a slight contribution from the intrinsic electrons at this temperature, but to a large extent the conductivity will be pretty constant until higher temperatures are reached, where the intrinsic contribution becomes significant and begins to increase measurably with temperature. Usually, one controls the dopant concentration so that the exhaustion region of the $\sigma$ versus $T$ plot will be in the vicinity of room temperature, because this is the temperature where most semiconducting devices operate, and a conductivity independent of temperature fluctuations is desirable. The conductivity curve does not follow exactly the number of electrons curve because of the mobility factor. The decrease in conductivity in the exhaustion region is associated with the decrease in mobility as the temperature increases.

*Example 10.4.* What is the conductivity of silicon in the exhaustion region when 10 parts per million (ppm) of arsenic is added to the silicon? The lattice parameter of the diamond cubic silicon is $5.4307 \times 10^{-8}$ cm.

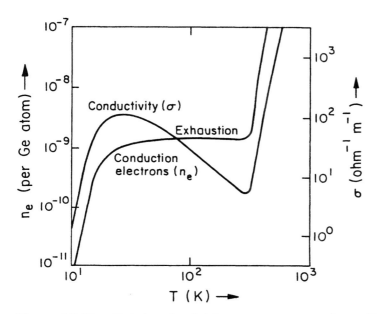

**Figure 10.12.** Variation of resistivity versus temperature for an *N*-type semiconductor. (Adapted from E. M. Conwell, *Proc. IRE,* Vol. 40, 1952, p. 1327; © 1952 IRE (IEEE).

Solution

$$\frac{\text{number carriers}}{\text{cm}^3} = N_d = \frac{1(\text{electron/As atom})(10^{-5}\ \text{As atoms/Si atom})(8\text{Si atoms/unit cell})}{(5.4307 \times 10^{-8})^3\ \text{volume of unit cell}}$$

$$N_d = \frac{8 \times 10^{-5}}{160 \times 10^{-24}\ \text{cm}^3} = 5 \times 10^{17}\ \frac{\text{electrons}}{\text{cm}^3}$$

$$\sigma \approx N_d q \mu_e = 5 \times 10^{16}\ \frac{\text{electrons}}{\text{cm}^3} \times 1.6 \times 10^{-19}\ \text{C} \times 1400\ \frac{\text{cm}^2}{\text{V} \cdot \text{s}}$$

$$\sigma \approx 11.2\ (\Omega \cdot \text{cm})^{-1}$$

P-Type Semiconductors

If instead of adding 5-valency dopant atoms we add 3-valency impurities (e.g., B, Al, Ga, and In), we obtain a positive hole bound to an acceptor impurity as illustrated in Figure 10.13. Since one of the bonds to the impurity atom is lacking, the hole acts as a positive charge. This corresponds in the band picture to a localized hole state. If the vacant state migrates away from the impurity atom, the latter finds itself associated with an extra electron density (i.e., it becomes negatively ionized). In the band picture (Figure 10.14) the impurity or acceptor

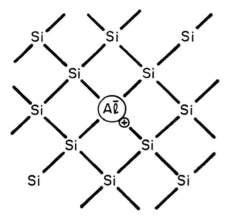

**Figure 10.13.** Positive hole bound to an acceptor impurity in a *P*-type semiconductor.

levels lie very close in energy to the filled valence band and the electrons can jump from the filled band to the acceptor levels by the room-temperature thermal energy of 1/40 eV. When the electron jumps from the filled band to the acceptor level, a positive charge or hole is left behind. The conductivity occurs by hole movement and becomes

$$\sigma = n_h q_h \mu_h \approx n_{0a} \exp\left(-\frac{E_a}{kT}\right) q_h \mu_h \qquad (10.17)$$

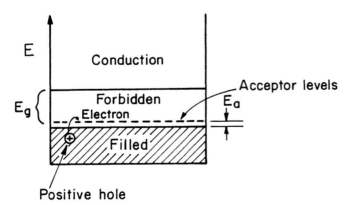

**Figure 10.14.** Energy band structure for a *P*-type semiconductor showing the empty energy acceptor levels near the filled valence band.

where $n_h$ = number of acceptors or holes

$\quad n_{0a}$ = constant

$\quad E_a$ = acceptor energy gap

Such extrinsic semiconductors are called acceptor or *P*-type. If we plot the conductivity versus temperature (Figure 10.15) a curve similar to that depicted in Figure 10.12 is obtained, except now the constant-temperature conducting region is called *saturation* and occurs when all the receptor levels are filled.

### 10.4.3   Silicon Devices and Their Processing

Modern alchemists are doing far better than the legendary ones who tried to turn a variety of metals into gold. Today, sand is turned into a product worth about $8500 per ounce. This process of conversion of sand to devices takes place in what is today called the microelectronic industry. Many of us now own more than 1 million transistors in the form of integrated circuits in computers, cars, appliances, and a multitude of other products. Although separate components, "discrete devices," such as transistors, capacitors, and resistors, were the initial circuit elements, today the semiconductor technologists put thousands of transistors on a small silicon wafer and incorporate diodes, resistors, capacitors, and so on, together with transistors to make integrated circuits. It is predicted that at some time in the 1990s there will be more than 10 million transistors on a silicon chip about 5 mm$^2$, compared to hundreds of thousands today. Hybrid circuits are a combination of integrated circuits and discrete components. The major portion of an electronic system is assembled

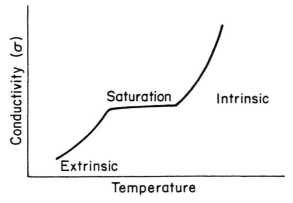

**Figure 10.15.** Variation of conductivity with temperature for a *P*-type semiconductor.

on a printed circuit board (PCB) with a typical mix of components as follows (figures from DM Data Inc.):

| | |
|---|---|
| Integrated circuits | 40% |
| Discrete transistors | 2% |
| Diodes | 5% |
| Resistors | 25% |
| Capacitors | 23% |
| Inductors | 2% |
| Other | 3% |
| Total | 100% |

A personal computer has several circuit boards.

With all of the different ways of manufacturing devices, and the multitude of steps involved in processing, it is impossible in a short section such as this to do more than just skim the surface of the subject. All integrated circuits are designed and processed using either a metal-oxide semiconductor (MOS) or a bipolar configuration, with the trend being toward MOS. As shown in Figure 10.16, either Ge, Si, or GaAs wafers can be used for integrated circuit fabrication, with Si being used in all but a few devices.

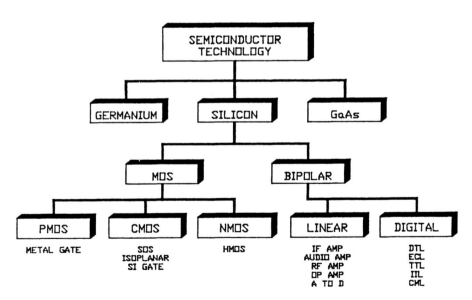

**Figure 10.16.** Semiconductor technology *tree.* Silicon devices account for well over 90% of the device market. (From *The Semiconductor Picture Dictionary,* DM Data, Inc., 1989, p. 9.)

Sand is chemically processed to produce trichlorosilane, which is subsequently treated with other chemicals and a lot of power to create polycrystalline silicon. Large single crystals (currently up to 8 in. in diameter) are grown by the Czochralski process, where a small silicon seed crystal is made to come into contact with the molten silicon. As the seed is slowly withdrawn while it is simultaneously being rotated, the molten silicon solidifies onto the seed, resulting in a giant crystal (Figure 10.17). This crystal is then sliced into thin wafers, lapped and polished to a mirror-finish surface, and supplied to circuit manufacturers for device fabrication (Figure 10.18).

The fabrication of semiconductor devices involves the creation of a series of *P-N* junctions located and patterned in such a manner as to form both active transistors and passive elements, such as resistor and capacitors. These elements are then interconnected by a metallization process, often involving several layers of metal, which most often are created by the sputtering process, although some vapor deposition of metal layers is still done. In the sputtering operation a target material of the desired layer is connected to a high positive potential in a vacuum chamber. A small quantity of argon gas is then introduced into the chamber where the atoms become ionized, forming a plasma. These ions bombard the target, dislodging atoms of this material which are deposited onto the substrate.

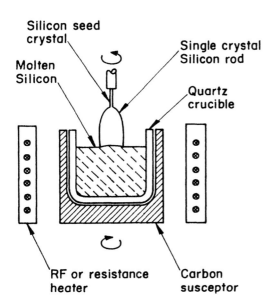

**Figure 10.17.** Silicon crystal growth developed by the Czochralski process.

**Figure 10.18.** Processed silicon wafer. A single chip contains hundreds of thousands of transitors of an area of about 5 mm square. (From *The Semiconductor Picture Dictionary*, DM Data, Inc., Scottsdale, Ariz., 1989, p. 17.)

The purified silicon is modified in a number of ways with $N$- or $P$-type dopants. The starting wafer is either $N$- or $P$-type silicon into which other dopant elements are diffused to create a junction. Dopants are also injected by ion implantation techniques, which has very rapidly been replacing most diffusion processes. Another key process available is the epitaxial growth of a single-crystal silicon layer onto the single-crystal wafer. This added layer can be doped to any concentration of either $N$- or $P$-type during the epitaxial growth process.

Other processing steps involve passivation, the forming of a thin $SiO_2$ surface layer by thermal oxidation of the wafer, and patterning the device structure by use of photomasks (often glass with a thin metal film) in conjunction with photoresist materials. Individual masks are aligned and placed in contact with the photoresist for the exposure cycle. To obtain high resolution, short-wavelength light or radiation can be used. Electron beams and x-ray equipment have been developed for special applications. Typical photomasking schemes are depicted in Figure 10.19. A set of masks on the order of 10 are necessary to form the interacting patterns for processing each layer of a circuit structure. Just to give an idea of such processing steps, the number of material layers

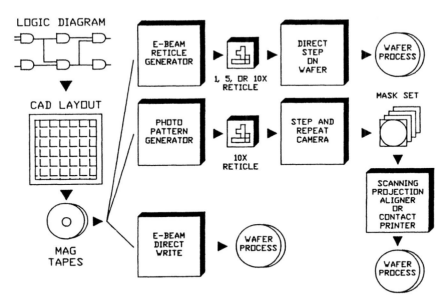

**Figure 10.19.** Typical photomasking approaches. (From *The Semiconductor Picture Dictionary*, DM Data, Inc., Scottsdale, Ariz., 1989, p. 25.)

involved in the construction of a typical CMOS integrated-circuit structure is illustrated in Figure 10.20. All of the masking steps have been omitted here. When the complex series of masking steps and their associated operations, including pattern definition, diffusion, ion implantation, sputter etching, oxidation have been completed, the final device cross section appears as shown later in Figure 10.27.

A photomask is a flat glass plate with either a photo emulsion or thin metal film on one surface. The patterned areas are opaque to ultraviolet light. Direct contact printing may be used. In newer systems, such as projection alignment, a mask is mounted in the system away from the wafer. This increases the life of the mask and reduces mask-induced defects. For electron beam systems, which are more expensive but are becoming more popular, masks are not used. The electron beams "paint" the exposure pattern on the photoresist. A typical photomask and the resulting printed pattern is illustrated in Figure 10.21.

*P-N* Junction

As the name implies, such a junction involves bringing a *P*-type semiconductor into contact with an *N*-type material. In the intrinsic silicon the Fermi level is at the center of the forbidden region. In extrinsic semiconductors this is not so.

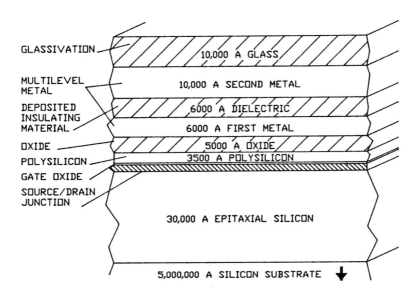

**Figure 10.20.** Material layers involved in the fabrication of a typical CMOS device. (From *The Semiconductor Picture Dictionary*, DM Data, Inc., Scottsdale, Ariz., 1989, p. 36.)

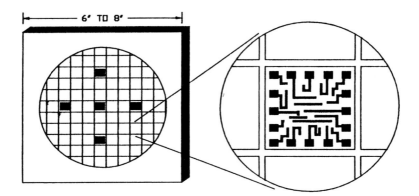

**Figure 10.21.** Example of a pattern formed by photomasking. (Courtesy of DM Data, Inc.)

At temperatures near and below the exhaustion and saturation regions, conduction comes from the impurity atom effects. The Fermi energy level must now lie near the impurity energy levels, since these are the levels from which the electrons are excited. When the *P* and *N* types are joined, the Fermi levels are at different positions; thus the electrons will move to the region of lowest energy in order to equalize the Fermi level. A similar flow of holes occurs by the same analysis. In essence we have electrons from the *N*-type and holes from the *P*-type flowing toward the junction. As soon as equilibrium is attained with a uniform Fermi potential, the flow stops. If we make the junction a part of a circuit by introducing a potential source (Figure 10.22), the electrons on the *N* side are continually removed and the Fermi level is decreased. This application of a potential creates a reverse-bias condition (i.e., the *N*-type material is connected to the positive terminal and the *P*-type to the negative side). The potential source causes a movement of majority carriers away from the junction, forming a depleted region which causes the current to cease flow. If the current is altered so that the positive potential is applied to the *P* side, we have a forward bias where both holes and electrons flow toward the junction and recombine. Electrons are continually drawn from the battery to replace those that are lost by recombination. A new hole is formed when an electron leaves the *P*-type material. Thus we have a rectifier in which current flows in one direction and not the other. This is called a rectifying diode, the same term that was used in the vacuum-tube days. A rectifying diode converts ac current

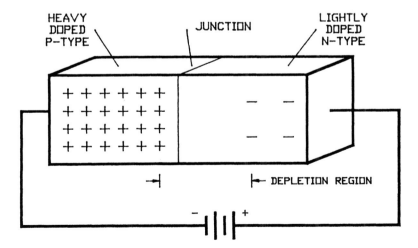

**Figure 10.22.**   Reverse-biased *P-N* junction diode. (Courtesy of DM Data, Inc.)

to a half-wave dc current. The current–voltage characteristics are shown in Figure 10.23.

Transistor

Transistors are devices that amplify electronic waves, or signals as they are sometimes called. Although most students today are not familiar with how the triode vacuum tube accomplished this feat, they are reminded of it here. It was a triode tube (i.e., three electrodes) in which the central electrode, called a grid (or base), accelerated electrons from the cathode (or emitter) to a plate (or collector). A small signal on the grid produced a large signal at the plate. The transistor does this via *N-P-N* junctions as depicted in Figure 10.24. A *P-N-P* junction configuration will accomplish this same feat. The *P* region, called the base in an *N-P-N* transistor, is a very thin section of a semiconductor (just as is the *N*-type layer in a *P-N-P* transistor) that is sandwiched between two larger *N* sections, one being the emitter and the other being the collector. The positive voltage on the collector collects the electrons emitted by the emitter. There must be some current flow into the base, in the form of a small signal, to turn on the major flow between the collector and emitter. The *P* base is thin, so that although some electron–hole recombination occurs, most of the electrons go right through the base to the collector. The ratio of the flow out of the collector to the input current is the gain of the device, and values typically range from 30 to several hundred. A discrete transistor can obtain higher performance (e.g., power, gain, or frequency) than that obtained in integrated-circuit form.

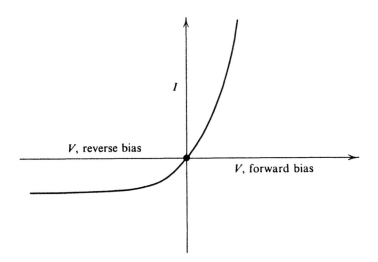

**Figure 10.23.** Current–voltage characteristic of a *P-N* junction.

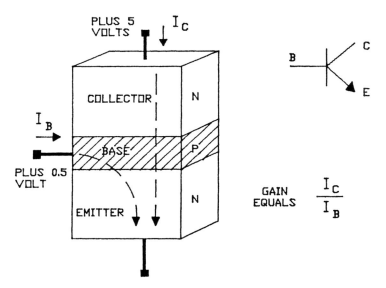

**Figure 10.24.** *N-P-N* bipolar transistor schematic showing how amplification occurs. (Courtesy of DM Data, Inc.)

This junction transistor is known as a bipolar transistor. The basic bipolar integrated-circuit element is the *N-P-N* transistor. A typical bipolar transistor in an actual plan-view picture (above) and in a plan-view schematic layout (center), together with the cross-section construction (below), is shown in Figure 10.25. The basic bipolar manufacturing process involves 33 separate steps requiring a seven-mask processing sequence.

The bipolar transistor was the first transistor developed by W. Shockley and of course was a discrete device. As pointed out earlier, only 2% of the components on a typical printed circuit board are discrete transistors. The metal-oxide semiconductor (MOS) emerged later and has replaced a lot of the bipolar transistor applications. The basic *N*-channel MOS transistor cross section is shown in Figure 10.26. One can also have a PMOS device (i.e., a *P*-channel MOS). When a chip contains both types it is referred to as a complementary metal-oxide semiconductor (CMOS), shown in Figure 10.27. MOS transistors are of the field-effect type, called a metal-oxide semiconductor field-effect transistor (MOSFET). They are used both as power devices and as integrated-circuit MOS devices and function in essentially the same manner in both circuits. To explain their amplification process, we refer to Figure 10.26. Two *N*-type regions are formed in a *P*-type substrate, one being the source and the other the drain. The conductor or gate is the third component. In contrast to the

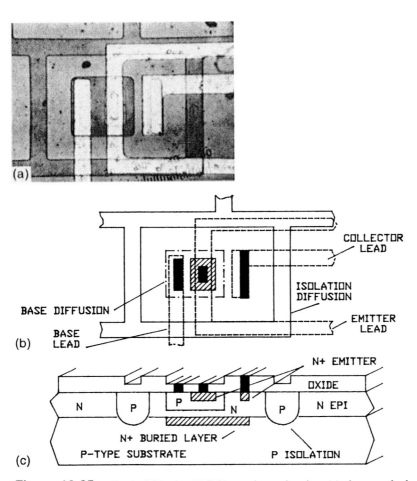

**Figure 10.25.** Typical bipolar *N-P-N* transistor showing (a) the actual plan-view device along with (b) the plan-view schematic and (c) the cross-section schematic. (Courtesy of DM Data, Inc.)

bipolar transistor, no current flows into the MOS device. A potential is applied to the positive gate electrode and the source and draws electrons toward the gate. But the gate is insulated from the source, so the electrons cannot enter it. The concentration of electrons around the gate makes this region more conductive, so that a small positive potential at the gate will produce a large electron flow between the source and the drain, thereby amplifying the input. A negative potential on the gate will drive charge carriers out of the channel and

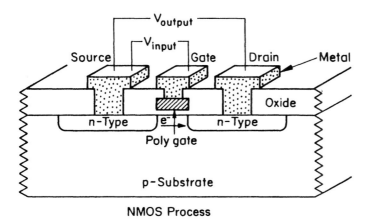

**Figure 10.26.** Cross section of an *N*-channel MOS field-effect transitor. A small input potential on the gate will produce a large output between the source and drain.

reduce the conductivity. In essence a voltage applied to the gate electrode is used to set up an electric field that controls the conduction between the heavily doped *N*-type source and drain regions. By changing the input voltage between the source and gate, a larger output potential is achieved between the source and drain. The gain in MOSFET devices is usually measured in terms of a voltage ratio instead of a current ratio as in the bipolar devices. The MOS devices

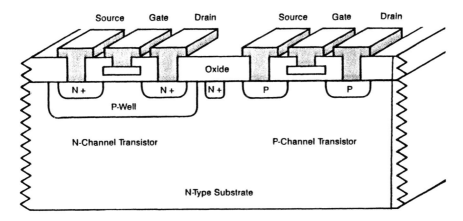

**Figure 10.27.** CMOS structure cross-section schematic. (Courtesy of DM Data, Inc.)

are preferred by computer manufacturers because they are smaller, require less power, and are less expensive to produce.

The MOS process technology is based on an 8- to 10-mask sequence with resolution in the range 1.5 to 2 $\mu$m. The key to this process is the use of a polysilicon gate, which has the advantage of *self-alignment* and an added conductor layer. With the use of polysilicon as a gate material, the poly is grown on the gate oxide, patterned, and then used as part of the masking to define the source and drain areas. However, for high-performance devices, an additional layer of a refractory metal on the polysilicon is used to reduce the resistance and increase the circuit speed.

The student should not be too concerned with all of these processing steps, however. New processes have already been proposed for both bipolar and CMOS devices as they attempt to scale down to the 1-$\mu$m level. By the time this book has been published, changes will have been made. These operations are difficult to teach via a textbook. I encourage engineering faculty to design an interdisciplinary laboratory-oriented course in microprocessing and failure analysis of electronic devices. The undergraduate materials engineer will never have sufficient electronic circuit knowledge and the corresponding electronic probing technique know-how to locate the source of circuit failure. Such failures usually can be traced to materials and materials processing errors. The region of failure on the device is today found by the electronics engineer, who then turns it over to the materials engineer for dissection and detailed materials analysis involving electron microscopic and electron microprobe analysis. Many failures have been traced to metallization voids (some from deposition errors and some from electromigration), corrosion, formation of intermetallic compounds, lack of step coverage in metallization, metallization adherence, electrical overstress, creep in solder joints, and bonding lead failures, to name a few.

### 10.4.4 Compound Semiconductors

Group 3–5 Compounds

When certain elements of 3-valency combine to form a crystalline compound with 5-valence atoms, the resulting crystal structure, net valence, and energy band structures are identical to those of the group 4 elements carbon, germanium, and silicon. The only difference between the pure elements and the compounds, shown in Figure 10.28, is that in the latter each atom has four nearest neighbors of unlike atoms, in contrast to like atoms in carbon, silicon, and germanium. Both structures are of the diamond cubic lattice type. GaAs, GaP, and InSb are such group 3–5 compounds that have been used as semiconductors, with GaAs being by far the most famous of the group. The band-gap energies and carrier mobilities of these compound semiconductors are given in Table 10.2.

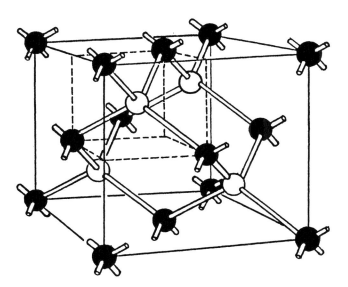

**Figure 10.28.** GaAs and GaP crystal structure. This structure is known as the zinc blende type (ZnS). Except for the unlike atoms it is the same structure as that found in diamond, silicon, and germanium.

GaAs and GaP were prepared in single-crystal form and used as light-emitting diodes (LEDs) in the early 1970s in applications such as the first hand-held calculators and wristwatches. These materials lost out to the liquid crystals for most displays many years ago. There are some applications where LEDs are still in use, however. This light emission occurs due to electron transitions between the conduction and valence bands or between impurity levels in either band and is a result of the recombinations of electrons and holes. The frequency, and hence the color of the light, are a function of the difference in the energy levels involved. Figure 10.29a shows how recombination could occur. This phenomenon is called *P-N* junction injection electroluminescence. By varying dopants and composition of the compounds, a number of different colors of light emission will result (Figure 10.29b). *N*-doped Ga(As,P) can emit green, yellow, amber, or red by varying the Ga–As–P ratios. In GaP green light is emitted from a nitrogen donor to the valence band, and red light from a ZnO impurity level to a valence-band transition. Both GaAs and the ternary AlGaAs are used to fabricate diodes that emit visible light in the range 500 to 700 nm. GaAs has also been used in lasers and photovoltaic cells. But the principal use for GaAs is for integrated-circuit transistors. It will never be a threat to the Si integrated circuits because of its high cost. But GaAs circuits switch at faster rates, pass signals more rapidly, consume less power at high

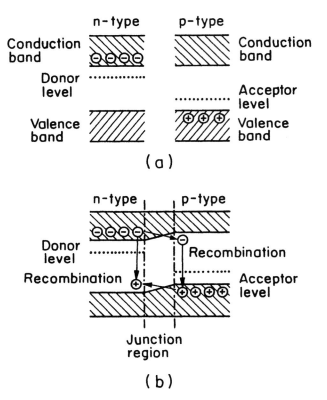

**Figure 10.29.** Energy band and dopant energy levels in a GaAs *P-N* junction injection light-emitting diode.

speeds, and operate over a wider temperature range than do similar Si circuits. GaAS is used in monolithic microwave integrated circuits (MMIC) as diodes or as bipolar or MESFET transistors. Since GaAs does not form an oxide, an MOS per se cannot be made. But metal-semiconductor field-effect transistors (MESFETs) can be and are made.

The base structure of a GaAs field-effect transistor is shown in Figure 10.30. It is an *N*-channel device formed from several layers of doped and undoped GaAs. A refractory metal gate of tungsten silicide modulates a thin channel of lightly doped *N*-type GaAs that connects the source and drain. GaAs has been used as discrete devices in microwave and satellite communication applications where high frequencies are required.

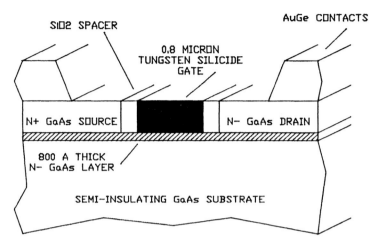

**Figure 10.30.** Cross-section schematic of a GaAs *N*-channel field-effect transistor. (Courtesy of DM Data, Inc.)

Group 2–6 Compounds

When atoms of 2-valency combine with atoms of 6-valency, we again have a net 4-valency. Many of these compounds have the diamond cubic structure shown in Figure 10.28. ZnS (zinc blende) was one of the earliest used compounds of this type, and in fact all of these diamond cubic compounds are referred to as having the zinc blende structures. ZnS has been used as a phosphor on television screens. Electrons bombard the coated screen and excite electrons across the 3.54-eV energy gap of the ZnS. Other group 2–6 semiconducting compounds include CdS, ZnTe, and CdTe.

## 10.5 DIELECTRIC MATERIALS

In Chapter 7 we discussed how a dielectric material, either via induced or spontaneously existing dipoles, can increase the charge on capacitor plates by virtue of its presence (Figure 7.24). The dielectric constant $K'$, which was expressed by equation (7.7) as follows

$$C = \frac{K'\epsilon_0 A}{d} \quad \text{or} \quad K' = \frac{Cd}{\epsilon_0 A} \tag{7.7}$$

is a material constant, but it varies with the frequency of the applied electric field. Dielectric constants are usually presented at 60 Hz, but these constants are also often stated at much higher frequencies. These constants can be found in many handbooks together with the dielectric strength (i.e., the volts/mil of thickness that will break down the dielectric material) and their resistivities. A number of such materials and their properties are listed in Table 10.3. In equation (7.7) the dielectric constant was defined in terms of the capacitance, permittivity, and capacitor plate dimensions. It can also be expressed simply as

$$K' = \frac{\epsilon}{\epsilon_0} \tag{10.18}$$

which is the ratio of the permittivity of the material to the permittivity in vacuum. It is a dimensionless quantity. We used $K'$ so as not to confuse it with the thermal conductivity $K$. The most useful dielectrics are those with high dielectric constants, such as the titanates. In these compounds (e.g., $BaTiO_3$) polarizations exist in the normal crystal since the $Ti^{4+}$ and $O^{2-}$ ions are displaced slightly from their face-centered positions, causing them to be permanently polarized (Figure 10.31a). When an ac alternating field is applied, the titanium ion moves back and forth between its two allowed positions (Figure 10.31b). Some amorphous polymers have sufficient mobility to polarize, although their dielectric constants are low compared to the titanates. The polarization is greater the larger the dielectric constant and can be expressed as $P = (K' - 1)\epsilon_0\xi$ where $\xi$ is the strength of the electric field in volts/mil.

Since dielectrics are used at high frequencies, their dipoles must switch often and thus give rise to a dipole friction energy loss. The more difficult the dipoles are to orient, the greater the loss. There are several types of polarization. Electronic polarization is that where electron rearrangement occurs in the

**Table 10.3**  Properties of Selected Dielectric Materials

| Material | Dielectric constant | | Dielectric strength ($10^6$ V/m) | Loss factor ($10^6$ Hz) | Resistivity ($\Omega \cdot m$) |
|---|---|---|---|---|---|
| | 60 Hz | $10^6$ Hz | | | |
| Polyethylene | 2.3 | 2.3 | 20 | 0.00023 | $10^{16}$ |
| PVC | 3.5 | 3.2 | 40 | 0.15 | $10^{10}$ |
| Soda-lime glass | 6.9 | 6.9 | 10 | 0.063 | $10^{13}$ |
| $BaTiO_3$ | — | 3000 | 12 | — | — |
| $Al_2O_3$ | 9.0 | 9.9 | 9.8 | 0.0004 | $10^9$–$10^{12}$ |
| Nylon | 8 | 4 | 12–24 | 0.14 | $10^{11}$–$10^{13}$ |
| Steadites | 6 | 6 | 8–14 | 0.007–0.025 | $> 10^{12}$ |
| Fosterites | — | 6.2 | 8–12 | 0.001–0.002 | $> 10^{12}$ |

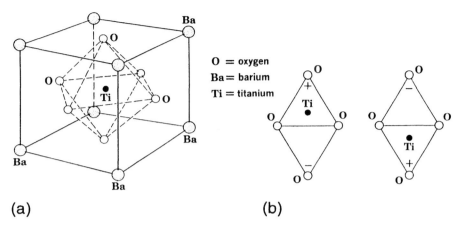

**Figure 10.31.** BaTiO₃ unit cell. The Ti⁺ ion can be in either the (a) Perovskite structure or (b) central structure position. This permanent dipole will switch from one position to another by the application of an electric field. (From C. W. Richards, *Engineering Materials Science*, Brooks/Cole, Monterey, Calif., 1961, p. 453.)

presence of an applied field. The electrons move to the side of the atom nucleus nearest the positive end of the field. Ionic polarization involves an elastic shift of ions in an ionic solid. In molecular polarization molecules or atoms shift in the presence of a field to form strong dipoles. Some materials have permanent molecular dipoles even in the absence of a field. Molecular polarization is the one of interest in our present discussion. Electronic polarization, which involves no atom shift, occurs easily at high frequencies and hence little energy loss. This is not so for molecular polarization. In a perfect dielectric the current will lead the voltage by 90°. However, due to losses, the current leads the voltage by 90° − δ, where δ is the *dielectric loss angle*. The power lost in the form of heat, called the *dissipation factor*, is given by

$$\text{dissipation factor} = \tan \delta \tag{10.19}$$

and the dielectric loss factor is

$$\text{dielectric loss factor} = K' \tan \delta \tag{10.20}$$

The total power loss $P_L$ is a function of the electric field, the frequency, the dielectric constant, and the dissipation factor and is given by

$$P_L = 5.556 \times 10^{-11} \, K' \tan \delta \xi^2 f V \tag{10.21}$$

where the electric field strength $\xi$ is given in V/m, the frequency $f$ in hertz, the volume $V$ in m³, and the loss in watts.

## 10.6 FERROELECTRIC AND PIEZOELECTRIC BEHAVIOR

A few materials have atom arrangements that affords polarization of the ions without the assistance of an external electric field. This characteristic is called *spontaneous polarization* and the materials are called *ferroelectrics*. They have nothing to do with ferrous materials. The ferroelectric name was selected to be analogous to *ferromagnetic* materials, which are materials that are spontaneously magnetized in the absence of an external magnetic field. Obviously, the field of magnetism was studied before the recognition that ferroelectric behavior existed. Magnetism is covered in the next section. The election to place ferroelectric behavior first was simply because polarization had been explained in Section 10.5.

In spontaneous polarization permanent dipoles exist within the crystal. The most widely studied and used material of this type is barium titanate ($BaTiO_3$), mentioned previously. It has the *perovskite* structure and displaced charges as we have shown in Figure 10.31 when we discussed dielectric materials. The position of the titanium ion can be in either of the two positions shown in Figure 10.31b. In both cases the unit cell has a permanent dipole because of the unequal distribution of the electric charge within the cell. The titanium ion can also occupy similar positions in the other principal directions in the cell, for a total six possible polar axes. Barium titanate is used as a dielectric material and also in the role of a ferroelectric material.

Ferroelectric materials form domains, in which the dipoles of adjacent cells are aligned in the same direction, as depicted in Figure 10.32. In $BaTiO_3$, there are six possible directions, but in some ferroelectric materials there may be only two such directions. The domains tend to form because this is a lower-energy state. This concept is explored in more detail in Section 10.7.2. When an external field is applied, the domains rotate toward the same direction, thereby increasing the polarization as depicted in Figure 10.32. As the electric field is increased, the degree of alignment increases until it attains a maximum saturation polarization value, denoted as $P_s$ on Figure 10.32. At this point all of the domains are aligned. If the field is reversed, the domains begin to fall back to their random arrangement, but do not quite attain it when the field is reduced to zero. To obtain random orientation of dipoles, the field must be reversed. If we continue this field reversal, a cycle will be completed which results in the hysteresis loop depicted in Figure 10.32. Ferroelectric domains of various orientations can be observed in barium titanate by chemical etching and viewing the resulting microstructure at a magnification of about $500 \times$. Ferroelectric materials possess a temperature called the *Curie temperature*, again analogous to the magnetic behavior, above which spontaneous polarization does not occur. In $BaTiO_3$ this temperature is 120°C, 233°C in $PbZrO_3$, and 490°C in $PbTiO_3$.

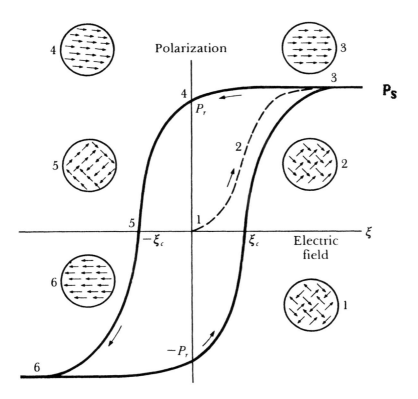

**Figure 10.32.** Ferroelectric hysteresis loop. $P_s$ represents the point where all domains are polarized in the same direction. (From D. R. Askeland, *The Science and Engineering of Materials*, 2nd ed., PWS-Kent, Boston, 1989, p. 696.)

Piezoelectric Effect

*Piezo* is derived from the Greek word *piezen,* which means *to press.* The piezoelectric effect is related to the relationship between the pressure applied to a material and the electric field created thereby. Compressive stresses reduce the dipole distance and accordingly, the dipole moment. This change in the dipole moment changes the charge and the voltage between the ends of the sample. The voltage is proportional to the strain. Conversely, the application of a voltage will cause a dimensional change, with the strain being proportional to applied voltage. Thus electric energy can be changed to mechanical, and vice versa. These actions are called the *piezoelectric effect.* All ferroelectric materials have piezoelectric properties, which has also been referred to as *electrostriction,* again because of a similar term and effect that exists in ferromagnetic materials. This pressure effect on ferroelectrics is illustrated schematically in Figure 10.33. Piezoelectrics are invaluable as electromechanical transducers because their large response makes them more efficient than magnetic transducers. Such devices are used for measuring pressure or strain, sonar, audio devices, ultrasonic vibrators, and radio tuners.

## 10.7 MAGNETIC MATERIALS

A moving charge creates a magnetic field. If this moving charge takes place as a current in a wire, the magnetic field flows around the wire. When the wire is formed into a coil the magnetic lines of flux reinforce each other and thereby increase the field within the coil (Figure 10.34a), which is given by the equation

$$H = \frac{4\pi NI}{10l} \qquad (10.22)$$

where $H$ = magnetic field strength
$\quad N$ = number of turns in the coil
$\quad I$ = current
$\quad l$ = coil length

In what used to be the more common cgs units, $H$ was expressed in oersteds when $I$ was in amperes and $l$ in centimeters. The SI units for $H$ is ampere turns

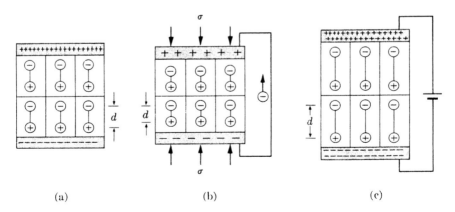

**Figure 10.33.** Piezoelectric effect. Application of pressure changes the net dipole moment and thus the voltage between the plates. Conversely, a voltage application produces dimensional changes. (From L. H. Van Vlack, *Elements of Materials Science and Engineering,* 3rd ed., Addison-Wesley, Reading, Mass., 1975, p. 279.)

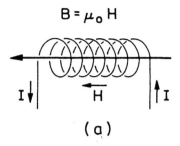

$$B = \mu_o H$$

(a)

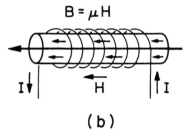

$$B = \mu H$$

(b)

**Figure 10.34.** Current passing through a wire creates a magnetic field around the wire. (a) When the wire is wound into a coil the magnetic flux density is reinforced. (b) When a core material with $\mu > 0$ is placed within the coil the flux is further increased.

per meter, more often expressed as amperes per meter (A/m). The units in magnetism are many in number and often confusing to the reader. A conversion chart and list of units used herein are presented in Table 10.4. When a material is placed within the coil (Figure 10.34b), the magnetic flux is increased if the material offers a lower-resistant path than does the air of Figure 10.34a. The magnetic field strength produces a magnetic flux density $B$ which is proportional to the field $H$, the proportional constant being $\mu$, the permeability of the material. In Equation form the $B$–$H$ relationship becomes

$$B = \mu H \qquad (10.23)$$

When $H$ is given in A/m, $B$ is expressed in webers/m$^2$ (tesla) and $\mu$ as henrys/ m; when $H$ is in oersteds, $B$ is in gauss. Most of us have heard of a gauss meter and perhaps a tesla meter or a tesla coil, but I doubt if we have run across a webers/m$^2$ meter. The gauss was originally thought of as the number of lines of magnetic force penetrating unit area. Perhaps the observance of the lines of iron filings when such filings were placed in a magnetic field had something to

**Table 10.4** Magnetic Property Conversion Units

---
$M$ (magnetization: 1 A/m $= 4\pi \times 10^{-3}$ Oe
$H$ (applied field): same as magnetization
$B$ (magnetic induction): 1 Wb/m$^2$ = 1 T = V-s/m$^2$ = $10^4$ G
$\mu_0$ (permeability in vacuum): $4\pi \times 10^{-7}$ H/m = Wb/m
$\mu_r$ (relative permeability): dimensionless = $\mu/\mu_0$
$\chi = M/H$ = susceptibility (dimensionless)

---

do with the lines of force concept (Figure 10.35). In vacuum, $\mu = \mu_0$. The relative permability $\mu_r$ then becomes dimensionless because

$$\mu_r = \frac{\mu}{\mu_0} \tag{10.24}$$

The permeability is somewhat like the dielectric constant in that the magnetic field is increased by the presence of a material (e.g., as a core material in a coil over that in air or vacuum). (Diamagnetic materials would decrease the field strength.)

Equation (10.23) is similar to the equation often used to express the *magnetization* of a solid, which is written as

$$B = \mu_0(H + M) = \mu_0 H + \mu_0 M \tag{10.25}$$

where $M$ has the same units as $H$. We can also write

$$M = \chi H \tag{10.26}$$

where $\chi$ is called the susceptibility. $M$ can be viewed as being related to the magnetic flux increase when a material of magnetization $M$ is placed in the field, because, except for diamagnetic materials, $\mu_0 H < \mu_0 M$ and for ferromagnetic materials, $\mu_0 H << \mu_0 M$. Equation (10.24) can also be written as

$$B = \mu_0 H(H + M) = \mu_0 \mu_r H \tag{10.27}$$

Therefore,

$$\mu_0 = \frac{H + M}{H} = 1 + \frac{M}{H} = 1 + \chi \tag{10.28}$$

Now that we have become sufficiently confused with units, let's return to the fundamentals of magnetism.

## 10.7.1 Sources of Magnetism in the Atom

An electron revolving in orbit around a nucleus is a moving charge and creates a magnetic field. In most atoms these orbits are randomly oriented and the

**Figure 10.35.** Iron filings reveal the magnetic field around a bar magnet. Note that the magnetic lines of force leave the bar at one pole and return to the other.

fields nearly cancel. But when an external field is applied to a material that derives its magnetic moment from orbital motion, and if the orbits reorient to oppose the field, and if this is the only effect of the field, the material is said to be *diamagnetic*. Cu, Bi, Zn, and the inert gases are diamagnetic. Their permeability is negative and about the order of $\mu \approx 1 - 1.00005$ (diamagnetic). When the orbital moments do not cancel and when they align with an applied field, and if this effect outweighs any diamagnetism in the material, the material is weakly paramagnetic and has a permeability of about $\mu \approx 1.005$ (weak paramagnetism). Many materials are weakly paramagnetic.

Electron spins create a magnetic field, and when there are an unequal number of parallel and antiparallel spins, either a somewhat strong paramagnetic or a *ferromagnetic* material exists. For the elements, these two types of magnetism occur in the transition element groups and result from the noncancellation of electron spins. The somewhat strong (relative to the weak paramagnetism) paramagnetic transition elements have permeabilities of approximately $\mu \approx 1.01$. But in the transition elements that are ferromagnetic, the permeabilities are very large and varied, but for now let's say that they are about $\mu \approx 10^6$ (ferromagnetic). In all of the following discussion of magnetic materials we will be dealing with ferromagnetism only, because these are the materials of commercial interest. It should be mentioned that there exists a nuclear magnetic effect that is small but useful for magnetic resonance applications, most notably in equipment used in the medical field.

## 10.7.2 Ferromagnetism

What is the source of ferromagnetism, and why are only a few of the transition elements ferromagnetic? It has already been mentioned that a necessary but not sufficient condition for ferromagnetism to exist in an atom is that it must have an imbalance of electron spins in the unfilled *d* energy band. This imbalance also occurs in the *f* band of higher-atomic-number elements. These include many of the rare earths. But we deal primarily with the first series of transition elements, which include Sc, of atomic number 21, through Ni of atomic number 28 (i.e., we will be concerned with the unbalanced spins in the 3*d* level of these elements). Ferromagnetism occurs in those transition elements that have an offset between $m_s = +\frac{1}{2}$ and $m_s = -\frac{1}{2}$ energy levels in their 3*d* band. Because of this effect, there arises an *exchange force* between adjacent atoms that causes their magnetic dipole moments to align in the same direction, and this occurs without the application of an external magnetic field (i.e., they possess the characteristic called *spontaneous magnetization*). The theories that explain this exchange force are rather complex and beyond our present consideration, but we do know that it is related to the ratio of the atomic diameter (or atom spacing) to that of the 3*d* shell diameter and we will simply express it as

$$\text{exchange force} = \text{function of} \left[ \frac{\text{atom diameter}}{3d \text{ shell diameter}} \right]$$

When this ratio is greater than 3, the exchange force is positive and causes the alignment of the magnetic moments of adjacent atoms. For the elements of interest here the number of the 3*d* and 4*s* electrons and their respective spins are shown in Figure 10.36. As the atomic number increases, the number of unbalanced spins increase from Sc through Mn. The 3*d* level can hold 10 electrons and becomes filled with Cu, of atomic number 29. But as the 3*d* levels are filled in these transition elements, their spins are in the same direction until we reach Fe, of atomic No. 26, and which has six electrons in the 3*d* level. The maximum number of unbalanced spins is five and is found in the elements chromium and manganese. (Chromium is an anomaly here since it contains only one electron in the 4*s* level.) One would thus expect that chromium and manganese would possess the largest ferromagnetic character. But when we examine the atom diameter/3*d* shell diameter ratio we find that only Fe, Co, and Ni possess a ratio greater than 3 (Figure 10.37). Thus they are the only

| | | 3d Electrons | 4s Electrons |
|---|---|---|---|
| | No. | Spin directions | |
| Sc | 1 | ↑ | 2 |
| Ti | 2 | ↑ ↑ | 2 |
| V | 3 | ↑ ↑ ↑ | 2 |
| Cr | 5 | ↑ ↑ ↑ ↑ ↑ | 1 |
| Mn | 5 | ↑ ↑ ↑ ↑ ↑ | 2 |
| Fe | 6 | ↑ ↑ ↑ ↑ ↑ ↓ | 2 |
| Co | 7 | ↑ ↑ ↑ ↑ ↑ ↓ ↓ | 2 |
| Ni | 8 | ↑ ↑ ↑ ↑ ↑ ↓ ↓ ↓ | 2 |

**Figure 10.36.** Spin directions of the 3d electrons in the first series of transition metals. Note the opposing spins in Fe, Co, and Ni.

elements of this series that possess a positive exchange force and as such are the only ferromagnetic elements of this series.

The spin magnetic dipole moment of a single electron is given by

$$\mu_b = \frac{eh}{4\pi m_e} \tag{10.29}$$

where $e$ = charge on electron
$h$ = Planck's constant
$m_e$ = mass of electron

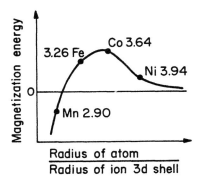

**Figure 10.37.** Magnetization energy as a function of the ratio atom diameter/3d shell diameter.

and is called a Bohr magneton. In SI units it has the value of $9.27 \times 10^{-24}$ A · m$^2$. Again one would expect from Figure 10.36 that Ni, which has three unbalanced spins would have the largest ferromagnetic character of this group of three elements. But such is not the case. In Table 10.5 the net Bohr magnetons are listed as 2.2, 1.7, and 0.6 for Fe, Co, and Ni, respectively. This is simply because not all electrons take on the character with regard to unbalanced spins depicted in Figure 10.36. Instead, we must deal with an average number of unbalanced spins. Thus Fe is the most ferromagnetic of the group. When its magnetization is saturated at 0 K, the net dipole moment is 2.2 Bohr magnetons per atom. This means that on the average 2.2 electron spins per atom are unbalanced. Not all atom moments align as shown in Figure 10.36. The ferromagnetic elements lose their ferromagnetic character as the temperature is increased. Thermal energy tends to break the spontaneous magnetic alignment of the moments of adjacent atoms until at a certain temperature, called the Curie temperature $\theta_c$, the exchange forces are no longer positive and the material become paramagnetic (Figure 10.38). The Curie temperature of some ferromagnetic elements are listed in Table 10.6. Gadolinium is ferromagnetic at temperatures below what we usually consider to be room temperature, 20°C (68°F).

Ferromagnetic Domains

There is still one characteristic of ferromagnetism that must be explained, and that is why certain ferromagnetic materials appear to be magnetic only in the presence of an external field. Actually, they are still ferromagnetic, but the internal fields of ferromagnetic materials can be so arranged that they cancel

**Table 10.5** Net Magnetic Moments per Atom for the Ferromagnetic Metals, Fe, Co, and Ni[a]

|  | Element | | |
|---|---|---|---|
|  | Fe | Co | Ni |
| Total number (s and d) electrons | 8 | 9 | 10 |
| Average number in s band | 0.2 | 0.7 | 0.6 |
| Average number in d band | 7.8 | 8.3 | 9.4 |
| Average number with parallel spins | 5.0 | 5.0 | 5.0 |
| Average number with antiparallel spins | 2.8 | 3.3 | 4.4 |
| Net moment per atom (number of Bohr magnetons) | 2.2 | 1.7 | 0.6 |

[a] These are not whole numbers as predicted by Figure 10.36.

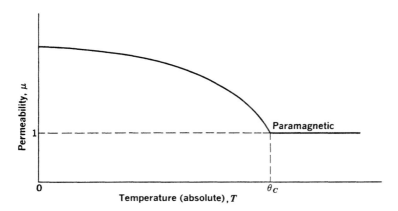

**Figure 10.38.** Decrease of the permeability of iron with increasing temperature. $\theta_c$ is the Curie temperature where the iron ceases to be ferromagnetic.

out, making it appear that the material is not magnetic. This phenomenon occurs because the aligned moments of individual atoms break up into regions called *domains*, where in one region the moments of the atoms are opposed to those of the moments in another region of the material.

Domain formation occurs because the presence of domains lowers the energy of the material. Domains in general are small in size (e.g., $10^{-2}$ to $10^{-5}$ cm across, smaller than most grains). ASTM grain size 7 represents a grain diameter of about 1/8 in. or approximately 0.3 cm. This lowering of the energy, called the *magnetostatic energy,* can be explained with the aid of Figure 10.39. The magnetostatic energy is the potential magnetic energy of a ferromagnetic material that can be produced by an external field. In Figure 10.39a, the moments of all the atoms are aligned in a single-crystal bar. However, this is not a desirable situation because the return magnetic flux path from the N back to the S pole is outside the bar. This path in air has a permeability

**Table 10.6** Curie Temperatures of Some Ferromagnetic Materials

| Element | Curie temperature (°C) |
|---------|------------------------|
| Fe | 780 |
| Co | 1125 |
| Ni | 365 |
| Gd | 16 |

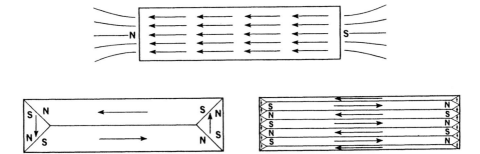

**Figure 10.39.** Single crystal of a ferromagnetic material subdividing into domains to reduce its magnetostatic energy. (From C. W. Richards, *Engineering Materials Science*, Brooks/Cole, Monterey, Calif., 1961, p. 431.)

on the order of one millionth ($10^{-6}$) of that in the ferromagnetic material. In this single-crystal bar of iron, all the magnetic dipoles prefer to align in a $<100>$-type direction. This is the direction of easy magnetization in iron, and when this occurs the energy is also reduced. This reduction in energy by virtue of the crystallography of the iron crystal is known as the *anisotropy energy.* Without it domains could not form and the *magnetostatic energy* could not be reduced. In BCC iron there are six such directions all along the cube edges. Thus the bar magnet breaks up into little domains where the moments align in more than one of the six directions (e.g., a [100] in one region while in another region they are aligned in a [010] direction at 90° to those in the other region). By so doing, all of the magnetic flux can find an easy path through the ferromagnetic bar of a high permeability of about $10^6$ without passing through the air or vacuum outside the bar. So a big reduction in the magnetostatic energy occurs when domains are formed. The crystal breaks up into small domains that are arranged such that the magnetic flux can stay completely within the bar. This situation is depicted in Figure 10.39b for this single crystal. In essence there exist small regions that act as individual magnets possessing N–S poles. We know from basic physics principles that magnetic poles of opposite signs attract. In Figure 10.39b the magnetic flux flows around the crystal in an internal circuit that consists of a series of N–S poles all of which are in a $<100>$-type direction in order to reduce both the anisotropy and magnetostatic energies.

Now Figure 10.39b is not the lowest-energy state. If the crystal forms more domains, the magnetostatic energy is further reduced. This situation is depicted in Figure 10.39c. There is, however, a new energy that must be considered, and that is the domain wall energy. The magnetic moments in the wall or boundary separating the domains are not aligned with those in either domain. Thus the domain wall is a region of higher energy. As the crystal subdivides into smaller domains its volume magnetostatic energy is reduced, but a gain in energy is experienced by the creation of more domain walls. When the point is reached where the decrease in volume energy is just balanced by the increase in energy due to the increase in domain wall volume, no further subdivision takes place.

In a polycrystalline metal each grain or crystal subdivides into domains just as described above for the single crystal, resulting in a large number of small flux paths all going around in small circuits in $<100>$ crystallographic directions. The magnetic fields of all of these randomly oriented crystals with their regions of small domains effectively cancel each other, so it appears as if the bar is not magnetic at all. When a magnetic field is applied, however, the domains shift their dipole moments in order to align with the field. This can occur in two ways. One method is by domain wall motion, as one domain becomes large at the expense of its neighbor. This occurs at low magnetic fields. First there is a reversible wall motion where the domain wall returns to its original position when the field is removed. The reversible displacement is followed by an irreversible displacement as the field is increased in strength (Figure 10.40). At higher fields the moments of individual atoms rotate. As a consequence the variation of $B$ with $H$ is nonlinear. Equation (10.23) is still used but $\mu$ is no longer constant but varies with $H$. Thus $\mu$ is analogous to the secant modulus of elasticity that we discussed in Chapter 8 for polymers.

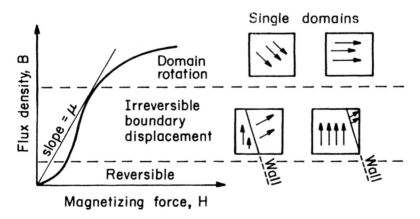

**Figure 10.40.** Effect of a magnetic field on domain alignment and resulting flux density $B$. Alignment occurs first by domain wall movement, but as the magnetizing force is increased, eventually the magnetic moments of adjacent atoms switch instantaneously to align in the same direction, thereby removing a domain wall.

Eventually, all domains will be aligned at a certain field strength. This is called the saturation magnetization $B_s$, shown in Figure 10.41. It should be noted that each domain by itself is always in a saturated condition, which is the spontaneous magnetization characteristic of ferromagnetism. When the field strength is reduced from the overall saturation the domains begin to return to their random orientations. But even though the field is reduced to a zero value, not all of the domains return to the original random orientation and the remaining magnetic field is called $B_r$, a residual flux called the *remanence* or remanent induction. A field, $H_c$, known as the *coercive force* is required to reduce the internal field to zero. During a complete cycle of the field $H$, the domains are forced to go through irreversible changes in direction, and as a result energy is expended that is in proportion to the area of the hysteresis loop.

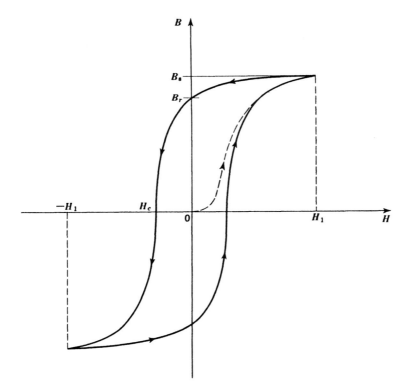

**Figure 10.41.** Hysteresis loop produced by reducing field from $H_{max}$ ($H_1$) to the same magnitude in the opposite direction, $-H_{max}$ ($-H_1$). Dashed line shows original magnetization curve.

Soft and Hard Magnetic Materials

It has been found that the area of this hysteresis loop (i.e., the energy dissipated per cycle of $H$) is very structure sensitive. The remanence, coercive force, and area of the hysteresis loop are higher for materials containing solute atoms, scond-phase particles, and dislocations. These types of defects tend to interfere with domain orientation and domain wall movement. Such materials are said to be *magnetically hard*, so named because many of the same factors that make a material mechanically hard also increased the loop size and make the material magnetically hard. Such materials are used as permanent magnets because they retain their overall alignment of domains to a large degree after the field is removed. Alloys of aluminum, nickel, and cobalt (Alnicos) have been developed that have very high coercive forces and $BH_{max}$ products, the terms often used for specifying permanent magnets. The best permanent-magnet materials today are the rare earth compounds, such as $SmCo_5$ and Sm–Pr–Co. Due to their relatively high cost, they are limited to small magnet applications such as in small dc motors. Permanent magnetic materials in general are developed for high induction, high resistance to demagnetization, and maximum energy content. The magnetic properties of some hard magnetic materials are compared in Table 10.7.

Magnetically soft materials are those with a small hysteresis loop. The loop sizes for soft and hard magnetic materials are compared schematically in Figure 10.42. Magnetically soft materials are used in transformer cores to minimize the hysteresis energy loss in the core. Pure iron makes a good transformer core. However, *eddy currents,* which are induced by the magnetic field, are another source of energy loss. Alloys of iron and silicon have been developed that have a high electrical resistivity to minimize eddy current loss, and at the same time have a grain alignment (texture) in sheet form such that the anisotropic easy direction of magnetization can be more fully utilized. The important characteristics of soft magnetic materials include high permeability, high saturation induction, low hysteresis energy loss, and low eddy current loss

**Table 10.7** Magnetic Properties of Some Hard Magnetic Materials

| Material | $H_c$ | | $BH_{max}$ |
| --- | --- | --- | --- |
| | $kA \cdot m^{-1}$ | Oe | $(kJ/m^3)$ |
| AlNiCo 5 | 50 | 620 | 42 |
| $SmCo_5$ | 690 | 8556 | 191 |
| Sm–Pr–Co | 805 | 9982 | 207 |
| PtCo | 355 | 4450 | 80 |

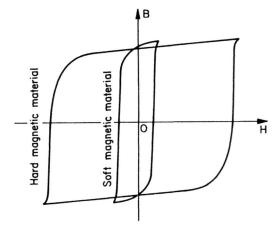

**Figure 10.42.** Comparison of the hysteresis loops for soft and hard magnetic materials.

in ac applications. The magnetic properties of some soft magnetic materials are listed in Table 10.8.

The spontaneous magnetization in a ferromagnetic crystal is accompanied by an elongation or contraction of the material called *magnetostriction*. It is associated with the electrical anisotropy of the crystal structure and the easy directions of magnetization. The saturation magnetostriction strain is very small. Nickel has the highest value, about $-40 \times 10^{-6}$. The property of magnetostriction is used in transducers, but as mentioned earlier, ferrostriction is larger and more often used for this purpose.

Antiferromagnetism and Ferrimagnetism

In some materials the exchange interaction prevents the magnetic moments of adjacent atoms from lining up as they do in a ferromagnetic material, but instead, the moments of neighboring atoms point in opposite directions (Figure 10.43). These materials are *antiferromagnetic*. The elements manganese and chromium exhibit antiferromagnetism at room temperature because the ratio of the atomic spacing to the diameter of their $3d$ shells is less than 3 (Figure 10.37).

In some ferrites, most notably the iron oxide $FeO \cdot Fe_2O_3$ (magnetite— usually written as $Fe_3O_4$), the magnetic moments point in opposite directions but are different in magnitude, a characteristic known as *ferrimagnetization*. These ferrites possess a high resistivity, to as much as $10^7$ $\Omega \cdot m$. Eddy currents resulting from alternating fields are therefore reduced to a minimum,

**Table 10.8**  Magnetic Properties of Some Soft Magnetic Materials

| Material | Maximum permeability | $B_s$ (kg) | Resistivity ($\mu\Omega \cdot cm$) | Applications |
|---|---|---|---|---|
| | | DC properties | | |
| 99.9% Fe | 8,000 | 21.5 | 10 | Flux carriers |
| Fe–1.1% Si–0.5% Al | 7,700 | 21.0 | 25 | Transformers |
| Fe–4.0% Si–0.5% Al oriented | 18,500 | 19.5 | 25 | Transformers |
| | | AC properties | | |
| Fe–(Si + Al)–1.9% ASTM A677 | 1,300 (60 Hz and 15 kG) | 20.7 | — | Rotating machinery |
| Fe–(Si + Al)–3.2% ASTM 876 oriented | > 1,800 at 10 Oe: | 20 | — | Power transformers |
| 79% Ni–5% Mo–15% Fe | 90,000 at 40 G | 7.9 | 16 | Shielding |

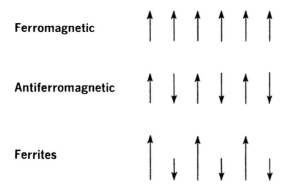

**Figure 10.43.**  Alignment of magnetic spin moments in ferromagnetic, antiferromagnetic, and ferrimagnetic materials.

and the range of electronic applications extend to high frequencies such as those found in microwave devices.

## 10.8 SUPERCONDUCTORS

In 1911 the Dutch physicist Kammerlingh Onnes discovered that the electrical resistivity of mercury disappeared at the temperature of liquid helium (4.2 K). The actual transition to essentially zero resistivity was found to be 4.15 K. This phenomenon has been termed superconductivity and the temperature at which a material becomes superconducting has been termed the critical temperature $T_c$. The resistivity–temperature relationships for a normal conducting material, Cu, and a superconducting metal, Pb, is shown in Figure 10.44. It is now known that about half of the metallic elements are superconductors at low temperatures, with niobium having the highest $T_c$, 9.3 K. Not too much attention was paid to the subject of superconductivity until the 1950s when $Nb_3Sn$ was found to be superconducting at a $T_c$ value of 18.5 K. Not only did $Nb_3Sn$ have a higher $T_c$, but it also had a high $J_c$, the critical current that the material could carry before becoming a normal conductor, and a high $H_c$, the magnetic field, which could also destroy the superconducting state. These are the three critical parameters for superconductors and are depicted in Figure 10.45. Within the shaded area the material is superconducting, but the magnitude of the superconductivity is a function of all three factors, $T$, $H$, and $J$. For $Nb_3Sn$, $J_c$ is around $10^5$ to $10^6$ A/cm$^2$ and $H_c$ about 11 T, all at about 4.2 K. A number of $Nb_3Sn$ superconducting magnets have been made for research use and for

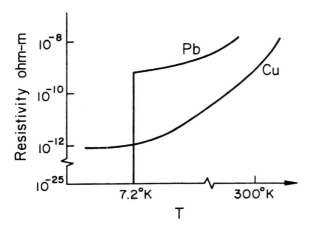

**Figure 10.44.**   Resistivity–temperature relationship for a normal (Cu) and a superconducting (Pb) metal.

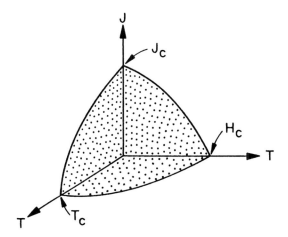

**Figure 10.45.** Three critical superconducting parameters, $J_c$, $H_c$, and $T_c$. At points within the shaded envelope the material will be superconducting.

nuclear magnetic imaging systems used for medical diagnosis, but other than that there has been little application for this material and its superconducting properties.

In the 1960s and 1970s superconductivity was found in some metal oxides, a rather unusual finding since oxides are ceramics and ceramics are not supposed to be conductors. The real excitement occurred in 1986 when scientists at IBM in Zurich, Switzerland found superconductivity in some rare earth containing oxides which had a $T_c$ value of 35 K. Since 1986 and continuing to the present there has been a flurry of activity in this field with the big goal that of finding a material that will be superconducting at room temperature ($\sim 293$ K). As of mid-1991 the highest $T_c$ value reported was 125 K, found in a compound of Tl–Ca–Ba–Cu–O (Herman and Zheng).

There are two types of superconductors, type I being those materials where a magnetic field does not penetrate the conductor until a critical field is reached (Meissner effect). These materials are completely diamagnetic in their super-conducting state and consist mainly of the pure metals. In these materials the current flows primarily in the outer layers and their critical current densities are rather low.

Type II superconductors have two critical fields, $H_{c1}$, below which the material is superconducting throughout its body, and $H_{c2}$, above which it possesses normal conducting behavior throughout the material. In between $H_{c1}$ and $H_{c2}$ the material is a mixture of both normal and superconducting regions (Figure 10.46).

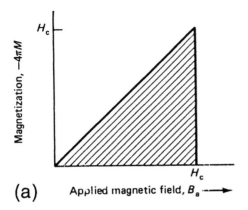

(a)

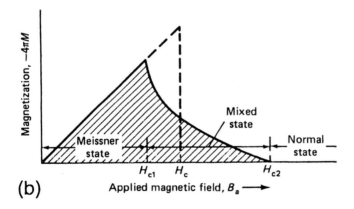

(b)

**Figure 10.46.** *B–H* relationship in type II superconductors. (a) Type I. This type exhibits perfect diamagnetism (Meissner effect). Above $H_c$ the material is a normal conductor. (b) Type II. The applied field begins to enter the sample at a field $H_{c1}$ that is lower than $H_c$. Superconductivity persists in the mixed state to a high field of $H_{c2}$, above which the material is a normal superconductor.

The more recently discovered oxide superconductors are of the type II variety, and despite their high $T_c$ values, they present other problems that must be overcome. They are very brittle and difficult to fabricate into wire or some other useful shape, and their critical current densities are on the order of $10^3$ A/cm$^2$, much less than that of the NbSn-type superconductors. There have been some useful thin-film devices fabricated from the high $T_c$ oxide superconductors with current densities in the vicinity of $10^5$ A/cm$^2$. The Tl–Ca–Ba–Cu–O superconductors also have shown some rather high current densities in the $10^6$-A/cm$^2$ region.

The theory of the superconducting state has still not been completely unraveled. The Bardeen–Cooper–Schrieffer (BCS) theory, for which a Nobel prize was granted, described the electrons as moving in pairs of opposite spin. The motion of the two electrons became correlated such that if one is scattered, the other responds in a way that causes a net zero effect (i.e., no scattering and no resistance).

## DEFINITIONS

*Acceptor level*: permitted energy levels that are formed in the forbidden region but near the filled valence of an extrinsic semiconductor when it is doped with 3-valence impurity atoms.

*Anisotropy energy*: energy by which a magnetic material is reduced when the field is applied in the direction of easiest magnetization.

*Base*: electrode in a bipolar transistor that controls the flow of charge.

*Bipolar*: type of transistor that has an emitter, a base, and a collector.

*Brillouin zone theory*: theory that determines energy levels in three-dimensional reciprocal lattice space at which electrons are allowed.

*CMOS*: complementary MOS device that contains both *p* (*PMOS*) and *n* (*NMOS*) channels.

*Coercive force*: magnetic field required to switch domains in a magnetic material back to the random arrangement that they had before the field was applied.

*Collector*: electrode that collects the electrons in a bipolar transistor.

*Diamagnetic*: material in which the magnetic dipole moments oppose an applied field and the magnetic permeability is less than unity.

*Domain*: region in a ferromagnetic material in which the magnetic moments of all atoms are aligned in the same direction.

*Donor level*: energy level in a semiconductor that is formed in the forbidden region of the semiconductor near the conduction band.

*Dopants*: impurity atoms added to an *intrinsic* semiconductor to make it an *extrinsic* type.

*Drain*:   electrode to which the electrons go in a MOS device.
*Emitter*:   electrode that emits the electrons in a *bipolar* transistor.
*Exchange force*:   force (or energy) that causes *spontaneous magnetization* in ferromagnetic materials.
*Extrinsic*:   semiconductor where the conduction is due primarily to intentionally added atoms (*dopants*).
*Ferroelectric*:   characteristic of a dielectric material that causes dipoles to be aligned spontaneously, even in the absence of a field.
*Ferromagnetic*:   characteristic of a magnetic material that causes magnetic electron spin moments to align even in the absence of a field.
*Ferrostriction*:   characteristic of ferroelectric materials whereby an application of a stress creates a shift of dipoles and an electric signal, and vice versa.
*Forbidden energy region*:   *See* Energy gap.
*Intrinsic semiconductor*:   semiconducting material whose conductivity is dependent and inherent to this material without additives or defects of any sort.
*Magnetostatic energy*:   reduction in the energy of a ferromagnetic material when domains are formed.
*Magnetostriction*:   characteristic of a magnetic material whereby a dimensional change affects the magnetic properties, and vice versa.
*Paramagnetism*:   magnetism that can be a result of either noncanceled orbital magnetic moments (weak) or of noncanceled spin moments (strong).
*Piezoelectric*:   characteristic of material whereby it can produce an electric signal via an application of pressure, or vice versa.
*Remanence*:   magnetization remaining in a ferromagnetic material when a magnetic field has been applied to reverse the spin moment alignment and this field has been reduced to zero.
*Source*:   electrode in an MOS device from which electrons are released.
*Susceptibility*:   proportionality constant between the magnetization $M$ of a material and the applied field strength $H$.

## QUESTIONS AND PROBLEMS

10.1.   Sodium has a lattice parameter of 4.289 Å at 20°C and a temperature coefficient of thermal expansion of $71 \times 10^{-6}$ per °C. Assuming that this coefficient is maintained (i.e., constant to 0 K), what lattice parameter will sodium have at 0 K?

10.2.   Compute the number of allowable energy levels in 1 cm³ of copper. The density of copper is 8.96 g/cm³.

10.3.   Estimate the probability that an electron in Sn can gain sufficient energy at 20°C to enter the conduction band. Compare your answer to Example 10.3 for Si.

10.4. Repeat Example 10.3 except for Ge.

10.5. What fraction of the charge in intrinsic Si is carried by the electrons? Compute the same fraction for a GaAs intrinsic semiconductor.

10.6. What is the resistance of a 0.001-in.-diameter gold wire of 10 ft in length at (1) $0°C$; (b) $100°C$?

10.7. Compute the conductivity of copper at $120°C$.

10.8. Sketch the energy band structure of semiconductors, metals, and insulators and explain their differences.

10.9. How many phosphorus atoms must one add to 1 mol of germanium atoms to give a resistivity of $10 \ \Omega \cdot cm$? What will be the concentration of phosphorus in parts per million (ppm)?

10.10. List all the factors that affect the conductivity of metals in order of the smallest to the largest effect.

10.11. At 300 K the electrical conductivity of intrinsic germanium is 22 $(\Omega \cdot m)^{-1}$. What is its conductivity at $300°C$?

10.12. Compare the advantages and disadvantages of bipolar versus MOS devices for (a) switching circuits in computers; (b) microwave appliances.

10.13. For a parallel-plate capacitor of 1 $cm^2$ area, a 1-cm separation of the plates, and a 10-V applied potential, what will be the capacitance for (a) a vacuum between the plates; (b) a material with a dielectric constant of 10?

10.14. Excluding cost as a factor, in what applications would GaAs devices be preferred over silicon devices? Consider both bipolar and MOS devices.

10.15. The intrinsic electrical conductivities for a semiconductor material at $20°C$ and $100°C$ are 0.5 and 5 $(\Omega \cdot m)^{-1}$, respectively. Estimate the $E_g$ gap width in eV for this material.

10.16. Explain the difference in the variation of conductivity with temperature between metals and semiconductors.

10.17. The *Handbook of Chemistry and Physics* gives the magnitude of a Bohr magneton as $9.27 \times 10^{-24}$ J/T. What is this value in the units of ergs/G?

10.18. The saturation magnetization of iron at 0 K is 2.2 Bohr magnetons. What is this value in ergs/G?

10.19. Several years ago copper was marketed for electrical conduction purposes as oxygen-free high-conductivity (OFHC) copper, yet chemical analysis showed oxygen contents of thousands of parts per million. But this oxygen content did not appreciably affect the electrical conductivity. How is this possible?

10.20. A *p*-type semiconductor is made by adding aluminum to silicon. The resistivity of such a doped silicon is measured to be $10^{-3}$ $\Omega \cdot$ m. Using the hole mobility from Table 10.2, what is the aluminum content in ppm?

10.21. How can a hole in a semiconductor act as a positive charge? Why does it have the same charge as an electron?

10.22. What are two ways in which Ohm's law can be stated? Which is preferred in the semiconductor industry, and why?

10.23. Why is the Fermi energy $E_F$ relatively unimportant in metals compared to semiconductors?

10.24. What causes the valence electron energy levels to broaden into bands as atoms are brought together, whereas the inner electron levels are essentially not affected by this particular equilibrium interatomic spacing?

10.25. List the possible dopants for $N$- and $P$-type semiconductors.

10.26. Why do most group 3-5 and 2-6 compounds behave as semiconductors? List the ones that have been used as such.

10.27. Why is the resistance of a metal at 20°C compared to that at 4.2 K a good indication of the metal's purity?

10.28. List and explain the reason for the wide difference in dielectric constants at 50 Hz for (a) polyethylene; (b) soda-lime glass; (c) $BaTi_2O_3$.

10.29. The maximum relative permeability for 99.91% pure iron and 99.95% pure iron is 500 and 180,000, respectively, yet the saturation magnetization for each is the same (2500 G). Explain.

10.30. Discuss all of the similarities (i.e., analogous characteristics) of ferroelectric and ferromagnetic behavior.

10.31. Why is the hole mobility always less than the electron mobility in semiconductors?

10.32. By what factor is the electrical conductivity of silicon increased in going from room temperature (300 K) to 473 K?

10.33. The resistivity of intrinsic silicon at 25°C is $5 \times 10^{-4}$ $(\Omega \cdot m)^{-1}$. Compute (a) the number of charge carriers/cm$^3$, (b) the value of the constant $n_0$ in equation (10.12).

10.34. What are the two requirements for a material to be ferromagnetic?

10.35. Why do domains form in a ferromagnetic material? What factor limits the minimum domain size?

10.36. What is the approximate ratio of the conductivity of metals at room temperature to that of a metal in the superconducting state?

## SUGGESTED READING

American Society for Metals, *Metals Handbook*, Vol. 2, 10th ed., ASM International, Metals Park, Ohio, 1990, pp. 761-782.

Dm Data, *The Semiconductor Picture Dictionary,* DM Data, Inc., Scottsdale, Ariz., 1989.

Hodges, D. A., and Jackson, H. G., *Analysis and Design of Digital Integrated Circuits,* McGraw-Hill, New York, 1983.

Smith, W. F., *Principles of Materials Science and Engineering,* 2nd ed., McGraw-Hill, New York, 1990.

# 11

# Environmental Degradation of Engineering Materials

## 11.1  INTRODUCTION

All materials will react to some degree with certain environmental constituents, but there is a vast difference in the degree of reactivity of the three major groups of materials: metals, polymers, and ceramics.

Metals are by far the most reactive of these three groups. That is why metals are found in nature in the form of compounds. Metals are extracted from their ores where nature put them. The metal compounds in the ores, oxides, sulfides, nitrides, and the like are very stable compounds. When the metal is removed from these other elements, they want to go back to where they came from. We take metals from their ores and refine them so that they are formable and easily fabricated into many desirable shapes. Metals are very useful materials, as evidenced in Figure 1.1. But we pay a price for this convenience. On their way back to their natural habitat, they may react with many elements that exist in whatever environments to which they may become exposed, including those that are not normally found in ambient atmospheres. These metal attackers include the acids; caustic bases; the gaseous hydrogen, oxygen, and nitrogen atmospheres; and many of the halides, chlorine, and fluorine in particular. The metals industry spends billions of dollars extracting metals from their compounds as found in nature, and then billions more to protect them from reactions that occur in their new environments. Estimates of losses in U.S. industries due to corrosion of steels fall in the range $70 billion to $80 billion per year. But some metals are not very reactive. These are the expensive precious metals. Gold does not have to be extracted from oxides and sulfides because it does not form such compounds. It is usually found (but not easily located) in nature as a pure elemental metal.

Metals are reactive because they have all those free electrons roaming around looking for something to which they can become attached, and in so doing lower the overall energy of the system. In polymers and ceramics the covalent and ionic bonds do not provide free electrons. Polymers are much more subject to environmental degradation then are ceramics. In polymers, light, particularly in the ultraviolet-wavelength region, can break the C-C bonds, a process called *scission*. Neutron radiation can also cause scission. At elevated temperatures the bonds of the atoms attached to the side of the chain of the carbon backbone can be broken, causing *charring*, also called *carbonization*. Ozone degradation of elastomers is a serious problem for the rubber industry. One of the larger effects of the environment on polymers is that caused by organic solvents. Whereas metals are attacked by electrochemical reactions, which requires electron conductivity, polymers are insulators. In polymers, the solvents diffuse into the polymer body and, without breaking bonds, cause plastization and swelling effects, processes generally referred to as *physical corrosion*. Organic liquids such as cleaning fluids, detergents, gasoline, lubricants, and sealants lead to the deterioration of polymer properties. Chemical attack by these solvents can cause swelling, scission, and depolymerization (unzipping) of the chain molecules. Some chemicals can cause crazing and alteration of the mechanical properties of polymers. In short, polymers are affected by sunlight, nuclear radiation, heat, oxygen, humidity, and organic solvents. But it must also be recognized that polymer degradation is very useful process when the need arises for the destruction of waste materials. There is an increasing demand for biodegradable self-destructing polymers for packaging, agricultural, and medical applications.

Ceramics are the materials least affected by their environment. But ceramics can be attacked by certain chemicals that lead to their degradation and loss of strength. The most obvious cases are the halides (e.g., NaCl), which are ceramics by most definitions and which are soluble in water and many other solvents up to certain concentrations. Some carbonates can be dissolved by weak acids. The carbides and nitrides are attacked by oxygen at elevated temperatures. Ceramic silicates are readily attacked by hydrofluoric acid, which is used to etch glasses. But the strongly bonded ceramics such as $Al_2O_3$ and $Si_3N_4$ are very stable and inert to attack by most aqueous solutions, including acids and alkaline solutions. A related ceramic corrosion problem is that facing highway engineers, where reinforced concrete bridges deteriorate due to deicing salts and marine environments. A January 1992 report by a committee of the National Academy of Sciences stated that NaCl is still the best way to clear ice and snow off roadways, even though its corrosive powers cost tens of millions of dollars to vehicles and bridges every year. The widespread use of salt substitutes, such as calcium magnsium acetate, are not warranted according to the 14-member panel of the National Research Council's transportation board because, even though the alternative deicers are less corrosive, they cost more than 20 times that of NaCl and are less effective. It appears to the writer that this is a fruitful field for exploration by some of our entrepreneurial scientists and engineers.

## 11.2  CORROSION OF METALS

The reaction of metals with their environment can be divided into two major groups: that with aqueous environments via the establishment of galvanic cells, which we call electrochemical corrosion, and that with other reactants, such as oxygen (oxidation), acids, and a host of other media. Galvanic corrosion is by far the most prominent degradation mechanism for metals and is the subject of this section. The others are discussed briefly in a subsequent section.

### 11.2.1  Laboratory Galvanic Cell

Before examining the overall galvanic cell corrosion behavior let's construct a large galvanic (electrochemical) cell using a piece of iron plate that is a perfect single crystal (i.e., it contains no grain boundaries, dislocations, vacancies, or impurity atoms). When this piece of iron is immersed in a solution of pure distilled water that contains nothing more than dissolved oxygen, as depicted in the beaker in Figure 11.1a, the following oxidation reaction begins:

$$Fe^0 \rightarrow Fe^{2+} + 2e^- \qquad\qquad (11.1)$$

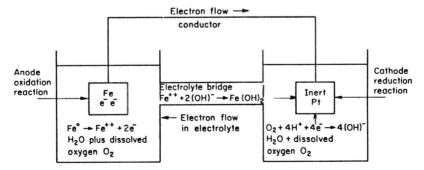

**Figure 11.1.** Laboratory galvanic cell.

Some of the neutral iron atoms go into solution as doubly charged positive ions, leaving electrons behind on the iron plate. But for every forward chemical reaction there is also a backward reaction which involves the ions returning to the plate and recombining with the electrons. After a short period of time these two reactions are equal, as denoted below by the two equal arrow lengths in the following:

$$Fe^0 \Longleftrightarrow Fe^{2+} + 2e^- \qquad (11.2)$$

At equilibrium there is no net loss of ions from the plate, and therefore the corrosion process ceases. Galvanic corrosion is defined as the loss of metal by a galvanic reaction. For the reaction to proceed, the balance must be upset. This can be done by removal of either the iron ions or the electrons. In a galvanic cell the electrons are removed as illustrated in Figure 11.1, where we have connected the iron plate via a conducting wire to another metal, the latter also immersed in the same kind of pure water as was used for the iron crystal. We consider the second metal to be inert, such as the precious metals platinum and gold. No metals are entirely inert, but these metals do approach that state. Now the electrons are drawn off the iron plate in the left beaker and into the right beaker, thus permitting the reaction in the chemical equation (11.1) to proceed. So now the corrosion process continues, but again not for long. In another short period of time the electron concentration on the inert plate will equal that produced on the iron plate. The reaction would again cease if it were not for the dissolved oxygen in the water. This permits the following reduction reaction to occur

$$O_2 + 4H^+ + 4e^- \longrightarrow 4(OH)^- \qquad (11.3)$$

This is called the *cathode reaction,* and that occurring at the iron plate, as denoted by equation (11.1), is the *anode reaction.* Electrons are produced at the anode and consumed at the cathode. So now the corrosion process can continue. But reaction (11.3) also has a backward reaction and when the forward and backward reactions equilibrate, there will be no net consumption of electrons at the cathode and again the corrosion process will cease. However, if we can mix the liquids in the beakers, for example by constructing an electrolyte bridge between the two beakers, another reaction will take place, because now the positive iron ions will be attracted to the negative hydroxyl ions:

$$Fe^{2+} \longrightarrow + 2(OH)^- + Fe(OH)_2 \qquad (11.4)$$

We call these liquids *electrolytes.* We could have mixed the liquids in the same container without the electrolyte bridge and the reactions would be the same. The electrolyte bridge was used only to illustrate that an electrolyte and the chemical reaction therein is required to complete the electric circuit that occurs within a galvanic cell. The overall reaction can be written as the sum of the equations above:

$$2Fe^0 + 2H_2O + O_2 \longrightarrow 2Fe^{2+} + 4(OH)^- \longrightarrow 2Fe(OH)_2 \qquad (11.5)$$

The compound formed is a white ferrous hydroxide. It is unstable and is further oxidized to ferric hydroxide, which has a reddish brown color called rust. The reaction is

$$2Fe(OH)_2 + H_2O + \tfrac{1}{2}O_2 \longrightarrow 2Fe(OH)_3 \quad (\text{rust}) \qquad (11.6)$$

Now our galvanic cell is complete. We have an anode, a cathode, an electrolyte, and a conductor. All four are required to produce a galvanic cell, and they must form a complete circuit. The electrons flow from the iron anode through the conductor to the cathode. Many of the electrons will be consumed at the cathode according to reaction (11.3) but some will flow through the electrolyte back to the anode. As long as more electrons are pumped out of the anode than are consumed at the cathode, the corrosion process will continue until all of the iron is consumed. With so many requirements for the corrosion reaction to proceed, the reader may wonder how so much corrosion can ever take place. We will pursue this further in a moment. But first note that what is depicted in Figure 11.1 is a battery. For many years, the automobile batteries consisted of a Pb anode and a $PbO_2$ cathode in an acid electrolyte, the latter for the purpose of providing hydrogen ions. The reactions are listed as follows:

anode:   $Pb + SO_4^{2-} \longrightarrow PbSO_4 + 2e^-$

cathode:   $PbO_2 + 4H^+ + SO_4^{2-} + 2e^- \longrightarrow PbSO_4 + 2H_2O$

A typical automobile battery contains several of these cells in series. The Pb anode corrodes (which in your car battery is labeled as the positive terminal) and in so doing produces electrons and $PbSO_4$. A corrosion cell and a battery are the same thing, only differing in that one results in destruction and the other in producing usable energy.

Now let's consider a piece of real iron. As shown in Figure 11.2, it contains grain boundaries, dislocations, and impurity atoms, mostly in the form of precipitates such as iron carbides or oxide inclusions left as a residue of the refining process. Any of these may become anodic or cathodic with respect to the main body of the iron. The atoms in the grain boundary, for example, can go into solution easier than the atoms within the grain proper, because the former are more disorganized and not as tightly bound together as are those in the perfect lattice. The terms *anodic* and *cathodic,* or *anode* and *cathode,* are strictly relative. If more atoms go into solution at the grain boundary than from the lattice proper, more electrons will be generated there and this boundary will become the anode. These excess electrons will be consumed in the interior

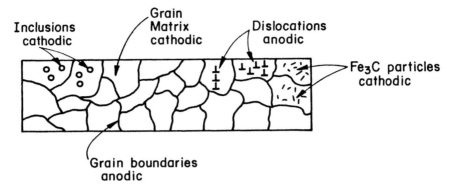

**Figure 11.2.**   Possible location of anodes and cathodes on a piece of polycrystalline iron in a humid atmosphere.

of the grain even though that region itself may be producing electrons by the dissolution of iron atoms, but at a lesser rate than the iron atoms at the grain boundary. So in the real piece of iron there are thousands of small galvanic cells established all over the surface. The rust will appear uniform to the eye (uniform corrosion) because we cannot see each little cell. In addition, these negative and positive electrodes shift position with time due to the changes in the concentrations of the chemical species involved in the reactions. This adds to the uniform corrosion appearance of the piece being destructed. A cold-worked iron or steel will corrode more than an annealed piece of iron because the dislocations are anodic with respect to the more perfect regions of the crystal. This characteristic of grain boundaries and dislocations being more reactive than the perfect crystal is used to reveal these defects by chemical attack with certain etchants. The grain boundaries are easily revealed by etching, and under certain controlled conditions, the dislocations will form etch pits when in contact with the etchant (Figure 5.8). Perhaps you have seen preferential rust formation at a bent or deformed region in a steel sheet caused by the higher dislocation density in the deformed regions.

It is not necesary to place the iron or steel in water to form rust. There is sufficient moisture in the air to provide a surface layer of electrolyte. Corrosion proceeds faster and is more extensive in humid atmospheres. The metal itself acts as the conductor, the humid atmosphere as the electrolyte, and the perfect lattice regions and defects become the cathodes and anodes, respectively. Furthermore, the ferric hydroxide rust flakes off the iron or steel surface due to the large coherency strains set up as a result of a mismatch in their respective lattice parameters. This flaking provides for removal of the corrosion products. The rusting will continue until the iron disappears. The rusty spike, shown in

Figure 11.3, has been attacked extensively over its entire surface. A considerable portion of its life has expired. It is a typical example of uniform corrosion.

## 11.2.2 Electromotive Force Series

Just as defects in iron are anodic with respect to the lattice proper, metals are anodic or cathodic with respect to each other. This is how giant battery cells are made: by putting two materials of widely different electrochemical characteristics into an electrolyte. One material will be the anode while the other is the cathode. The Pb–PbO$_2$ battery, mentioned previously, is the best example of one of the good guys: a galvanic cell that is put to work to generate power rather than to destroy the anode material as in the case of a corrosion reaction. As the battery produces current both plates are covered with lead sulfate as a result of the chemical reactions taking place in the electrolytes. Recharging this battery is accomplished by feeding current in the reverse direction, which decomposes the lead sulfate. The life of a battery is snuffed out when the amount of this lead sulfate and assorted corrosion products flake off the electrodes and fall in between them. The battery can then no longer be recharged. It is not the consumption of the anode on which the life of the battery is dependent. The anode can always be restored by current reversal as long as there is lead sulfate or lead compounds available to place the Pb back onto the anode during current reversal (recharging). Note also that the dry cell battery that we use in our flashlights is not really dry. In the common zinc anode–carbon cathode "dry" cell the electrolyte is a wet paste containing moist aluminum chloride.

Returning now to metals, their corrosion tendency is determined by measuring the electrons that they generate versus those of a standard electrode. The

**Figure 11.3.**   Rusty spike: uniform corrosion.

hydrogen electrode is universally used for measuring the standard electromotive force (emf) of metals. Since each metal is measured against a hydrogen electrode, they are written as half-cell values since the potential of the hydrogen standard electrode is taken arbitrarily as zero. If the metal electrode measured against the hydrogen electrode is negative, it has a greater tendency to give up electrons than does the hydrogen electrode, and this metal is anodic with respect to the hydrogen standard half-cell. Metals with high negative potentials (e.g., Li and Mg) are very reactive metals. These metals and most other metals characteristically give up electrons in what is called an *oxidation reaction*. Metals that measure a positive potential with respect to the standard hydrogen electrode give up fewer electrons than the standard cell and are more cathodic. These include the more noble precious metals (e.g., Au, Pt, and Pd). The hydrogen electrode (Figure 11.4) consists of an inert metal like platinum immersed in a 1 $M$ solution of $H^+$ ions into which hydrogen gas is bubbled. The metal to be measured is placed in a 1 $M$ solution of its own ions. The platinum provides a surface for the oxidation reaction of hydrogen gas to hydrogen ions ($H_2 \longrightarrow 2H^+ + 2e^-$) to occur. The electrolytes of the two half-cells are allowed to be in electrical contact but are not permitted to mix with one another. The voltage of each metal is measured versus the hydrogen half-cell at 25°C. These voltages are usually ranked in descending order of positive values until the arbitrarily defined zero potential of hydrogen is reached. Those metals that generate fewer electrons than the hydrogen electrode are listed as positive,

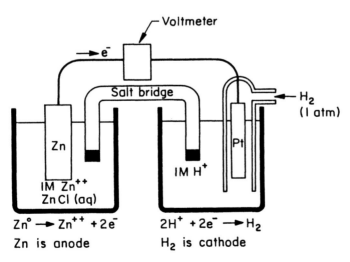

**Figure 11.4.** Hydrogen reference half-cell (on the right), better known as the hydrogen electrode is being used to measure the emf of zinc.

and those that generate more electrons are considered to be negative. The electromotive force voltages resulting from these measurements are listed in Table 11.1, and this listing is known as the emf series for metals. Any metal below another metal in this listing is anodic with respect to the one above it. The anodic one is also more reactive. If we form a cell (battery) between any two

**Table 11.1** Electromotive Force Series of Metals

| Electrode reaction | Standard potential at 25°C (77°F) (volts versus SHE) |
|---|---|
| $Au^{3+} + 3e^- \rightarrow Au$ | 1.50 |
| $Pd^{2+} + 2e^- \rightarrow Pd$ | 0.987 |
| $Hg^{2+} + 2e^- \rightarrow Hg$ | 0.854 |
| $Ag^+ + e^- \rightarrow Ag$ | 0.800 |
| $Hg_2^{2+} + 2e^- \rightarrow 2Hg$ | 0.789 |
| $Cu^+ + e^- \rightarrow Cu$ | 0.521 |
| $Cu^{2+} + 2e^- \rightarrow Cu$ | 0.337 |
| $2H^+ + 2e^- \rightarrow H_2$ | 0.000 |
| | (Reference) |
| $Pb^{2+} + 2e^- \rightarrow Pb$ | −0.126 |
| $Sn_2 + 2e^- \rightarrow Sn$ | −0.136 |
| $Ni^{2+} + 2e^- \rightarrow Ni$ | −0.250 |
| $Co^{2+} + 2e^- \rightarrow Co$ | −0.277 |
| $Tl^+ + e^- \rightarrow Tl$ | −0.336 |
| $In^{3+} + 3e^- \rightarrow In$ | −0.342 |
| $Cd^{2+} + 2e^- \rightarrow Cd$ | −0.403 |
| $Fe^{2+} + 2e^- \rightarrow Fe$ | −0.440 |
| $Ga^{3+} + 3e^- \rightarrow Ga$ | −0.53 |
| $Cr^{3+} + 3e^- \rightarrow Cr$ | −0.74 |
| $Cr^{2+} + 2e^- \rightarrow Cr$ | −0.91 |
| $Zn^{2+} + 2e^- \rightarrow Zn$ | −0.763 |
| $Mn^{2+} + 2e^- \rightarrow Mn$ | −1.18 |
| $Zr^{4+} + 4e^- \rightarrow Zr$ | −1.53 |
| $Ti^{2+} + 2e^- \rightarrow Ti$ | −1.63 |
| $Al^{3+} + 3e^- \rightarrow Al$ | −1.66 |
| $Hf^{4+} + 4e^- \rightarrow Hf$ | −1.70 |
| $U^{3+} + 3e^- \rightarrow U$ | −1.80 |
| $Be^{2+} + 2e^- \rightarrow Be$ | −1.85 |
| $Mg^{2+} + 2e^- \rightarrow Mg$ | −2.37 |
| $Na^+ + e^- \rightarrow Na$ | −2.71 |
| $Ca^{2+} + 2e^- \rightarrow Ca$ | −2.87 |
| $K^+ + e^- \rightarrow K$ | −2.93 |
| $Li^+ + e^- \rightarrow Li$ | −3.05 |

metals, the voltage generated will be the difference in the emf values of the two metals. For Li–Mn this voltage would be $-3.05 - (-1.18) = -1.22$ V. For Li–Cu the cell voltage would be $-3.05 - (+0.52) = -2.53$ V. Many emf series, including that of the *Handbook of Chemistry and Physics,* list Li at the top with a $-3.05$ V and gold at the bottom as a $+1.50$ V. This is the same listing in both sign and value that we show in Table 11.1 except that in the latter listing the more reactive metals are placed at the bottom of the list. Others, however, reverse the signs and put gold at the top with a $-1.50$ V and Li at the bottom with a $+3.05$ V. This difference is merely one of convention. Both conventions are still used, although this matter was supposedly settled at the International Union of Pure and Applied Chemistry Meeting in Stockholm in 1953. According to the Stockholm convention, a negative sign indicates a trend toward oxidation and corrosion in the presence of $H^+$ ions. Thus Li has a negative sign when it is undergoing an oxidation reaction.

Actually, corrosion engineers use the galvanic series, which includes alloys as well as pure metals, the latter seldom existing in most product designs. This series includes metals that form passive films and list the most negative (anodic) material at the top and the most noble (cathodic) at the bottom. The measurements are also usually conducted in about a 3.5% saltwater solution, comparable to seawater, and often measures the voltages, which are frequently not listed, against a graphite electrode. In practical corrosion problems galvanic coupling between metals in equilibrium with a 1 $M$ solution of its own ions is seldom observed. The galvanic series is a more accurate prediction of galvanic reactions than is the emf series. The galvanic series as determined by the LaQue Center for Corrosion Technology is given in Table 11.2. This series was obtained by measuring the specific metals listed against a calomel half-cell reference electrode, which is composed of mercury and calomel (mercurous chloride) in a solution of potassium chloride. The design engineer tries to avoid placing metals that are far apart in the series physically close together in an attempt to avoid large galvanic cells.

Most electrolytes in the real world are not 1 $M$ but of a lesser concentration of ions. In this case the driving force to corrode the anode is greater since there are fewer of its ions in solution. The effect of ion concentration is given by the Nernst equation (after W. H. Nernst) as

$$E = E° + \frac{0.0592}{n} \log C_{\text{ion}} \tag{11.7}$$

where $E$ = measured potential
$\quad\quad E°$ = standard emf
$\quad\quad n$ = number of electrons transferred per metal ion (valency)
$\quad\quad C_{\text{ion}}$ = molar concentration of ions

*Example 11.1.* If 200 g of $Mg^{2+}$ ions are present in 1000 g of water, what is the electrode potential of this half-cell?

**Table 11.2**   Galvanic Series of Metals

# CORROSION POTENTIALS IN FLOWING SEAWATER (8 TO 13 FT./SEC.) TEMP RANGE 50° - 80°F

VOLTS: SATURATED CALOMEL HALF-CELL REFERENCE ELECTRODE

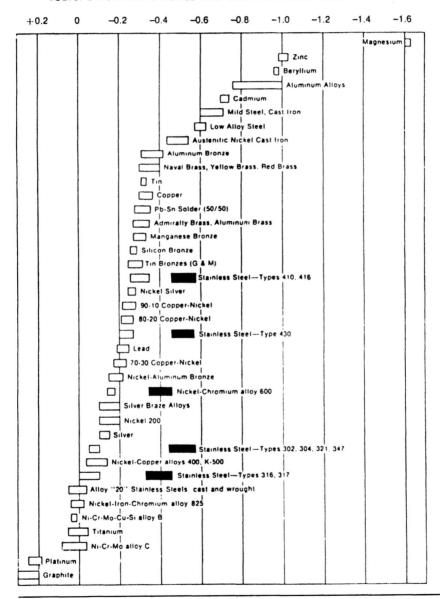

*Source*: Courtesy of the LaQue Center for Corrosion Technology, Inc., Wrightsville Beach, N.C.

Solution.   A 1 $M$ solution contains 1 mol of $\dot{M}g^{2+}$ ions to 1000 g of water. The atomic mass of magnesium is 24.312 g/g-mol. The concentration for 200 g of $Mg^{2+}$ becomes

$$C_{ion} = \frac{200}{24.312} = 8.22 \ M$$

$n = 2$ since the Mg ion has a $+2$ charge. $E°$ is obtained from Table 11.1 as $-2.37$ V. Thus

$$E = -2.37 + \frac{0.0592}{2} \log(8.22) = -2.4 \ V$$

## 11.2.3   Electrode Reactions and Their Associated Galvanic Cells

In our discussion of the corrosion of iron via electron production at the anode and electron consumption at the cathode, which were stated in equations (11.1) and (11.3), respectively, the anode reaction is always the same, that is

$$M° \rightarrow M^{n+} + ne^- \qquad (11.8)$$

where $n$ is the valency of the metal. However, at the cathode several different reactions may occur, all being reduction reactions, the exact type being dependent on the solution or electrolyte involved. One of the more common cathode reduction reactions is that which occurs via the corrosion of metals in acid solutions, where a high concentration of hydrogen ions is present as represented by the following chemical equation:

$$\text{hydrogen generation: } 2H^+ + 2e^- \rightarrow H_2 \ \uparrow \qquad (11.9)$$

This reaction occurs when the cathode is another metal and the electrolyte is acidic. This reaction is sometimes referred to as the plating out of $H_2$, and the upward arrow indicates that the hydrogen escapes as a gas. Unfortunately, not all of this hydrogen escapes. When metals are electroplated, hydrogen is also plated out according to equation (11.9) simultaneously with the metal at the plated part. Some of the hydrogen ions enter the plated metal. These ions have been the cause of many failed plated parts because hydrogen embrittles metals. There are many other sources of hydrogen in metal processing which lead to hydrogen embrittlement. This phenomenon is discussed more fully in the hydrogen embrittlement section.

$$\text{Water formation: } O_2 + 4H^+ + 4e^- \rightarrow 2(H_2O) \qquad (11.10)$$

In acid solutions containing dissolved oxygen, water is produced at the cathode instead of the $(OH)^-$ of equation (11.3), the latter occurring when the water is neutral or basic. The reaction that predominates depends on the electrolyte and the ion concentration. As pointed out above, more than one reaction may be occurring at the same time. But the predominate or controlled reaction determines the type of cell involved in the corrosion process. Metal ions in solution may also be reduced since these ions can exist in more than one valence state. The most common of these reactions in the corrosion process is the reduction of ferric to ferrous ions:

$$Fe^{3+} + e^- \longrightarrow Fe^{2+} \tag{11.11}$$

Galvanic Cells Between Two Metals

This is the type of cell that generates a large amount of corrosion when two metals that have a large separation between them in the galvanic or emf series are placed in direct contact within an electrolyte. This is also the type of cell that could be used as a battery. The Ni–Cd battery is another example of such a cell. The electrolyte is a solution of potassium hydroxide. Cadmium metal changes to cadmium oxide and the nickel to nickelous hydroxide on discharging, and the reverse reaction occurs on charging. The reactions are very complex and still not fully understood. The potassium hydroxide solution does not change in composition and strength and does not really become involved in the reactions. It is said to be a nonvariant electrolyte. The Ni–Cd cell lasts longer than the Pb–PbO$_2$ acid battery and requires less care. But most large galvanic cells between two dissimilar metals result in unwanted corrosion. The connecting together of copper pipes with a silver braze alloy is a typical example. Note how far apart these two materials are in the galvanic series of Table 11.2.

Concentration Cells

Concentration cells occur when the concentration of identical ions differ from one region to another in the system. One of the common cells of this type is the oxygen concentration cell. Consider the cell of Figure 11.5, where we have two iron electrodes in electrolytes of different oxygen concentrations. The electron-producing anode reaction and the cathode consumption reaction are the same as before:

anode:   $Fe^0 \longrightarrow Fe^{2+} + 2e^-$

cathode:   $O_2 + 2H_2O + 4e^- \longrightarrow 4(OH)^-$

Now both of these reactions are the same ones that occurred in our laboratory cell and the electrode reactions were expressed by equations (11.1) and (11.3). But our electrodes are of the same metal now. Since the cathode reaction requires electrons and also oxygen, it must have a higher concentration of oxygen than that at the anode. In an oxygen concentration cell the region low in oxygen will produce more electrons and become anodic with respect to the high oxygen region. The metal can be a single piece. Two separate metal electrodes, as shown in Figure 11.1, are not needed. One example of this type of cell is that of waterline corrosion. Immediately above the waterline the oxygen

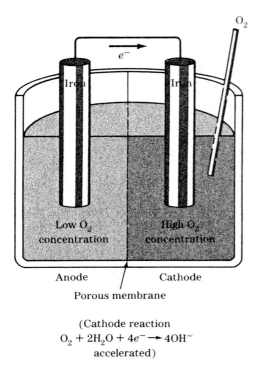

**Figure 11.5.** Oxygen concentration cell. The region of high oxygen concentration is cathodic to the region of low oxygen concentration. (After J. Wulff, Ed., *The Structure and Properties of Materials,* Vol. 2, Wiley, New York, 1964, p. 165.)

in the air will be of a higher concentration than that of the dissolved oxygen in the water below. Thus the latter region becomes anodic with respect to that on the same metal above the waterline. All of the requirements for a galvanic cell exist: the anode and cathode, the electrolyte, and the metal conductor. In this cell the corrosion is accentuated in the region of low oxygen concentration. Pitting often occurs immediately below the waterline in the oxygen-starved layer. Similar oxygen concentration cells can also form on metallic surfaces at localized spots where debris such as grease spots or dirt decrease the oxygen content underneath the accumulated junk. The same cell type could occur in cracks where the oxygen concentration might be low. This type of accelerated localized corrosion can lead to pit formation.

*Water Reduction.* In the absence of all other reactions water will be reduced as follows:

$$2H_2O + 2e^- \longrightarrow H_2 + 2(OH)^- \tag{11.12}$$

which is equivalent to (11.9) assuming dissociation of the water.

*Electroplating.* This reaction, although not considered as a corrosion reaction, obeys the same principles and occurs when there is a high concentration of metallic ions usually intentionally introduced to coat the cathode with metal atoms. The reaction is

$$M^0 + ne^- \longrightarrow M^{n+} \tag{11.13}$$

### 11.2.4 Corrosion Rates

Michael Faraday (1791–1867) established many of the basic principles of electricity and magnetism but is probably best known for his work on ion migration in an electric field. *Faraday's law,* which is applicable for both electroplating and corrosion is expressed as

$$W = \frac{ItM}{nF} \tag{11.14}$$

where $W$ = weight of metal plated or corroded per unit of time $t$ (g/s)
  $I$ = current flow (A)
  $M$ = atomic mass of metal (g/mol)
  $n$ = number of electrons produced or consumed
  $F$ = Faraday's constant = 96,500 C/mol or 96,500 A · s/mol.

*Example 11.2* For electroplating nickel from a nickel sulfate solution a current of 5 A was employed to plate a 1-mm thickness of nickel onto a steel surface of 10 cm$^2$ area. How much time was required to complete this operation?

Solution. The quantity of Ni = 10 cm$^2$ × 0.1 cm = 1 cm$^3$. Nickel has a density of 8.9 g/cm$^3$. Thus 8.9 g of Ni is required. In this reaction Ni has a valency of +2 since it goes into solution as follows: Ni° $\longrightarrow$ Ni$^{2+}$ + 2e$^-$.

$$W = \frac{ItM}{nF}$$

$$t = \frac{WnF}{IM} = \frac{8.9 \text{ g} \times 2 \times 96,500 \text{ A} \cdot \text{s/mol}}{5 \text{ A} \times 58.7 \text{ g/mol}} = 5097 \text{ s} = 1.1416 \text{ h}$$

Polarization

When metals are electropolated the current density is controlled and the plate thickness for a specific time can be accurately predicted. During corrosion reactions it is difficult to predict corrosion rates because of polarization, which may occur at either the anode or cathode, and thereby alter the potential at these electrodes. As corrosion progresses the corrosion rate will change due to changes in the electrolyte and oxides or other surface films. These changes are termed *polarization*. When the electrode potentials change due to polarization, the corrosion rate changes. Electrochemical reactions at both electrodes depend on the availability of the electrons involved in the reactions. During metallic corrosion, in the absence of polarization, the rate of production of electrons at the anode equals the rate of consumption at the cathode. If electrons are made available to the cathode too fast, the cathode becomes more negative and cannot accommodate all of these electrons in the cathode reaction. This is called *cathodic polarization*. On the other hand, a deficiency in the electrons required to sustain dissolution of the anode, which leaves its electrons behind, is called *anodic polarization*. As the deficiency becomes greater the driving force to produce more electrons becomes greater and the anodic dissolution increases. As the surface potential becomes more positive (i.e., fewer electrons available), the oxidizing power of the solution increases because of the larger anodic polarization. Polarization can be divided into two classes: *activation* and *concentration*.

*Activation polarization* is that related to the metal–electrolyte interface, where a sequence of steps is involved and the rate of one step in the process is the rate-controlling step of the entire process. The plating out of hydrogen is a good example to use here because it is also applicable to the hydrogen embrittlement problem to be discussed later. When hydrogen is evolved at the cathode, the reaction proceeds by the following steps:

1. Physical adsorption, where the gaseous molecules are attached to the surface by weak van der Waals forces.
2. Chemisorption, where the molecules become dissociated and are bound to the surface as atoms and/or radicals of the molecules (e.g., $H^+$ ions).
3. Electron transfer from the metal to form a hydrogen atom which combines with another H atom to form the $H_2$ molecule.
4. $H_2$ gas evolves as bubbles.

The slowest of these steps, probably step 1, determines the rate of the overall reaction. An activation energy barrier is associated with the rate of the limiting step. (Incidentally, many of the $H^+$ ions do not combine with electrons as in step 3 above but diffuse into the metal and cause hydrogen embrittlement).

*Concentration polarization* exists when the reaction rate is limited by diffusion in the electrolyte. If the number of $H^+$ ions is small, the reduction rate may be controlled by the diffusion of $H^+$ ions to the metal surface. Concentration polarization is usually not a factor in anodic dissolution of metals. Increasing the agitation of the corrosive medium will increase the corrosion rate only if the cathodic reaction process is controlled by concentration polarization. When both cathodic and anodic reactions are controlled by activation polarization, agitation will have an influence.

When the surface area of the cathode in a galvanic cell is large in comparison to the anode, the anodic current density is large and this anodic polarization leads to more pronounced anode dissolution. Conversely, a smaller cathode slows the galvanic action due to the predominate cathodic polarization.

*Passivation* is a phenomenon that affects the corrosion rate very significantly and always in a favorable direction. Most metals that have a high degree of passivity are also those that form tight adherent protective oxide layers on the surface of the metal, although it is believed that the passitivity mechanism is not simply that of an oxide barrier mechanically separating the metal from the electrolyte. Stainless steel is passive because of its chromium content. When this chromium content exceeds about 11 wt %, a complete surface oxide ($Cr_2O_3$) is formed in a few seconds on exposure to air. Similar oxide films form on aluminum ($Al_2O_3$), titanium ($TiO_2$), and silicon ($SiO_2$). Nickel and iron show some degree of passivity but not to the same extent as the others mentioned above. There is some controversy as to the nature of this barrier, but it is believed by many to be some type of thin hydrated oxide layer, on the order of tens of angstroms in thickness.

Passivation is usually demonstrated by polarization curves, where the electrode potential is plotted versus the log of the current density (Figure 11.6). For a normal active metal the current and hence the electrode dissolution rate increase exponentially with increasing positive potential on the electrode. This is termed the *active* state of the metal and correspondingly, the active region of the polarization curve. But the metals that possess the ability to become passive show a marked decrease in the corrosion rate, by a factor of $10^3$ to $10^6$ when the potential exceeds some critical value denoted as $E_p$ on the polarization curve. The corresponding critical current is known as $I_c$. As the electrode

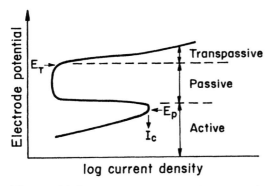

**Figure 11.6.** Schematic polarization curve for a metal that displays an active-passive transition.

potential exceeds the $E_p$ value, the current density suddenly decreases to a constant value that does not change with increasing electrode potential. This is the passive state of the metal and is noted on the polarization curve as the *passive* region. But when the electrode potential attains a sufficiently high value ($E_t$) the metal again becomes active. This part of the curve is termed the *transpassive* region.

The state of passivity can be altered by the environment. Chloride ions are noted to be detrimental to most oxide films, particularly that on aluminum. Breaks in this film lead to localized corrosion. The damage to bridge structures by NaCl, which was mentioned earlier, is to some extent related to the destruction of passivity of the metal reinforcing bars.

### 11.2.5    Types of Corrosion

The ability to identify the type of corrosion of concern is of particular importance to selection of the kind of corrosion protection that could be employed. The types usually experienced can be identified by the appearance of the metal surface.

Uniform Corrosion

As the term *uniform corrosion* suggests, this type of corrosion appears visually the same over all of the surface, as shown in Figure 11.3 for the rusted spike. There are many small galvanic cells on the surface, and these may shift position from time to time, causing the corrosion to spread over the entire surface. Weathering steels and some copper alloys generally show uniform attack (Figure 11.7). The rusting of iron and steel and the formation of patina on copper

**Figure 11.7.**   Uniform atmospheric corrosion of a structural steel member from the Golden Gate Bridge after 50 years of service. The irregular-shaped holes are where the corrosion penetrated the entire thickness of the member. (Courtesy of Prof. R. H. Heidersbach, Cal Poly.)

occur readily in moist atmospheres and are some of the best examples of uniform corrosion.

Uniform corrosion is one of the more common forms and can be controlled in a number of ways, such as by protective coatings, inhibitors, reduction in the moisture where possible, cathodic protection, or by alloy selection. Small additions of copper, nickel, and phosphorus have proven to be very beneficial in retarding atmospheric corrosion in steels. Aqueous environments, including seawater, will usually cause uniform corrosion on steel parts other than the stainless steels and aluminum, where pitting is more likely to be found. In certain industries chemical dissolution by acids, bases, and chelants frequently cause uniform corrosion and require alloys designed specifically for these environments. (*Note*: Uniform corrosion is sometimes referred to as *general corrosion*.)

## Pitting Corrosion

Localized corrosion is generally classified as pitting corrosion when the diameter of the pit is less than the depth. At certain depths the pits may branch out into several directions. In pitting corrosion the anodic and cathodic sites do not shift but become fixed at certain points on the surface. Pitting often requires a certain amount of initiation time but the pit grows faster with time as conditions within the pit become more aggressive. Oxygen concentration cells can initiate a pit and the difference in oxygen concentration between the pit and the surface will accelerate pit growth rates. In sea water, positively charged anodes at the bottom of the pit attract negative chloride ions. The reaction at the bottom of the pit is the usual anode reaction of $M° \rightarrow M^n + ne^-$, while the cathode reaction takes place at the surface forming $(OH)^-$ ions as in equation (11.3) [i.e., $O_2 + 2H_2O + 4e^- \rightarrow 4(OH)^-$]. The combination of these two reactions causes a buildup of acidic chlorides according to the following reactions:

$$M^+Cl^- + H_2O \rightarrow MOH + 2Cl^- + H^+ \tag{11.15}$$

The acid buildup at the bottom of the pit increases the reaction rate, which now becomes an autocatalytic reaction (Figure 11.8). A 6% $FeCl_3$ solution is a common testing medium for pitting and crevice corrosion. Pits may be initiated at surface inclusions, breaks in the inhibitor, or passive film coverage and surface deposits. Examples of pitting corrosion in stainless steels is shown in Figure 11.9.

## Crevice Corrosion

Crevice corrosion is a localized corrosion type that occurs in narrow gaps or small openings, where two materials, either metal to metal or metal to nonmetal, come into contact. In the latter case the second material may be debris,

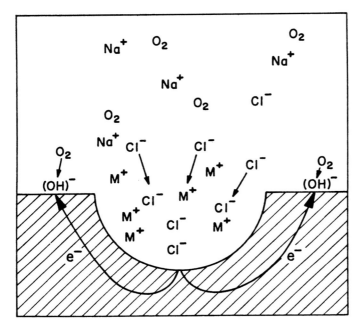

**Figure 11.8.** Schematic of an autocatalytic corrosion pit formation in seawater.

such as grease or mud, gaskets, and packing materials, and as a result this type of crevice corrosion has been called *deposite* or *gasket corrosion*. In the former situation cracks and seams may be the point of corrosion initiation even between two regions of the same metal. The actual mechanism varies from material to material and with the environment, but what is common in all crevice corrosion is the presence of a crevice where stagnant fluids collect. Furthermore, it is always a localized form of attack. Usually, there will exist some type of concentration cell such as that depicted in Figure 11.10. But the chemical reactions are believed to be more involved than those in simple concentration cells. Metal ions are released in the anodic crevice following equation (11.8) when the potential in the crevice exceeds some critical value. The cathodic reaction in the regions surrounding the crevice occurs according to equation (11.3). However, in the situation for iron alloys and the presence of chloride ions, the following reaction may occur:

$$Fe^{2+} + 2H_2O + 2Cl^- \longrightarrow Fe(OH)_2 + 2HCl \qquad (11.16)$$

Such crevice corrosion will proceed autocatalytically as more chloride is attracted to the crevice, much as in pitting corrosion.

(a)

(b)

(c)

**Figure 11.9.** Variation in stainless steel pitting corrosion resistance in model SO₂ scrubber environment: (a) type 304 in acid condensate; (b) type 316 in acid condensate; (c) type 304 in a limestone slurry zone. (Courtesy of the LaQue Center for Corrosion Technology, Inc., Wrightsville Beach, N.C.)

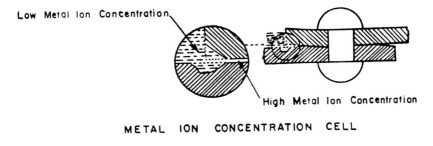

METAL   ION   CONCENTRATION   CELL

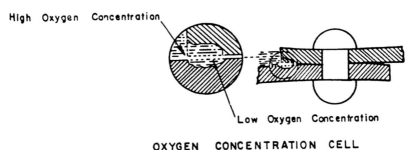

OXYGEN   CONCENTRATION   CELL

**Figure 11.10.**   Crevice   corrosion   due   to   concentration   cells.   (From   P.   A. Schweitzer,   Ed.,   *Corrosion   and   Corrosion   Protection   Handbook,*   2nd   ed.,   Marcel Dekker,   New   York,   1989,   p.   11.)

For stainless steels and in general the passive alloys, the concentration of acid and chloride ions become sufficiently aggressive to break down passive films. Examples of crevice corrosion are shown in Figure 11.11. The occurrence of crevice corrosion in copper alloys is believed to be of the more simple metal ion concentration type.

A Teflon or similar sealant can be used around mechanical joints to keep out the electrolyte. However, one must be careful to avoid crevices between metallic and nonmetallic parts such as gaskets and seals. The avoidance of regions where stagnant solutions will accumulate is a must. Increasing the alloy content with Cr, Ni, Mo, and nitrogen increases the resistance to both crevice and pitting corrosion. The duplex stainless steels discussed in Chapter 6 contain a few percent of Mo specifically for this purpose. To be effective, these alloying elements must all be in solid solution.

Stray-Current Corrosion

This type of corrosion occurs in the presence of some external direct current, most commonly encountered in soils containing water. It is more frequently a

(a)

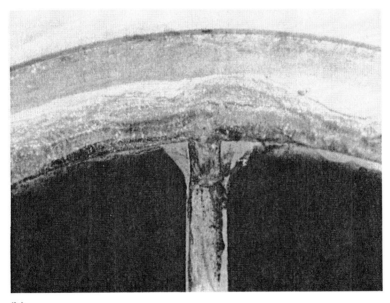

(b)

**Figure 11.11.** Examples of crevice corrosion: (a) metal-to-metal crevice site formed between components of a type 304 stainless steel fastener in seawater; (b) at a nonmetal gasket site on an alloy 825 heat exchanger. (Courtesy of the LaQue Center for Corrosion Technology, Inc., Wrightsville Beach, N.C.)

problem in underground structures than in underground piping. Protective coatings and insulation are two possible preventive methods that may be used for this type of corrosion.

## Intergranular Corrosion

Intergranulator corrosion consists of selective attack at the grain boundaries. The fact that chemical etchants reveal grain boundaries is evidence of their susceptibility to chemical attack and dissolution at these boundaries.

One of the most common examples of grain boundary corrosion is that termed *sensitization.* It occurs in austenitic stainless steels, some superalloys, and perhaps in other high-chromium alloys that have been heated into certain temperature ranges. For the austenitic stainless steels this range is from 500 to 600°C (950 to 1450°F). The chromium in this particular temperature range precipitates out of the solid solution as chromium carbides. The chromium is no longer available to form the passive chromium oxide ($Cr_2O_3$) protective film. The alloy is said to be in the *sensitized condition,* or in more simple words, is now susceptible to galvanic corrosion and the related rust formation. The austenitic stainless steels alloys 321 and 347 contain small amounts of Ti and Nb, respectively, which are introduced to tie up the carbon and reduce the precipitation of chromium carbides. But this is not a cure-all. 316L and 304L austenitic stainless steels of low carbon content have also been used, particularly where welding is involved. Welding operations often heat the adjacent metal sections into the critical temperature range for the chromium carbides to precipitate. A typical microstructure of a sensitized alloy is shown in Figure 11.12.

The other major types of intergranular corrosion—stress corrosion cracking and hydrogen-assisted cracking—are treated in a later section.

## Selective Leaching

This type of corrosion occurs when a particular element in an alloy is preferentially removed with respect to the major one. Dezincification of brass is a typical example where zinc is removed from brass in certain environments. Similar processes have been observed for other alloys in which Al, Fe, Co, and Cr are preferentially removed.

## Erosion–Corrosion

*Erosion–corrosion* describes a situation where metal removal occurs by the combined action of corrosion (chemical attack) and erosion (the wearing away of the metal by fluid motion). Soft metals are sensitive to this form of attack. Grooves and gullies will appear. Corrosion tests conducted under static conditions are not valid for predicting corrosion rates when a moving fluid is involved.

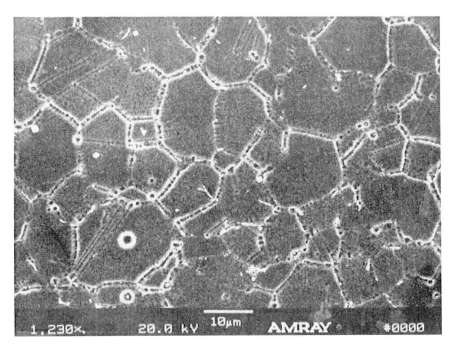

**Figure 11.12.** Example of sensitization in X750 Inconel after 2 h at 1400°F plus 5 h at 1100°F. Note the carbides at the grain boundaries. (Courtesy of Roxanne Hicks, Cal Poly.)

Corrosion Fatigue

*Corrosion fatigue* is a subject that, as its name suggests, involves two mechanisms, both of them very complex. We can simplify this phenomenon somewhat by separating the two mechanisms prior to joining them together. But first let's define corrosion fatigue as that type of fatigue failure that occurs during application of a cyclic stress in the presence of a corrosive environment or corrosive action. In Chapter 2 we described how fatigue crack nucleation would occur by the to-and-fro movement of dislocations, but that most fatigue cracks propagate from surface defects. One of the primary roles of corrosion in the corrosion fatigue process is that of providing surface defects for crack formation and propagation. A corrosion pit, for example, can concentrate stresses by factors of 10 to 100 or more. There is little doubt that fatigue cracks frequently start at corrosion pits. Crack propagation can also be enhanced by a corrosive process in some alloys whereby the energy required to break atom bonds and provide two new surfaces is reduced by the corrosive media. The breaking of surface films may also be involved. The interrelation of crack propagation in the presence of cyclic stress and corrosive environments is far beyond the scope of this book. The ASM *Atlas* (see the Selected Reading) presents a considerable amount of data and insight on this phenomenon. For the present we leave this subject to specialists in this field.

### 11.2.6 Stress Corrosion Cracking and Hydrogen Damage

When discussing stress corrosion cracking and hydrogen damage we encounter some difficulties, in part due to semantics and in part to the controversy that exists in the mechanisms involved in each process. Persons from different industries label the same embrittlement process by different names. So let's begin with the semantics problem by listing the various names that one could encounter in the literature for these embrittlement phenomena. First it must be mentioned that as far back as 1959, A. R. Troiano and co-workers concluded that much of what had been previously called stress corrosion cracking (SCC) was related to hydrogen effects, commonly referred to at that time as hydrogen embrittlement (HE). Today, it is generally considered that one form of HE is a mechanism of SCC.

*Stress corrosion cracking (SCC)*: phrase developed early in the twentieth century that has generally been defined as the result of the combined action of static stresses and corrosion, the stresses being either residual or externally applied.

*Hydrogen-assisted cracking (HAC)*: phrase, mostly used by metallurgists, to describe the principal type of HE, where a few ppm of hydrogen in the form of $H^+$ ions causes extreme embrittlement of metals, resulting in loss of ductility and often strength. HAC is the form of HE mentioned above that is now considered to be one mechanism of SCC.

*Hydrogen-induced cracking (HIC)*: phrase that to many engineers means the identical process as HAC. However, in the petroleum industry this phrase is used to describe blister formation on the surface of steels, where the hydrogen, in the form of $H_2$ gaseous molecules, builds up sufficient internal pressure at points near the surface to cause the formation of bumps (blisters). A 1989 article in the journal *Corrosion* states: "HIC normally observed in pipeline steels includes not only hydrogen blistering but also internal stepwise cracking parallel to the pipeline." It appears that even those in the petroleum industry do not always agree on the meaning of this term.

*Sulfide stress cracking (SCC)*: type of cracking identical to HAC, but the term was coined by the petroleium industry to describe cracking when hydrogen sulfide environments generate the hydrogen, in contrast to the term HAC used in other industries, where the source of hydrogen is not defined.

*Sulfide stress corrosion cracking (SSCC)*: same process as SCC.

*Hydrogen embrittlement (HE)*: term used synonymously with HAC by many metallurgists but by others to include all forms of hydrogen damage.

Despite the confusion in semantics, certain forms of SCC can, we believe, be separated from hydrogen effects, and vice versa. It will assist in clarifying the picture if SCC is considered as a phenomenon that consists of two mechanisms, one that we label as anodic dissolution, which occurs independently of hydrogen, and the other as some type of HAC mechanism, where only a few ppm of hydrogen ions is required for embrittlement.

Stress Corrosion Cracking

The first acknowledgment or recognition of SCC was that which occurred early in the twentieth century when cartridge shells made of 70% Cu–30% Zn were found to crack over a period of time. It was later realized that ammonia from decaying organic matter in conjunction with the residual stresses in the brass was responsible for the cracking of these shells. The presence of high humidity also promoted SCC. This phenomenon was termed *season cracking* because it was more prominent in warm, moist climates (or seasons). Stress-relief annealing of the brass reduces the residual stresses without significantly reducing the strength, and at the same time reduces the susceptibility of the brass to SCC. SCC of certain nonferrous metals and the responsible environments are listed in Table 11.3. Anodic dissolution is the most likely cause of this type of SCC. Notice that steels have been omitted from this table intentionally, because in steels the cracking mechanism of SCC phenomenon is probably that due to hydrogen effects. The mechanisms of the forms of SCC in Table 11.3 are not completely understood, although anodic dissolution of the metal by a localized galvanic cell accompanied by tensile stresses is responsible for crack growth. Crack nucleation may begin at surface pits or similar discontinuities. The crack grows in a plane perpendicuar to the applied stress (this is also the frequent mode of crack propagation in the absence of corrosion). It has been stated *that if either the corrosion or the stress is removed, the crack stops growing* provided that the crack is below a certain critical length for the stress intensity present at the crack tip. This is the stress intensity that is a function of the applied stress and the crack length as described in Chapter 2 [equation (2.17)] and becomes the $K_{1c}$ material constant at the critical crack length for propagation to fracture. In corrosion we often see the term $K_{1SCC}$ used for the critical stress intensity to propagate a crack in the presence of a corrosive medium. This $K_{1SCC}$ is not a true material constant but is a useful measure of the relative susceptibility of materials to cracking in certain environments. The role of the tensile stresses is not completely understood but is believed to be associated with the rupture of surface films, which thereby promotes further anodic dissolution and an increase in stress concentration as the crack grows via the anodic dissolution process. A tensile stress is necessary for propagation of the cracks

**Table 11.3**   Environments That Produce Stress Corrosion Cracking in Metals
(Partial Listing)

| Nonferrous metals that show SCC | Environment |
| --- | --- |
| Copper alloys | Ammonia |
| Aluminum alloys | Most halide ion solutions |
| Magnesium alloys | Chloride solutions |
| Titanium alloys | Most halide ion solutions |
| Zirconium alloys | $FeCl_3$ or $CuCl_2$ solutions |
| High-nickel alloys | Aerated hydrofluoric or hydrofluorosilicia acid vapor |

in addition to the rupturing of surface films. SCC that occurs by anodic dissolution often occurs along grain boundaries and results in intergranular cracking (IGSCC), although considerable transgranular cracking has also been observed.

SCC by Hydrogen

In this book the SCC by hydrogen (HAC) type of hydrogen embrittlement is reserved for that commonly encountered in steels, although it has also been observed in Al, Ti, Ni, and possibly other nonferrous alloys, where only a few parts per million of hydrogen can cause embrittlement. This is also the type of cracking referred to above as a subclass of SCC. It appears in many cases that it can occur without residual or applied stresses. Applied stresses after introduction of the hydrogen are usually necessary to observe the embrittlement effect, which changes the fracture from a ductile to a brittle type as a result of the small hydrogen content. In a few cases a ductile failure has been observed accompanied by a reduction in ductility. Some have speculated that an applied stress, subsequent to or during the introduction of hydrogen, allows dislocations to move, and during their movement they collect hydrogen and dump it at interfaces such as grain boundaries and phase boundaries, and as such, stress is required. Stress may also enhance the diffusion of hydrogen ions. The belief that stress was required is probably the result of early tests where stress applied in hydrogen atmospheres produced curves such as that shown in Figure 11.13. This curve shows stress versus time to fracture, and where the stress levels off, the material appears to be immune to the effects of hydrogen. Because of the close similarity to the *S–N* curves of fatigue (Figures 2.19 and 2.20), this phenomenon has been labeled *delayed fatigue*. This term is misleading, and hopefully will eventually disappear from the literature. But there is ample evidence that stress for HAC in many cases is not really required, other than that for observing the effect. A few isolated cases have been reported

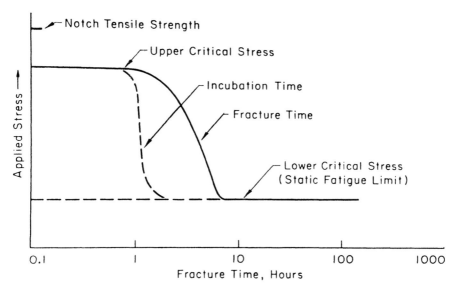

**Figure 11.13.** Schematic representation of delayed failure characteristics of a hydrogenated high-strength steel. (From A. R. Trioano, *Trans. ASM,* Vol. 52, 1960, p. 54.)

where cracking occurred just due to the introduction of the hydrogen ions alone (not $H_2$ gas), and many results can be found in the literature where cracking and fracture of previously hydrogen-charged specimens occurred when stresses as low as 30% of the yield stress were subsequently applied to hydrogen-charged specimens (Figure 11.14). Thus the hydrogen ions have drastically reduced the fracture stress. It is doubtful if stress-induced hydrogen migration plays a significant role at such low stresses. For low hydrogen contents the ductility is reduced, while at higher hydrogen contents both strength and ductility are severely reduced. Figure 11.15 shows a brittle ring in a PH stainless steel tensile test specimen that had been electrolytically charged with a few ppm of hydrogen prior to testing to failure. The boundary between the brittle and ductile failure regions in Figure 11.15 is that where the concentration of hydrogen was below some very small critical level. SEM fractographs (Figure 11.16) show brittle intergranular failure in the brittle ring and a dimpled ductile microvoid coalescence core in the center of the specimen. In a few cases the brittle fracture was transgranular, possibly due to hydrogen trapping at carbide–matrix interfaces. The crack, of course, will follow the path of least resistance, and this is often the grain boundary. In many cases it is difficult to

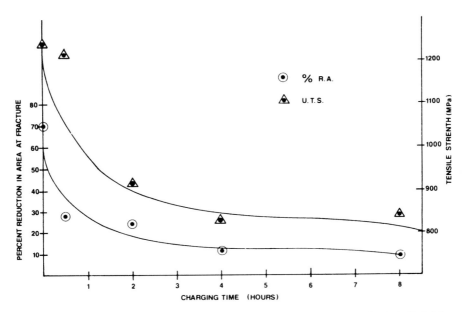

**Figure 11.14.** Effect of charging time on strength and ductility of a PH 13-8 Mo stainless steel (H 1050 condition). (From G. T. Murray et al., *Corrosion,* Vol. 40, 1984, p. 147.)

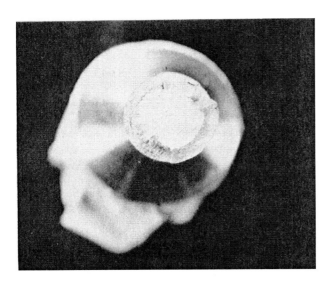

**Figure 11.15.** Brittle ring on the exterior of PH 13-8 Mo stainless steel (H1050 condition) that was fractured in a tensile test after being electrolytically charged with hydrogen at a current density of 9 mA/cm$^2$ for 4 h. 10×. (From G. T. Murray et al., *Corrosion,* Vol. 40, 1984, p. 147.)

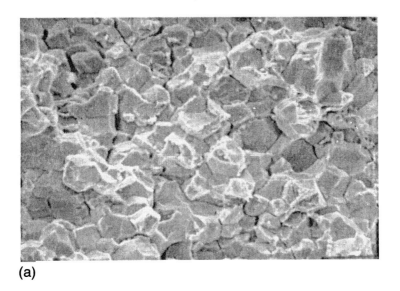

(a)

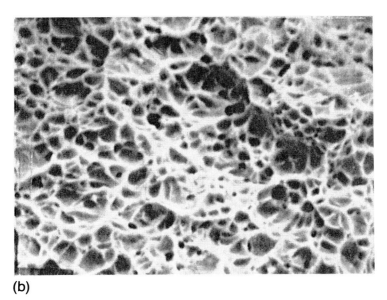

(b)

**Figure 11.16.** SEM fractographs of (a) brittle hydrogen failure taken from the brittle ring in the specimen of Figure 11.15, and (b) ductile failure taken from central portion of specimen in Figure 11.15. (From G. T. Murray et al., *Corrosion*, Vol. 40, 1984, p. 148.)

distinguish the anodic dissolution form of SCC from HAC since both often show branched cracking (Figure 11.17) and both can be either intergranular or transgranular. The classic form of anodic dissolution SCC, such as that observed in brass, often shows more branched cracking than does that due to hydrogen. The separation of the two mechanisms is further complicated by our inability to ascertain what chemical reactions may be occurring at the crack tip. The chemistry at the crack tip may be entirely different from that taking place on the external surface of the metal part.

There are many sources of hydrogen that can cause HAC, including $H_2S$, water (particularly salt water), surface cleaning acids, electroplating, welding, and contact with hydrogen gas. Surface protection, particularly oxides on steel, will retard hydrogen penetration. Another preventive measure is to determine the critical stress level and ensure that this stress is not exceeded. Heating the metal after hydrogen exposure (baking) will remove a large percentage of the hydrogen, but in many cases this operation will not prevent HAC.

Hydrogen Damage (Other Than SCC)

There are certain forms of hydrogen damage that are definitely distinguishable from SCC. These hydrogen effects are described in the following:

1.  *Hydrogen attack.* This type of embrittlement is a result of hydrogen combining with carbon to form a gas, usually methane. The gas pressure builds up to the point where it can cause cracking of the metal. Surface decarburization by the hydrogen occurs at temperatures above 540°C, while internal decarburization can occur at temperatures as low as 200°C. Closely related to this effect is the decreasing solubility of $H_2$ in molten metal with decreasing temperature. Upon solidification $H_2$ pressure builds up sufficient to cause cracks to form. These cracks were observed many years ago, and due to their shiny fracture surfaces, visible without microscopes, were called *fisheyes, flakes,* and an assortment of similar names.

2.  *Hydrogen blistering.* This term covers a multitude of causes. Hydrogen gas blistering has been found to occur in plated surfaces and in the petroleum industry, where it is caused by $H_2S$ reacting with the metal. In the latter case it would be called HIC.

3.  *Hydride formation.* Hydrogen cracking via brittle hydride formation occurs in metals that readily form hydrides (e.g., Ta, Ti, Zr, and Nb). Hydrogen concentrations greater than 100 ppm are usually required. The cracks form in the brittle hydrides rather than the metal matrix.

In summary, there are many ways in which hydrogen can enter and embrittle or cause cracks, blisters, and so on, to form in metals. The vocabulary of hydrogen damage is large and sometimes overlapping and confusing. Strain rates and temperature have an effect. Some mechanisms are better understood than others. The HE (HAC) caused by a few ppm hydrogen content is believed

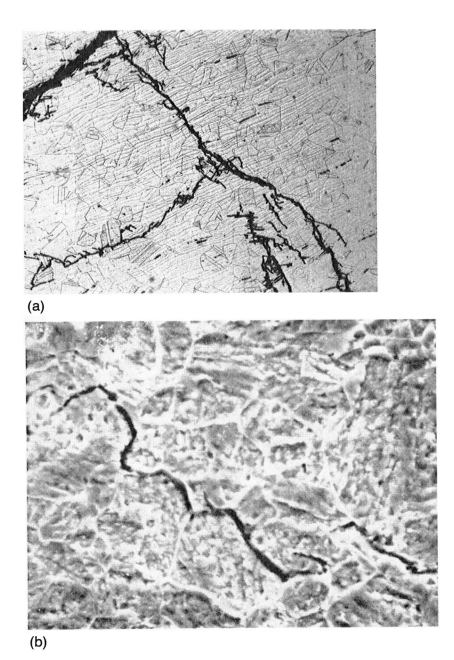

(a)

(b)

**Figure 11.17.** (a) Branched crack formation, typical of stress corrosion cracking, found in an austenitic stainless steel bracket on a sailboat. Both transgranular and intergranular cracking are evident. (Courtesy of Eric Willis, Cal Poly). (b) Intergranular cracking in a hydrogen-charged PH stainless steel.

to be related to some type of decohesion (i.e., the energy to form two new surfaces is reduced by the presence of hydrogen ions).

## 11.2.7 Microbiological Corrosion

This subject has been given more attention recently, although its intrusion into the field does not present any real new form of corrosion process nor does it affect any of the fundamental electrochemical equations stated earlier. There are two ways in which it plays a role: (1) by the introduction into the system of new chemical entities such as acids, alkalides, sulfides and other aggressive ions, and (2) by entering into one or more of the electrochemical reactions at the surface of a part, thereby affecting the kinetics of the reaction. Hydrogen ingress has been reported to occur at some microbiological deposites. Microorganisms covering a wide range of species are encountered in corrosion problems, but in the common ones, sulfur and/or its compounds play a vital role. Sulfate-reducing bacteria (SBRs) reduce sulfates to sulfides, which appear as curved rods. They are aerobic; that is, they require oxygen for growth. On the other hand, the most serious corrosion problems involving microorganisms occur on iron alloys in the absence of oxygen (anerobic conditions). Also, some rather serious corrosion of concrete pipes, cooling towers and building stones have been associated with this type of corrosion. Prevention methods include inhibitors, which kill the organisms; elimination of stagnant conditions where growth proceeds at a rapid rate; and protective coatings. A typical form of this corrosion is shown in Figure 11.18.

## 11.2.8 Prevention of Metallic Corrosion

Although some preventive methods were mentioned where appropriate in previous sections, we will briefly review a number of common preventive methods. General or uniform corrosion is perhaps the easiest form to prevent because of the numerous prevention methods available. Coatings, in the form of paint, oil, oxides, inhibitors, and cathodic protection, have all proven to be effective methods for slowing the action, or preventing the formation, of the trillions of galvanic cells on the metal surface. *Cathodic protection* is accomplished by attaching an extremely anodic (active) metal such as zinc or magnesium on the metal surface at close intervals. Cathodic protection uses the more anodic material to supply electrons to the metal structure to be protected. These electrons flow to the cathode, allowing the electrical circuit to be completed without drawing electrons from the more relatively less anodic metal that is being protected. Cathodic protection of a buried steel pipline using a magnesium *sacrificial anode* is depicted in Figure 11.19. In this pipeline the magnesium corrodes preferentially to the steel. Cathodic protection using sacrificial anodes has been applied to ship structures and hot-water tanks to

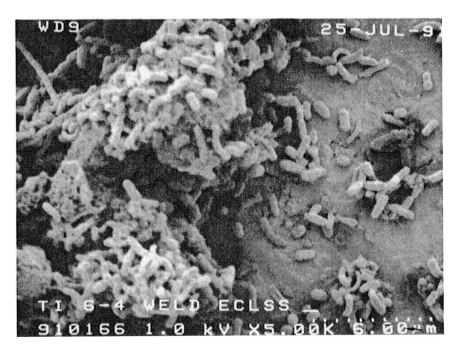

**Figure 11.18.** Microbiological corrosion: deposits of rod-shaped sulfate-reducing bacteria (SBR) on titanium after exposure to tap water. (Courtesy of Prof. Dan Walsh, Cal Poly.)

name a few applications. To determine the size and spacings of the sacrificial anode, the energy content and anode efficiency must be known. For magnesium, the most common sacrificial anode metal (approximately 10 million pounds per year), the theoretical energy content is 1000 A-h/lb with 50% efficiency. The number of pounds of Mg required to provide 1 A for a year can be computed as follows:

lb Mg/A-yr = (8760 h/yr)/(500 A-h/lb) = 17.52

Equations have been developed for protection of underground structures based on soil resistivity, number of anodes, anode spacing, and size (see the book by Schweitzer in the Suggested Reading). The resistance of a system in terms of the variables above is expressed by the Sunde equation as follows:

$$R = \frac{0.005P}{NL} \left[ 2.3 \log \left( \frac{8L}{d-1} \right) + \frac{2.3}{S(\log 0.656N)} \right] \qquad (11.17)$$

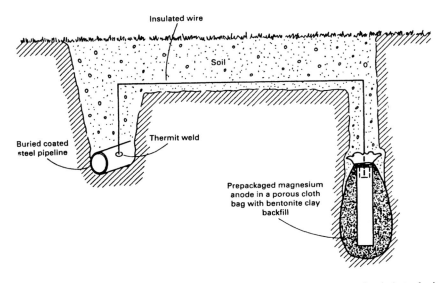

**Figure 11.19.** Sacrificial anode of magnesium used to protect a buried steel pipeline. (Courtesy of Prof. R. H. Heidersbach, Cal Poly.)

where $R$ = resistance of system ($\Omega$)
$\quad\quad P$ = soil resistivity ($\Omega \cdot$ cm)
$\quad\quad N$ = number of anodes
$\quad\quad L$ = anode length (ft)
$\quad\quad d$ = diameter of anode (ft)
$\quad\quad S$ = anode spacing (ft)

Another protection technique, somewhat analogous to the sacrificial anode method, is to impress a current with an external dc power supply connected to an underground tank or pipeline. The negative terminal of the power supply is connected to the metal to be protected, and the positive terminal to an inert anode such as graphite, located some distance away from the structures to be protected. For underground structures the resistivity of the soil will determine the required applied voltage. If the leads are properly insulated, current passes to the metallic structures and suppresses the corrosion reaction that would otherwise take place by metallic ions of the protected metal going into solution and leaving electrons behind. This sort of reverses the reaction stated in equation (11.1). To provide cathodic protection a current density of a few milliamperes per square foot is required. Insulating coatings are also helpful. One magnesium anode is capable of protecting about 100 ft of bare pipeline compared to about 5 miles of coated pipeline.

For prevention of giant galvanic corrosion cells where large pieces of dissimilar metals are placed in contact with each other, there are a number of possible approaches. First, try to avoid dissimilar metal contact by design, or at least minimize it by choosing two metals or alloys that are close together in the galvanic series. When it is not possible or practical to avoid high anodic–cathodic combinations, try to find a way to insulate these two metals (i.e., prevent their contact). Polymer washers, couplings, and bolt heads have been used for this purpose. Where metallic junctions cannot be avoided, cover the junction with paint or some other protective medium that will assist in preventing the electrolyte from contacting the dissimilar junction.

Crevice corrosion prevention is another form of corrosion that must be approached from the design standpoint, the most important being to avoid crevices where possible. Avoid areas where stagnant liquids might collect. Again, surface protection can be used in most cases.

Pitting corrosion prevention must be approached in the same manner as that for crevice corrosion. Some materials pit more than others (e.g., austenitic stainless steels, aluminum, and titanium), so material selection is important. Aluminum is particularly susceptible to pitting in chloride environments. The $Cl^-$ ion tends to break down the protective passive film. Both hydrogen and chloride ions promote dissolution of metals and may start a pit that feeds on itself by collecting more $H^+$ than $Cl^-$ ions. These environments must not come into contact with the bare metal. The selection of the duplex stainless steels is a possible solution that should be seriously considered for pitting environments.

Sensitization of stainless steels can be minimized by selection of a stabilized 321 or 347 or a low-carbon stainless steel. For other forms, consult tables (see the Suggested Reading) that show the metal and the media where intergranular attack can be expected to occur.

For many years the most publicized and touted method for prevention of SCC has been that of a *stress-relief anneal,* which is simply an anneal or heat treatment of a cold-worked metal that relieves much of the elastic stresses resulting from the cold work without sacrificing most of the strength intentionally introduced by the cold-working operation. In theory this causes a realignment or disentanglement of the dislocations introduced by, and responsible for, the increase in strength caused by the cold-working operation. Whether or not the theory is correct, a stress-relief anneal does appear to reduce the susceptibility of the material to the SCC of brass. In the case of the HE form of stress corrosion cracking, it has been observed that high dislocation density promotes cracking.

Inhibitors will reduce the rate of anodic oxidation or cathodic reduction, or both processes. Inhibitors are physically or chemically absorbed, the latter involving transfer or sharing of charge. Inhibitors can be broken down as passivation, organic, or precipitation inhibitors. Passivation inhibitors are the most effective and usually consist of chromate or nitrite substances that are absorbed into metal surfaces in the presence of dissolved oxygen. Chromates are inexpensive and widely used in automobile radiators engine blocks, and cooling towers. A wide variety of organic inhibitors [e.g., sodium benzoate, ethanoline phosphate (used in ethylene glycol cooling systems)] are available, which are absorbed on to the surface and provide a protective film, thus retarding dissolution of the metal in the electrolyte. Precipitation inhibitors are those that provide a protective film in the form of a solid precipitate in the metallic surface. Sodium phosphate will cause precipitation of a calcium or magnesium orthophosphate onto a metal surface.

### 11.2.9 Other Types of Metallic Degradation

Metals are attacked by many environments that result in deterioration of their desirable properties, even in the absence of galvanic cell formation. The most notable of these effects are those due to oxygen. Oxidation can occur by the oxygen penetrating a porous, previously formed oxide film such that the oxidation reaction at the metal–oxide interface continues. Oxygen may also diffuse atomically through a nonporous oxide film and react at the metal–oxide interface, a somewhat slower process of oxidation. Similarly, the metallic cations may also diffuse through the nonporous oxide layer and react with the air–oxide or oxygen–oxide outer surface.

The oxide may form a thin protective layer, as in the passivation process, or continue to grow in thickness until it flakes off the metal, exposing a new fresh metal surface, a process that continues until the metal is entirely consumed. The Pilling–Bedworth ratio $R$, which was first presented in the *Journal of the Institute of Metals* in 1923, can be used to predict the tendency to form a protective oxide coating. $R$ is expressed as follows:

$$R = \frac{Md}{amD} \tag{11.18}$$

where $M$ = molecular weight of oxide of formula $M_aO_b$
$d$ = metal density
$m$ = metal atomic weight

For a value of $R > 1$, the oxide tends to be protective. In addition to the passive metals mentioned earlier, Cu, Mn, Be, and Co tend to form protective oxides according to the Pilling–Bedworth criterion.

The reactivity of metals to other environments is far too broad a subject to be treatd here. The interested reader should consult the ASM *Atlas* listed in the Suggested Reading.

## 11.3 ENVIRONMENTAL DEGRADATION OF POLYMERS

Polymers are inert in many aqueous solutions and as such are frequently used to replace metal parts. Typical examples include ABS, polypropylene, and PVC as piping, tanks, scrubbers, and columns in process equipment. Teflon has been used in heat exchangers, and nylons and acetals for small parts and gears. In the chemical industry polymers have been used alone and as metallic coatings, the latter being where more strength is desired. It is because of all of these applications of polymers that their reactions with certain environments must be considered.

Chemical attack of polymers can be broadly classified as follows:

1. Degradation of a physical nature such as permeation or solvent action
2. Oxidation, the breaking of bonds
3. Hydrolysis-bond splitting by the addition of water (e.g., attack of ester linkages)
4. Radiation

One could include thermal degradation to this list, but since this is not a chemical attack it will not be included in this chapter (see Chapter 7). The theory involved in all such attacks is beyond the scope of this book. Suffice it to say that softening, embrittlement, crazing, dissolution, and blistering are typical results. There is a significant, and in some cases vast, difference in the reaction of polymers to specific environments.

The most chemically resistant polymer developed to date is tetrafluoroethylene (Teflon or Halon). It is practically unaffected by acids, alkalies, and organics at temperatures to 260°C (500°F). It is an excellent material for gaskets, O-rings, and seals. It has also been used as linings for tanks and ductwork. Polyethylene is a low-cost polymer that has excellent resistance to a wide variety of chemicals. Because of its low strength and low cost, it is used in large quantities as thin sheet liners in metal containers and for agricultural purposes. It is temperature limited and is used primarily at ambient temperatures. PVC materials have excellent resistance to oxidizing acids, other than nitric and sulfuric, and in all nonoxidizing acids. Many of the synthetic rubbers have good resistance to nonoxidizing weak acids but not to oxidizing acids. Epoxy resins have excellent resistance to weak acids and to alkaline materials but poor resistance to strong oxidizing acids. Many polyesters, particularly bisphenol polyester, have good general chemical resistance. The bisphenol polyesters have been the workhorses of the chemical industry. These materials are easy to fabricate, have superior acid resistance, and moderate resistance to alkaline and most solvents, such as the alcohols, gasoline, and ethylene glycol. The chemical resistance and applications of the more common and commercially used polymers are listed in Table 11.4.

**Table 11.4**   Summary of the Chemical Resistance of the Common Polymers and Their
Applications

| Type of plastic | Normal upper service limit | | Typical areas for use |
|---|---|---|---|
| | °C | °F | |
| Polyvinyl chloride | 60 | 140 | Piping: water, gas, drain, vent, conduit, oxidizing acids, bases, salts; ducts: (breaks down into HCl at high temperatures); windows plus accessory parts and machine equipment in chemical plants; liners with FRP overlay |
| Chlorinated PVC | 82 | 180 | Similar to PVC, but upper-temperature limit is increased |
| Polyethylene | 60 | 140 | Tubing, instrumentation (laboratory): air, gases, potable water, utilities, irrigation pipe, natural gas; tanks to 12-ft diameter |
| High-density polyethylene | 82 | 180–220 | Chemical plant sewers, sewer liners, resistant to wide variety of acids, bases, and salts; generally carbon filled; highly abrasion-resistant; can be overlaid with FRP for further strengthening |
| ABS | 60 | 140 | Pipe and fittings; transportation, appliances, recreation; pipe and fittings is mostly drain, waste, vent; also electrical conduit; resistant to aliphatic hydrocarbons but not resistant to aromatic and chlorinated hydrocarbons; formulations with higher heat resistance have been introduced |
| Polypropylene | 82–104 | 180–220 | Piping and as a composite material overlaid with FRP in duct systems; useful in most inorganic acids other than halogens; fuming nitric and other highly oxidizing environments; chlorinated hydrocarbons cause softening at high temperatures; resistant to stress cracking and excellent with detergents; flame-retardant formulations make it useful in duct systems; further reinforced with glass fibers for stiffening, increases the flex modulus to $10^6$ and deformation up to 148°C (300°F) |
| Polybutylene | 104 | 220 | Possesses excellent abrasion and corrosion resistance; useful for fly ash and bottom ash lines or any lines containing abrasive slurries; can be overlaid with fiberglass for further strengthening when required |

**Table 11.4** Continued

| Type of plastic | Normal upper service limit | | Typical areas for use |
|---|---|---|---|
| | °C | °F | |
| General-purpose polyesters | 50 | 120 | Made in a wide variety of formulations to suit the end-product requirements; used in the boat industry, tub-showers, automobiles, aircraft, building panels |
| Isophthalic polyesters | 70 | 150 | Increased chemical resistance; used extensively in chemical plant waste and cooling tower systems plus gasoline tanks; also liners for sour crude tanks |
| Chemical-resistant polyesters | 120–150 | 250–300 | Includes the families of bisphenol, hydrogenated bisphenols, brominated, and chlorendic types; a wide range of chemical resistance, predominantly to oxidizing environments; not resistant to $H_2SO_4$ above 78%; can operate continually in gas streams at 148°C (300°F); end uses include scrubbers, ducts, stacks, tanks, and hoods |
| Epoxies (wet) | 150 | 300 | More difficult to formulate than the polyesters; more alkaline-resistant than the polyesters, with less oxidizing resistance; highly resistant to solvents, especially when postcured; often used in filament-wound structures for piping; not commonly used in ducts; more expensive than polyesters |
| Vinyl esters (wet) (dry) | 93–140 | 200–280 | Especially resistant to bleaching compounds in chlorine-plus-alkaline environments; wide range of resistance to chemicals, similar to the polyesters; used extensively in piping, tanks, and scrubbers; modifications developed to operate continuously up to 140°C (280°F) |
| Furans (wet) | 150–200 | 300–400 | Excel in solvent resistance and combinations of solvents with oxidizing chemicals; one of the two resins to pass the 50 smoke rating and 25 fire-spread rating; carries about a 30–50% cost premium over the polyesters; excellent for piping, tanks, and special chemical equipment; does not possess the impact resistance of polyesters or epoxies |

*Source*: P. A. Schweitzer, Ed., *Corrosion and Corrosion Protection Handbook,* 2nd ed., article by J. H. Mallison, Marcel Dekker, New York, 1989, pp. 348–350.

## 11.4  ENVIRONMENTAL DEGRADATION OF CERAMIC MATERIALS

Since ceramic materials are basically insulators, the type of uniform corrosion experienced in metals cannot occur because of the lack of a conductor, one of the four requirements for the formation and existence of a galvanic cell. However, other forms of chemical attack are possible. Almost all materials can be attacked by some environments. Hydrofluoric acid is used to etch glass and other ceramic materials. The degradation of ceramics by their environment has not been categorized nearly to the same extent as that for metals and polymers. Many ceramic materials will decompose at high temperatures, and in many cases this decomposition may be accelerated or altered by the environment to which they may be subjected.

It is more common in ceramics to group corrosion and other environmental effects as *dissolution corrosion,* which may include some of the same metallic characteristics of grain boundary or stress corrosion, and even galvanic cell corrosion. Ceramics contain many components (e.g., bonding agents, mineralizers, grain growth inhibitors, etc.), many of which are capable of providing a conductive path for electrons to flow in a galvanic cell raction. Corrosive attack often centers around minute quantities of such constituents. Hot pressing, reaction sintering, and chemical vapor deposition have removed the need for many of these minor constituents and hence less concern for corrosive effects.

Considerable work has been done on the galvanic corrosion of refractories containing a glassy phase. Refractories that establish an electrical potential relative to glass of around 0.5 V are fairly resistant to this type of corrosion, but when this potential exceeds 1.0 V, the corrosion resistance is poor. In many ceramics a glassy phase surrounds all other phases. Refractories that have a negative potential relative to glass should not be used.

The corrosion of glasses by atmospheric conditions is referred to as *weathering.* It is basically a result of a water vapor reaction and is believed to be related to tensile stresses set up by an ion exchange of the alkali by hydrogen ions. Soda-lime silica glasses lose considerable strength due to atmospheric corrosion by water vapor. Glass corrosion by liquids is much more frequent than that by vapors. The release of PbO and other toxins is currently of much interest and is being studied intensely. Low-pH solutions release Pb more slowly than the higher-pH liquids. Glasses are soluble in liquids over a wide range of pH values from acids to bases, including water to a slight extent. The high-silicate glasses such as the borosilicates and alumina silicates of about 98% $SiO_2$ content have excellent corrosion resistance compared to the somewhat water-soluble sodium silicates. In the soda-lime glasses the $H^+$ ion from the water replaces an alkali ion which goes into solution while the $(OH)^-$ ion destroys the Si–O–Si bonds forming nonbridging oxygens. The reaction sort of feeds on itself since the nonbridging oxygen reacts with the $H_2O$ molecule forming another nonbridging oxygen–hydrogen bond plus another $(OH)^-$ ion. The hydrolysis of glasses determines the service life from weathering and corrosion. The rate of hydrolysis of the alkali glasses increases according to the modifier element in the following order: $Cs > K > Na > Li$.

Glass fibers are much more subject to corrosion than are the bulk glasses due to the higher surface/volume ratio. Humid environments lower the strength of E-glass fibers. For this reason many glass fibers are produced with a polymer coating.

The resistance of ceramics to chemical attack can be related to the bond strength. Whereas weakly bonded ionic salts, nitrates, oxalates, chlorides, and sulfates are dissolved in water and weak acids, the more strongly bonded $Al_2O_3$, $Si_3N_4$, and $ZrSiO_4$ are resistant to many types of chemicals, including aqueous solutions, strong acids, bases, and liquid metals and hence can be used as crucible materials. Many of the oxides are used for furnace linings, metal refining, and glass processing. These oxides are also resistant to acids and bases and are used either to replace metals or as a coating for them.

Elevated-temperature environmental reactions of ceramics are of more concern now since those materials are being considered seriously for use in gas turbines. Oxides and silicates are more stable in high-temperature oxide environments than are the carbides, nitrides, and borides. Oxygen pressures of 1 mmHg will form a protective layer of $SiO_2$ on SiC and $Si_3N_4$ rendering them as passive materials. It is essential that the latter be free of impurities, or else low-melting silicates may be formed. Porosity is also a factor since oxidation can occur along interconnected pore channels. For heat engines, $Si_3N_4$ and SiC appear to be the leading candidate materials since they possess good stability, thermal conductivity, low thermal expansion, and good high-temperature strength. In addition, they can be protected to some extent by $SiO_2$ layers.

## 11.5 CORROSION OF COMPOSITES

Steel rusts, polymers swell in moisture and change color and strength in sunlight, rubber ages faster under the effects of ozone, wood splits because of loss of water, and ceramics are attacked by a host of chemicals. Since the properties of composites are a function of their constituents, one could say that the environmental effects would sort of follow the rule of mixtures. To a large extent this is true. Any of the environmental effects mentioned previously on the individual constituents would be expected to have similar effects on these respective constituents, provided that there exists a path whereby the damaging ingredients can reach the individual constituents. For example, it is now generally accepted that steel in concrete is protected by the passivity provided by the highly alkaline nature of the porewater, and persists as a result of the presence of the water and oxygen. However, if the concrete fails to provide a barrier to the ingress of certain substances, the reinforcing steel bars can corrode. Carbon dioxide and chloride can cause depassivation of the steel. One of the problems is to prevent crack formation in the concrete such that such substances can enter.

In the more sophisticated composites such as the organic matrix–fiber composites, other problems arise. The most commonly feared is the deterioation and breaking of interfacial bonds. Water can cause swelling and plastization of resins and thereby permit the ingress of water or other liquids by capillary action along the fiber–matrix interface. Deterioration of PMCs by moisture or chemical absorption is of great concern. The strength of these composites can be reduced if the chemical species in the environment attack the polymer, the reinforcement material or the interface between the two. To some extent these problems can be circumvented by judicious selection of the materials used in the composite construction. Similar problems can arise in metal and ceramic matrix composites as well. MMCs are not as susceptible to moisture deterioation as are PMCs. Of more concern for MMC applications is the temperature stability of the matrix and reinforcement constituents in the environment of interest. Damage from the external environment can be minimized if the reinforcement material is not exposed. Galvanic corrosion has been observed in some MMCs, particularly Al–B, Al–SiC, and Mg–graphite. The coupling of electrochemically dissimilar materials can occur in these composites when electrolytes are available. The compatibility of certain alloys and intermetallic compounds with hydrogen atmospheres is also of concern. But the state of the art in many of these composites has not advanced to the point where many corrosion data are available. For the resin composites a number of studies have been made, but for now the corrosion of composite materials will be considered beyond the scope of this book.

## DEFINITIONS

*Anode*:   electrode in a *galvanic cell* that is attacked by the electrolyte, thereby removing ions and producing electrons.

*Cathode*:   electrode in the galvanic cell that consumes electrons by a variety of chemical reactions.

*Composition cell*:   galvanic cell composed of two different materials.

*Concentration cell*:   type of galvanic cell that establishes anodes and cathodes because of a difference in concentration of ions on or near the metal surface. The *oxygen concentration cell* is the best known of this type.

*Crevice corrosion*:   corrosion occurring where stagnant fluids collect in crevices. The corrosion involves a number of reactions depending on the materials and the environment.

*Electrolytes*:   liquids in the galvanic cells that act as media for the combination of the ions resulting from anode and cathode reactions.

*Electromotive force series (emf)*:   ranking of the elements according to their tendency to corrode.

*Erosion-corrosion*:   combination of corrosion and fluid flow, where the latter may accelerate the former.

*Galvanic series*:   ranking of metals and alloys according to their corrosion tendency.

*Hydrogen attack*:   type of hydrogen damage where a gas, often methane, is generated via a chemical reaction of hydrogen with $Fe_3C$. The gas builds up sufficient pressure to cause cracking.

*Hydrogen electrode*:   electrode against which metals are measured to establish their position in the emf series. It consists of the bubbling of hydrogen into 1 $M$ solution of $H^+$ ions. Its voltage is arbitrarily set at zero.

*Hydrogen embrittlement*:   form of hydrogen damage and SCC where a few ppm of hydrogen causes decohesion. Sometimes called *hydrogen-assisted cracking* (HAC).

*Hydrogen-induced cracking*:   type of cracking usually associated with $H_2S$ in petroleum products that produces hydrogen gas sufficient to cause blistering.

*Inhibitors*:   complex ion formation on the surface of a metal that protects it in some way from the electrolyte. (*See also* Passivation.)

*Intergranular corrosion*:   type of anodic dissolution of the grain boundary. A form of *stress corrosion cracking*.

*Nernst equation*:   equation that expresses the effect of metal ion concentration on the standard emf of a metal half-cell, the latter being set at a 1 $M$ solution of the metal ions.

*Passivation*:   state existing on the metal surface where the corrosion rate does not change with increasing electrode potential. It is a characteristic of metal surfaces that usually occurs when thin adherent metal oxides are formed, although inhibitors are also believed to affect passivity by being absorbed into the metal surface.

*Pitting corrosion*:   localized corrosion that forms a pit where the surface diameter is less than the pit depth.

*Polarization*:   *activation polarization* affects the corrosion rate by a rate-controlling reaction at the metal–electrolyte interface. *Concentration polarization*: affects the corrosion rate by a rate-controlling step in diffusion within the electrolyte.

*Sacrificial anode*:   highly anodic metal, such as Mg, that is attached to a less anodic metal and corrodes in preference to this metal.

*Selective leaching*:   preferential removal of one element from a solid alloy by corrosion process.

*Sensitization*:   precipitation of chromium carbides at the grain boundaries in austenitic stainless steels and other high-chromium-content alloys that render them sensitive to galvanic corrosion.

*Stray corrosion currents*:   corrosion by current flow other than the intended path and often found in soils.

*Sulfide stress cracking*:   type of hydrogen embrittlement where the $H_2S$ is the principal source of hydrogen.

## QUESTIONS AND PROBLEMS

11.1.   What is the maximum voltage that one could obtain from any two pure metals arranged as two electrodes in a galvanic cell?

11.2.   What chemical reaction always occurs at a metallic anode in a galvanic cell?

11.3.   Write balanced chemical equations for three types of reactions that could occur at the cathode of a galvanic cell. What types of electrolytes are involved in each case?

11.4.   What is the difference between the electromotive force and the galvanic series?

11.5.   What are the four requirements for the operation of a galvanic cell?

11.6.   Describe and give examples of the following corrosion types: (a) uniform corrosion; (b) crevice corrosion; (c) pitting corrosion.

11.7.   Determine the electrode potential with respect to hydrogen for a copper electrode in a solution containing 10 g of $Cu^+$ ions/liter.

11.8.   A $TiO_2$ layer that forms on a sheet of titanium passivates the metal. How thick should this layer be to limit the current density to $10^{-4}$ A in a cell that produces 4 V? Assume the electrical resistivity of $TiO_2$ to be $10^{18}$ $\Omega \cdot m$.

11.9.   Why is concentration polarization seldom rate controlling for oxidation reactions?

11.10.   Gold and cooper both have a valency of $+1$, and both are good conductors of electricity. Why is copper so much more aggressively attacked in ambient environments, whereas gold is essentially stable? Would iron be more reactive than copper?

11.11.   Suppose that it was necessary in a certain construction to connect a 1020 steel tube to a Ni–Co Monel K alloy. Suggest a method to prevent the formation of a galvanic composition cell.

11.12.   What is meant by the term *sensitization* of austenitic stainless steels? How can this phenomenon be avoided in the welding of such steels?

11.13. Cd shows $-0.403$ V and nickel $-0.250$ V on the emf series. Why is the voltage of this galvanic cell $+0.146$ V rather than $-0.403 - (-0.250) = -0.153$ V?

11.14. What is the corrosion rate in g/s for a Zn plate in combination with a Cu plate of the same area in a galvanic cell that is producing 1 A of current flow?

11.15. What is the meaning of a standard half-cell oxidation–reduction potential?

11.16. Copper concentrations of 0.05 and 0.1 $M$ occur in an electrolyte at 25°C at opposite ends of the copper wire. Which end will corrode?

11.17. Within a mild steel in a humid atmosphere, what region or constituents will become anodic with respect to the matrix material? What regions or constituents will be cathodic with respect to the matrix?

11.18. What is the purpose of galvanizing steel? With respect to corrosion prevention, how does galvanizing differ from tin plate?

11.19. Define the term *passivity* and how it is measured. Name four passive-type metals and the film that causes their passivity.

11.20. Do inhibitors cause metals to become passive? Explain how they differ from oxide films.

11.21. List and describe four types of hydrogen damage.

11.22. In what ways are SCC and hydrogen effects similar? In what ways are they different?

## SUGGESTED READING

American Society for Metals, *Metals Handbook*, Vol. 13, 9th ed., ASM International, Metals Park, Ohio, 1987.

Fontana, M. G., *Corrosion Engineering*, 3rd ed., McGraw-Hill, New York, 1986.

Jones, D. A., *Principles and Prevention of Corrosion*, Macmillan, New York, 1992.

McEvily, A. J., Jr., Ed., *Atlas of Stress-Corrosion and Corrosion Fatigue Curves*, ASM International, Metals Park, Ohio, 1990.

Schweitzer, P. A., Ed., *Corrosion and Corrosion Protection Handbook*, 2nd ed., Marcel Dekker, New York, 1989.

Schweitzer, P. A., Ed., *Corrosion Resistant Tables*, 3rd ed., Marcel Dekker, New York 1991.

# 12

# Comparative Properties

## 12.1 INTRODUCTION

When design engineers are constructing their drawings, these engineers, engineers who specialize in this particular arena, or materials engineers must specify the materials for constructing this object or system being designed. Large corporations frequently have materials engineers who work closely throughout the design process with the design engineers. In smaller companies the design engineer often specifies the materials. In the latter case the design engineer will frequently rely on vendors' suggestions or use materials from existing designs of similar objects or systems. In some cases materials handbooks may even be used, and in a small percentage of cases a materials consulting engineer may be hired.

In selecting materials for designs, we can separate the products into two large categories:

1. The mundane components, machinery, standard automobile parts, containers, and so on, of the mature technologies: In most of these designs the person selecting materials selects those that have been used for years and have a proven record of satisfactory performance. But even for these established products, cost savings can often be made by switching to more recently developed materials whose properties have been well documented and certified by the producers. This is particularly true in the automobile industry, and perhaps these engineers are the most eager of those in the mature industries to risk

using relatively new materials and processes after careful materials evaluation and testing in prototype systems.

2. The high-tech systems of the aerospace, electronics, nuclear, medical, and some chemical industries: these designs frequently require new or recently developed materials, many of which are *engineered*. The aerospace industry leads the way in development of new materials, some that eventually filter down to the more established technologies. A good example of the latter are the boron and silicon carbide fibers and their composites that were first used in the space shuttle, other aircraft, and missiles that are now being widely used to manufacture sports equipment: skis, golf clubs, tennis rackets, fishing poles, boats, and the like.

In the materials selection process for either of the categories listed above, assuming that we are considering materials that are available or can be made available on a reasonably short notice, it helps to know in what ballpark the materials of interest for a particular application might be found. For example, if the materials are going to experience temperatures in excess of 500°C, should we waste our time and effort by considering polymers and polymer matrix composites? Let's narrow our field of consideration as quickly as possible. After all, we have many thousands of materials from which to choose. Our first objective is to reduce this number to a few hundred. This is sort of analogous to the block diagram approach used for detecting failures in electronic equipment. We must first isolate the block of interest. It is the intent of this chapter to establish some ballpark categories that should assist us in narrowing the field of possible materials for any given application.

In this chapter the cost of the materials is not included. These figures frequently change and for many of the newer materials are often unknown. Comparative cost figures for established materials are provided in Appendix B.

## 12.2  COMPARATIVE MECHANICAL PROPERTIES

The mechanical properties of paramount interest in most designs are those of strength, hardness, ductility, fatigue, and fracture toughness. For other than room-temperature applications these properties must also be known in the temperature range of the intended application. For higher-temperature applications, we must add the creep and stress–rupture properties. In many room-temperature uses and in all elevated-temperature applications, the effect of the environment must be considered. Although in some electronic materials applications the mechanical properties of housings, substrates, and other components of electronic equipment designs are of some significance, usually the electronic and thermal properties are far more important in this industry. Consequently, the mechanical properties of electronic components are not considered in this section. Metals, polymers, ceramics, and composites are the materials that will be judged and compared on the basis of their mechanical properties in this chapter.

## 12.2.1 Comparison of Room-Temperature Mechanical Properties of Metals, Polymers, Ceramics, and Composites

In Figure 12.1 the ultimate tensile strengths of typical materials of the four categories are shown. To conserve space we have elected to use ksi (100 psi) instead of both systems of units. To obtain SI units of MPa, one can multiply the ksi figures by 6.9. Representative materials from each category have been selected. Space does not permit examination of a larger number of materials. For other materials, and for more detailed data, handbooks or computerized data bases must be used. One could use average values of a large number of materials in each category. But there are many processing variables involved, and such details affect the average values. By choosing specific materials, at

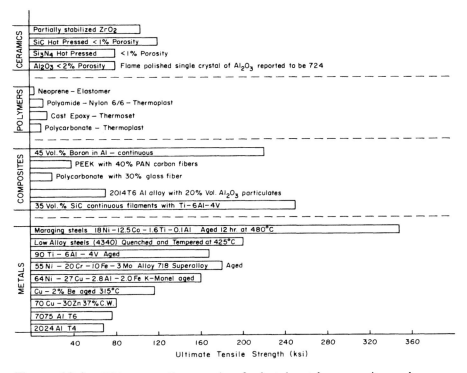

**Figure 12.1.** Ultimate tensile strengths of selected metals, composites, polymers, and ceramics.

least some of the processing variables, such as heat treatment, porosity, and fabrication methods, can be included. The ductilities and moduli for similar materials are depicted in Figures 12.2 and 12.3, respectively.

Fracture Toughness

In Section 2.2.4 fracture toughness measurements by both the impact test, where the toughness is reported in ft-lb of energy absorbed during fracture of a standard ASTM Charpy V-notch test specimen, and by the more modern fracture mechanics measurement method were discussed. In the latter test method the results are reported as $K_{1c}$ values, which are true material constants and are usually expressed in the SI units MPa$\sqrt{m}$ or the English units of ksi$\sqrt{in}$.. The conversion factor between the two units is 1.1, so for all practical purposes they are interchangeable. The $K_{1c}$ value is a function of the applied stress and length of the crack at failure. The material must exceed a certain critical thickness as described in ASTM specification E399 in order for the measured $K_{1c}$ to

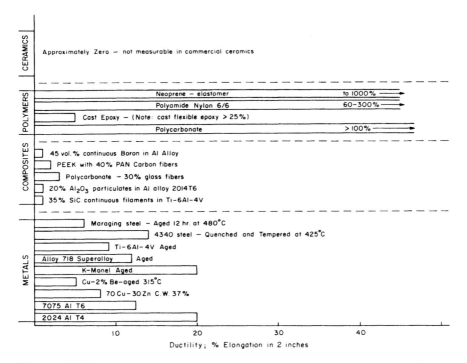

**Figure 12.2.** Ductilities of selected metals, composites, polymers, and ceramics.

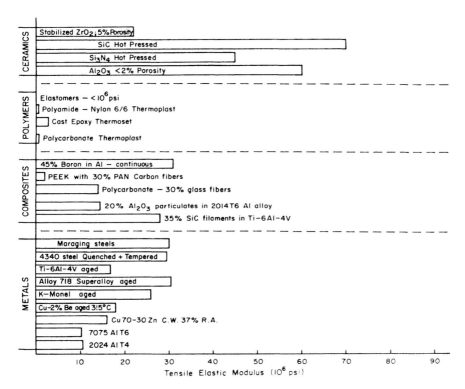

**Figure 12.3.** Tensile moduli of selected metals, composites, polymers, and ceramics.

be a true material constant. In comparing the fracture toughness characteristics of the various materials we must choose either the impact test or the fracture mechanics method. Unfortunately, most polymers have been measured by the impact test method, since the ductile ones, such as polycarbonate, require thickness on the order of 12.7 mm (0.5 in.). Nevertheless, there have been a sufficient number of $K_{1c}$ values reported on polymers to permit us to choose this more up-to-date method for comparison of the fracture toughness of the various groups of materials. Most polymers have rather low fracture toughness. The glass-reinforced thermosets, which are really composites, have some of the higher toughness values. (Impact data on polymers were presented ·in Table 8.6.) In Figure 12.4 representative materials of each category were selected for toughness comparisons. It must be realized that there exists a wide variation in the fracture toughness of metals, depending on the heat treatment and other processing variables. Many designs will specify that $K_{1c}$ exceed a certain value, for example 25 MPa√m. Other designs may specify lower or higher values, depending on the application. The data listed in Figure 12.4 are room-temperature values.

Some comments on the data of Figure 12.4 are in order here. The fracture toughness values for PMCs have not been reported in $K_{1C}$ terms. Impact data show values ranging from 32 to 43 J/cm notch. In the ceramic–ceramic composites category the highest value of 25 MPa√m was reported for SiC fibers in a SiC matrix. The SiC–Al alloy particulate composites are usually in the range 10 to 30 MPa√m. The high value of 60 MPa√m was reported for an aluminum sheet composite, which probably had some directional variation in its fracture toughness. Perhaps this value should be treated with some caution until it has been well established and all of the conditions clearly understood. There are many metals that have fracture toughness values higher than those reported in Figure 12.4. One quenched and tempered steel of medium-carbon content (ASTM A533B) was reported to have a 200 MPa√m fracture toughness. Low-carbon steels also have very high values of fracture toughness, particularly in the annealed condition. Even the high-strength maraging steels have $K_{1C}$ values on the order of 150 MPa√m.

There are certain generalities that are noticeable in Figure 12.4. First, the metals category, on the average, has the highest fracture toughness of the four classes of materials listed. There are many metals, including the copper and

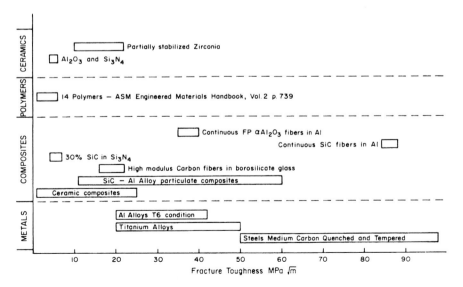

**Figure 12.4.** Fracture toughness of selected metals, composites, polymers, and ceramics.

nickel alloys, that were not included. The nickel alloys were not shown since the Monels (Ni–Cu alloys) have fracture toughness values above those needed for structural purposes. Monel alloy K-500 has an impact strength of 37 ft-lb at room temperature and the other Monels possess even a higher fracture toughness. Many of the other nickel-based alloys consist of the superalloys used at elevated temperatures where fracture toughness is far more than adequate for these applications. Fracture toughness values for the copper alloys, including the high-strength copper–beryllium alloys, are not usually reported, even in the well-known handbooks. They are seldom, if ever, excluded from design purposes because of their fracture toughness numbers.

The continuous fiber SiC–metal composites have good fracture toughness, while the particulate SiC–Al composites have more than adequate toughness in most cases (the higher the SiC content, the lower the toughness). Polymers and polymer composites, together with ceramics and most ceramic composites, have very low fracture toughness and their usage is currently limited by this factor. In the composites category, however, we will probably see higher fracture toughness values in the near future, especially in the SiC–SiC fiber composites.

### 12.2.2 Comparison of Elevated-Temperature Properties of Metals, Ceramics, and Composites

What is meant by *elevated* in terms of a temperature scale? Since atoms only recognize the Kelvin absolute temperature scale, the temperature of 77 K (liquid nitrogen) is an elevated temperature with respect to that of liquid helium (4.2 K). In our comparison we will be somewhat more practical and arbitrarily define temperatures above 500°C as elevated temperatures. This will rule out all polymers and most of the reinforced polymer matrix composites. Furthermore, the elevated-temperature properties must include creep and stress-rupture data because lifetime is a very important design parameter in the high-temperature range. But just for the purpose of comparison, Figure 12.5 shows the variation of the strength with temperature of representative members of all groups except for the polymers. Time was not a factor here; that is, the tests were conducted in a period of a few minutes at the usual tensile test strain rates of 0.05 to 0.5 per minute. In ceramic materials one finds a considerable amount of scatter in the data, being particularly large for silicon nitride. Also, ceramic strengths are most often determined in bend tests, which give flexural rather than tensile strengths. In Figure 12.5 the flexural strengths for the bend tests are so noted. The strength of the other materials reported are those obtained in tensile tests. The data of Figure 12.5 were obtained from a number of sources, most listed in the Suggested Reading, and must be considered as typical rather than precise values. Of particular note is the high strength of the superalloy 718 at room temperature and the sudden drop in strength at about 800°C. This

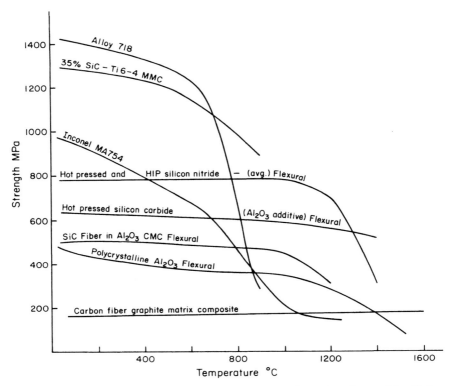

**Figure 12.5.** Ultimate strengths versus temperature for selected materials. Unless otherwise noted (e.g., flexural), the values were determined in tensile tests.

is, of course, the reason behind the development efforts for nonmetallic materials for high-temperature applications. There is a considerable amount of high-temperature tensile strength, creep, and fatigue data on ceramics and composites now being generated at Oak Ridge National Laboratories (Figures 12.6 and 12.7). Returning to Figure 12.5, the ceramics, the MMCs, and CMSs, and particularly, the carbon–carbon composites, retain their strength to much higher temperatures, even though their room-temperature strengths are somewhat lower than the room-temperature strength of the superalloys. The oxide-dispersioned superalloys (e.g., MA 754, Ma 6000, and MA 956) retain their strengths to higher temperatures than do the other superalloys. Their advantages will show more clearly in the stress-rupture tests.

The most important function of materials for elevated-temperature applications is their ability to withstand a particular stress and temperature for a reasonable length of time: that of the lifetime of the particular design of interest. In many instances it is not necessary for the life of the component to be the

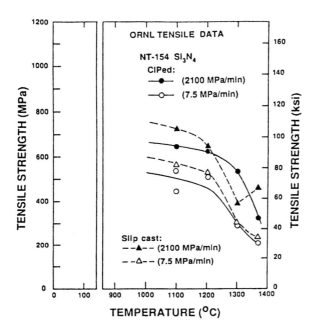

**Figure 12.6.** Tensile strength versus temperature for CIP and slip cast $Si_3N_4$. (From K. C. Liu and C. R. Brinkman, *Proceedings of the Automotive Technology Department Contractors Meeting*, Paper 243, Society of Automotive Engineers, Warrendale, Pa., Oct. 1990, p. 238.)

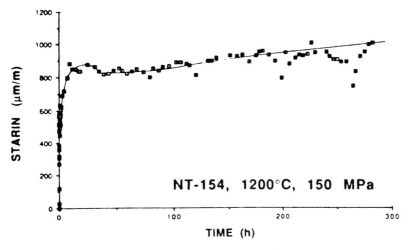

**Figure 12.7.** Creep of HIP $Si_3N_4$ at 1200°C and a stress of 150 MPa. (From K. C. Liu, H. Pih, C. O. Stevens, and C. R. Brinkman, *Proceedings of 27th Automotive Technology Department Contractors Meeting*, Paper 230, Society of Automotive Engineers, Warrendale, Pa., Apr. 1990, p. 217.)

same as that of the entire system of the design. Certain components are replacable. The life of a turbine blade in an aircraft jet engine is far less than the life of the aircraft as a whole. The turbine blades can be and are replaced many times. But we need to know with some degree of certainty the lifetimes of all components under the conditions of stress, temperature, and environment to which they will be exposed.

Following the argument above leads us to the subject of *creep* and *creep-rupture* or as it is sometimes called, *stress-rupture*. The three stages of creep were discussed in Chapter 2 and presented in Figure 2.21. Creep and stress-rupture were discussed, but actual data were not presented. Now we must consider these properties. As a reminder, creep and stress-rupture were properties that were of concern only at temperatures above about one-half of the melting point in kelvin (i.e., $0.5T_m$). Remember that atoms recognize only the absolute temperature scale. At about $0.5T_m$ the atom vibrations are of such amplitude that atom-vacancy exchanges become very probable. Diffusion of atoms occurs. Dislocations can move readily with the assistance of very low stress, far below the short-time yield stress at that particular temperature. Dislocation movement creates plastic deformation, albeit a very slow process compared to that in a short-time tensile test. Nevertheless, plastic deformation does occur, and over a long period of time, substantial dimensional changes take place which we call creep. Eventually, the creep strain will be sufficient to cause rupture, called creep-rupture. The term *stress-rupture,* which has the same meaning as *creep-rupture,* denotes the constant stress to which the material was subjected over this long period of time (from days to years, depending on the stress and temperature). So for elevated-temperature usage, we are not particularly interested in the strength, as determined in tensile, compression, or bend tests, which may span a period of a few minutes. We are more interested in the creep rate, since perhaps the design specifies a maximum dimensional change during its lifetime, or, more often, the time to rupture at a specific constant stress. We will emphasize the latter (i.e., stress-rupture data) in our material comparison picture.

The time to rupture for a typical superalloy (Inconel 617) and two oxide dispersion-strengthened superalloys (Inconel MA 6000 and MA956) and a ceramic material ($Si_3N_4$) are shown in Figure 12.8. (Many polymers were illustrated in Figures 8.12 and 8.13. The temperatures considered for these polymers were all below 500°C.) Creep data for several ceramic materials were reported in Figures 7.16 and 7.17. Unfortunately, there is only a limited amount of data available on the stress-rupture of ceramic materials and high-temperature composites, but the data that are available indicate that ceramics possess the ability to withstand stress at temperatures much higher than that for superalloys. With the considerable experimental work now being conducted on

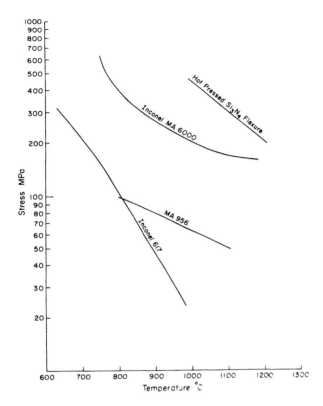

**Figure 12.8.** 1000 hour stress–rupture curves for superalloys and hot-pressed Si₃N₄. (Superalloy data taken from Inco Alloys Int., Inc., and the Si₃N₄ from G. Quinn, *J. Am. Ceram. Soc.*, Vol. 73, 1990, p. 2374.)

ceramics and composites for high-temperature applications, it is expected that more stress–rupture data will soon be forthcoming.

In the above-mentioned figures, data is presented for certain stresses and temperatures. If one desires a different set of stress-temperature combinations, methods have been developed for extrapolating from one set of conditions to another over a wide range of temperatures and stresses by the use "master curves" for a given alloy. These master curves have been developed for numerous superalloys and for some stainless steels, and can be found in the published literature. Unfortunately they do not exist for composites and ceramics.

These curves are not difficult to use or interpret. Generally they involve subjecting the alloy to certain stresses at high temperatures, most often at higher temperatures than the intended application of the alloy, in order to enforce failure in a short period of time. This data is then used to extrapolate to lower temperatures of application where this particular alloy will be used for much longer periods than is practical to conduct stress-rupture tests. These methods are referred to as parameter methods, the best known and most widely used being the Larson-Miller (L-M method) developed in 1952. Other parametric methods have also been developed but we will explore only the L-M method. The answers by the other methods do not vary widely because they are all based on the Arrehenius relationship where a certain process, in the present case either creep strain rate or the time to rupture, is an exponential function of temperature. In the case of the time to rupture, $t$, it becomes

$$t = A \exp \left[ \frac{Q}{RT} \right] \tag{12.1}$$

Taking logarithms of both sides we have

$$\ln t = \ln A + \frac{Q}{RT} \tag{12.2}$$

If $A$ and $Q$ are functions of stress only, as assumed by Larson and Miller, this equation is linear in $\ln t$ and $1/T$ for any given stress. By studying a large amount of data, Larson and Miller concluded that the lines for different stresses representing equation 12.2 converged at a common point on the $\ln t$ axis. This point then becomes the intercept $\ln A$ and the slope of the lines then become $Q/R$. The L-M parameter is then defined as

$$T(\ln A + \ln t) \tag{12.3}$$

where $T$ is in degrees Kelvin and $t$ is in hours.

The master curves are constructed by obtaining the time to rupture at various temperatures and stresses. This process can be best explained by referring to the actual data obtained on Inconel 718 illustrated in Fig. 12.9. In graph (a) the conventional log of stress versus log of time to rupture is depicted for several temperatures. In (b) constant time to rupture intercepts are determined, as denoted by the dashed lines, for various values of stress and temperature. Using these intercepts the log $t$ versus $1/T$ curves of Fig. 12.9 (c) are constructed. The common intercept would be $\ln A$ in the linear equation 12.2 if the curves had been plotted in terms of degrees Kelvin. However, it is common practice in industry to use the Rankine temperature scale and log to the base 10 in the parameter methods of presenting data. Following this convention, equation 12.2 now becomes

$$\log t = 2.3 \log A + 2.3 \frac{Q}{RT} \tag{12.4}$$

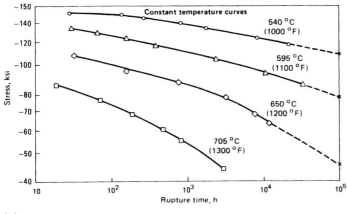

**(a)**

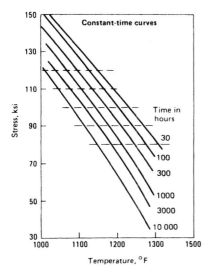

**(b)**

**Figure 12.9.** Method of constructing a master curve using data for Inconel 718. (From *ASM Int. Metals Handbook, Vol. 3, 9th ed.*, 1980, p. 239.)

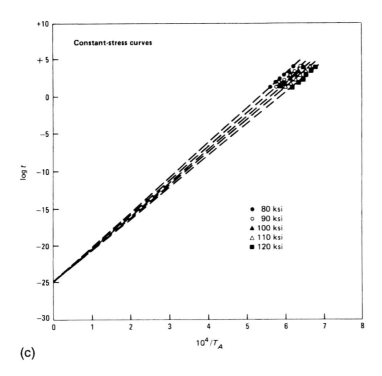

(c)

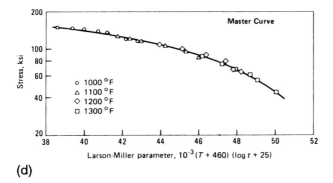

(d)

**Figure 12.9.** Continued.

which can be rewritten as

$$\log t + C = m \frac{1}{T} \tag{12.5}$$

where $m$ = slope = $2.3Q/R$ or as

$$T(\log t + C) = m = \text{L-M parameter } P \tag{12.6}$$

where $C = 2.3 \log A$. The parameter $P$ is usually defined as

$$P = T(\log + C) \times 10^{-3} \tag{12.7}$$

where $T$ is in degrees Rankine. $C$ is now the constant determined by the intercept on the $\log t$ axis in Fig. 12.9 (c). This constant $C$ is found to fall in the range of 15 to 25, and many engineers will use a value of 20 without doing the cross-plotting described above. Each point in the time, stress, and temperature family curves of Fig. 12.9 (a) can now be used to obtain the master curve of Fig. 12.9 (d). Now any combination of time and temperature that equals the parameter $P$ in equation 12.7 will fail at the same stress.

*Example 12.1*    Using the L-M master curve of Fig. 12.9 (d) determine the time to rupture of Inconel 718 for an applied stress of 80 Ksi and a temperature of 600°C.

Solution    The temperature of the stress application must be converted to the Fahrenheit scale which is found to be 1112°F. Next a horizontal line is constructed from 80 Ksi until it intersects the master curve. The abscissa at the point of intersection is the value of the parameter $P$, which for 80 Ksi is 47. In this particular master curve it is not necessary to convert $T$ to degrees Rankine since the parameter $P$ is stated as

$$P_{\text{L-M}} = 10^{-3}(T + 460)(\log t + 25) = 10^{-3}(1572\,°\text{R})(\log t + 25)$$

where $T$ is in °F

$$\log t = \frac{47}{1.572} - 25 = 29.9 - 25 = 4.9$$

$$t = 79{,}433 \text{ h} = 9.1 \text{ years.}$$

If the L-M parameter is used in terms of the minimum creep rate instead of time to rupture the parameter $P$ becomes

$$P = T(C - \log \text{strain rate}) \tag{12.8}$$

where $T$ is in degrees Rankine.

Fatigue data for some representative materials are shown in Figure 12.10. Steels and the continuous fiber composites are the better materials with respect

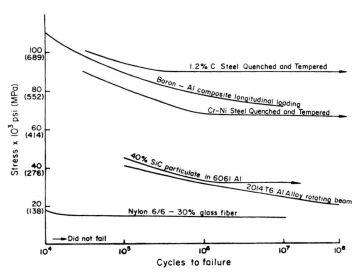

**Figure 12.10.** Fatigue behavior of selected composites and alloys.

to fatigue life. The fatigue life of these composites will, however, vary considerably with fiber direction and loading direction. Because of their anisotropic characteristics, composite materials have very complex fatigue failure mechanisms. Fatigue tends to occur by extensive damage throughout the specimen rather than by a single crack or a few cracks that propagate to failure as in the case of metal fatigue. The most common fatigue failures occur by delamination and interfacial debonding, although fiber fracture has also been observed. But the latter could be a result of the debonding and delamination. Cross-ply laminates tend to show a gradual reduction of strength until failure without a well-defined endurance limit, while the unidirectional laminates show little strength loss prior to failure. The aluminum alloys that contain particulates have a somewhat better fatigue behavior than do the nonreinforced aluminum alloys.

## 12.3  COMPARATIVE PHYSICAL PROPERTIES

We have already treated the electrical conductivity of metals, insulators, and semiconductors (Figure 2.22). Because there exists such a large variation among alloys, the electrical resistivity of metals is treated in a separate graph in Figure 12.11. The univalent and divalent metals and their alloys have lower resistivities than do the transition metals and their alloys. Probably the physical property of next most importance and interest is the thermal conductivity. These data for the various categories are summarized in Figure 12.12. Again, typical materials in each group have been selected. The values for composites vary considerably, as do all composite properties. For many composites physical property data are difficult to locate.

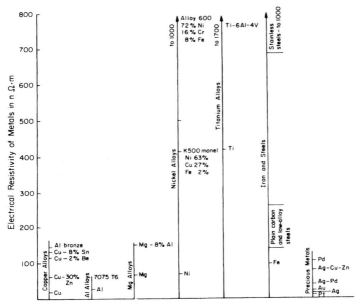

**Figure 12.11.** Electrical resistivities for selected metals. The variation is quite large, being smaller in the univalent, divalent, and precious metals than in the transition metals and their alloys.

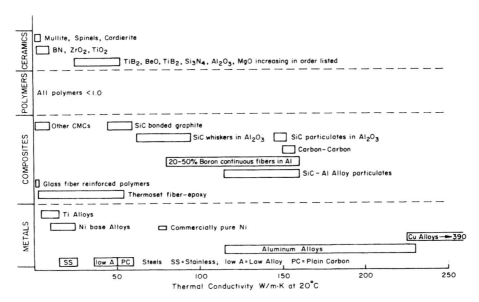

**Figure 12.12.** Thermal conductivities of selected materials. The polymers and ceramics have significantly lower values than do most metals and composites.

Most metals, particularly those with high electrical conductivities, because of their large number of free electrons, also have the highest thermal conductivities, while the covalently bonded polymers and ceramics have the lowest values of thermal conductivity. The thermal conductivities of ceramics tend to decrease with increasing temperature, while those for metals remain reasonably constant (see Figure 7.18). The thermoset fiber–epoxy composites, particularly the HMS graphite type, possess reasonable thermal conductivities. The metal matrix composites, both particulates, discontinuous and continuous fiber type, generally have good thermal conductivities. As a result, many of these, especially the boron–aluminum composites, are found to be desirable for electronic packaging applications. Their low density and high modulus are additional benefits for this purpose.

The coefficient of thermal expansion must be considered in many designs where the materials experience large variations in temperature. Just as in creep, only certain dimensional changes can be permitted. Furthermore, dimensional changes created by thermal expansion can cause localized stress gradients due to nonuniform thermal conductivities. And if large members are restricted in movement during changes in temperature, stresses will be set up since the change in temperature and the corresponding desire to expand will place these members in compression. The coefficients of thermal expansion for a number of materials are listed in Table 12.1. The polymers appear to have the largest thermal expansion coefficients, but these materials do not usually experience wide temperature fluctuations since they are already temperature limited in their applications. Some ceramics have surprisingly significantly high coefficients of thermal expansion, especially those with a high degree of ionic bonding, and along with their low thermal conductivity and high elastic moduli account for their susceptibility to thermal shock. Composites have sort of a mixed bag in the way of thermal expansion coefficients much as they behave with other properties. Some of those with low-thermal-expansion coefficients and high thermal conductivities are desirable electronic device packaging materials.

## QUESTIONS AND PROBLEMS

12.1. If a steel rod of 10 in. in length with a CTE of $10^{-6}$ $K^{-1}$ is restricted in its motion by fixed nonexpandable members, how much stress is established if the rod is heated to 300°C above room temperature (20°C)? Assume a modulus of elasticity of $30 \times 10^6$ psi and that the CTE is constant over the temperature range 20 to 300°C.

12.2. Locate, by the use of handbooks, vendor's literature, and data bases, materials in each category that have higher strength, higher ductility, and higher moduli than those possessed by the materials shown in Figures 12.1, 12.2, and 12.3.

**Table 12.1**  Coefficient of thermal Expansion at 20°C ($10^{-6}$ K$^{-1}$)

| Metals | CTE | Composites | CTE | Polymers | CTE | Ceramics | CTE |
|---|---|---|---|---|---|---|---|
| Al alloys | 20–24 | SiC whiskers in Al alloy | 16 | ABS | 53 | BeO | 7.6 |
| Cu alloys | 17–20 | SiC particulates 30% in Al alloy | 12 | Nylon 6/6 | 40 | MgO | 12.8 |
| Plain C steels | 11–14 | Continuous boron fibers in Al | 6 | PEEK | 26 | $SiO_2$ | 22.2 |
| Low-alloy steels | 12–15 | $Al_2O_3$ | 6.7 | PET | 15 | $ZrO_2$ | 2.1 |
| Stainless steels | 11–19 | Si whiskers 30% in $Al_2O_3$ | 6.7 | PMMA | 34 | SiC | 4.6 |
| Ti alloys | 7–8 | Carbon–carbon | 0 | Bakelite | 16 | $Al_2O_3$ | 7.6 |
| | | Glass fibers in thermoplasts | 34–63 | PPS | 30 | Mullite | 5.5 |
| | | Glass fibers in thermosets | 9–32 | Polysulfone | 31 | | |
| | | Graphite–epoxy thermosets | 0.6 | | | | |
| | | Boron–epoxy | 4.5 | | | | |

12.3.   Convert the value of 150 W/m · K thermal conductivity to the units cal/cm · s · °C.

12.4.   Convert the value of 200 ksi to MPa.

12.5.   Using handbook data, find the ductility for the highest-strength maraging steel commercially available. Would this ductility limit is usage? What compromise materials would you select for optimum strength and ductility?

12.6.   Why is Ce–Be much stronger than other copper alloys? For what applications is Cu–Be a suitable alloy?

12.7.   Would a SiC continuous-fiber Al alloy composite be favored over a boron continuous-fiber Al alloy (a) for strength at 20°C; (b) for high-temperature strength at 500°C (assuming that 500°C is in the all-alpha range of the aluminum alloy)?

12.8.   What are the advantages and disadvantages of the two composites above with respect to each other?

12.9.   Assuming similar fatigue *S–N* curves for both a steel and a boron–Al continuous-fiber composite, which would you choose (a) where strength-to-weight ratio is important; (b) where fatigue is the only consideration?

12.10.  What advantage do the particulate composites have over the continuous-fiber composites in the SiC–metal composites?

12.11.  Why do the titanium, steel, and nickel alloys have such high electrical resistivities compared to the other alloys of Figure 12.8?

## SUGGESTED READING

American Society of Metals, *Advanced Materials and Processes: Guide to Selecting Engineering Materials,* ASM International, Metals Park, Ohio, 1990.

American Society of Metals, *Engineered Materials Handbooks,* Vols 1–4, ASM International, Metals Park, Ohio, 1992.

American Society of Metals, *Engineered Materials Reference Book,* ASM International, Metals Park, Ohio, 1989.

American Society of Metals, *Metals Handbooks,* Vols. 1 and 2, 10th ed., ASM International, Metals Park, Ohio, 1990.

Richerson, D. W., *Modern Ceramic Engineering,* Marcel Dekker, New York, 1992.

See also data sheets from the following vendors.

Inco Alloys International
Textron Specialty Materials, Textron, Inc.
Duralcan USA
DuPont Lanxide Composites
DuPont Kevlar
GTE Laboratories and GTE WESGO

# 13
# Materials Selection

## 13.1 INTRODUCTION

Materials selection for a particular project or design is a problem that most engineers encounter at some point in their professional career. Many nonmaterials engineers who are alumni of our school have called for opinions on what materials should be used for the project in which they were involved. That is one method of attacking the problem—call in a qualified consultant. For some inquiries, some ballpark answers may be suggested over the phone, but most projects require a more detailed study. First, it is seldom that the caller has all of the information required for a proper answer. Often, more questions are raised than are answered. The engineer is not always aware of all the parameters required in the materials selection process. My experience has been that the mechanical engineers encounter the material selection problems more often than do other engineers. MEs are more involved with all types of designs and materials and often do not have access to materials experts. Even materials engineers who have been confined to a narrow field during their careers cannot always respond to such questions without doing some rather extensive study and review of the literature.

Probably aeronautical engineers are involved in materials problems to the same extent as are the mechanical engineers. But their needs are often so specialized that they work very closely with materials engineers on designs which often involve new or recently developed materials. Nevertheless, there are also some rather mundane or standard materials that must be specified in the aircraft industry. They use fasteners, springs, insulation, and similar common materials throughout their work. Even electronic and electrical engineers are not immune to materials problems. Parts as common as electrical contacts involve wear resistance, spring characteristics, and electrical conductivity, and all of these properties must be considered. In thin-film microcircuits it has been estimated that 75 to 80% of device failures can be traced to materials problems. The creep of solders, electrical migration of atoms, interdiffusion between thin layers of metals, metal deposition *whiskers,* and other defects are just a few of the materials problems encountered in electronic devices. Certainly, civil engineers work with materials of construction and must stay up to date on new materials and their properties, especially their corrosion resistance. Of course, industrial engineers are always concerned with the cost of materials. In many situations this can be one of the most important factors.

With the development of data bases, which will be considered later, there have been attempts to develop heuristic expert systems that reduce the materials selection process to a set of numbers called *weighting factors.* Digital logic analysis techniques are used in the evaluation, which results in the elimination, or at least minimizes, the level of subjectivity in the assignment of weighting factors. This is a noble and futuristic approach because all materials engineers have their favorite materials, often because of familiarity and past successes, and as such cannot make an objective evaluation of all candidate materials. The only problem with this approach is the determination of the weighting factors that must be assigned to each material. Cost data can easily be obtained. Upper and lower limits of all properties, both mechanical and physical, of a suitable material can be reasonably well established. Then one must also establish acceptable limits for wear, corrosion resistance, creep and stress–rupture, and fatigue in many cases. Weighting factors, which may vary considerably, must be assigned to all materials under consideration. Separate weighting factors are assigned for each property: hardness, cost, fracture toughness, and so on. The use of nondimensional weighted property index techniques provides a basis for accelerating and standardizing the process of evaluating candidate materials, which must first be identified and screened by examining service conditions. This approach has its merits for well-documented materials. Jebb and McCollough* have used digital logical analysis to assign weighting factors to gas turbine blades, cutting tools, roller bearings, and artificial-hip-joint materials. Their weighting factors for cutting tools to achieve metal removal, well-established metals, and for gas turbine blades are listed as follows:

---

*A. Jebb and J. M. McCullough, *Chart. Mech. Eng.,* Vol. 34, No. 10, 1987, p. 49.

| Cutting tools | | Turbine blades | |
|---|---|---|---|
| Operating-temperature hardness | 0.40 | Specific rupture strength | 0.27 |
| Room-temperature hardness | 0.25 | Thermal fatigue resistance | 0.23 |
| Toughness | 0.15 | Specific strength | 0.20 |
| Liability to fracture | 0.10 | Cost | 0.13 |
| Cost | 0.15 | Oxidation resistance | 0.17 |
| | 1.00 | | 1.00 |

But they also concluded that ceramic materials cannot yet be selected by such procedures. The variability in their properties is still too large and they suffer from a limited amount of accurate data. Yet weighted property index methods used with suitable material data bases will eventually emerge as an important and useful method for the selection of engineering materials.

What we attempt to do in this chapter is to look into the principles involved or knowledge needed in the materials selection process, the use of handbooks and data bases, and present some case histories of materials selection. In the latter situation the process of selection, the materials considered, and why each was selected were the questions for which answers were sought from manufacturers. Many manufacturers in competitive businesses treat such information as proprietary. In other cases the answer often received was: "Well, this is what they all use in this industry." I am sure that in many situations encountered in this survey, this was a true answer, but in others it may have been a way of protecting confidential data.

## 13.2 DETERMINATION OF THE MATERIALS FOR CONSIDERATION

It is probably wise to ignore cost in making the first cut of materials that can be eliminated from consideration. The cost factor must be considered, but only after the list of materials that will fulfill the other requirements has been narrowed for technical reasons. There are several ways to approach the elimination process. The following order is only a suggested list. The actual order will depend on the requirements for each specific application.

1. *Temperature.* There is really no point of considering polymers, polymer composites, and low-melting-point metals for elevated-temperature usage. For temperatures above 500°C we can narrow our list of thousands of possible materials to some number in the neighborhood of a few hundred or so. If the application is one of room temperature (e.g., −10 to 50°C), the selection process becomes more complicated simply because of the much larger number of

materials that must be considered. The design engineer must remember that at high temperatures (and also where fluctuating stresses are encountered) the yield strength has no bearing on the selection process (except for short-time applications). Only creep and stress–rupture values are of interest.

2. *Stress.* Once the temperature range of usage has been established, stress and ductility are the next most prominent mechanical properties that must be examined. Here the first decision that must be made is whether we are dealing with yield or ultimate tensile strengths and whether these strengths should be those determined in tension, torsion, or compression. Can some plastic deformation and dimensional changes be tolerated? Most designs are based on yield strength. The type of loading, however, must also be considered. The yield strength in torsional loading is only about one-half of that in uniaxial tension. Another example of loading that affects the properties is that found in most brittle materials, such as cast iron, ceramics, and graphite. Brittle materials are generally much stronger in compression than in tension. In brittle materials tensile stresses readily cause crack propagation, while compressive stresses tend to close cracks. Cracks formed in compression are less likely to propagate to failure under compressive stresses than are those cracks subjected to tensile stresses. This factor of compressive versus tensile strengths for brittle materials is an entirely different consideration from that of fracture toughness. The latter deals with the rate of loading and the resistance of the material to crack propagation. The types of loading for fracture toughness tests is specified by ASTM E399. Impact tests were described in Chapter 2. The rate of loading also affects the ductility, and in metals the yield strength may be altered by the loading rate, but not extensively. A factor of safety must always be introduced, but this is not a materials selection problem per se. The factor of safety must be ascertained prior to selecting a material, although this factor of safety depends on the type of material (i.e., the confidence level of the materials properties). When weight is an important consideration the specific strengths and specific moduli (i.e., the strength or modulus/density ratio) become important.

If cyclic stresses are suspected, the fatigue behavior must be known. Again the yield strength can be thrown out of the window. Only fatigue data can be used, and these data must be applied in a statistical fashion. It must be determined in advance what probability of failure can be accepted (see Figures 2.19 and 2.20).

3. *Ductility.* The ductility is usually taken into consideration at the same time that the stress requirements are being evaluated because in general the higher the strength, the lower the ductility of the material. If both properties are important, some compromises must be made. For metals, the strength can be increased considerably by reducing the grain size without a significant reduction in ductility. But few other options are available. In composites one can play around with the volume fraction of the various components and their

arrangements in order to improve the ductility with a minimum reduction in strength. Their are probably more compromises that can be made on the strength–ductility relationship in polymers than in the other categories, because all polymers are relatively weak compared to the other materials, yet their exists a wide variation in their ductilities and glass temperatures. At present, ductile ceramics do not exist, yet theoretically, they are our strongest materials.

4. *Toughness.* The fracture toughness must also be considered in addition to the ductility, particularly if shock loading is believed to be an operating condition. If it is not clearly known whether shock or impact loading could occur, then err on the conservative side. Today, many designs specify fracture toughness values, even in the absence of shock loading, either in terms of ft-lb of impact energy absorbed during fracture at a certain temperature, or $K_{1c}$ values, the latter becoming more popular now because they are true material property constants that can be used for design purposes, whereas impact energy measurements or of the go/no go type (i.e., they merely define the temperature above which the material can be used). $K_{1c}$ values will vary with temperature just as do the strength, ductility, and modulus values.

5. *Modulus of elasticity.* The modulus is an important design parameter, and where weight is a consideration, the specific modulus (i.e., the modulus/density ratio) is a more useful number. It must be realized that one cannot alter the inherent modulus of a material without changing its composition. For metals the modulus of a soft metal is identical to that of the same strain-hardened metal. The moduli of polymers are very small and not well defined. Perhaps composites are the only materials that can be designed to a desired modulus. Ceramics have the highest moduli of all groups because of their strong covalent–ionic bonds.

6. *Physical properties.* The coefficient of thermal expansion and the thermal and electrical conductivities are the most prominent physical properties that one must usually know. It should be kept in mind that these properties do vary with temperature and one must select values that were determined over the temperature range of interest. Density and specific heats are physical properties that are of lesser importance, although the density must be considered in terms of the specific strength and specific modulus, as pointed out above.

7. *Corrosion resistance.* This is one of the most difficult material characteristics with which one must deal. First, the data available are reported in many different ways, some being of a qualitative nature only. Second, the environments to which the part will be exposed are not always known. The corrosion resistance of metals to many environments have been compiled and discussed in Vol. 13 of the *Metals Handbook,* 9th ed. Another good data source for metals consists of the National Association of Corrosion Engineers (NACE) series of handbooks, which began publication in 1982. NACE also has a corrosion-resistance data base, which we discuss in the next section. Some

corrosion data sources for polymers and reinforced polymers include the *Engineered Polymer Source book* by R. B. Seymour, published by McGraw-Hill; the annual *Guide to Selecting Engineering Materials,* published as a special issue of the *Advanced Materials and Processes Journal* published by ASM International; the McGraw-Hill *Plastics Encyclopedia;* and the *Corrosion Protection Handbook* by P. E. Schweitzer, published by Marcel Dekker. I strongly advise consulting a corrosion expert for specifying materials that will be exposed to corrosive environments. Vendor data, when available, should also be taken into consideration.

8. *Reliability.* The reliability of the component itself is defined in terms of the probability that the component or device will perform its intended function for the intended period of the design. But for this section we are referring to the reliability of the property data discussed in the foregoing sections. The reliability of the life of the component is based on the combined reliability of all the properties discussed above. As mentioned earlier, fatigue data must be treated in a statistical fashion. One could make a similar statement for the other mechanical properties. However, for metals, the strengths, ductilities, fracture toughness, and moduli are known to a rather high degree of certainty, and for room-temperature properties a conservative factor of safety will assure a high probability of survival of the part. At elevated temperatures the data on metals are not as precisely known as those for lower temperatures. Creep and stress-rupture tests require long periods of time. The usual method of handling such tests is to enforce failure in a short time at elevated-temperature tests and then extrapolate back to lower temperatures. These so-called parameter methods (the Larson–Miller parameters being the most common) were discussed in Chapter 12. The parameter curves for most superalloys can be found in the ASM *Metal Handbooks* (e.g., Vol. 1, 10th ed., p. 627, 1990). These curves are not difficult to understand and use with a little study or tutoring by persons working in this field.

The reliability of the data on polymers and glass-reinforced polymers is also reasonably well established and again can probably be used with conservative safety factors. But the mechanical property data on ceramics, metal matrix composites, and ceramic matrix composites must be treated with caution. The scatter and nonreliability of data on these materials are restricting their use at present. In many such materials the availability of data is insufficient to permit application of a statistical treatment.

The term *factor of safety* can be associated with uncertainties in property data and manufacturing processes and uncertainties in loading and service conditions. Brittle materials have much greater variation in properties than do ductile ones and hence require a larger factor of safety (which is always greater than 1). The nominal strength $S$ and the allowable strength $S_a$ are related by

$$n_s = \frac{S}{S_a} \qquad (13.1)$$

where $n_s$ is the factor of safety. The factor of safety $n_1$ related to loading conditions is given by

$$n_1 = \frac{L}{L_a} \qquad (13.2)$$

where $L$ is the maximum load that one could use based on normal working conditions and $L_a$ is the allowable load.

The overall factor of safety now becomes

$$n = n_s \times n_1 \qquad (13.3)$$

One must also assign factors of safety to the other parameters—toughness, stress–rupture, creep, and so on—depending on the intended application. Factors of safety range from about 1.2 to 20, where the highest values involve fatigue conditions (i.e., cyclic stress applications) where values as low as 20% of the ultimate tensile strength are occasionally used. Of course, one could use the fatigue strength and apply a smaller factor of safety. In the absence of cyclic stresses the more common safety factors range from 1.5 to 10, and here the more brittle materials require the higher factors of safety.

In many applications the designer is required to follow established codes which include an allowable working stress, $S_w$, based on certain derating factors. In some situations, particularly where fatigue and corrosive environments are involved, $S_w$ is determined by multiplying these derating factors, which are less than unity, times the nominal strength $S$ as follows:

$$S_w = Sd_1, d_2, \ldots, d_i \qquad (13.4)$$

Factors of safety and derating factors are then combined to obtain the following:

$$S_w = \frac{Sd_1 d_2 \cdots d_i}{n_1 n_2 \cdots n_i} \qquad (12.5)$$

There are established standard statistical treatment procedures of data involving standard deviations, distribution curves, and probability calculations with which we are not concerned herein. The interested reader may consult the books listed at the end of the chapter. Some of these approaches become rather involved when many components of different materials are used in one complex system.

## 13.3  COST ANALYSIS

Introduction of the cost factor must be done when it has been established that a number of materials meet all the other requirements. The cost factor includes more than the cost of the material itself. Additional manufacturing processes, such as machining, forming, heat treatments, ease of assembly, number of components to be manufactured, and reliability must also be considered. One of the more important decisions to be made is whether the cost of the material is a major factor in the final product cost. In products in which precious metals such as gold, platinum, and palladium are used, the cost of the raw metal is often 90 to 95% of the cost of the final product. This will vary from jewelry and dental alloy manufacturing, where alloys and intricate die designs are involved, to the electronic industry, where the precious metals are sold as bonding wires, evaporation wire and slugs, and sputtering targets for the deposition of thin films. In the electronic industry the price to the customer is often fixed at the precious metal price on the day the order is placed, plus a fabrication charge. Otherwise, both parties are playing the precious metal market game, which is not much different than betting on which way the stock market will go. Where there is room for design factors, one can afford to gamble, and in fact must do so, on the price fluctuations in what the stockbrokers refer to as the *metals* market. On the other hand, in many of the common metals, such as copper, aluminum, and nickel, the price of the raw material, particularly for small parts and small lots, is often a small percent of the selling price. Unless weight is a factor, one can overdesign these products.

In Appendix B we list the prices of a number of materials obtained in February 1992. Many metals are quoted by the American Metals Market daily newspaper and some can also be found in *Iron Age* magazine and the McGraw-Hill *Metals Price Report*. Because prices vary considerably with form and quantity, a vendor's quotation is essential where material cost is a prominent factor for consideration. In Appendix B the cost of a few of the primary metals (i.e., the starting metals as they emerge from the refiner) can be compared to the cost of the metal in alloy or fabricated form. Using such comparisons, one can obtain an idea of the possibility for profit by considering another metal. Titanium, the wonder metal of the 1940s, is still lagging the other metals in quantity used because of the high price of the titanium metal ingot. The price gap between titanium and the more common metals is decreasing because the consumption of titanium is increasing (which drives down the extraction and refining costs) and because of the technical improvements being made on the extraction and refining end of the process of ingot formation. Continuous casting of electron-beam-melted ingots has provided considerable cost reduction plus improvement of metal quality.

When we consider alloys such as the superalloys, especially the single-crystal superalloy turbine blades, processing and manufacturing costs far exceed the raw materials cost. This ratio increases the more complex the alloy and the less the usage quantity. When we consider the composite materials, particularly the more exotic metal–boron and SiC fiber composites, the cost of producing the fibers themselves becomes a major factor, not the cost of the boron or SiC raw materials. The same argument goes for carbon–carbon composites and boron– and SiC–epoxy composites. Many of these composites have been used only in special products, although this situation could change significantly within the next few years. This is sort of the trickle-down reasoning applied by the defense industry in their funding requests.

Now let's return to the real life and what could be more realistic than the automobile industry. We can almost use the automobile as a yardstick for inflation. The ratio of the starting salaries of engineers to automobile prices has been fairly constant over many decades. In the automotive industry the cost of many parts are computed to two or three decimal places. Now the cost of the raw material becomes a vital part of the total cost and accounts in part for the demise of our steel industry. In summary, the importance of materials cost varies widely from industry to industry and should be seriously considered only when it is a vital part of the total cost. This determination must be made on a case-by-case basis.

## 13.4 CASE HISTORIES INVOLVING MATERIALS SELECTION

### 13.4.1 Solar Arrays for Spacecraft

Candidate materials are as follows:

The candidate materials for solar cells are limited to silicon, GaAs, GaSb, GaAlAs, and GaAsP at the present time. The number of semiconductor materials for solar devices will undoubtedly increase very rapidly within the next few years.

Substrates for the solar cell devices could include lightweight beryllium sheet, a variety of polymer films, and lightweight PMCs (e.g., glass-fiber-reinforced polymers).

Housings could consist of carbon–carbon composites, boron and SiC MMCs, and titanium or aluminum alloys.

Solar arrays are devices for collecting the sun's energy and transforming it into electricity. Ideally, the structure should weigh nothing for spacecraft applications. Silicon has been the semiconductor solar cell material of choice for many years, but gallium arsenide is about to take over. GaAs cells are used in outer-space vehicles because they are more efficient than silicon cells. These cells are manufactured by Applied Solar Energy Corporation and Spectrolab, a unit of Hughes Aircraft. GaAs cells cost about 10 times that of silicon, but are 40% more efficient, which mean smaller arrays and less weight. Boeing Aerospace and Electronics has reached efficiencies (i.e., the percent light striking its surface to that converted to electricity) of more than 30% by stacking GaAs and GaSb cells in tandem. The GaSb captures infrared light just outside the range of GaAs. GaAlAs cells are also being considered for use in tandem to capture other "wasted wavelengths" of light. By such arrangements, 40% efficiency devices are on the horizon. Today, the cells are often mounted on lightweight beryllium hot-rolled sheet metal and housed in a lightweight carbon–carbon composite housing.

Lockheed Missiles and Space is developing big solar arrays of a total of 16,400 cells, presumably for space stations, which will produce far more power than the 84-panel 12.5-kW system carried aloft on *Discovery*'s maiden voyage. The new arrays will be mounted on DuPont Kapton polyimide film only 1 mil thick, which is coated with protective silicon dioxide to protect it from atomic oxygen impingement.

Rigid arrays have some advantages, and in these applications high-modulus graphite–epoxy composites are being considered as substrates. One of the problems yet to be solved is finding composite materials thin enough for solar cell surfaces. Hercules is continuing to develop ultra-high-modulus materials to compete with pitch-based carbon–fiber composites.

### 13.4.2 Materials for the Cooler Regions of Jet Engines

In the cooler regions of the jet engine we are considering temperatures below 300°C (570°F). Aluminum alloys will overage at these temperatures. Most resins are limited to about 150°C (302°F), but bismalemides and polyimides are useful to about 325°C (617°F). Obviously, lightweight materials, preferably of a high strength/density ratio, are preferred. Candidate materials are:

Beryllium sheet
Titanium alloys
Graphite–epoxy composites
Bismalemides as a resin or with a variety of reinforcements such a glass, carbon fibers, and SiC fibers
Polyimides with the same type of reinforcement as for bismalemides
Boron–Al and SiC–Al composites (very expensive)

Advanced composite materials are now making their way into the production of jet turbine engines in both the military and commercial sectors. Graphite–epoxy cases for the space shuttle solid rocket motors (SRMs) were being manufactured by Hercules in the mid-1980s. The following is a partial listing of such applications in jet engines:

Nacelles and thrust reversers are made routinely with PMCs.

Ducts, cowls, and fans—also PMCs.

Graphite–epoxy outlet guide vanes.

Rolls-Royce; polymer composites for noise spinners and compressor core, farings on RB-12 engines, and graphite–epoxy bypass ducts on Rolls Tay engines used in the Fokker 100 and Gulfstream 100 aircraft.

Forward fan blades for the GE90 power plant of the next-generation jets (in progress).

Pratt and Whitney will use a variable geometry graphite–polyimide guide vane inlet for the Advanced Tactical Fighter.

A GE–Rohr joint ventures uses a graphite–epoxy core cowl now installed on a Lufthansa Airbus A310.

Kevlar aramid fiber containment casings have been standard materials on the GE CF6 engines since 1980 and outlet guide vanes have been made of graphite–epoxy composites since 1985.

Companies are now working on ceramic and metal matrix composite materials for the hotter engine components. Service temperatures in the range 1600 to 1800°F are anticipated for the first materials, although these materials have the potential for much higher temperature usage. Alumina fibers from Sumitomo and an alumina–silica–boric oxide fiber called Nextal from the 3M Company are involved in these studies. (Source: *Aerospace America*, July 1991).

### 13.4.3  Surgical Implants

Body implants for replacement of bone structures are known as *prostheses*. Various metal devices in the form of wires and pins, mostly made from the precious metals, have been used since the 1860s. The relatively low strengths plus their high cost and, surprisingly, some infections have limited their use. Experiments in laboratory animals have shown that the common alloys of copper, low-alloy nickel–chromium, and plain carbon steels have both biochemical and biomechanical limitations. Nevertheless, certain steels were introduced early in the twentieth century but were later replaced by the higher-chromium-containing alloys. These passive alloys were relative inert, much less expensive, and far stronger than the precious metals. They began to be used in the late 1920s. Stainless steels were introduced several years later.

Most of the modern implants have centered around repairing long bones and joints. Hip joint replacement is a very common procedure today, probably greater than 300,000 per year worldwide. Many persons past the age of 50 have developed calcium deposits in these and other body regions that are not only painful under most limb movements, but in some cases virtually incapacitate the person. In the 1950s and into the 1960s, metals used for such implants were the cobalt-based alloys, which had first been used for highly stressed gas-turbine blades in 1941. Similar alloys, known by the trade name Vitallium, had been used for dental implants (e.g., bridges and partial dentures) even prior to 1941. But the primary use for these alloys and their modifications had been in the piston-type aircraft and later for bladeings and turbines in the first jet aircraft. They were known as Stellites or Haynes-Stellites, developed by a division of Union Carbide Corp. (now part of the Cabot Corporation). They usually contained from 23 to 27% Cr for high-temperature oxidation resistance with small additions of other elements to the cobalt base. Later versions contained approximately 15 to 20% nickel. It was their good corrosion resistance, because of their high chromium content, and the experience that had been acquired over several years in precision casting that brought them to the attention of orthopedic surgeons. They also possessed good wear resistance and the ability to be polished to a smooth attractive surface. These alloys were a logical choice for prostheses at that time. Their main drawback was their brittleness. Most were used as precision castings, but later it was found that with some alloy modifications, such as by the additions of nickel, molybdenum, and tungsten, these alloys could be fabricated by forging and other mechanical working operations. At about the same time the wrought surgical stainless steels of the 316L type were being examined. They could be fabricated and were strong and ductile. But they were found to be susceptible to pitting corrosion and did not possess the strength of the cobalt-based alloys. In the 1950s and early 1960s titanium and zirconium were introduced because alloys existed that had been developed for aerospace and nuclear applications. Their passive nature suggested that they could be compatible with body fluids and tissue. Titanium alloys have been an important implant material since that time. But also, the nonmetallics, ceramics, carbon, and high-density polyethylene were gaining interest because of the inertness and strength. In the 1970s bioactive substances were considered for the in vivo bonding to bone, and considerable emphasis was given to the interactions between synthetic biomaterials and tissues. Ceramics are being used more and more in hip replacements, and their usage will probably increase within the next few years, despite their brittleness. Friction and wear studies have shown that alumina–alumina surfaces are superior to alumina–polyethylene, which surpasses the metal–polyethylene combination. The catastrophic failure of ceramic prosthetic components is a constant fear. But ceramics can be tested to loads far exceeding those to be experienced, by at least a factor of 2.5. Thus ceramic devices can be individually and nondestructively tested and thereby eliminate the statistical scatter normally experienced in other ceramic applications. Partially stabilized zirconia, because of its toughness, strength, and biocompatibility, may replace alumina in the near future. Calcium phosphate ceramics interact with bone and thereby form an unusually attractive bond. But calcium phosphate has limited strength in bulk form and as a result is being considered more seriously as a coating for other materials, including metals.

The mechanical requirements of implant materials are not nearly as stringent as is their biocompatibility with body tissue and fluids. Bones, for example, have compressive strengths of about 105 MPa and elastic moduli around 15 GPa, values easily attained by most materials, including reinforced polymers. Also, bones are brittle. They have low fracture strength and ductility. Fatigue properties of implant materials are important, however, because fatigue loading is several times the body weight for hip implants, and the frequency is on the order of a few million cycles per year or tens of millions per lifetime, considerably beyond that of most conduct fatigue tests.

A sketch of the end of a femur (the thighbone) is depicted in Figure 13.1a. Figure 13.1b shows the reconstructed bone containing a metal stem and head. The shaded area is the prostheses, the shaft or stem being inserted into the medullary canal of the femur bone and held in place by a cement.

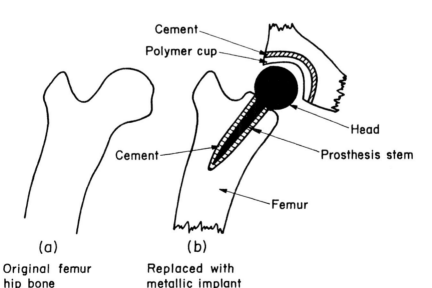

**(a)**
Original femur
hip bone

**(b)**
Replaced with
metallic implant

**Figure 13.1.** Sketch of the femur bone: (a) before reconstruction; (b) with implants.

The total hip joint prostheses consists of several parts, each having different requirements. The head, which replaces the femur bone head, requires fatigue strength in excess of 200 MPa. Almost any material meeting this requirement will also meet the other strength requirements. Other properties of significance are wear resistance, elastic compatibility with bone structure, and of course, corrosion resistance. Materials that meet these requirements include the cobalt-based alloys previously mentioned, 300 stainless steels, with cold-worked 316 being a current favorite material, the titanium alloys (Ti–6% Al–4% V), ceramics for the head that meshes with the cup or socket, and some reinforced epoxy composites. The cobalt-based alloys have excellent corrosion resistance and biocompatibility, but low toughness and variability in mechanical properties caused their replacement with more desirable materials, particularly the titanium alloys. The anchorage stem, which is placed within the medullary of the long bone and the head, was formally, and in some cases still is, cast or forged as one piece. The socket bearing surface on the femur head replacement must provide lubrication and wear resistance between these two components. The socket is almost invariably a nonmetallic material. Ultrahigh-density polyethylene and polyoxymethylene are the favorite polymers for this application. The friction between alumina and polymers is much less than that for the metal–polymer interfaces, so the trend today is to bond a metallic stem to an alumina head by cements, usually a two-component mixture acrylic bone cement. PMMA bone cements are a recent consideration. It is interesting that during the past 10 years the Co–Cr alloys have staged a comeback, particularly in implants other than the hip joint. The main reason for their resurgence is their good wear resistance plus some minor improvements in their toughness and consistency. HIPing of Co–Cr powders have achieved considerable improvement over the cast parts. Although titanium is still preferred for the stem that is inserted into the bone medullary canal, the head, which requires high wear resistance, now often consists of an alumina ceramic ball bonded to the titanium alloy stem rather than a one-piece stem and head titanium alloy. Material advancements in bioengineering are proceeding at a rather rapid pace. Hydroxyapatite is a relatively new bioceramic coating under study. This ceramic material, which is regarded as synthetic bone, readily allows bone in-growth and could be a possible coating for implants. In the 1980s and into the present, synthetic biomaterials are being engineered for implants and are predicted to play a major role for future implants.

Another example of a frequently employed surgical implant is that of the rotating hinge knee shown in Figure 13.2. On the left is an ultrahigh-density polyethylene tibial sleeve that fits onto the top of the hinged knee implant shown on the right. The long stem that fits onto the hinged section is machined from a Ti–6% Al–4% V alloy, while the hinge distal itself is machined from a

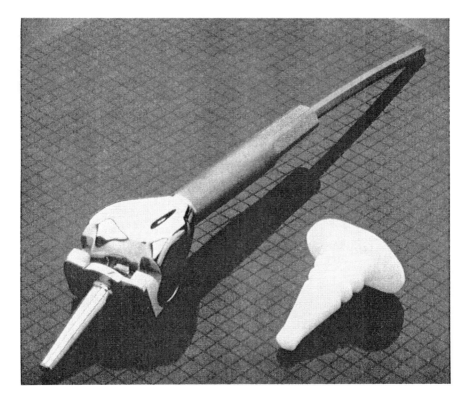

**Figure 13.2.** Rotating hinged knee constructed from a cobalt–chromium alloy used in conjunction with a titanium alloy stem and hinges and ultrahigh-density polyethylene bearings and sleeve. (Courtesy of Techmedica Corp.)

cobalt–chromium alloy. The hinges, which are rotated with the assistance of polyethylene bearings, are of the same titanium alloy as the stem section.

### 13.4.4 The Space Shuttle

The space shuttle flew its first orbital journey in April 1981. The most difficult materials problem for this first reusable space vehicle was the ability of the exterior materials to withstand the ascent and reentry temperatures to which these surfaces would be subjected as a result of the frictional heating caused by the earth's atmosphere. These reentry temperatures versus exposure time for the critical shuttle components are depicted in Figures 13.3 and 13.4. The maximum temperature approached 1400°C (2552°F). Although the time span

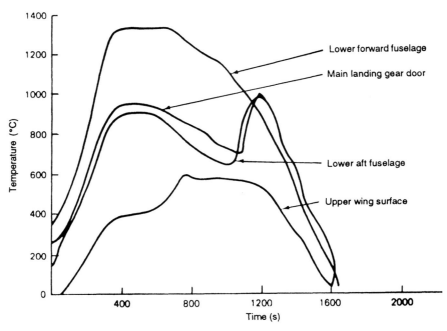

**Figure 13.3.**   Ascent and reentry temperatures versus exposure time for various sections of the space shuttle.

within the range 1000 to 1400°C (1832 to 2552°F) was only about ¼ h per trip, there were no commercially available metallic alloys at that time, nor at the present, that could meet this temperature requirement. Inconel X750 had been used on the X-15 aircraft to temperatures of about 900°C (1652°F). X750 has a yield strength of 400 MPa at 800°C (1472°F) compared to 700 MPa for alloy 718, but X750 has better stress-rupture properties. Some refractory metal alloys, particularly the niobium-based alloys, have been used for rocket nozzles, leading edges, nose caps for hypersonic flight vehicles, and guidance structures for reentry vehicles to temperatures of 1400°C (2552°F), where moderate strength is required. Silicide coatings have been developed for niobium that extend their operating temperatures and have been used for aerospace applications but are not recommended for sharp edges or applications where spot welding and riveting fabrication operations are required.

The candidate metallic materials for the lower temperatures included the aluminum alloys [to about 177°C (350°F)] and the titanium alloys [500°C (932°F)]. As we progress upward in temperature the superalloys could be used to 900°C (1652°F) and the niobium silicide-coated alloys to temperatures of 1482°C (2700°F). In the nonmetallic area the graphite–epoxy and Kevlar–

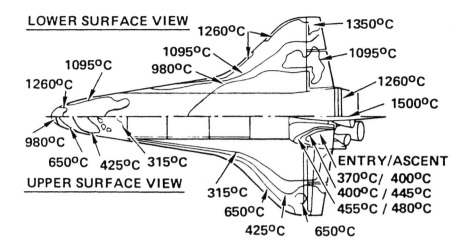

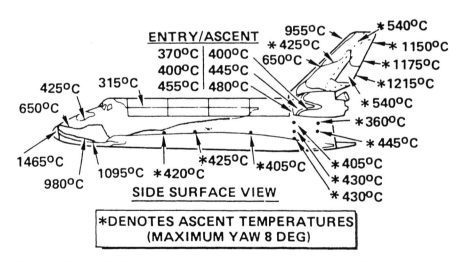

**Figure 13.4.** Typical (though not the most severe) thermal profile of the space shuttle orbiter. (From L. Korb et al., *Ceram. Soc. Bull.*, Vol. 60, No. 11, 1981, p. 1188.)

epoxy composites were the leading candidates for the lightweight low-temperature use [< 150°C (302°F)]. For the thermal protection system (TPS), which would experience temperatures in excess of 1600°C (2912°F), carbon–carbon was the likely material for protecting the nose cone and leading edges, but it was not practical to use carbon–carbon composites to protect the main orbital structure from the high heat encountered in ascent and reentry. Ablative systems, hot radiative metallic materials, and regenerative cooled metallic structures were all considered for the main body thermal protection system. Ablators protect simply by becoming detached as they absorb heat. Ablators were not acceptable because of their aerodynamic effects, and they did not provide a TPS that could withstand the 100-h mission requirement with a minimum 160-h turnaround refurnishment requirement.

The materials finally selected for the first shuttle *Columbia* (there were some modifications for subsequent shuttles) can be divided into three major categories: composites, metals, and the nonmetals used for the TPS.

Composites

1. *Carbon–carbon.* Carbon fibers in a graphite matrix were used in the hottest regions that were covered by tiles. These regions could attain temperatures on the order of 1600°C (2912°F) and included the nose cone and leading edges.

2. *Carbon–epoxy.* Carbon fibers in an epoxy matrix were used for the huge (about 225 ft$^2$ in area each) cargo bay doors. These doors, which consisted of eight separate doors ganged together, required a strong lightweight material. They were designed to be nonoperable at atmospheric pressures but could be opened in space. This same composite was also employed for the orbital maneuvering system (OMS) pods.

3. *Boron fiber-epoxy-impregnated tape.* This composite tape was used to cover Ti-6 Al-4 V alloy tubing in the aft section structures that supported the engine. These support systems were of both round and square shape.

4. *Boron–aluminum composite structural tubes.* These composites, which were described in Chapter 9, were manufactured in the form of structural tubes that were used as stabilizing support members for the cryogenic tanks and also as stabilizing members for the main structure as struts (similar to I-beams). Several layers of these tubes were diffusion bonded to titanium alloy end members.

5. *Kevlar–epoxy.* This polymer matrix composite was widely employed for (a) large air ducts for purging or ventilation in order to equalize pressure, (b) honeycomb sandwich structures for cabin equipment containers, and (c) overwrapping of pressure vessels lined with titanium alloys or Inconel 718. The composite carried about 80% of the load.

6. *Fiberglass composites.* These were used in honeycomb sandwich structures.

Metallic Materials

1. *Aluminum alloys.* The orbiter structure for the most part was made from aluminum alloys in a skin-stringer, integrally stiffened machined panel, or as adhesive-bonded honeycomb sandwich structures. 2124 T6 Al alloy (4.4%

Cu, 1.5% Mn) was employed in the basic structure because it had better toughness than other aluminum alloys in 1-in.-thick sections. It could be used to 177°C (350°F) without more than a 10 to 15% loss in strength in a 100-h period. The stronger 7075 alloy overages faster than alloy 2124. In the cabin structure and for flat bulkheads 2219 (6% Cu, 0.3% Mn) was employed with different tempers, depending on the location. T86 was common for bulkheads. 2219 alloy was also used for liquid helium storage.

2. *Titanium alloy 6Al–4V.* In addition to the boron–epoxy-wrapped tubing, this alloy was employed for the structural beams, some being on the order of 22 ft in length. They were initially machined and bolted together to attain these lengths but were later forged to near final shape.

3. *Beryllium.* This metal played a more important function in the design of the shuttle than was ever publicized, even in the materials journals. It is a very lightweight (1.85 g/cm$^3$), high-modulus [303 × 10$^6$ MPa (44 *times*10$^6$ psi)], brittle, expensive metal. Its first claim to fame was its use in nuclear reactors and radiation measuring instruments. It is practically transparent to most types of radiation. But its specific modulus is substantially greater than that of Al, Ti, or Mg. It is usually prepared by hot pressing of powders, which can subsequently be hot worked to obtain other shapes. Its first real structural use was in the aerospace industry. In the shuttle it was used as structural doors for oxygen fuel tanks. Both CIP and HIP beryllium beams were employed as structural members for stiffening the structure between windows and also as window frames, as well as other stiffening applications involving gyro areas and star tracker arms.

Thermal Protection System (TPS)

Ablators and hot metallic radiative structures were not considered for the TPS. Despite extensive experience with silicide-coated niobium rocket nozzles, the multitude of parts needed for the TPS required a high degree of manufacturing complexity and coating reliability problems. Local loss of the coating could result in burn-through or embrittlement, leading to failure of the niobium substrate.

The TPS materials eventually selected are listed in Table 13.1. The basic tile system was composed of four key elements: the ceramic tiles, a nylon felt mounting pad, a filler bar, and a room-temperature vulcanized rubber. The tiles were treated with a high-emittance borosilicate glass. The felt mounting pad isolated the tile from the thermal and mechanical strains of the substructure. The filler bar, also a nylon felt material coated with silicone rubber, protected the tile-to-tile gaps from overheating. The adhesive bonded the tiles to the isolator pads and the filler bar to the structure.

Two different-strength tiles were employed, both containing high-purity amorphous silica fibers but of different densities (Table 13.2). Most of the tiles used were of the low-density (1.44 g/cm$^3$) type, but where higher strength was

**Table 13.1**   Materials used in the Space Shuttle TPS

| Material | Temperature capility[a] [0°C (°F)] | Areas of use |
|---|---|---|
| Carbon fibers in a carbon matrix coated with SiC | 1650 (3000) | Nose cone, wing leading edges, forward external tanks separation panels |
| $SiO_2$ borosilicate glass coating with $SiB_4$ added | 650–1260 (1200–2300) | Lower surfaces and sides, tail leading and trailing edges, tiles behind carbon |
| $SiO_2$ borosilicate glass coating | 400–650 (750–1200) | Upper wing surfaces, tail surfaces, upper vehicle sides, OMS pods |
| Felt surface insulation–nylon felt–silicone rubber coating | 400 (750) | Wing upper surfaces, upper sides, cargo bay doors, sides of OMS pods |

*Source*: L. Korb et al., *Ceram. Soc. Bull.*, Vol. 60, No. 11, 1981, p. 1188.
[a] Based on 1000 missions.

**Table 13.2**   Densities of Some Selected Materials

| Material | Density $(g/cm^3)$ | Material | Density $(g/cm^3)$ |
|---|---|---|---|
| Aluminum | 2.7 | Beryllium | 1.85 |
| Boron | 2.34 | Carbon (graphite) | 2.25 |
| Iron | 7.87 | Magnesium | 1.74 |
| Nickel | 8.9 | Cobalt | 8.0 |
| Titanium | 4.51 | Niobium | 8.6 |
| 7075 Al | 2.7 | 2219 Al | 2.83 |
| 2014 Al | 2.80 | 2024 Al | 2.75 |
| Ti–6% Al–4% V | 4.43 | Inconel 718 | 8.19 |
| Niobium alloy FS-85 | 10.60 | 301 stainless steel | 8.03 |
| Al–Li alloy | 2.685 | Nickel superalloy | 7.9 |
| SiC filament | 3.0 | Boron filament | 2.57 |
| Al–20% boron | 2.63 | Boron–epoxy | 2.0 |
| 60% Carbon–epoxy | 1.66 | SiC hot pressed | 3.2 |
| SiC fibers–Ti–6% Al–4% V | 3.86 | SiN hot pressed | 3.2 |
| 40% SiC fibers–6061 Al | 3.0 | Carbon–carbon | 1.95 |
| High-density polyethylene | 0.96 | PH 17-4 stainless steel | 7.8 |

required, high-density ($3.52$ g/cm$^3$) tiles were employed. Those tiles on the underside were black because of the addition of silicon tetraboride for high emittance at high temperatures. The tiles were about 93% void, thus possessing good thermal conductivities, in the range 0.017 to 0.052 W/m·K. Their low coefficient of thermal expansion and low modulus practically eliminated thermal stresses and the corresponding thermal shock problems.*

### 13.4.5 Golf Clubs

Sporting goods have been the big recipients of space-age fall-outs, which include golf club shafts, tennis rackets, skis, boat hulls, fishing rods, and bicycles. In the case of golf club shafts the number of candidate materials now number close to those for golf club heads. The poor golfer struggling to improve his game is confronted by advertisements of how to do this by choosing certain products, including the golf balls (a simple lead pencil with an eraser could sometimes be of more help). But despite all this advertising hype there is some foundation to the claims that certain materials will perform better than others for the intended objective. Of course, the design of the golf club head is important but we will not confront that problem here except to mention that the lighter the shaft weight, the more weight that can be placed in the head. Also, the head can be enlarged to give a larger ball striking area. We begin with the golf club shaft.

Candidate materials for shafts are:

Boron fiber-reinforced graphite
Titanium
Graphite fiber–epoxy filament wound
Tape-wound graphite–epoxy shafts
Stainless steels
Chrome-plated steels

The graphite composite shafts, which consist of graphite fibers in an epoxy matrix, have emerged for the present as premier golf club shafts. Graphite is strong and light weight, which allows reduction in the weight of the shaft, permitting more weight in the club head for the same overall swing weight. Generally, they are made of graphite fibers in an epoxy matrix. More recently, boron fibers in an epoxy matrix have been added at the small tapered end of the shaft. The average graphite composite shaft which contains the boron–epoxy section weighs 6 g less than the all graphite–epoxy shaft. Boron fibers are 10 to

---

*The author is indebted to L. Korb, retired from Rockwell Int., for supplying some unpublished information.

15% heavier than carbon fibers, but their higher strength permits a smaller total volume of fiber. Boron fibers are considerably more expensive than graphite (carbon) fibers. Graphite shafts weigh about 70 to 85 g, compared to 100 to 125 g for the steel shafts. The boron–epoxy composites are made by Textron, and a large percentage of the shafts are manufactured by Aldila Corp. A photomicrograph of the boron–epoxy composite construction was shown in Figure 9.27. Figure 13.5 shows a photomicrograph of the cross section of a graphite–epoxy shaft with an outer layer of boron–epoxy composite tape,

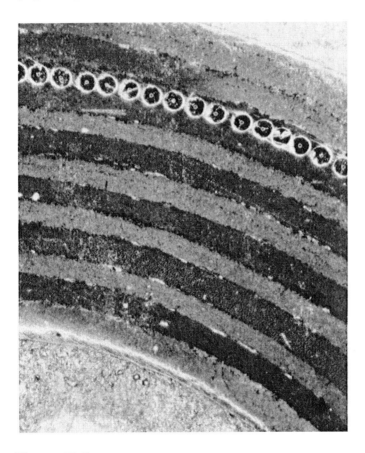

**Figure 13.5.** Structure of a graphite–epoxy golf club shaft with an outer layer of boron–epoxy composite material. The difference in shading of the graphite–epoxy layers is a result of the difference in orientation of alternate plies. The larger-diameter boron fibers are placed in two outer layers. The individual graphite fibers are barely discernible at this $50 \times$ magnification. (Courtesy of Aldila Corp.)

which has been added at the lower 5 to 8 in. of the shaft length. The basic physics is somewhat complicated, but a few comments are in order. Lighter shafts enable greater club head speed, which equals greater distance. Golf Laboratories, an independent equipment testing company that tested these shafts, concluded that the boron–graphite shafts outhit the all graphite–epoxy shafts by 14 yards, steel shafts by 13 yards, and titanium shafts by 16 yards. (From Golf Magazine, Jan. 1992). In addition, during the swing the shafts twist and untwist, which we call torque. The extra high stiffness (modulus) of the boron shaft allows the graphite layers to be angled to reduce the torque. A lower torque means more torsional stiffness, allowing a player with a faster swing to meet the ball squarely, causing a more accurate shot. The boron reduces the torque to about 2° when used with high-modulus graphite (some claim less torque), similar to that of steel. Normally, the boron is added only in the bottom portion of the shaft. Boron fibers are placed in the bottom end to limit breakage around the hosel, to reduce the torque, and also to provide for certain flex points. The flex point is where the shaft does most of its bending. A low flexpoint gives a little more "kick" than a higher one.

Graphite–epoxy shafts are made by either table rolling or filament winding. In table rolling the graphite fiber–epoxy tapes are laid side by side to form thin sheets that are rolled around steel mandrels (all shafts are made in tubular form). After the graphite is baked and hardened, the mandrel is removed. In filament winding, now the preferred method, individual fibers are wound around the mandrel, resulting in a seamless shaft.

Boron is lighter and stronger than steel and 10 to 15% heavier than graphite. Sometimes boron is laid atop the graphite fibers and then table rolled. Figure 13.6 shows the positions of the various layers near the tapered end section of the shaft. The boron–epoxy preimpregnated tape is shown in Figure 13.7. The finished shaft is slightly stiffer, stronger, and heavier than the graphite-only shaft. Boron experts say that a full two-layer boron wrap around the shaft circumference is necessary to realize the full benefit of boron. Textron specialty materials, America's only supplier of boron, identifies these shafts by a boron-certified sticker. In 1990, graphite-type shafts were attached to 18% of the woods (includes metal woods) and about 5% to iron heads. Estimates for 1991 point toward about one-third of all shafts being graphite and/or graphite with boron composites.

Titanium is about 50% lighter than steel but considerably heavier than graphite or boron. It is also more expensive than steel and has a higher torque. Some golfers, however, believe that the higher the cost of the golf club, the better it is and accordingly the lower the score. Materials experts doubt that the titanium shafts will effectively compete with the graphite and graphite–boron shafts. Incidentally, aluminum shafts appeared on the market for a short time in the 1960s. Perhaps they are collectors items by now.

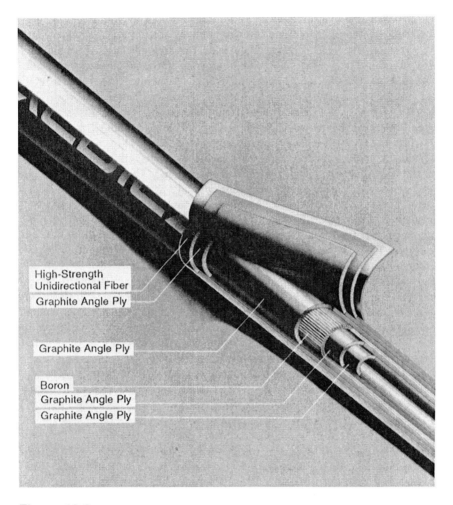

High-Strength
Unidirectional Fiber
Graphite Angle Ply

Graphite Angle Ply

Boron
Graphite Angle Ply
Graphite Angle Ply

**Figure 13.6.** Layered construction of a graphite–epoxy/boron–epoxy golf club shaft. (Courtesy of Aldila Corp.)

For golf club heads, we have an ever larger number of materials from which to choose. The candidate materials are:

Graphite–epoxy
Cast aluminum
Cobolt–chromium alloys
Titanium alloys

**Figure 13.7.**   Preimpregnated boron–epoxy tape. (Courtesy of Textron Corp.)

Maple-laminated woods
Kevlar-reinforced ABS copolymer
Plain carbon steel, chrome plated
Cast or forged Al–20% SiC fibers (Langert)
Zirconia (Coors Cerasports)
Stainless steel (PH 17-4, 18-8 austenitic, 431 martensitic)
Lexan (GE polycarbonate)
Persimmon woods
Be–Cu alloy
Kevlar-reinforced graphite
Cast iron

One will have many reasons for choosing a certain type of club. Most of them can be attributed to advertising hype (not realized by the golfer), but in some cases there are some real beneficial effects of certain materials. For the graphite composites and titanium heads, high strength and low weight are again touted. Solid woods are the traditional choice for feel, while metal woods allow a larger head with a larger ball striking area because the head is hollow, permitted by the higher strength of the metal. This hollow head also promotes control via weight distribution in the club head, reducing the twisting of the club head at impact. The nongolfer may be confused by the term "metal woods." Prior to about 1975, the longer-shaft, longer-driving-distance clubs that were used from the tee and for longer fairway shots had heads that were traditionally made of wood, and the shorter-length clubs were called "irons" since the club heads were made of steel, originally of plain carbon and later most often of the stainless type. When manufacturers began constructing drivers and long-fairway-shot clubs with metal heads, they were ironically and ambiguously referred to as "metal woods," a term that has struck despite efforts to call them something else (e.g., a metal driver).

The majority of metal wood heads are precision cast (hollow) from PH 17-4 stainless steel. There are about seven other precipitated hardenable stainless steels, all of which, like the PH 17-4, can be heat treated to obtain strengths within the range 115 to 220 ksi (794 to 1518 MPa), and correspondingly, hardness values ranging from 24 to 45 Rockwell C. The cast PH 17-4 steels used in golf club heads have hardness values of about Rc 38, corresponding to a tensile strength of about 185 ksi (1276 MPa). These heads are not usually heat treated, which means that some precipitation of second-phase particles occurs during cooling from the molten state because the solution-annealed condition has a lower hardness, about Rc 34. There are several advantages of the PH stainless steel club heads. The high strength gives the designer considerable flexibility in club head shape, including a larger ball striking area. The PH stainless steels also have very high ductilities, being on the order of 10% elongation in 2 in. (44% R.A.). But why 17% Cr–4% Ni instead of PH 13–8 and 15-5 stainless steels. They all have similar properties which can be adjusted by head treatment. Although it is reported that the 17-4 polishes well and has good wear resistance, the 17-4 has been much more widely used for other products, and manufacturers prefer to stick with the better known materials. The 17-4 had been used to a large extent in the aerospace industry, and some of these engineers eventually found their way into casting golf club heads. Their experience in investment (precision) casting of the 17-4 steel prompted them to stay with this alloy. Prior to this transfer of technology most metal club heads had been made from 431 martensitic stainless steels, which in the normal tempered condition possess strengths on the order of 115 ksi (795 MPa) with 15% ductility by elongation, although lower tempering temperatures can produce considerably higher strengths. This steel is still employed for the shorter-distance "irons" and compete with the common 18-8, and also PH 17-4, stainless steels for the "iron" club heads. The lower-priced "iron" heads are sometimes made of chromium-plated carbon steel, which was a very common "iron" club head many years ago. There is a reason for using 18-8 stainless heads, because their higher ductility permits bending of the "hosel" section into which the shaft fits. This permits last-minute alignment during assembly. The stronger stainless steels and Be–Cu alloys do not have sufficient ductility for this maneuverability. But one gives up hardness and strength when choosing the 18-8 stainless over the other stainless steels. Incidentally, Yamaha Corp. now offers these "iron" heads made from graphite and boron composites.

Like graphite, titanium has lighter weight but more strength than graphite composite heads. On a strength/weight ratio, titanium comes out ahead of all the metals. This again means that club heads can be made in larger sizes. Titanium and ceramics were the new club head materials in 1991, although titanium had been used in shafts for a few years. The Coors ceramic head is actually a zirconia insert on the ball-striking area of a wooden head. Many people think that Coors only makes beer, but the company has been manufacturing ceramic products for decades, perhaps even before entering the brewing business.

The Cu–Be alloys (commonly referred to as beryllium–copper, although the beryllium content is only about 2 wt %) have been featured for a few years as the new metal for "iron" heads. It is touted for its soft feel and gorgeous yellow-gold-like appearance. With respect to appearance, some prefer the black nitride oxidized or proprietary molten salt bath "melonite" black-colored steels that have been used for years on gears and crankshafts for wear and corrosion resistance. There are two other metal heads that have recently appeared on the market, namely cast aluminum metal–SiC fiber heads for the driver and cast-iron "iron" heads for the shorter-length clubs.

With respect to the nonwood "metal" wood heads, the Kevlar-reinforced compression-molded graphite, Kevlar-reinforced ABS copolymer, and Lexan polycarbonate have recently emerged. The entry of polymers into the golf club head business is very new but expected by most materials engineers because the polymer manufacturers are not going to miss any opportunities to replace metal, wood, concrete, or any other material with their products. And, of course, that is the way that a free-enterprise system should function. Lexan has been described as indestructible, providing good shock absorption and impact strength. Similar descriptions have been provided for Kevlar-reinforced ABS club heads.

Armed with all this information, what material does one choose? There is no obvious answer. In the March 1992 issue of *Golf* magazine the results and opinions were reported of a club test by 16 serious golfers with a large number of different woods and irons. There were no clear-cut winners, although it appears that the light graphite and boron composite shafts and metal wood heads are going to replace a lot of the traditional stainless steel shafts and wood heads for the drivers and long fairway woods. One manufacturer offers both woods and irons made with the low-torque reinforced graphite shafts matched to a graphite head.

### 13.4.6 The Mountain Bike

Mountain bikes frames have been constructed of a number of materials within the past decade, with each being replaced in rather rapid succession by stronger lightweight materials (i.e., higher strength/density ratios). Candidate materials are:

Cr–Mo low-alloy steels
Aluminum alloys
Titanium alloys
Metal matrix composites

Traditionally, mountain bike frames have been made with Cr–Mo steel alloys or 6061 T6 Al alloy. But just when manufacturers were beginning to find ways to process economically the high strength/weight ratios of the titanium alloys, along came the new metal matrix composites. The metal matrix composite that has excited the bike enthusiasts the most is the 6061 Al alloy–10% alumina particulate composite made by Duralcan. This material, which can be cast, forged, and machined, is 40% stiffer than the 6061 T6 aluminum alloy and 5 to 10% stronger. It is also no more difficult or expensive to produce than the Ti–6% Al–4% V tubes (bike frames are usually of tubular construction), but the costs are somewhat higher than those of 6061 T6 aluminum alloy. This bike (built by Specialized Corp; see Figure 13.8) still has the traditional Cr–Mo steel alloy fork welded to a Cr–Mo steel stem. The frame is made of the Duralcan metal matrix composite. Other parts include an inner stainless steel chain ring, a direct-drive titanium alloy pedal, a titanium-alloy frame seat, and titanium-alloy direct-drive bottom bracket.

On the subject of bikes, Tioga Corp. has used 19 layers of carbon mesh fiber bonded to a two-ply glass fiber core to produce a high-strength handle bar. A Cr–Mo steel sleeve is placed at the center and the entire fiber construction expanded to form a reinforced center-bulge handlebar.

### 13.4.7 Materials for Angioplasty Devices

Coronary artery disease (CAD) affects millions of people per year. CAD is caused by the gradual buildup and hardening of fatty deposits (plaque) within the walls of coronary arteries. There are three basic ways to treat CAD: medication, coronary artery bypass surgery, and percutaneous transluminar coronary angioplasty (PCTA). *Percutaneous* means that the procedure is done through the skin and *transluminar* means that it is done through the artery.

**Figure 13.8.** First mountain bike to employ a metal–matrix composite. The tubular frame is constructed of Duralcan SiC–6061 Al composite. (Courtesy of Specialized Corp.)

PTCA is a technique of widening arteries which have restricted blood flow, and several materials used in the devices associated with this procedure are the subject of this case history.

Angioplasty basically involves the temporary placement of a balloon catheter into one of the coronary arteries via either the femoral artery located near the groin or less commonly via the brachial artery located near the underarm. First, a flexible guide wire is steered through the particular coronary artery until its tip extends beyond the point of narrowing caused by plaque buildup (Figure 13.9). A hollow catheter with a balloon at its distal end is then moved over the guide wire into the narrowed segment of the artery and filled with a radiopaque fluid so that it may be viewed on the x-ray monitor. The balloon is typically inflated for about 90 s, and it may be inflated and deflated several times until the artery remains opened to the appropriate size. Some exterior widening of the artery may also occur.

The materials used in devices for PCTA are selected in close cooperation with medical personnel and engineers involved in the device design and manufacture. Although these devices are in contact with body fluids for only a short period of time, nonreactive, biocompatible materials are still required. For the guide wire the logical choice is austenitic stainless steel. For strength and resiliency the wire is drawn to a high tensile strength and subsequently processed so that its diameter tapers down at the distal end. This end is covered with wound wire coils to provide a final assembly that has a uniform outside

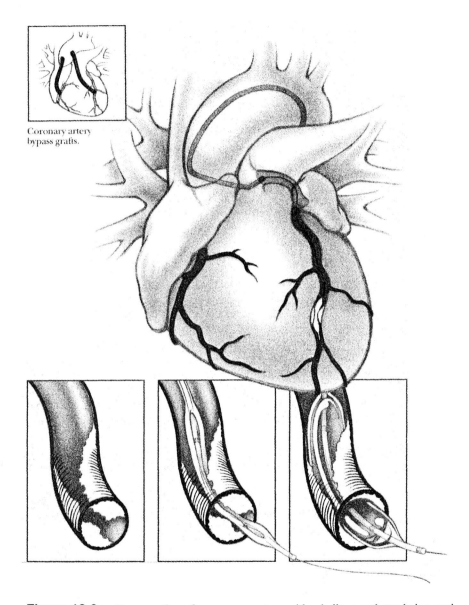

Coronary artery
bypass grafts.

**Figure 13.9.**   Cross section of a coronary artery with a balloon catheter being positioned in the narrowed plaque-buildup section. (Courtesy of Advanced Cardiovascular
Systems, Inc.)

diameter but whose flexibility varies along its length. The result is a combination of superior torque transmission and steerability, with adequate resiliency to avoid unintentional plastic deformation during normal use.

Another interesting example of materials selection for angioplasty devices lies in the wound wire coils at the tip of the guide wire. Since the guide wire is used to cross the narrowed portion of the artery and then to guide the balloon catheter into position, it is desirable for the guide wire tip to be visible on an x-ray monitor. Platinum has much greater radiopacity then stainless steel, and consequently, platinum alloys, which can be drawn into high-strength wire for coil resiliency, are commonly used. In this case the material selection is based on a combination of strength and radiopacity. A 0.014-in. high-torque guide wire is shown in Figure 13.10.

In balloon catheters, the balloons are made of materials that can withstand high inflation pressures, yet when deflated have low profiles to permit initial crossing of the narrowed artery. Special formulations of polyethylene (PE) or polyethylene terephthalate (PET) are commonly used. Balloons in catheters manufactured by Advanced Cardiovascular Systems, Inc. are irradiationed hardened by electron bombardment so that they will expand to a precise size which remains nearly constant over a broad range of inflation pressures. The balloons in such catheters are manufactured to a variety of specific sizes which are selected by the physician according to the particular artery of interest.

These unique examples further illustrate the necessity for the materials, design, and manufacturing engineers to be well aware of the materials requirements and options available for the very precise control demanded by the medical industry. To meet such demanding performance requirements, many materials processing details are involved. Most of these details are rightfully considered proprietary by the manufacturers.

## 13.4.8 Cargo Loaders for Aircraft

This case history portrays the use of metals in an established industry. The main deck bridge of a cargo loader designed for loading cargo onto all of the

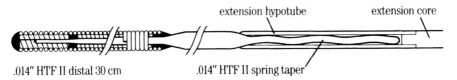

extension hypotube    extension core

.014" HTF II distal 30 cm    .014" HTF II spring taper

**Figure 13.10.** A 0.014-in.-diameter high-torque guide wire used in angioplasty. The wire-wound coils surrounding the guide wire at its distal end are shown in cross section. (Courtesy of Advanced Cardiovascular Systems.)

large jets (e.g., Boeing 747, MD-11, DC-10, and all other wide-body aircraft main decks up to 18 ft high) is shown in Figure 13.11. This loader, manufactured by the Lantis Corporation of Salinas, California, is a dual-platform loader (two levels) and has a load capacity of 15,000 lb for each platform. Since the structural members can be as heavy as one may desire, materials cost becomes an important factor in the selection process. On a cost–strength basis,

**Figure 13.11.**   Cargo loader for wide-body aircraft. (Courtesy of the Lantis Corp.)

steels are still the best buy for many structural applications. The steels currently used for this cargo loader are listed as follows:

1. ASTM A36 is used for the majority of plates, rectangular bars, channels, and angles. ASTM A36 is a common structural steel priced at about $0.30 to 0.40 per pound, depending on form and quantity. The specifications include a ultimate tensile strength of 400 to 500 MPa (58 to 80 ksi), yield strengths in the range 220 to 250 MPa (32 to 36 ksi), and a minimum elongation of 20% in a 2-in. gauge length. It falls in the general composition range of a 1022 SAE steel and is used here in the hot-rolled or hot-finished condition. The selection process reduces primarily to one of cost. One could use an ASTM 514 low-alloy steel of structural quality which is often quenched and tempered. It has a higher strength and better hardenability (depth of hardening) but a higher price because to attain this hardenability requires the addition of chromium (up to 1.5%). It may also involve a heat treatment operation which would add to the production cost and also subject the steel to undesirable microstructural changes if welded. A more expensive microalloyed steel (e.g., ASTM 572) would provide more strength and could be welded without serious property deterioration, but again at a higher materials cost.

2. Cold-drawn 1018 steel is used in rectangular bars for structures requiring a strength higher than the A36 can provide. The bridge scissors utilize cold-drawn 1018 for stiffeners of the longitudinal square tubes. This is a less expensive route than either the use of ASTM 514 or 572 mentioned above. Cold-drawn 1018 has a yield strength on the order of 275 to 345 MPa (40 to 50 ksi), depending on the amount of cold work. Its composition is similar to A36 but has slightly better machinability.

3. ASTM A500 grade B is used for the rectangular and square tubing structural components. This steel has a minimum tensile strength of 310 MPa (58 ksi) and a corresponding yield strength of 228 MPa (33 ksi), along with an elongation in 2 in. of 23%. It is a general-purpose structural steel used in buildings and bridges. It can be obtained in cold-formed welded or seamless tubular form as well as in round bars and other shapes. It contains 0.3% maximum carbon, 0.063 maximum sulfur, and 0.18 minimum Cu.

4. ASTM A514 grade C is a high-strength steel [690 to 895 MPa (100 to 130 ksi) UTS and a yield strength of 620 MPa (90 ksi)] used in plate form for parts that require very high strength and limited welding. It is used in the elevator lift mechanism (not shown in the picture). This steel, containing only about 0.20 C, attains its strength via the quench and temper process.

5. An ASTM A311 class B is a high-strength steel used as the pivotal pins that allows the use of smaller pin diameters to support the load. It is comparable to a 1044 SAE type with 828 to 900 MPa (120 to 130 ksi) yield strength and good machinability, the latter being due to a sulfur content of 0.3 wt %, about 10 times normal. It is a resulfurized cold-drawn steel and is the least

expensive steel, with yield strengths greater than 100 ksi. This cost factor plus a reduced machining cost makes it a logical selection for the intended application. One should be aware that this steel has a low fatigue strength, due in part to the high sulfur content.

### 13.4.9 Miscellaneous Applications of Advanced Materials

The F-18 Navy fighter aircraft uses high-strength 7000 series alloy extrusions and forgings for the spars, ribs, and bulkheads. The fuselage skins and ribs employ honeycombed structures, while the wing skin segments are fabricated from 2000 and 6000 series aluminum alloys. High-strength maraging steels form the landing gear assemblies and titanium alloys are used for certain high stress points.

In 1986, President Reagan announced a nationwide research to develop the National Aero-Space Plane (NASP), sometimes called the X-30 program. The plane was to be designed to take off from a runway, accelerate to orbital velocities, and descend back to land in a conventional manner. It could travel to Tokyo, for example, in less than 2 h. Two engines will probably be employed, a scramjet to attain speeds several times that of sound, followed by a rocket that would send it into orbit. One material that has been selected for this plane is the SiC–$\beta$-titanium alloy composite developed by Textron. Timet 21S is the candidate titanium alloy. This composite will be used for the airframe and fuselage and will be protected on the outside with a mechanically attached carbon–carbon composite. An inside fuel tank containing liquid hydrogen will most likely be made of a graphite–epoxy composite. Other materials being considered include the intermetallic aluminides, probably in some type of composite form.

An ultrahigh-molecular-weight polyethylene fiber developed by Allied Signal has strength many times that of most steels and is being explored for use in resin composite form for lightweight personal armor helmets and vests.

A new high-temperature [370°C (700°F)] polymer known as AFR 700B has been invented—*finalized* may be a better word—by Tito Serafine of TRW Ballistic Missiles Division. Serafine also developed the most widely commercialized polyimide that could be used to temperatures of 343°C (650°F) known as PMR-15. The AFR 700B material has a $T_g$ of 400°C (752°F), and at these temperatures it is impinging on the performance of titanium, a much heavier material. It could be used to replace titanium in the turbine compressor stage static structures and could also be used to reduce the weight of the aft fuselage. It is based on a six-fluorine structure backbone. Right now its main drawback is its high cost, about $400 per pound.

Rigid-rod polymers (RRP) are a family of liquid-crystal polymers that have a tendency to align in certain solutions and thus reinforce themselves. They are ram-rod straight, and extremely stiff and strong, with a tensile strength $>600$ ksi and a tensile modulus $>400 \times 10^6$ psi, yet their density is only 1.69 g/cm$^3$.

Allied Signal has developed a phenolic–triazene (PT) copolymer for use in applications that now require bismaleimides. The $T_g$ is 300°C and it can be heated to as high as 450°C before decomposition begins.

An industry, university, and government team in Canada has developed a $Si_3N_4$ ceramic diesel engine valve that has withstood exhaust temperatures in excess of 500°C at 1000 rpm.

Researchers at McDonnell Aircraft are studying an improved steel for landing gears that may replace the standard 300M maraging steel. The modified AF1410 steel has a higher toughness ($K_{1c}$ of 143 MPa·m$^{1/2}$) than the 300M steel and the necessary strength. The nominal composition of AF1410 is 10% Ni–14% Co–2% Cr–1% Mo–0.16% C. The carbon content was increased from 0.16 to 0.2 wt %. Improvements in this steel resulted in a strength of 1790 MPa (260 ksi) and improved resistance to hydrogen embrittlement (SCC). Titanium matrix–SiC fiber composites are also being evaluated for landing gears.

Microalloyed–steel forgings are creating new design opportunities due to recently developed new grades and the introduction of a new ASTM specification ASTM A909. This specification applies to forgings with a maximum 4-in. section thickness and a maximum forge weight of 150 lb. Crankshafts in both diesel engines for trucks and gasoline engines for automobiles, yokes for drive shafts, connecting rods, front axles, engine mounts, spindles, and suspension and steering components are some of the applications of these forgings in the automotive industry.

Textron's three-dimensional carbon–carbon composite material is used in the integral throat entrances on the Peacekeeper, Trident II, and NASA's advanced solid rocket motor. PAN-based carbon fiber is used as a brake material for aircraft.

## 13.5 USE OF DATA BASES FOR MATERIALS SELECTION

Probably the most desirable achievement of data bases, short of artificial intelligence, is one projected in the following scenario. The design engineer, after careful computation of stresses involved, including those at the desired temperature and cyclic fatigue stress (if such a situation exists) and the environments that may be involved for the particular design of interest, shoots this information over to the purchasing office. The purchaser has cost information, including machining and forming costs, and with this information and the design engineer's request, he or she punches a few buttons on the PC, and pronto—the materials, their costs, availability, and possible vendors emerge on the printout sheet. If the availability or cost is unacceptable, a few more buttons are punched and a list of possible alternative materials are printed. Now perhaps a meeting with co-workers and the boss is in order, or the engineer simply returns an answer to select B or C instead of the desired A materials.

Unfortunately, data bases do not work this way, although one could envision such a happening at some time in the not–to-distant future. Data bases are a good fast way to report facts (i.e., numbers), but are not so well developed that they can produce running commentary on the relative merits of a material. For most established materials, their properties and behavior can be predicted to a high degree of certainty, except perhaps for fatigue and corrosion. But even the materials engineer must make some guestimates. Materials properties have a range of values, which the data bases should show, but yet their may be a need for some additional information, which sometimes can be supplied by a vendor or handbooks that list more than just numbers. Those handbooks that include only the facts, such as the *Handbook of Chemistry and Physics,* and others, such as *Smithells' Metals Reference Book,* are very suitable for conversion to data bases. But others (e.g., the ASM International handbooks) supply a lot of information involving processing, comparative materials, and some of the most recently developed materials whose data may not be considered appropriate for data bases. For these reasons data bases will still require a certain amount of interpretation and "guestimating" beyond the numbers that appear on the computer screen. But even at the present state of the art, data bases have proven to be valuable sources of information. Some of the more popular materials data bases are listed herein.

1.  *Materials Property Data (MPD) Network.* MPD is a private nonprofit organization established in 1984 by the Materials Property Council of the U.S. government.

(a)  Consists of about 13 separate data bases.
(b)  Will be accessible worldwide on the STN International system.
(c)  MPD files to be searched as a single unit with a menu-driven interface, including an interactive thesaurus.
(d)  The user menus provide for a choice between a menu for experienced searches and a guide for beginners.
(e)  The following data bases will be included (many are already available):
AAASD: Aluminum standards and data
AIFRAC: Aluminum fracture toughness
MARTUF: Toughness of steels for marine structures
METALS DATAFILE of ASM International
MIL-HDBK-5: Military standardization handbook
MPD SEARCH: Directory of materials property data bases
PLASPEC: Plastic materials selection data base

STEELTUF: Pressure vessel steels
IPS: International plastics selector
CDA: Copper alloys data base
POLYMAT: Plastics data base
PLASNEWS: Business news for the plastic industry
COR-SUR: A data base on corrosion by NACE

(f) The MPD network will have numeric search capabilities (i.e., the ability to locate a specific numeric value or a range of values).

(g) Permits search in six major sets of units.

(h) MPD works in a complementary and cooperative partnership with many technical societies, including ACS, ASM, ASME, AWS, and NACE, trade associations, and government laboratories.

2. *MAT.DB: ASM Center of Materials Data.* This system was designed for IBM and compatible computers using DOS version 3.0 or later. It requires 640 K of memory and a hard disk. The Mat DB system consists of MAT.DB, MAT.DB data, and En Plot.

(a) *Mat.DB.* This is the data base manager and is capable of maintaining, searching, and displaying textural, tabular, and graphical information on engineered materials.

(b) *Mat.DB data.* These are the data bases that contain property data on engineered materials. Mat.DB "reads" Mat.DB data bases in much the same way that a cassette player reads cassettes. The basic Mat.DB program comes with a "starter" data base. More detailed data bases on specific alloys and other classes of materials can be purchased separately and added to the starter data base. Data bases now available for purchase include alloy steels, stainless steels, structural steels, aluminum, copper, titanium, and magnesium alloys, and plastics. Other materials data bases are in progress of being developed.

(c) *En Plot.* This is an analytical graphics program that enables the creation of publication quality graphics on a PC.

3. *STN Int.* This is a scientific and technical information network that has an on-line system with direct access to some of the world's best scientific and technical data bases in all fields of science, engineering, and math, including patents. A special front-end package allows the formulation of search strategies off-line at your own PC. STN is operated in North America by CAS, a division of the American Chemical Society; in Europe by FIZ Karlsruke; and in Japan by the Japan Informational Center for Science and Technology (JICST). All three are not-for-profit scientific organizations. The STN Software "messenger" supports searching in all traditional bibliographic access points: subject terms, titles, author names, and additional bibliographic information as provided in the data bases. Using any microcomputer or terminal that is ASCII

compatible, one can search and display text and structures. For a completed listing of data bases and their producers, contact:

STN International
254 Olentangy River Road
P.O. Box 3012
Columbus, OH 43210–0012
Phone: 800–848–6533 in North America
Telefax 614–447–3713

4. *Worldwide Standards Service by IHS information services.* This service provides a comprehensive data base on over 100,000 standards from 375 developing bodies.

5. *IHS DoD standardization service.* This service provides PC access to more than 60,000 active U.S. military specifications and DoD standardization documents.

6. *Copperselect.* This is a computerized system for selecting copper alloys developed by the Copper Development Association, Inc.

7. *Corrosion.* Produced by Marcel Dekker, this data base includes behavior of alloys, carbon, plastics, rubbers, and glass in 600 corrosive environments.

8. *M/VISION.* Produced by PDA Engineering; several M/VISION data bases are available which use their own software systems for a PMC-90 databank (i.e., for advanced polymer–matrix composites). In addition, M/VISION includes MIL-HDBK-5E, a databank for Metals, and MIL-HDBK-17A, a databank for common structural composites.

9. *Orbit Search Service.* This search service provides access to more than 100 computerized data bases that are heavily concentrated in the areas of science, technology, and patents.

10. *COR.AB.* Computerized abstracts of the worlds corrosion literature are provided by the National Association of Corrosion Engineers (NACE).

11. NACE has other software, namely Cor-Sur, which provides computerized reference data and Cor-Sur 2 provided in conjunction with the National Institute of Standards and Technology (NIST). Requires IBM compatible, DOS version 2.0 or later, and 320K RAM, one disk drive.

12. *Structural Ceramics Database (SCD) Version 1.0.* Issued by NIST, this data base requires 640K of memory and DOS version 2.1 or higher.

13. *Plastics SELECTOR.* From D.A.T.A. Business Publishing, this data base requires IBM or IBM-compatible PCs. It searches over 12,700 grades of molding and extruding plastic materials.

14. UNS Version 2.0. This SAE PC software, uses the unified numbering system jointly created by SAE and ASTM and includes up-to-date information on over 400 metals and alloys. This system requires IBM, XT, AT, PS2, or

IBM compatibles, DOS version 2.0 or later, hard disk, 640K of RAM, and 8 megabytes of free storage.

    15. *NIST Standard Reference Data Program.* This program provides reliable well-documented data to scientists and engineers. This program is more than just a materials data base. It includes analytical chemistry, chemical kinetics, molecular structures, and thermodynamical data.

    16. *SAE Global Mobility Database.* Materials and structures are just one part of this data base, which includes electronics, engines, aviation, and several other subjects.

    17. *MPR Publishing Service.* From Shrewsberg, England, this service provides a software package for powder metal purchase.

    18. *LAP.* Life analysis program of fatigue data is produced by MTS System Corp.

    19. *Engineering Design Database (EDD).* This data base is produced by G.E. Plastics of Pittsfield, Mass.

    20. *Polyfacts.* This data base was developed by DuPont.

    21. *Cenbase.* This data base was developed by CenCad services; includes thousands of polymeric thermosets, thermoplastics, and elastomers, and over 3000 composite materials.

    22. *HTMP.* This is a list of high-temperature materiels properties by CINDAS of Purdue University in conjunction with the U.S. Department of Defense.

    A worthwhile book for materials information was published by ASM International in 1986, entitled *A Guide to Materials Engineering Data and Information.* Subsequent editions are planned.

## QUESTIONS AND PROBLEMS*

  13.1.  For an application temperature of 1000°C and a static 1000-h stress–rupture strength of 200 MPa, what class of materials should one consider? Of those listed, select the one that you consider best suited for these conditions, taking into account their reliability and cost.

  13.2.  If the material above was to be subjected to a cyclic stress with a $\sigma_a$ of ±20 MPa superimposed on the static 200-MPa stress, what materials would you select?

  13.3.  Approximately at what temperatures must one consider using either metals or MMCs rather than PMCs?

  13.4.  At what temperatures might MMCs be favored over the superalloys?

---

*Answers to some of these problems will be found in other chapters of the book.

13.5.   At what temperatures might ceramics or CMCs be favored over MMCs?

13.6.   For a dead-bolt lock on a residence dwelling, what materials are highly recommended?

13.7.   What tool steel would you select for a room-temperature punch-press die?

13.8.   What extrusion die material would be useful for the extrusion of (a) zircalloy, a zirconium alloy; and (b) for forging Ti–6% Al–4% V?

13.9.   What are some of the more common applications of aluminum bronze? What comparative material might be considered?

13.10.  Rank the refractory metal alloys in order of increasing strength.

13.11.  For a common ceramic material such as $Al_2O_3$, how many different room-temperature strength values can you find? List property value, source, test method, and any processing variables listed along with the strength.

13.12.  List five applications of advanced ceramics.

13.13.  List as many applications as you can find for the PH stainless steels.

13.14.  What aluminum alloy has the highest strength/density ratio at room temperature?

13.15.  What are the advantages of the particulate MMCs over the continuous-filament type?

## SUGGESTED READING

Farag, M. M., *Selection of Materials and Manufacturing Processes for Engineering Design*, Prentice Hall, Englewood Cliffs, N.J., 1989.

Gladius, L., *Selection of Engineering Materials*, Prentice Hall, Englewood Cliffs, N.J., 1990.

Johnson, L. G., *Statistical Treatment of Fatigue Experiments*, Elsevier, Amsterdam, 1974.

Park, J. B., *Biomaterials Science and Engineering*, Plenum Press, New York, 1984.

Petty, W., Ed., *Total Joint Replacement*, W.B. Saunders, Philadelphia, 1990.

# Appendix A
## Conversion Units

### Length

1 m = 3.28 ft
1 cm = 2.54 in.
1 mm = 0.0394 in.
1 km = 0.622 mile

1 Å = $10^{-8}$ cm
1 $\mu$m = $10^{-6}$ m
1 nm = $10^{-9}$ m
1 mil = 0.0254 mm

### Area

1 $m^2$ = 10.76 $ft^2$
1 $cm^2$ = 0.155 $ft^2$

1 $yd^2$ = 0.837 $m^2$
1 $ft^2$ = 0.093 $m^2$

### Volume

1 $cm^3$ = 0.0610 $in.^3$
1 $m^3$ = 35.32 $ft.^3$

1 $in.^3$ = 16.39 $cm^3$
1 $ft.^3$ = 0.0283 $m^3$

### Mass

1 kg = 2.205 lb
1 g = 2.205 $\times$ $10^{-3}$ lb
1 lb = 453.6 g

1 oz = 28.34 g
1 troy oz = 31.1 g

## Density

$1 \text{ kg/m}^3 = 0.0624 \text{ lb}_\text{m}/\text{ft}^3$

$1 \text{ g/cm}^3 = 0.0361 \text{ lb}_\text{m}/\text{in.}^3$

$1 \text{ lb}_\text{m}/\text{ft}^3 = 16.02 \text{ kg/m}^3$

$1 \text{ lb}_\text{m}/\text{in.}^3 = 27.7 \text{ g/m}^3$

## Force

$1 \text{ N} = 0.2248 \text{ lb}_\text{f}$

$1 \text{ N} = 10^5 \text{ dyn}$

$1 \text{ lb}_\text{f} = 4.44 \text{ N}$

$1 \text{ dyn} = 10^{-5} \text{ N}$

## Current

$1 \text{ A} = 1 \text{ C/s}$

$1 \text{ C} = 3 \times 10^9 \text{ esu of charge}$

## Stress

$1 \text{ N/m}^2 = 1 \text{ Pa}$

$1 \text{ kg/m}^2 = 9.8 \text{ Pa}$

$1 \text{ g/mm}^2 = 1.422 \text{ psi}$

$1 \text{ dyn/cm}^2 = 0.10 \text{ Pa}$

$1 \text{ Pa} = 0.145 \times 10^{-3} \text{ psi}$

$1 \text{ ksi} = 6.89 \text{ MPa}$

$10^6 \text{ psi} = 6.89 \text{ GPa}$

$1 \text{ psi} = 0.703 \text{ g/mm}^2$

## Viscosity

$1 \text{ poise} = 1 \text{ g/cm} \cdot \text{s}$

## Impact Energy

$1 \text{ ft-lb}_\text{f} = 1.355 \text{ J}$; divide net value of energy by specimen width to obtain J/m or ft-lb/in.

$1 \text{ ft-lb/in.} = 1.355 \text{ J/m}$

## Fracture Toughness

$1 \text{ MPa} \sqrt{\text{m}} = 0.91 \text{ ksi} \sqrt{\text{in.}}$

$1 \text{ ksi} \sqrt{\text{in.}} = 1.1 \text{ MPa} \sqrt{\text{m}}$

## Energy

$1 \text{ J} = 1 \text{ N} \cdot \text{m}$

$1 \text{ cal} = 4.18 \text{ J}$

$1 \text{ cal} = 3.97 \times 10^{-3} \text{ Btu}$

$1 \text{ J} = 0.738 \text{ ft-lb}_\text{f}$

$1 \text{ eV} = 1.6 \times 10^{-19} \text{ J}$

$1 \text{ Btu} = 1054 \text{ J}$

$1 \text{ erg} = 10^{-7} \text{ J}$

$1 \text{ Btu} = 252 \text{ cal}$

## Power

$1 \text{ W} = 1 \text{ J/s}$

$1 \text{ cal/s} = 14.29 \text{ Btu/hr}$

## Thermal Conductivity

$1 \text{ W/m} \cdot \text{K} = 2.39 \times 10^{-3} \text{ cal/g} \cdot \text{K}$
$1 \text{ Btu/lb}_m \cdot {}^\circ\text{F} = 1.0 \text{ cal/g} \cdot \text{K}$
$1 \text{ cal/cm} \cdot \text{s} \cdot \text{K} = 24.8 \text{ Btu/ft} \cdot \text{hr} \cdot {}^\circ\text{F}$

## Specific Heat

$1 \text{ cal/g} \cdot \text{K} = 1.0 \text{ Btu/lb}_m \cdot {}^\circ\text{F}$       $1 \text{ cal/g} \cdot \text{K} = 4.184 \text{ J/g} \cdot \text{K}$

## Constants

| | | |
|---|---|---|
| Avogadro's number | $N_0$ | $6.023 \times 10^{23} \text{ mol}^{-1}$ |
| Atomic mass unit | $amu$ | $1.661 \times 10^{-24} \text{ g}$ |
| Electron mass | $m_e$ | $9.110 \times 10^{-28} \text{ g}$ |
| Electronic charge (magnitude) | $e$ | $1.602 \times 10^{-19} \text{ C}$ |
| Planck's constant | $h$ | $6.626 \times 10^{-34} \text{ J} \cdot \text{s}$ |
| Gas constant | $R$ | $1.987 \text{ cal/mol} \cdot \text{K}; 8.314 \text{ J/mol} \cdot \text{K}$ |
| Boltzmann's constant | $k$ | $8.620 \times 10^{-5} \text{ eV/K}$ |
| Permittivity constant | $\epsilon_0$ | $8.854 \times 10^{-12} \text{ C}^2\text{N} \cdot \text{m}^2)$ |
| Permeability constant | $\mu_0$ | $4\pi \times 10^{-7} \text{ T} \cdot \text{m/A}$ |
| Bohr magneton | $\mu_B$ | $9.274 \times 10^{-24} \text{ A} \cdot \text{m}^2$ |
| Faraday | $F$ | $0.6485 \times 10^4 \text{ C/mol}$ |
| Gravitational acceleration | $g$ | $9.806 \text{ m/s}$ |

# Appendix B
## Material Costs

## B.1.1 Composites

| Material | Form | Price (Jan. 1992) ($/lb)* |
|---|---|---|
| Boron filament–epoxy | Impregnated tape | ~400 |
| Boron filament–Al matrix | Impregnated tape | 400–500 |
| SiC filament–Al 6061 | Impregnated tape | Not quoted > boron–Al |
| SiC particulate 20% vol.–Al 6061 | Ingot | 3.30 |
| $Al_2O_3$ particulate 10% vol.–Al 6061 | Ingot | 3.15 |
| Kelvar-reinforced epoxy | Molding type | 50.00 |
| Glass-fiber-reinforced epoxy | Varies | 5.00 |

*Depends on form and quantity; prices above are for small quantities.

## B.1.2 Polymers

| Material | Form | Price (Jan. 1992) ($/lb)* |
|---|---|---|
| Polyethylene HD | 0.5-in.-thick sheet | 1.73† |
| Polycarbonate | 0.5-in.-diameter rod | 15.00† |
| PVC | 0.5-in. sheet | 3.60† |
| Polycarbonate + 30% glass fibers | 0.5 in. rod | 37.00† |
| Acrylics | 0.125 in. sheet | 3.00† |
| Polyethylene HD resin | Raw material | 0.40‡ |
| Polystyrene resin | Raw material | 0.30‡ |
| PET resin | Raw material | 0.60‡ |
| Polyamide resin | Raw material | 1.75‡ |
| PEEK | Raw material | 25.00‡ |

*Depends on form and quantity.

† Price for small quantity.

‡ Price for tonnage quantity.

Relative Cost of Polymers Based on Density (Polyethylene = unity)

| | | | |
|---|---|---|---|
| Polyethylene HD | 1.0 | Nylon 66 30% glass filled | 7.2 |
| Polyproplyene | 1.1 | Nylon 66 | 6.6 |
| Polystyrene | 1.7 | Nylon 6 | 4.9 |
| Polystyrene 30% glass filled | 1.8 | Polycarbonate | 5.6 |
| PVC | 1.6 | PET 30% glass filled | 6.8 |
| ABS | 2.4 | Polysulfone 30% glass filled | 15.6 |
| ABS 30% glass | 3.4 | Polyamide-imide | 59.2 |
| Acrylic | 3.6 | PEEK | 88.0 |

## B.1.3 Metals

| Material | Form | Price (Jan. 1992) ($/lb)* |
|---|---|---|
| Low-alloy steel | Plate | 0.28 |
| Plain carbon steel | Hot-rolled sheet | 0.28 |
| Al alloy 384 | Ingot | 0.59 |
| Al alloy 319 | Ingot | 0.62 |
| 2024 Al T6 | Plate | 1.75 |
| 7075 Al T6 | Plate | 1.75 |
| 304 Stainless steel | Coiled sheet | 1.47 |
| 316 Stainless steel | Coiled sheet | 1.99 |
| 310 Stainless steel | Coiled sheet | 3.10 |
| 304 Stainless steel | Plate | 1.32 |
| 316 Stainless steel | Plate | 1.73 |
| 310 Stainless steel | Plate | 3.03 |
| 410 Martensitic stainless steel | Bar | 1.50 |
| 17-4 PH | Bar | 1.80–2.00 |
| Copper | Wire-bar | 1.10 |
| Cu–Be alloy | Strip | 8.40 |
| Cu–Zn (brass) | Rod | 2.29 |
| Nickel 200 (high purity) | Rod | 14.00 |
| Nickel | Melt stock | 4.00 |
| Chromium | Electrolytic | 4.62 |
| Titanium | Sponge | 4.40 |
| Zinc | As refined | 0.29 |
| Vanadium | As refined | 9.50 |
| Zirconium | — | 11.00–17.00 |
| Aluminum | Electrolytic | 0.65 |
| Ti–6% Al–4% V | Plate | 8.00 |
| Monel 400 | — | 5.00 |
| Monel K500 | Rod | 10.00 |
| Tin | Melt stock | 2.43 |
| Magnesium | — | 1.43 |
| Lead | — | 0.35 |
| Cobalt | Cathode | 12.50 |

*100- to 1000-lb quantities.

# Index